Molecular- and Nano-Tubes

Oliver Hayden · Kornelius Nielsch
Editors

Molecular- and Nano-Tubes

Springer

Editors
Oliver Hayden
Siemens AG
Corporate Technology
Erlangen, Germany
oliver.hayden@siemens.com

Kornelius Nielsch
University of Hamburg
Institute of Applied Physics
Hamburg, Germany
kornelius.nielsch@mail.physnet.uni-hamburg.de

ISBN 978-1-4419-9442-4 e-ISBN 978-1-4419-9443-1
DOI 10.1007/978-1-4419-9443-1
Springer New York Dordrecht Heidelberg London

Library of Congress Control Number: 2011931467

Printed on acid-free paper

Springer is part of Springer Science+Business Media (www.springer.com)

Preface

Why for heaven's sake a book on nanotubes or nanopore? Just another summary of nanoscale gimmicks without any hope for future engineering? The answer is in fact yes and no. Nanotubes and nanopores are probably the most fascinating materials on the nanoscale due to one simple reason: We can barely fabricate tubes and pores on the nanoscale top-down with a perfection such as carbon nanotubes and we are still puzzled by the elegant functionality of biological ion channels. At the same time we have clever bottom-up synthesized materials though it is not clear how we can solve the tremendous engineering problems associated for rational device architectures. When it comes to real-world applications the tubes and pores have obviously tremendous potential though only a few start-ups have currently products on the market. Limitations for new tube products are often related to the lack of sufficient bulk quantities and purity of the materials. For example in the case of carbon nanotubes the scattered intellectual property on material synthesis is one of the reasons why nanotube products are still in their infancy.

The idea to a book on nanotubes and nanopores was the result of two symposia at the Material Research Society (MRS) conference held at Boston and San Francisco which the editors organized and where we realized that working interdisciplinary is still more of a buzzword in the community. Furthermore, we have been able to organise an exploratory workshop funded by the European Science Foundation (ESF) on nanowires and again experienced that the highly heterogeneous mix of scientists was one of the most fruitful experiences to exchange ideas across scientific borders though it is difficult to keep track on the exponential increase if applications. Thus, it was time to write something down not for all low-dimensional materials but for the least accessible materials from an engineering point-of-view: nanotubes and nanopores.

Within this book we tried not to cover all aspects of tubes, which would go beyond the scope of the editors, but rather introduce the readers to tubular structures from biological, organic and inorganic materials as well as their functionality. Some nanotubes are synthesized by ingenious chemistry and others are spontaneously formed by physical processes which we often do not understand in detail. With respect to functionality we tried to cover the most interesting aspects of tubular materials, which should allow the readers to evaluate the potentials for these new materials and derive some rule of thumb for their own research or interest. Last, we have been adding chapters on top-down technologies which can be used to fabricate rational tubular structures on the nanoscale to give the reader an impression how creative scientists and engineers start to be when it comes to small holes. In most chapters the readers will find some critical theory to understand the physics as well as detailed descriptions of the chemistry applied. Furthermore, the authors of the chapters were asked to be critical about their own work and to explain critical experiments thoroughly.

However, one should not expect to have a student textbook in his hand but a highly interdisciplinary book covering material synthesis, electronics, optics, and membrane science where basic understanding of the physics and chemistry is required to understand the content. This might also be the hint for the decision making process why to purchase our book on tubes and pores. Anyone who is interested in both applied science and engineering will probably benefit most.

Last, we would like to thank all authors who have put much effort in their chapters. It was quite difficult to convince the principal investigators in academia and industry to participate. Some of the new exciting topics could not be covered such as sequencing with nanopores. Nevertheless, we believe that the selected topics are a good starting point for readers to think about material and engineering issues with nanotubes and nanopores.

Only recently one of the PI's, Prof. Dr. Ulrich Goesele, director of the MPI for Microstructure Physics at Halle, Germany, passed away. His outstanding contributions to materials physics and chemistry which regularly led to key innovation for industrial applications will be remembered. Our book is a tribute to him.

Erlangen, Germany — Oliver Hayden
Hamburg, Germany — Kornelius Nielsch

Contents

Contributors

Olaf S. Andersen Department of Physiology and Biophysics, Weill Cornell Medical College, New York, NY 10065, USA, sparre@med.cornell.edu

J. Appenzeller School of Electrical and Computer Engineering, Purdue University, West Lafayette, IN 47907, USA, appenzeller@purdue.edu

W.M. Arnold Industrial Research Limited, Lower Hutt, New Zealand; The MacDiarmid Institute for Advanced Materials and Nanotechnology, Lower Hutt, New Zealand, m.arnold@irl.cri.nz

Francisco Bezanilla Department of Biochemistry and Molecular Biology, The University of Chicago, Chicago, IL 60637, USA, fbezanilla@uchicago.edu

F. Bonaccorso Department of Engineering, University of Cambridge, Cambridge CB3 0FA, UK, fb296@cam.ac.uk

M.F. Broom Izon Science, Christchurch 8053, New Zealand, murray@izon.com

Mihaela Daub Max Planck Institute of Microstructure Physics, 06120 Halle, Germany, mihaela.daub@gmx.net

Paul Dommersnes MSC, Université Paris Diderot, F-75205 Paris, France, paul.dommersnes@univ-paris-diderot.fr

A.C. Ferrari Department of Engineering, University of Cambridge, Cambridge, CB3 0FA, UK, acf26@eng.cam.ac.uk

Ulrich Gösele (deceased) Max Planck Institute of Microstructure Physics, D-06120 Halle, Germany, goesele@mpi-halle.mpg.de

Nicole Grobert Department of Materials, University of Oxford, Oxford OX1 3PH, UK, nicole.grobert@materials.ox.ac.uk

T. Hasan Department of Engineering, University of Cambridge, Cambridge CB3 0FA, UK, th270@eng.cam.ac.uk

Lloyd P. Horne Department of Chemistry, Center for Research at the Bio/Nano Interface, University of Florida, Gainesville, FL 32611-7200, USA, lloyd.horne@gmail.com

M.L. Jansen Industrial Research Limited, Lower Hutt, New Zealand, m.jansen@irl.cri.nz

Aldo Jesorka Department of Chemical and Biological Engineering, Chalmers University of Technology, SE-41296 Göteborg, Sweden, aldo@chalmers.se

J. Knoch Faculty of Electrical Engineering and Information Technology, RWTH Aachen University, Templergraben 55, 52056 Aachen, Germany, knoch@iht.rwth-aachen.de

Roger E. Koeppe II Department of Chemistry and Biochemistry, University of Arkansas, Fayetteville, AR 72701, USA, rk2@uark.edu

Jan Lagerwall Faculty of Chemistry and Physics, Institute of Chemistry – Physical Chemistry, Martin-Luther-Universität Halle-Wittenberg, D-06108 Halle/Saale, Germany, jan.lagerwall@lcsoftmatter.com

Cheol Jin Lee School of Electrical Engineering, Korea University, Seoul, Korea, cjlee@korea.ac.kr

Tatsiana Lobovkina Department of Chemical and Biological Engineering, Chalmers University of Technology, SE-41296 Göteborg, Sweden, tatsiana@chalmers.se

Andreas Langner Max Planck Institute of Microstructure Physics, D-06120 Halle, Germany, alangner@mpi-halle.mpg.de

Charles R. Martin Department of Chemistry, Center for Research at the Bio/Nano Interface, University of Florida, Gainesville, FL 32611-7200, USA, crmartin@chem.ufl.edu

Frank Müller Max Planck Institute of Microstructure Physics, D-06120 Halle, Germany, frank.mueller@mpi-halle.mpg.de

Kornelius Nielsch Max Planck Institute of Microstructure Physics, 06120 Halle, Germany; Institute of Applied Physics, University of Hamburg, 20355 Hamburg, Germany, knielsch@physnet.uni-hamburg.de; knielsch@mpi-halle.de

Björn Önfelt Department of Cell Physics, Royal Institute of Technology, 10691 Stockholm, Sweden; Department of Microbiology, Tumor and Cell Biology, Karolinska Institute, 10691 Stockholm, Sweden, bjorn.onfelt@ki.se

Owe Orwar Department of Chemical and Biological Engineering, Chalmers University of Technology, SE-41296 Göteborg, Sweden, orwar@chalmers.se

Maja Remskar Jozef Stefan Institute, SI-1000 Ljubljana, Slovenia, Maja.Remskar@ijs.si

John Robertson Engineerings Department, University of Cambridge, Cambridge, UK, jr@eng.cam.ac.uk

Siegmar Roth School of Electrical Engineering, Korea University, Seoul, Korea; Sineurop Nanotech GmbH, Stuttgart, Germany, siegmar@korea.ac.kr; sgmrroth@aol.com

A.G. Rozhin Department of Engineering, University of Cambridge, Cambridge CB3 0FA, UK, A.Rozhin@ston.ac.uk

V. Scardaci Department of Engineering, University of Cambridge, Cambridge CB3 0FA, UK, vittorio.scardaci@hp.com

Sigrid E. Schmutzer Department of Chemistry and Biochemistry, University of Arkansas, Fayetteville, AR 72701, USA, sigrid@hayden.at

Lindsay T. Sexton Department of Chemistry, Center for Research at the Bio/Nano Interface, University of Florida, Gainesville, FL 32611-7200, USA

Toshimi Shimizu Nanotube Research Center (NTRC), National Institute of Advanced Industrial Science and Technology (AIST), Tsukuba, Ibaraki 305-8565, Japan, tshmz-shimizu@aist.go.jp

Martin Steinhart Institute for Chemistry, University of Osnabrück, 49069 Osnabrück, Germany, martin.steinhart@uos.de

Z. Sun Department of Engineering, University of Cambridge, Cambridge CB3 0FA, UK, zs244@eng.cam.ac.uk

P.H. Tan Department of Engineering, Cambridge and State Key Laboratory for Superlattices and Microstructures, University of Cambridge, 912 Beijing, China, phtan@red.semi.ac.cn

G.R. Willmott Industrial Research Limited, Lower Hutt, New Zealand; The MacDiarmid Institute for Advanced Materials and Nanotechnology, Lower Hutt, New Zealand, g.willmott@irl.cri.nz

R.M. Young Industrial Research Limited, Lower Hult, New Zealand, r.young@irl.cri.nz

Chapter 1
Ion Channels, Nanotubes in Living Cells

Francisco Bezanilla

Abstract Living cells are surrounded by a lipid bilayer that separates the internal from the external media. As the very hydrophobic nature of the core of the lipid bilayer prevents the exchange of charged species between the interior and exterior of the cell, there are specialized structures inserted in the lipid bilayer that carry out the exchange of ions. These structures are integral membrane proteins that may be classified as transporters, ion pumps and ion channels. Of special interest here are the ion channels which are proteins specialized to conduct ions across the membrane with the distinguishing characteristic that the ionic flow is driven exclusively by the electrochemical gradient of the conducted ionic species. These channels are found in the cell surface membrane and also in membranes of internal cell compartments.

Different types of ionic channels. Following the classification of the Transport Classification Database of the University of California at San Diego (http://www.tcdb.org) the major subclasses within the channels/pore class of transporters are the 1. α-Type channels, 2. β-barrel porins, 3. pore-forming toxins, 4. Non-ribosomally synthesized channels and 5. Holins. The α-Type channels are transmembrane proteins found in all type of organisms from bacteria to higher eukaryotes and they consist mainly of α-helical membrane spanning segments. An example of this large group is the KcsA channel shown in Fig. 1.1. The β-barrel porins are found in the outer membrane of Gram-negative bacteria, mitochondria, plastids and possibly in acid-fast Gram-positive bacteria. The Pore-forming toxins are normally synthesized by one cell and after secreted to the medium they can insert into other cells and form pores that, in general produce transport of electrolytes and small molecules that kill the cell. Hemolysin is an example of a pore-forming toxin and is shown in Fig. 1.1.

Although many molecules of type 2 and 3 have been studied in detail and some have been modified to be used as sensors, we will be mainly concentrate here on the α-type channels that play fundamental roles in cell homeostasis and dedicated function. The α-type channels have been characterized extensively in their function, genetics and structure.

F. Bezanilla (✉)
Department of Biochemistry and Molecular Biology, The University of Chicago,
Chicago, IL 60637, USA
e-mail: fbezanilla@uchicago.edu

O. Hayden, K. Nielsch (eds.), *Molecular- and Nano-Tubes*,
DOI 10.1007/978-1-4419-9443-1_1, © Springer Science+Business Media, LLC 2011

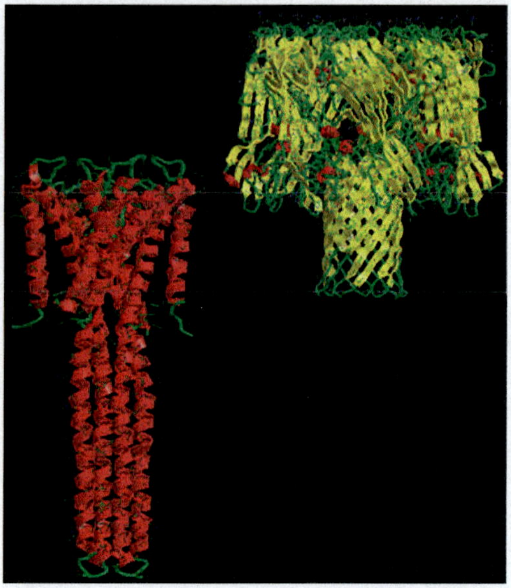

Fig. 1.1 On the *left* is the full structure of KcsA [20] an α-Type channel and on the *right* is hemolysin [21] a pore forming toxin that is mainly a beta-barrel structure. The two *white lines* represent the approximate boundaries of the lipid bilayer

A brief history of ion channel research. The study of ionic conductances that were intimately connected to the desire to understand the origin of the nerve impulse using the squid giant axon, may be considered the first steps in ion channel research. The ion channel as a conducting pore with properties of permeation, selectivity and gating was introduced in the 1970s, before any data on ion channel molecular structure was available, based on electrophysiological evidences. During that time channels were incorporated to artificial lipid bilayers and the first recordings of the tiny currents through one isolated channel were obtained. The number of biological preparations was expanded enormously when the patch clamp was introduced that also allowed to see, for the first time, the opening and closing of the channels in their native membrane. After that, molecular cloning initiated our understanding of ion channels at the molecular level and it was further expanded by the crystal structures and spectroscopic methods since the 1990s. A more detailed account may be found in [1].

It is important to note that in ion channel research the source of the channels studied has essentially spanned the entire biological world. Ion channels are fundamental for ion exchange and they play a fundamental role at all levels of the biological scale, from resting potential to cell homeostasis. In several cases we have learned about the structure of a bacterial channel before its eukaryotic counterpart because it has been more amenable to over expression and crystallization but in the process we have learned that there is strong conservation within families across distantly related species. The objective of understanding how the function of a channel is carried out by the structure has driven a large part of the research but also practical applications, such as drug design have been a very strong driving force in ion channel research.

Common features of ion channels. The fact that the driving force of ion conduction is the electrochemical gradient, suggests a simple picture whereby the protein has a conducting tube or pore, mostly hydrophilic, that connects both sides of the membrane. This minimum picture is normally complicated by specialized features that make channels selective, rectify and gate open and close. In addition, the minimum picture is complicated by the large number of families of ion channels that differ in their basic structure making it difficult, if not impossible, to talk about a prototypic ionic channel (see previous paragraph). The pore or the conducting tube, however is still the common theme in all these families but the pore is formed in very different ways in different families and this aspect is still the subject of active research giving us clues as of how conductance, selectivity and gating is achieved in different types of channels. Thus we should mention that are ion channels formed anywhere from one subunit (or homologous domain) to multiple subunits. Examples are: a single subunit per pore such as the Cl channel and the proton channel (although dimerized, forming two channels), three subunits (acid sensing ion channel), four subunits (or domains) (K^+ channels, voltage gated Na^+ and Ca^{2+} channels), five and seven subunits (ligand-gated and stretch activated channels).

Ion channels are very efficient ion transport machines: typically they can carry about 10^8 ions per s and many of them they do so by excluding all other ionic species for which they are not selective. There is no external source of energy to carry out the ion transport, except for the electrochemical gradient of the ions being transported. Ion channels do not transport ions against the electrochemical gradient. The size of the channel varies greatly between different families and loosely depends on how complicated their function might be. For example, KcsA, a bacterial K^+ channel that opens at low pH has a molecular weight of about 67 KD while a voltage gated Na^+ channel is about 230 KD. The number of ion channels in cells depends on the cell type and the channels may be located in different densities in different parts of a cell depending on the specific cell function. For example, the number of Na channels can be as large as 30,000 in the node of Ranvier which has a very small area compared to the internodal region that has only a few channels. On the other hand in a non-myelinated nerve the Na channel density may vary between 6 and 300 channels per μm^2 of cell surface area.

Most of the ion channels gate. This means that the conduction pathway is not always available because it goes from closed to open or to some intermediate conducting state. In most cases this gating process is controlled by chemical and physical interactions of the channel protein and its surroundings. The role and function of ion channels had a sudden advance when the basic description of the ion conductance changes that are responsible for the generation and propagation of the nerve impulse were unveiled [2, 3]. Although the level of understanding in these studies was at the macroscopic level with only hints that the ionic conduction was through discrete entities [3] these studies revealed that the conductance was regulated by the membrane voltage. Once the recording of the currents through only one channel (single channel recording) was possible [4] we learned that the ionic conduction through the conducting pore is interrupted by periods when the pore does not conduct at all. The periods of conduction are controlled by the membrane potential in the case of the voltage-gated channels or by the concentration of the neurotransmitter in the case of the ligand-gated channels. Regardless of the mechanism, even channels that do not seem to be controlled by external physical or chemical forces, show interruptions in the ion conduction in a process that has been called gating. The molecular basis of the gating process has been clarified in several cases and varies from a conformational change of part of the channel protein that physically obstruct the conduction pathway to occlusion of the conduction pore by polyvalent cations driven by the electric field.

Selectivity. The reason for the gating process becomes clear when one realizes that different types of channels exchange different ions and that the cell must regulate what ions must go in and out depending on the cell function and its metabolic state. Selectivity means that the channel will have preference for a particular ion over others even if the others are in excess. The classical example where selectivity plays a major role in cell homeostasis is the resting and active nerve. In nerve cells there are many types of ion channels but of particular interest for the nerve conduction are the sodium selective channels and the potassium selective channels. Most eukaryotic cells have a much larger concentration of K^+ inside than outside and the reverse is true for Na^+ ions (see Fig. 1.2). In the resting state, the cell has a voltage of about –70 mV (negative inside with respect to the outside). This membrane potential is mainly given by K^+ channels and the membrane potential is very close to the equilibrium potential of K (see Fig. 1.2a). It is clear that if other ion channels, such as Na^+ channels were in the conducting state, the resting potential would collapse. In fact, as a few Na^+ channels are open, the membrane potential is not quite at the equilibrium potential of K^+ and there is a small unidirectional flow on K^+ leaving the cell and Na^+ entering the cell that the sodium potassium pump has to correct by transporting those ions against their electrochemical gradient using the chemical energy of ATP. (see Fig. 1.2b).

Because Na^+ channels are selective to conduct Na^+ ions, when the Na^+ channels open the membrane becomes more selective to Na^+ and the membrane potential shifts temporarily towards the Na^+ equilibrium potential which is positive inside generating the upstroke of the nerve impulse which is terminated by an increase of the number of active K^+ channels and decrease of conducting Na^+ channels. The cycle of resting potential and nerve impulse illustrates how important is ion channel selectivity and its regulation by changing the number of channels that are open by external events, such as the membrane potential. Calcium channels are another example of an important player in the homeostasis in cells. Normally Ca^{2+} concentration is maintained at very low values inside the cell by keeping the Ca^{2+} channels closed. However, multiple fundamental processes are regulated by the increase in ionic Ca^{2+} in the cytoplasm such as the release of synaptic vesicles or muscle contraction and those changes in concentrations are carried out by opening Ca^{2+} channels either in the surface of the membrane or in Ca^{2+} channels located in membranes that delimit intracellular Ca-containing compartments.

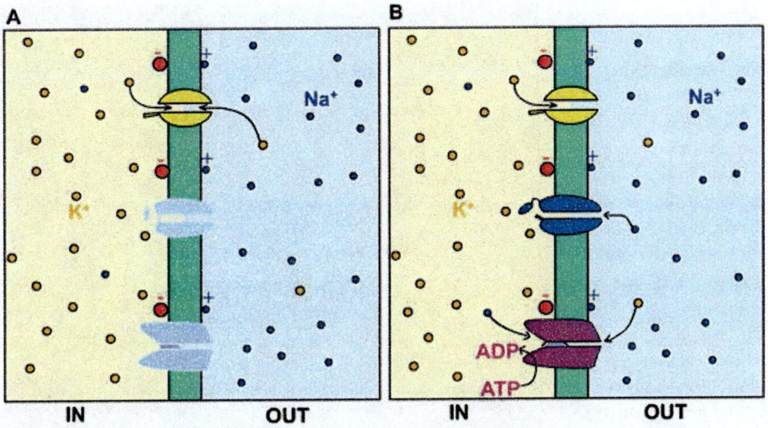

Fig. 1.2 Schematic representation of the ionic gradients, fluxes and the origin of the resting potential. The K ions (*brown*) are about 10 times more concentrated inside than outside the cell, while Na ions (*blue*) are 10 times more concentrated in the outside than inside. Anions are not pictured except for the ones that are in excess in one side with respect to the other side (*red*). (**a**). Case where the K channels (*brown*) are the only open channels. In this case the system is in equilibrium and the flux is zero because the chemical gradient is balanced by the electrical potential that is negative inside (see ionic dipoles across the membrane) according to the Nernst equation [22]. (**b**). In the more realistic case, in the resting state for about 10 K channels that are open one Na channel is open, therefore there is some exchange of the internal K for external Na but, as the membrane is mainly selective to K, the resting potential is still negative inside and its value can be approximated by the Goldman-Hodgkin and Katz equation [22]. To maintain the gradients in the presence of these fluxes, the Na/K pump (*violet*) transports Na and K against their gradients using the energy from ATP hydrolysis

The molecular basis of conduction and selectivity. It is clear from the above discussion that the ability of an ion channel to discriminate between different ionic species (selectivity) is a fundamental property that allows multiple cells functions. In fact ions channels are loosely classified according to their ionic selectivity. The most important players are cationic selective or anionic selective channels with a few examples of channels that have almost no selectivity. How is selectivity implemented in ion channels has been a question that has been asked using electrophysiological techniques in combination with site directed mutagenesis but the real breakthrough did not occur until the crystal structure of KcsA channel was unveiled [5].

KcsA is a bacterial channel that is K^+ selective and is made of four identical subunits each having two transmembrane helices organized around a central conducting pore (see Fig. 1.1). This basic pore structure has been found in several other K^+ channels that have been crystallized, including the voltage-gated channels Kv1.2 and its chimera with Kv2.1 (Fig. 1.3). The central pore has an intracellular region lined by the inner helices and an extracellular region that is formed by a reentrant amino acid loop where all the backbone carbonyls point toward the pore lumen. This region has been demonstrated to be the seat of K^+ selectivity in what

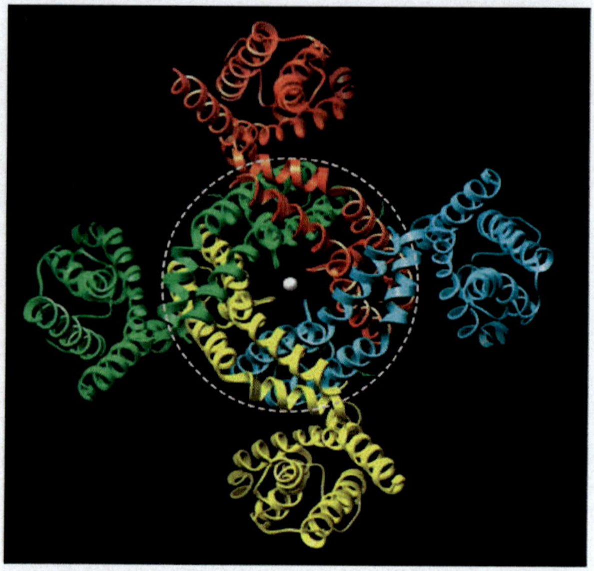

Fig. 1.3 Bird's eye view of a Kv channel from the intracellular side from the crystal structure of a chimera of Kv1.2 and Kv2.1 [12]. Subunits are color coded. The region enclosed in the *circle* is the pore region that shows a K^+ ion in the *center*. The *inner helices* are in the open state, broken at the PVP motif. The channel is expected to be in the inactivated state because it has been depolarized for a long time. The *helices* outside the *circle* form the voltage sensor

is called the selectivity filter. The carbonyls are at the correct distance to replace the water of hydration of the K^+ ions (see Fig. 1.3) with almost no change in free energy. Therefore conduction occurs after the K^+ ion loose its hydration shell and gets 'hydrated' by the backbone carbonyls of the pore while progressing from one carbonyl site to the next. With the atomic structure it has been possible to compute the flow of potassium through the KcsA channel [6] giving a physical representation of the transport through the pore as a multisite jumping process that reproduces the unidirectional flux exponent found experimentally [7].

The question of how the pore can select K against Na has been explained as a consequence of the different ionic radii of Na and K. As the K ionic radius is a perfect fit in the carbonyl cage, the smaller Na radius would not fit as well and it would be excluded from the permeation pathway. The problem with this notion is that due to thermal vibration one would expect that the carbonyl cage could get smaller to accommodate the Na^+ ion. However, this does not occur because the carbonyl dipole moment repulsion prevents them to become closer to each other making the radius-based exclusion hypothesis plausible [8].

The question of how selectivity is achieved in voltage gated Na^+ channels has not been answered at the same level because there is no crystal structure available but comparison of the amino acid sequences of K^+ and Na^+ channels indicate the importance of two negatively charged residues (aspartate and glutamate) that change the local electrostatics favoring Na^+ over K^+. In the case of the voltage-gated Ca^{2+} channels there are four negatively charged residues, as expected by the increase in the positive charge of the Ca^{2+} ion with respect to Na^+. Multiple proposals to explain selectivity of Ca^{2+} channels have been made taking in account the negatively charged amino acids in the pore [9] and using the K^+ channel structure as a basis but without a crystal structure they must be considered as models. However, it is important to note that in fact just the presence of the negative charges and how the ions are crowded in the pore are enough to explain Ca^{2+} selectivity [10].

KcsA has also been important in understanding the molecular basis of gating. The four inner helices cross each other blocking the access of ions to the pore in the closed state. The channel becomes conducting when these helices break at a level of a glycine residue in the helix allowing a rotation while at the same time all four helices spread apart opening the passage of the ions [11]. The probability of going to the open state is increased by lowering the internal pH, making this channel a pH-gated channel. Similar channels that are also gated by pH in a more narrow range exist in eukaryotic cells and are important in maintaining resting potential. In the case of the voltage-gated K^+ channel, the inner helices of the pore bend in the PVP motif [12]. Recently, it has been found that the bundle crossing of the inner helices is not the only gate but that the selectivity filter also acts as a gate [13]. This means that the pore has two gates in series and conduction will occur only when both gates are open. The selectivity filter gate explains a phenomenon called slow inactivation (the channel stops conducting when it has been activated to open for a long time) that occurs in most of the voltage-gated Na^+ and K^+ channels.

When the diameter of the conducting pore is larger, the selectivity is decreased and this is expected when the pore is central of an increasing number of subunits. This is the case of the ligand gated activated channels such as the well studied acetyl-choline receptor that are made of five subunits surrounding a central pore. These channels do not select well between Na^+ and K^+ ions and the larger sized pore is confirmed in the available structures [14]. An even larger increase of the number of subunits decreases selectivity even more as has been shown for the stretch activated channels MscL and MscS that have a central pore surrounded by five and seven subunits respectively.

Asymmetry is also possible. The prototype of the anion channel is the Cl^- channel that in fact is a family compose of several members. In this case the pore is not a central cavity surrounded by several subunits but is a tortuous channel lined by multiple helices of an extremely complex structure that forms a dimer with one of these pores in each subunit [15]. A complication has arisen in the interpretation of conduction based on the crystal structure because it was found that the protein that was crystallized is not a channel but a transporter where Cl^- is simultaneously moved with protons [16]. The studies been carried out presently with Cl^- channels and their relation to the structure are of great interest because they start to blur the line that separates channels and transporters and may help to understand both types of membrane conductances at a detailed atomic level.

The proton channel [17] is another example of a pore that is not formed by the association of subunits of domains around a central region. This channel is a dimer [18] but it seems that each monomer contains its own pore. Different to the case of Na^+, K^+ and Ca^{2+} channels, the central conducting pore structure is not found in the proton channel although it has all four transmembrane segments that form the voltage sensor of voltage gated channels. The actual pathway for the translocation of the protons has not been identified but a possibility is that it is formed by the sensor itself as it was found in a mutated voltage sensor of the Shaker K^+ channel [19].

Relation to synthetic nanotubes. One of the main challenges facing the understanding of biological ion channels is the correlation of structure and function at the atomic level. This has only been possible in a very limited number of cases such as the K^+ selectivity in KcsA. One obstacle is the determination of the three-dimensional structure of ion channels because membrane proteins are more difficult to obtain in crystal form. But even when this obstacle is overcome, the structure obtained by solving three dimensional crystals is of one particular state of the channel and in an environment that does not resemble their native surrounding: the lipid bilayer. By having a better knowledge of the structures these problems are decreased in the case of synthetic nanotubes and the correlation of their structure with their function can also be expanded by molecular dynamics. The similarities of ion channels with nanotubes (voltage-gated ion channels are molecular transistors; ions channels are selective) should help in elucidating the basic mechanisms underlying their function.

But there is another aspect where the synergism of simultaneous research on biological ion channels and synthetic nanotubes can be extremely productive. Many ion channels are very selective and also respond to nanomolar amounts of specific transmitters or toxins making them excellent nanoprobes. By combining our knowledge of how the biological ion channels acts as probes and the powerful techniques for synthesis of nanotubes we should be able to build better and more robust chemical and biological probes that could be used in health care and chemical detection.

Acknowledgments Many thanks to Dr. R. Latorre for reading the manuscript. Supported by NIH grant GM030376.

References

1. F. Bezanilla, *Neuron* **60** (3), 456 (2008).
2. K.S. Cole, H.J. Curtis, *The Journal of General Physiology* **22**, 649 (1939).
3. A.L. Hodgkin, A.F. Huxley, *The Journal of Physiology* **117** (4), 500 (1952).
4. G. Ehrenstein, R. Blumenthal, R. Latorre et al., *The Journal of General Physiology* **63** (6), 707 (1974); E. Neher and B. Sakmann, *Nature* **260** (5554), 799 (1976).
5. D.A. Doyle, J. Morais Cabral, R.A. Pfuetzner et al., *Science (New York, NY)* **280** (5360), 69 (1998).
6. S. Berneche, B. Roux, *Proceedings of the National Academy of Sciences of the United States of America* **100** (15), 8644 (2003).
7. A.L. Hodgkin, R.D. Keynes, *The Journal of Physiology* **128** (1), 61 (1955).
8. S.Y. Noskov, S. Berneche, B. Roux, *Nature* **431** (7010), 830 (2004).
9. W.A. Sather, E.W. McCleskey, *Annual Review of Physiology* **65**, 133 (2003).
10. D. Boda, M. Valisko, D. Henderson et al., *The Journal of General Physiology* **133**, 497 (2009).
11. Y. Jiang, A. Lee, J. Chen et al., *Nature* **417** (6888), 515 (2002); E. Perozo, D.M. Cortes, and L.G. Cuello, *Science (New York, NY)* **285** (5424), 73 (1999).
12. S.B. Long, X. Tao, E.B. Campbell et al., *Nature* **450** (7168), 376 (2007).
13. J.F. Cordero-Morales, V. Jogini, A. Lewis et al., *Nature Structural & Molecular Biology* **14** (11), 1062 (2007).

14. R.J. Hilf, R. Dutzler, *Nature* **452** (7185), 375 (2008); N. Unwin, *J Mol Biol* **346** (4), 967 (2005).
15. R. Dutzler, E.B. Campbell, M. Cadene et al., *Nature* **415** (6869), 287 (2002).
16. C. Miller, *Nature* **440** (7083), 484 (2006).
17. I.S. Ramsey, M.M. Moran, J.A. Chong et al., *Nature* **440** (7088), 1213 (2006); M. Sasaki, M. Takagi, and Y. Okamura, *Science (New York, NY)* **312** (5773), 589 (2006).
18. H.P. Koch, T. Kurokawa, Y. Okochi et al., *Proceedings of the National Academy of Sciences of the United States of America* **105** (26), 9111 (2008).
19. D.M. Starace, F. Bezanilla, *Nature* **427** (6974), 548 (2004); F.V. Campos, B. Chanda, B. Roux, and F. Bezanilla, *Proceedings of the National Academy of Sciences of the United States of America* **104**, 7904 (2007).
20. S. Uysal, V. Vasquez, V. Tereshko et al., *Proceedings of the National Academy of Sciences of the United States of America* **106** (16), 6644 (2009).
21. L. Song, M.R. Hobaugh, C. Shustak et al., *Science (New York, NY)* **274** (5294), 1859 (1996).
22. B. Hille, *Ion channels of excitable membranes*, 3rd ed. (Sinauer, Sunderland, MA, 2001).

Chapter 2
Gramicidin Channels as Cation Nanotubes

Roger E. Koeppe II, Sigrid E. Schmutzer, and Olaf S. Andersen

Abstract The linear gramicidins constitute a family of peptide antibiotics produced by the soil bacterium Bacillus brevis. The first antibiotics to be used in clinical practice, the linear gramicidins exert their antibacterial activity by forming bilayer-spanning channels that increase the monovalent cation permeability of target bacterial plasma membranes. Gramicidin channels are synthesized by non-ribosomal peptide synthesis on large protein complexes and contain both D- and L-amino acid residues; they were the first channels of known chemical composition to be studied. The channels effectively serve as cation-selective organic nanotubes that span lipid bilayer membranes and provide a basis for examining many aspects of ion-channel function and channel-lipid bilayer interactions. The nanotube properties can be tuned by means of mutations or chemical changes to the subunit architecture, as well as by altering the channels' bilayer environment (e.g., the bilayer thickness). Indeed, many analogue sequences within the extended peptide family have been prepared by semi-synthesis or total synthesis. Diverse applications of gramicidin channels have enhanced our understanding of the microphysics of ion permeation, lipid-protein interactions and membrane protein function.

2.1 Introduction

The linear gramicidins constitute a family of peptide antibiotics produced by the soil bacterium *Bacillus brevis* [1, 2]. Analogue sequences within the extended peptide family also can be prepared by semi-synthesis or total synthesis [3–6]. The linear gramicidins were the first antibiotics to be used in clinical practice [7]. They exert their antibacterial activity by forming bilayer-spanning channels [8] that increase the monovalent cation permeability of the target bacterial plasma membranes [9].

Among the ion-selective channels that span lipid bilayer membranes, gramicidin channels can be considered as robust organic nanotubes. The linear gramicidins are peptides containing both D- and L-amino acids that are linked in specific sequences within bacteria such as *Bacillus brevis* by means of a non-ribosomal synthesis in which the sequence is specified by the order of the of the domains that couple

R.E. Koeppe II (✉)
Department of Chemistry and Biochemistry, University of Arkansas, Fayetteville, AR 72701, USA
e-mail: rk2@uark.edu

O. Hayden, K. Nielsch (eds.), *Molecular- and Nano-Tubes*,
DOI 10.1007/978-1-4419-9443-1_2, © Springer Science+Business Media, LLC 2011

formyl-LVal[1]-Gly[2]-LAla[3]-DLeu[4]-LAla[5]-
-DVal[6]-LVal[7]-DVal[8]-LTrp[9]-DLeu[10]-LTrp[11]-
-DLeu[12]-LTrp[13]-DLeu[14]-LTrp[15]-$NHCH_2CH_2OH$.

Fig. 2.1 Amino acid sequence of gramicidin A, the primary linear gramicidin produced by the soil bacterium *Bacillus brevis*. The N-terminal is blocked with a formyl group; the C-terminal is blocked with an ethanolamide group. Because the sequence strictly alternates between residues of L- and D-chirality, with Gly^2 considered as an "honorary" D-residue, the sequence can fold to accommodate a β-helical fold, The four tryptophans at positions 9, 11, 13 and 15 serve as "anchors" at the membrane/water interface to orient the respective subunits of a bilayer-spanning dimeric channel

amino acids to the nascent chain [10, 11]. The prototypical member of the gramicidin family is gramicidin A ("gA"), whose sequence includes glycine-2 (which is achiral) and otherwise all D-amino acids at the even-numbered sequence positions, interspersed between all L-amino acids at the odd-numbered sequence positions (Fig. 2.1). The strictly alternation between residues of L and D chirality confers resistance to proteases and allows the native channel structure to fold as a helix with a secondary structure similar to β-sheets, in which the side chains are on the outside of the helix and the peptide backbone lines the ion permeable pore.

The N-formyl blocking group of gramicidin A is added biosynthetically [10] and is required for channel activity [12]. By contrast, the C-aminoethanol blocking group, which derives biosynthetically from glycine, can be modified rather extensively with retention of channel activity [13–17].

In lipid bilayer membranes of appropriate thickness, namely those whose phospholipid acyl chains contain 12–18 carbons, the linear gramicidins fold into single-stranded dimeric membrane-spanning channels. In these channels, the formyl-L-valines meet at the bilayer center and the Trp indole rings anchor the subunits to the respective membrane/water interfaces (Fig. 2.2). The channel lumen is about 0.4 nm in diameter and permits the single-file passage of water molecules and alkali cations [18]. Anions are rejected because the peptide backbone residues do not solvate anions well, as can be deduced from bulk phase free energies of transfer [19] and computational studies on gA channel models [20]. Divalent cations are rejected because the electrostatic barrier for ion entry into the pore becomes forbidding, e.g. [21]. When the Trp interfacial interactions are broken or stressed – namely in organic solvents [22], or in thin or very thick membranes (those whose lipid acyl chains have less than 10 or more than 20 more carbons) [23, 24], or by selected substitution of Trp residues by less polar/amphiphilic residues [25] – the gramicidin dimer refolds away from the single-stranded, functional channel structure into alternative double-stranded dimer conformations (Fig. 2.3) [26], All of the folded dimers are β-helices, with a central backbone, with side chains projecting to the outside [26, 27], as depicted for the single-stranded dimeric "nanotube" in Fig. 2.2 and the double-stranded structure in Fig. 2.3. Usually the double-stranded conformations (also known at π helices [13]) are inert, i.e. not measurably permeable to

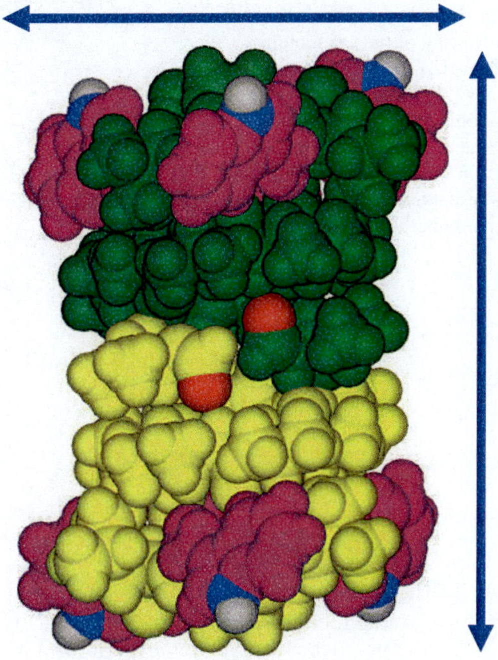

Fig. 2.2 Dimensions of the gramicidin channel, depicted using a CPK model. The calibration *arrows* represent 2.6 nm (*vertical*) and 2.0 nm (*horizontal*). The membrane-spanning channel surrounds a water-filled, cation-selective pore of about 0.4 nm diameter (see also Figs. 2.3, 2.4, and 2.5)

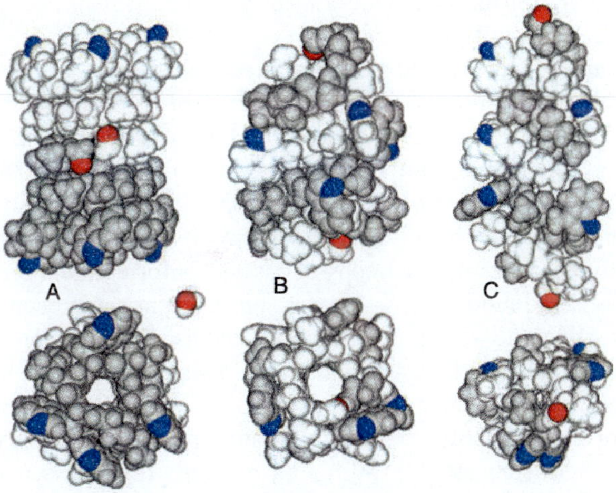

Fig. 2.3 Folding of the gramicidin channel. Side views (*upper*) and end views (*lower*) showing: (**a**) the membrane-spanning channel, $\beta^{6.3}$-helical dimer with 6.3 residues per helical turn; and (**b**, **c**) two members of the set of double-stranded dimer conformations that are observed in organic solvents. Formyl oxygens are *red* and Trp indole NH groups *blue*. A water molecule is shown for reference. Structures **a** and **b** each enclose a pore of about 0.4 nm diameter, whereas structure **c** does not

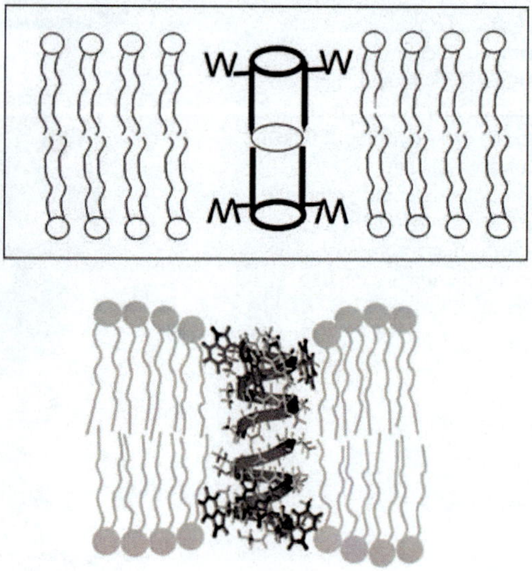

Fig. 2.4 Tryptophan anchoring of the gramicidin channel. The Trp indoles rings ("W" in the upper cartoon) anchor each subunit of the membrane-spanning, cation-conducting channel to its respective membrane/water interface

ions [28], but selected combinations of subunit sequences can sometimes exhibit double-stranded channel activity [29, 30]. In the single-stranded channel conformation, the tryptophans of each gramicidin subunit serve as anchors to the membrane/water interface (Fig. 2.4), holding each respective subunit within the lipid leaflet on the side of the membrane to which it was added [31]. In the interfacial attachment-stabilized conformation, the folded β-helical backbone surrounds a pore of about 4 Å, allowing for the single-file transport of monovalent cations and water molecules (Figs. 2.3 and 2.4) [32].

Gramicidin channels have served as prototypical channels in the development of many physical approaches toward understanding the structure and function of membrane proteins [33]. The main physical approaches have included: (a) single-channel analysis using, e.g., the bilayer punch ("patch clamp") method [34, 35], (b) circular dichroism (CD) spectrosopy to detect the channel's secondary structure [22], (c) size-exclusion chromatography [25, 36] to distinguish weak from strong subunit interactions, (d) fluorescence spectroscopy to assess conformational preference and tryptophan positions and motions [37–39], (e) solution magnetic resonance (NMR) spectroscopy for complete structure determination using detergent-encapsulated (membrane-mimetic) samples [40, 41], and (f) solid-state NMR for complete structure determination using bilayer-incorporated samples [42, 43]. Some of the issues concerning conformational polymorphism and structural heterogeneity have been summarized elsewhere [28, 39, 44]. Although methods for the structural characterization of monomers have remained elusive, an in-plane X-ray scattering study of a gramicidin analogue for which dimer formation is blocked[1] has indicated that membrane-incorporated monomers assume a similar β-helical fold as the individual subunits within the native dimeric channels [45].

---

[1]When the N-formyl group is replaced by N-*t*-butoxycarbonyl, the dimer/monomer equilibrium is shifted by $\sim 10^{-5}$ [45].

When considered as cation nanotubes, the canonical single-stranded gramicidin channel can be used in diverse ways to investigate the properties of lipid bilayer membranes as well as the physics of ion transport through narrow tubes. This chapter will address fundamental properties and selected practical applications of these nanotubes which assemble from a pair of head-to-head single-stranded gramicidin molecules. The properties of these peptide nanotubes can be tuned by altering the amino acid side chains [46]. Furthermore, the two subunits that compose a given nanotube can be different (in heterodimers) or identical (in homodimers). Each of these assemblies offers advantages for particular applications, and it is possible to measure the energetic consequences of altering the amino acid sequence at the subunit interface.

2.2 Tuning the Channel Properties

Cation-conducting gramicidin channels assemble in lipid bilayer membranes when subunits from opposing bilayer leaflets diffuse laterally, to "line up" and then dimerize; the bilayer-spanning channels are stabilized by six intermolecular hydrogen bonds [31]. Dimer formation (and disappearance) is detected as a step change in the current across a high-resistance "black" lipid[2] membrane (Fig. 2.5) [8, 35]. That is, gramicidin channels do not "open" and "close" – except in a few special

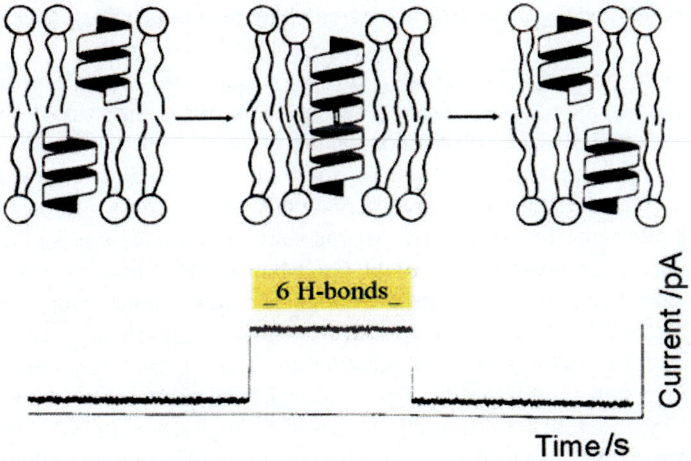

Fig. 2.5 Formation of a gramicidin channel. When subunits in opposing lipid leaflets dimerize, the dimer is stabilized by six hydrogen bonds, and a step change in the transmembrane current can be observed. The step size is typically about 3 pA for Na^+ or about 10 pA for Cs^+ permeation through a channel of gramicidin A (1.0 M salt, diphytanoylphosphatidylcholine/*n*-decane bilayers, 200 mV applied potential, 25°C). The channel lifetime is typically about 1 s. See also Figs. 2.7 and 2.9

[2] A "black" lipid membrane or "bilayer" lipid membrane ("BLM") is only one lipid bilayer thick and is optically black [80].

cases [47, 48] – but rather appear and disappear by a process of transmembrane association and dissociation. In the right-handed β-helical channel, the six hydrogen bonds link the carbonyl oxygens of L-residues 1, 3 and 5 to the backbone NH groups of L-residues 5, 3 and 1 on the respective subunits [49]. This hydrogen-bonded subunit arrangement was first proposed by Urry [27], albeit for a left-handed channel. The right-handed helix sense was established using magnetic resonance (NMR) methods [40, 50, 51].

Gramicidin channels are remarkably conductive. Even though the small pore radius restricts ions and water to move in single file [32, 52], the rate of ion movement through a channel is rapid: in 1.0 M NaCl the single-channel conductance is only about fivefold less than would be predicted for free diffusion through a 2.6×0.4 nm pore (Fig. 2.2) that offers no "excess" resistance [53]. Because gramicidin channels are seemingly ideally selective for monovalent cations [54], single-channel measurements provide direct information about the net cation flux through the channel. Gramicidin channels therefore are not just water-filled pores. Favorable short-range ion-channel interactions effectively compensate for the electrostatic barrier for ion movement through the low-dielectric bilayer core. Like other membrane-spanning ion channels, gramicidin channels catalyze ion movement across a lipid bilayer by providing a reaction path that obviates the ion's passage through the lipid bilayer hydrophobic core per se. For this reason, gramicidin channels (and other ion channels) belong to a special class of enzymes in which no covalent bonds are made or broken during the catalytic cycle.

Both the single-channel conductance and the mean channel lifetime of gramicidin channels can be regulated by engineered changes in the amino acid sequence [46].

A. *Conductance*: The cation conductance of gramicidin channels is sensitive to the introduction of many different side chains into the amino acid sequence (Fig. 2.6). For example, the substitution of amino acids with more polar sulfur- or fluorine-containing polar side chains for the hydrophobic valine at position one decreases the single-channel conductance. Notably, substituting trifluorovaline at position one alters the ion selectivity, as this substitution causes a 3-fold reduction in current in the presence of 1.0 M CsCl but a 6-fold reduction for 1.0 M NaCl[3] [55]. Replacing any of the four tryptophan indole dipoles with a nonpolar (and non-dipolar) side chain, such as Phe, decreases the single-channel conductance by about 25% [56] – with the deeper substitutions causing the larger conductance changes. Conversely, enhancing the Trp dipole by 5-fluorination will increase the conductance by about 20% [57]; in this instance through-space ion/5-F-Trp dipole interactions over a distance of several nm serve to attract cations toward the channel entrance and lower the energy barrier for ions to pass the bilayer center [58]. Even the aliphatic "spacer" residues between the tryptophans are important, as replacing

[3]Gramicidin A channels are about 3-fold selective for K^+, Rb^+ or Cs^+ over Na^+ [54]. Amino acid substitutions can serve to "tune" this cation selectivity [55]. Experiments with different cations are chosen for mechanistic studies.

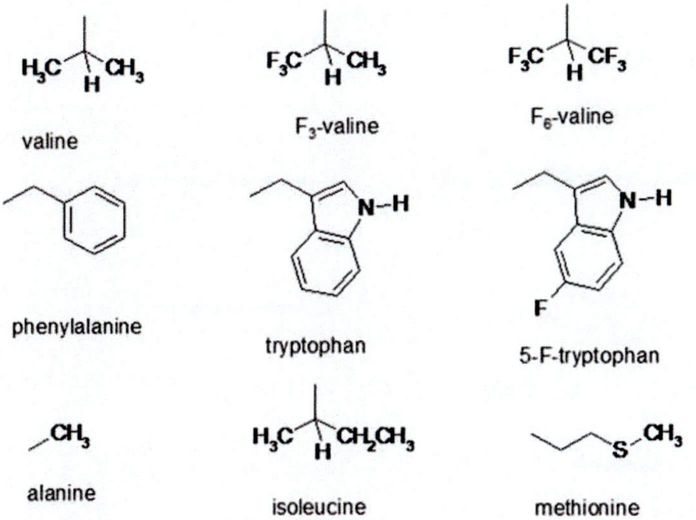

Fig. 2.6 Selected side chains of amino acids that have been introduced into gramicidin channels

D-Leu^{10,12,14} by D-Ile^{10,12,14} or D-Ala^{10,12,14} decreases the single-channel conductance for 1.0 M CsCl by 25–33% [59]. Furthermore, the cation conductance can be tuned by more subtle alterations in the amino acid sequence [49, 55]. These conductance changes occur within the framework provided by the β-helical fold of the membrane-spanning channel, which depends on maintaining the pattern of alternating residues of L- and D-chirality within the gramicidin channel sequence [4, 47, 57, 60].

To illustrate how subtle changes can modulate channel function, Fig. 2.7 shows the influence of glycine and alanine replacements at positions 1 and 2 on the properties of gramicidin channels [49]. [Gly^1-Gly^2]gA channels conduct, with a single-channel current of about 2 pA in the presence of 1.0 M NaCl (and 200 mV applied potential), and exhibit a mean channel lifetime of about 70 ms (Fig. 2.7a, b). Replacing Gly^1 by Ala^1 causes an approximate doubling of the Na^+ current and a 2.5-fold increase in lifetime, whereas a Gly→Ala substitution at position 2 causes no change in the single-channel current but a >10-fold increase in the lifetime (Fig. 2.7c, d). Similar changes in single-channel lifetimes are observed when Cs^+ is the permeant ion, though the current changes are less [49]. Remarkably, the effects of alanines on the channels' ion permeability and lifetimes at the two positions are largely uncoupled.

B. *Channel lifetimes*: As noted above, gramicidin channel lifetimes also are sensitive to the nanotube's amino acid sequence and to the overall channel length, specifically how well the length of the channel's hydrophobic exterior matches the thickness of the host lipid bilayer (Fig. 2.8), a concept known as *hydrophobic matching* [61, 62]. The results shown in Fig. 2.8 illustrate that neither the absolute tube length nor the absolute lipid bilayer thickness, but rather their difference, regulates

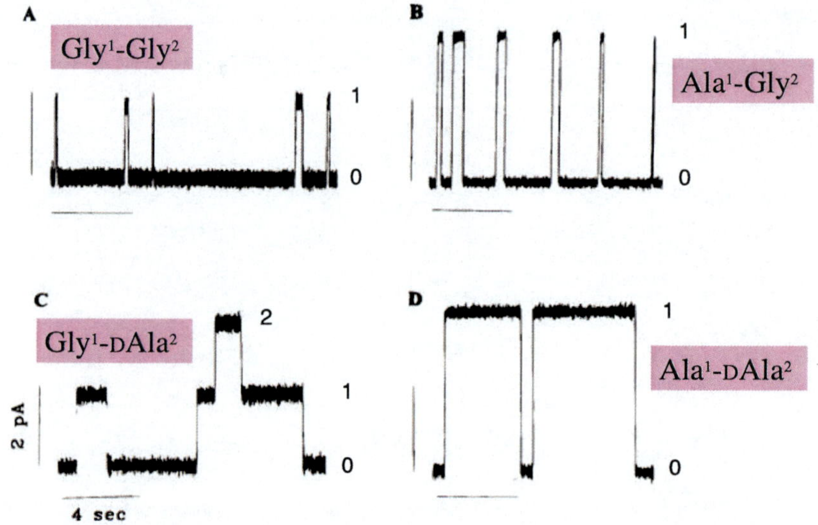

Fig. 2.7 Examples of single-channel current transitions that accompany the dimerization of gramicidin subunits that have glycine or alanine at positions one and two. Current level "1" or "2" denotes one or two conducting channels, whereas level "0" denotes the baseline when there are zero conducting channels. Diphytanoylphosphatidylcholine/*n*-decane bilayers, 1.0 M NaCl, 200 mV applied potential, 25°C. Modified from [49]

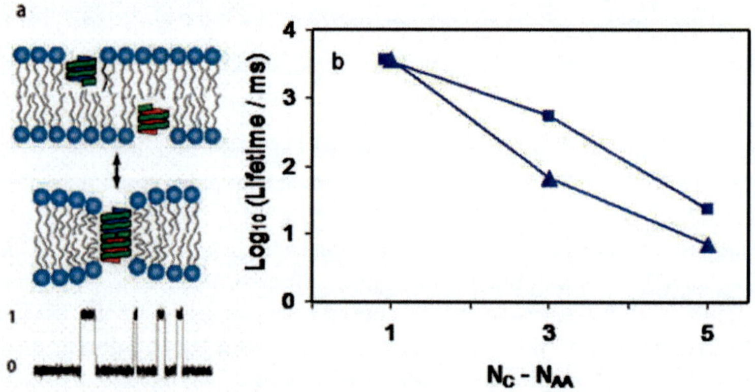

Fig. 2.8 Effect of a bilayer-channel hydrophobic mismatch on gramicidin channel lifetimes. (**a**) Gramicidin channels form by means of transmembrane dimerization of nonconducting subunits (cf. Fig. 2.5), which is observable as discrete current transition between two levels: 0, in which there is no conducting channel, and 1, in which there is one conducting channel. Because channel formation causes a local bilayer thinning, the average channel lifetimes vary as a function of the hydrophobic mismatch. (**b**) Gramicidin channel lifetimes as a function of a hydrophobic mismatch parameter ($N_C - N_{AA}$), where N_C denotes the number of carbon atoms in the acyl chains of the bilayer-forming lipids (monounsaturated phosphatidylcholines), and N_{AA} denotes the number of amino acids in one gramicidin subunit. The *squares* (■) denote experiments in which N_{AA} was varied while the lipid acyl chains had 16 carbons. The *triangles* (▲) denote experiments in which N_C was varied at constant N_{AA} of 15 amino acids per subunit

Table 2.1 Variation of gramicidin channel lifetimes as function of the nanotube amino acid sequence[a]

Residue 1	Residue 2	Dimer lifetime (ms)
Val	Gly	810
Norval	Gly	220
Leu	Gly	210
Norleu	Gly	280
Met	Gly	200
F_3-Val	Gly	180
F_6-Val	Gly	50
Phe	Gly	770
Gly	Gly	70
Ala	Gly	190
Gly	D-Ala	1,100
Ala	D-Ala	2,200

[a]Data from [55, 65] and [49]. Experimental conditions: 1.0 M NaCl, diphytanoylphosphatidyl choline/n-decane, 200 mV, 25°C. Day-to-day variations of ±10% are observed for the mean channel lifetimes [65]. Residues 1 and 2 are located at the subunit/subunit junction in a gramicidin dimer (Fig. 2.2). Channel lifetimes also vary significantly if the tryptophans near the C-terminal are substituted [56, 57]. These tryptophans are located away from the subunit junction but are critical anchors of the channel subunits to the bilayer/electrolyte interface [31] and therefore also influence channel lifetimes, often in highly dramatic ways [56].

the channel lifetime. The concept holds true regardless of whether the tube length is varied within a fixed bilayer or, conversely, a fixed-length tube is examined in bilayers of different thicknesses (Fig. 2.8b) [63].

Substitutions in the amino acid sequence also influence the channel lifetime. For example, polar residues at position one or two – at the subunit interface and near the bilayer center – decrease the channel lifetime (Table 2.1). Remarkably, branched hydrocarbon side chains (Val, Phe) confer significantly longer lifetimes than do straight side chains (Norval, Norleu, Met). Furthermore, D-Ala^2 yields significantly longer channel lifetimes than does Gly^2, independent of the identity of residue one (Table 2.1) [49]. Away from the subunit interface, tryptophans anchor the subunits to the respective membrane water interfaces. Substituting one or more of the Trp indole rings by phenylalanine or glycine, and even varying the intervening "spacer" residues, also have dramatic effects on the channel lifetime [56, 59, 64].

2.3 Applications of Heterodimers

Heterodimer experiments involve the random and independent associations of subunits when two chemically different gramicidin subunits are added to the same lipid bilayer membrane (Fig. 2.9). It thus becomes possible to measure the energetics of different subunit combinations [65, 66]. In varying experimental protocols, the

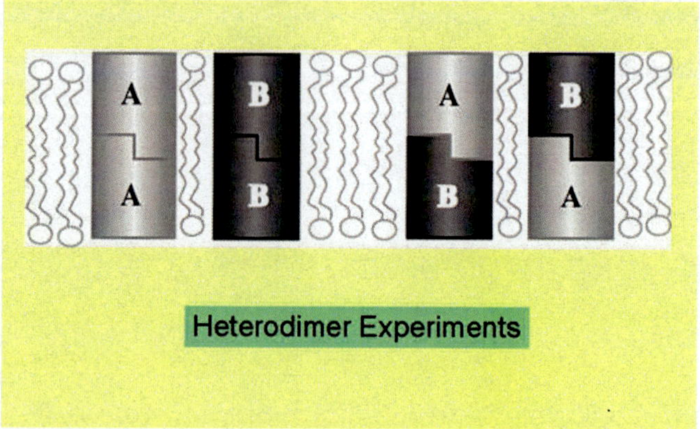

Fig. 2.9 Heterodimer formation. When two chemically different gramicidin subunits A and B are added to both sides of a bilayer, four types of channels may be able to assemble

respective subunits can be added to the same unique side of a membrane, to different sides, or to both sides. The Trp-anchored single-stranded subunits that constitute standard gramicidin channels tend to remain within the leaflet on the side of the membrane to which they are added initially [31]. This "anchoring" by the Trp indole rings at the membrane/water interface is effective for at least 30 min after single-sided addition of gramicidin subunits to only one side of a lipid bilayer membrane [31], meaning that gramicidin channels usually are observed only after adding gramicidins to both sides of a bilayer. If one or more of the Trp side chains are progressively substituted by other aromatic rings such as Phe, naphthyl or 1-Me-indole, the modified gramicidins have an increased probability of forming channels when added to only one side of the bilayer indicating that individual subunits can more easily cross the membrane [25, 67]. In contrast to the canonical single-stranded channels, the double-stranded channels that are formed only in rare cases tend to assemble when both subunits come from the same side of the bilayer membrane [24, 30].

Figure 2.10 shows results from a heterodimer experiment with single-stranded subunits. In such an experiment, subunits A and B (sequence analogues of gramicidin, in this case [Ala^1-D-Ala^2]gA and gA, respectively) were added to both sides of a planar lipid bilayer. The single-channel current trace (top of Fig. 2.10) shows current steps resulting from the formation of homodimeric AA and BB channels, and heterodimeric (or hybrid) channels. The heterodimeric channels will exist in two different orientations (relative to the applied potential), one in which the current is in the A$\rightarrow$B direction and one in which the current is in the B$\rightarrow$A direction. These two orientations are, in principle, distinguishable because the potential of mean force for ion movement will differ for the two channel orientations [68]. In this case, however, the hybrid channel orientation does not influence the channel properties because the A$\rightarrow$B and B$\rightarrow$A channels contribute to the same peak – of intermediate magnitude – in the current transition amplitude histogram (middle of

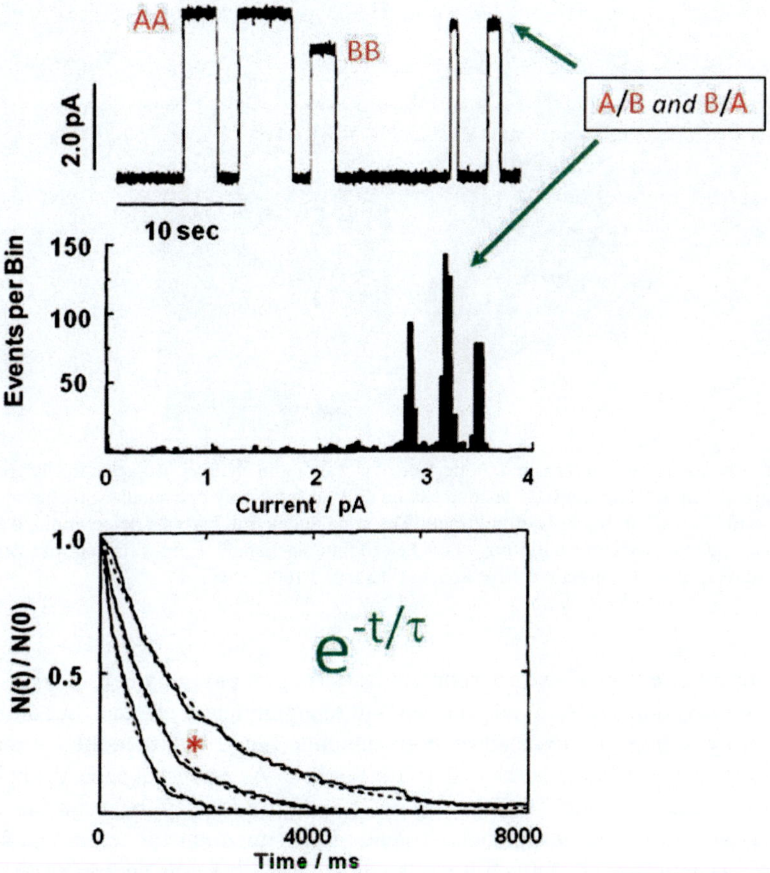

Fig. 2.10 Examples of heterodimeric channels. From top to bottom, the figure depicts a single-channel current trace, a current transition amplitude histogram, and a lifetime survivor plot that depicts the probability of channel disappearance as a function of time after formation. In this example, the heterodimers A→B and B→A exhibit nearly identical single-channel currents (*) and lifetimes (*) which, in both cases, are intermediate between the corresponding properties of the homodimeric AA and BB channels. Subunit A is [Ala^1-D-Ala^2]gA; subunit B is gA. Diphytanoylphosphatidylcholine/*n*-decane bilayers, 1.0 M CsCl, 200 mV applied potential, 25°C. Modified from [49]

Fig. 2.10). (The case where the two orientations can be distinguished will be illustrated below.) The hybrid channels also exhibit a single exponential probability of dissociating and a mean channel lifetime that is intermediate between those of the respective homodimers in a channel lifetime histogram or survivor plot (* in lower section of Fig. 2.10).

Heterodimer experiments reveal the (energetic) consequences of interrupting the pattern of alternating L- and D-residues that constitute the gramicidin channel backbone. For example, the pattern can be interrupted by deletion or insertion of one

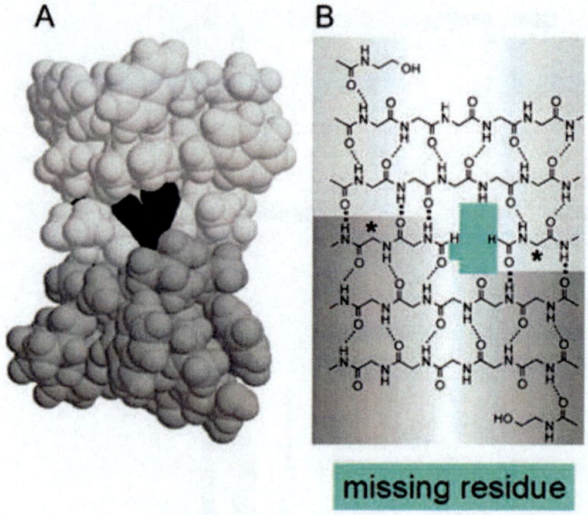

Fig. 2.11 Missing residue at the subunit interface of a gramicidin channel. Adjacent to Gly^2 (*black* in structure **a**; * in net diagram **b**), a residue can be deleted from only one subunit of a gramicidin channel (with "repair" of the N-formyl group). The residue deletion leaves a defect in the wall of the channel and lowers the mean lifetime of the heterodimeric channel, from which one can deduce that the "gap" incurs an energetic cost of about 10 kJ mol^{-1} [66]

residue, or by insertion of extra carbon atoms (CH_2 groups) into the backbone. If a single residue is deleted from only one subunit of a gramicidin channel, but the crucial N-formyl groups are retained on both subunits (Fig. 2.11), then there remains an effective "gap" or defect in the wall of the channel. As a consequence, the hybrid channel conductances and lifetimes are no longer intermediate between those of the respective homodimeric channels. Instead of the situation depicted in Fig. 2.10, the hybrid channels exhibit much *lower* single-channel currents and much *shorter* mean channel lifetimes than either of the corresponding homodimers (c.f. figures 3 and 4 in [66]). The consequence is that, when two gramicidin subunits differ in length by ± one residue, the channel's backbone structure and helix sense are not affected, but the resulting hybrid channels are destabilized by ~10 kJ mol^{-1} [66]. The destabilization of the heterodimeric nanotube is caused by the de facto "gap" – and loss of one intersubunit hydrogen bond – in the peptide backbone at the junction between the two subunits. This "missing residue" or "gap" analysis gives remarkably consistent results, regardless of whether a residue is added or removed from the N-terminal of gramicidin. (The symmetry of a dimer dictates that the initial residues where the respective subunits meet each other should be of the same chirality, while the β-helical fold dictates that residue chirality should alternate thereafter.) This consistency suggests that the transition state for dimer dissociation is reached when two hydrogen bonds are broken, when the subunits move apart in a rotating/sliding motion [66, 69, 70].

Gramicidin channels also can be used to examine general questions relating to the structural and functional consequences of inserting non-genetic amino acids into the amino acid sequence of a well-folded structure. What are the consequences of

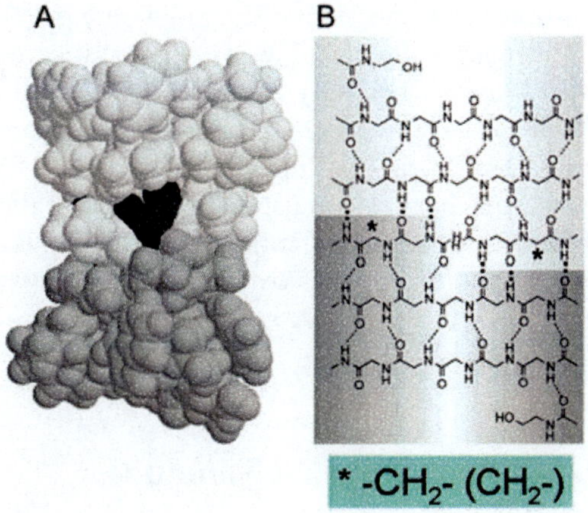

Fig. 2.12 Peptide backbone modification at the subunit interface of a gramicidin channel. Adjacent to Gly^2 (*black* in structure **a**; * in net diagram **b**), one (or two) extra CH_2 group(s) can be inserted into the backbone of a gramicidin channel. See text for details

inserting "extra" atoms into the channel backbone structure? Figure 2.12 illustrates the concept, with Gly^2 shown in black in panel A. Next to the backbone CH_2 of Gly^2 (* in Fig. 2.12b), we have incorporated one or two extra CH_2 groups. Remarkably, the extra methylene groups in the backbone of the nanotube are well tolerated, such that the subunits retain the ability to assemble into conducting channels. Figure 2.13

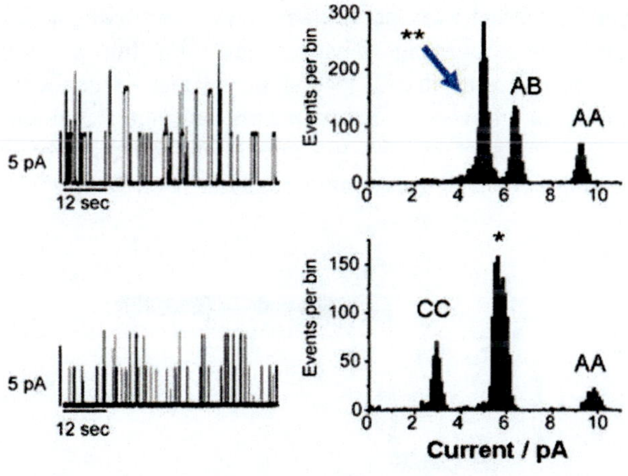

Fig. 2.13 Current reduction due to extra methylene groups at the subunit interface in a gramicidin channel. Subunits A, B and C have 0, 1, or 2 extra $-CH_2-$ groups, respectively, inserted next to the $-C_\alpha H_2-$ of Gly^2 in the native gramicidin sequence. *Left*: single-channel current transitions for mixtures of A/B subunits (*upper*) or A/C subunits (*lower*). *Right*: the respective current transition histograms. In the upper histogram AA homodimers and AB heterodimers exhibit distinct currents, whereas the BB homodimers and BA heterodimers exhibit the same average current (**→ peak). In the lower histogram, AA and CC homodimers exhibit distinctive high and low currents, while the oppositely oriented AC and CA heterodimers exhibit the same intermediate current (* peak). Experimental conditions: 1.0 M CsCl, diphytanoylphosphatidyl choline/n-decane, 200 mV, 25°C

illustrates that one extra CH_2 group (subunit B; upper panels) and even two extra CH_2 groups (subunit C; lower panels) still permit conducting channels to assemble. Interestingly, when only one extra CH_2 is present in only one subunit, the hybrid channel orientation – B→A or A→B – makes a difference, such that the current is different in the two directions. In one of the orientations, the current through the hybrid channel is indistinguishable from the current through the symmetric BB channels (peak ** in upper right of Fig. 2.13). By contrast, the heterodimers in which one subunit incorporates two extra CH_2 groups exhibit no such asymmetry, and the currents through the CA and AC channels are indistinguishable (peak * in lower right of Fig. 2.13).

2.4 Double-Barreled Nanotubes

As noted above, gramicidin channel formation is associated with a local bilayer deformation that varies with the (hydrophobic) mismatch between the bilayer thickness and channel length (see above). Further information about the importance of hydrophobic mismatch can be gained by forming covalent crosslinks between adjacent subunits in a bilayer leaflet, such that double-barreled channels may assemble [71]. Figure 2.14 illustrates the result of adding such tandem subunits – linked at their C-terminals, using a flexible hydrophilic linker that extends from the membrane/water interface out into the aqueous solution. When a channel forms, one first observes some initial "bursting" activity (red ellipse in Fig. 2.14), as the first barrel makes several apparent attempts to open, and then finally succeeds. Almost concerted with the ultimate opening of barrel one – or within a few milliseconds thereafter [71] – one observes that a second barrel also is conducting a cation current. The opening of the second barrel occurs without a prelude of bursting activity and is likely cooperative with the opening/stabilization of the first barrel. Together, the double-barreled assembly remains open for, typically, minutes

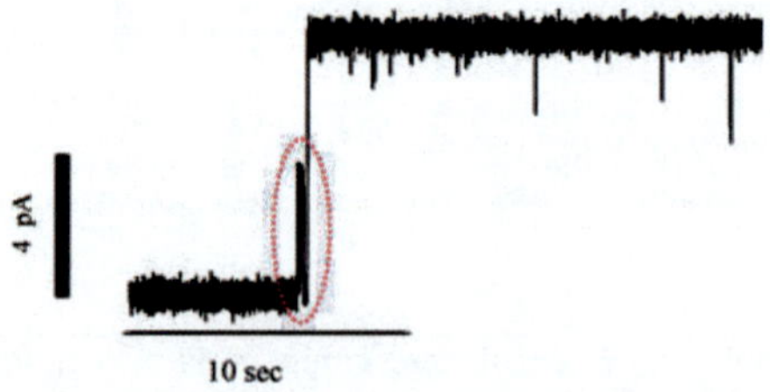

Fig. 2.14 Formation of a double-barreled gramicidin channel. Single-channel current trace obtained with tandem gramicidin subunits that are pairwise linked at the C-terminals. Channel formation is preceded by bursting channel activity (*dotted ellipse*), and is very quickly followed by an apparently concerted or cooperative formation of the second barrel of a double-barreled gramicidin channel. The double-barreled ensemble then remains stable for many tens of seconds (or even minutes). For further details, see Goforth et al. [71] Diphytanoylphosphatidylcholine/*n*-decane bilayers, 1.0 M CsCl, 200 mV applied potential, 25°C

(Fig. 2.14), much longer than the ~0.8 s lifetimes of standard gramicidin channels having unlinked subunits. The tandem dimers are stabilized some 100,000-fold relative to two independent channels [71].

The decisive results with the double-barreled channels suggest strongly that there indeed is a bilayer deformation energy is associated with gramicidin channel formation (because the channel length is somewhat less than the bilayer thickness). One infers that the bulk of the energy cost is associated with the formation of the first barrel (witness the associated bursting behavior), that the cost is much less for forming the second barrel, and that the double-barreled assembly is remarkably stable within the lipid bilayer membrane [71].

2.5 Energetics of Channel-Bilayer Hydrophobic Coupling

The properties of the double-barreled nanotubes provide additional experimental evidence for the notion of an energetic coupling between gramicidin channel formation and lipid bilayer deformation [69, 72–74]. Further experiments have been designed to measure the actual deformation energy, or rather changes in bilayer deformation energy. These types of experiments typically employ gramicidin subunits of different lengths [75]. A design concept is illustrated in Fig. 2.15. When the hydrophobic length (l) of a bilayer-spanning gramicidin nanotube differs from the average thickness (d_0) of the host bilayer (Fig. 2.15), the bilayer thickness will vary locally in the vicinity of the channel dimer in order to "match" the length of the nanotube's hydrophobic exterior to the thickness of the bilayer hydrophobic core. Such bilayer deformations incur an energetic cost, the bilayer deformation energy ($\Delta G^0_{\mathrm{def}}$), which will vary as a function of the tube shape, the inclusion/bilayer hydrophobic mismatch (d_0–l), the lipid bilayer elastic properties and the lipid intrinsic curvature [72, 73, 76, 77].

To explore these issues further, we have developed single-molecule methods to measure $\Delta G^0_{\mathrm{def}}$ and to probe changes in bilayer elastic properties using gramicidins as molecular force transducers [75]. The basis for these experiments is to construct nanotubes of different lengths by using gramicidin subunits of different lengths. Three fundamental approaches to measuring the deformation energy include: first, measurements of changes in channel lifetimes and appearance rates as the lipid bilayer thickness or channel length are varied; second, measurements of the equilibrium distribution among channels of different lengths, formed by homo- and heterodimers between gramicidin subunits of different lengths; and third, measurements of the ratio of the appearance rates of heterodimer channels relative to parent homodimer channels formed by gramicidin subunits of different lengths [75]. Each of these methods contributes an independent measure of the bilayer deformation energy. The best estimates to date (for hydrocarbon-containing bilayers) are 10–20 kJ mol^{-1} nm^{-2} for monoglyceride-based bilayers and 35–45 kJ mol^{-1} nm^{-2} for phosphatidylcholine-based bilayers [69, 75].

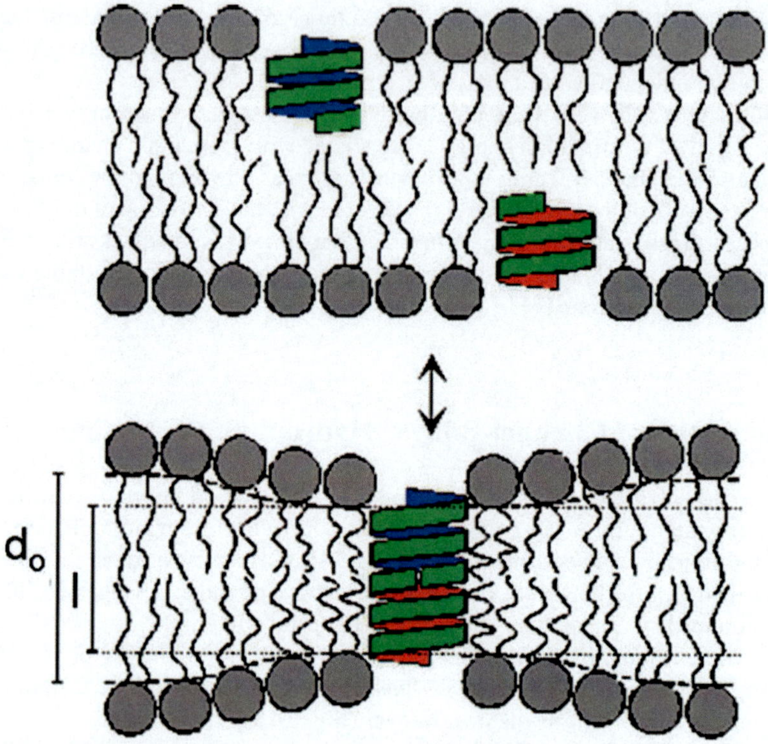

Fig. 2.15 Bilayer deformation energy. gA channel formation by the transbilayer dimerization of subunits may incur an energetic cost. When the channel's hydrophobic length (l) differs from the average thickness of the unperturbed bilayer hydrophobic core (d_o), channel formation will be associated with a bilayer deformation [78, 79], due to the compression and bending of the two bilayer leaflets, which is caused by the hydrophobic couple between the bilayer-spanning channel and the host bilayer. Figure from [53]

2.6 Summarizing Perspective

Gramicidin channels were the first channels of known chemical composition to be studied [8]. These channels serve as cation-selective nanotubes that span lipid bilayer membranes, and continue to provide a robust basis for examining many different aspects of ion channel function and channel-lipid bilayer interactions [39]. In the present context, the nanotube properties can be tuned by means of mutations or chemical changes to the subunit architecture. Diverse applications of gramicidin channels, in proof-of-principle studies and in their own right, have enhanced our understanding of the microphysics of ion permeation, lipid–protein interactions and membrane protein function.

Acknowledgments This work was supported in part by NIH grants GM21342 and RR15569. We thank Denise Greathouse, Gwen Mattice, Robyn Goforth, Haiyan Sun and Claus Nielsen for helpful discussions.

References

1. Dubos, R.J. Studies on a bactericidal agent extracted from a soil bacillus. I. Preparation of the agent. Its activity in vitro. J. Exp. Med. **70**, 1–10 (1939).
2. Hotchkiss, R.D. Gramicidin, tyrocidine, and tyrothricin. Anv. Enzymol. Relat. Areas Mol. Biol. **4**, 153–199 (1944).
3. Bamberg, E., Apell, H.J., Alpes, H., Gross, E., Morell, J.L., Harbaugh, J.F., Janko, K., and Läuger, P. Ion channels formed by chemical analogs of gramicidin A. Fed. Proc. **37**, 2633–2638 (1978).
4. Morrow, J.S., Veatch, W.R., and Stryer, L. Transmembrane channel activity of gramicidin A analogues: effects of modification and deletion of the amino-terminal residue. J. Mol. Biol. **132**, 733–738 (1979).
5. Heitz, F., Spach, G., and Trudelle, Y. Single channels of 9, 11, 13, 15-destryptophyl-phenylalanyl-gramicidin A. Biophys. J. **40**, 87–89 (1982).
6. Greathouse, D.V., Koeppe, R.E., II, Providence, L.L., Shobana, S., and Andersen, O.S. Design and characterization of gramicidin channels. Methods Enzymol. **294**, 525–550 (1999).
7. Herrell, W.E. and Heilman, D. Experimental and clinical studies on gramicidin. J. Clin. Invest. **20**, 583–591 (1941).
8. Hladky, S.B. and Haydon, D.A. Discreteness of conductance change in bimolecular lipid membranes in the presence of certain antibiotics. Nature (London) **225**, 451–453 (1970).
9. Harold, F.M. and Baarda, J.R. Gramicidin, valinomycin, and cation permeability of Streptococcus faecalis. J. Bacteriol. **94**, 53–60 (1967).
10. Kessler, N., Schuhmann, H., Morneweg, S., Linne, U., and Marahiel, M.A. The linear pentadecapeptide gramicidin is assembled by four multimodular nonribosomal peptide synthetases that comprise 16 modules with 56 catalytic domains. J. Biol. Chem. **279**, 7413–7419 (2004).
11. Schracke, N., Linne, U., Mahlert, C., and Marahiel, M.A. Synthesis of linear gramicidin requires the cooperation of two independent reductases. Biochemistry **44**, 8507–8513 (2005).
12. Szabo, G. and Urry, D.W. N-acetyl gramicidin: single-channel properties and implications for channel structure. Science **203**, 55–57 (1979).
13. Bamberg, E., Apell, H.J., and Alpes, H.J. Structure of the gramicidin A channel: discrimination between the pi-L,D and the beta-helix by electrical measurements with lipid bilayer membranes. Proc. Natl. Acad. Sci. U.S.A. **74**, 2402–2406 (1977).
14. Koeppe, R.E., II, Paczkowski, J.A., and Whaley, W.L. Gramicidin K, a new linear channel-forming gramicidin from *Bacillus brevis*. Biochemistry **24**, 2822–2826 (1985).
15. Cifu, A., Koeppe, R.E., II, and Andersen, O.S. On the supramolecular structure of gramicidin channels. The elementary conducting unit is a dimer. Biophys. J. **61**, 189–203 (1992).
16. Woolley, G.A., Jaikaran, A.S., Zhang, A., and Peng, S. Design of regulated ion channels using measurements of cis-trans isomerization in single molecules. J. Am. Chem. Soc. **117**, 4448–4454 (1995).
17. Woolley, G., Zunic, V., Karanicolas, J., Jaikaran, A., and Starostin, A. Voltage-dependent behavior of a "ball-and-chain" gramicidin channel. Biophys J. **73**, 2465–2475 (1997).
18. Andersen, O.S. and Koeppe, R.E., II. Molecular determinants of channel function. Physiol. Rev. **72**, S89–S158 (1992).
19. Cox, B.G., Hedwig, G.R., Parker, A.J., and Watts, D.W. Solvation of Ions. XIX. Thermodynamic properties for transfer of single ions between protic and dipolar aprotic solvents. Aust. J. Chem. **27**, 477–501 (1974).
20. Sung, S.S. and Jordan, P.C. The interaction of Cl- with a gramicidin-like channel. Biophys. Chem. **27**, 1–6 (1987).
21. Allen, T., Andersen, O., and Roux, B. Molecular dynamics – potential of mean force calculations as a tool for understanding ion permeation and selectivity in narrow channels. Biophys. Chem. **124**, 251–267 (2006).

22. Wallace, B.A., Veatch, W.R., and Blout, E.R. Conformation of gramicidin A in phospholipid vesicles: circular dichroism studies of effects of ion binding, chemical modification, and lipid structure. Biochemistry **20**, 5754–5760 (1981).
23. Greathouse, D.V., Hinton, J.F., Kim, K.S., and Koeppe, R.E., II. Gramicidin A/short-chain phospholipid dispersions: chain length dependence of gramicidin conformation and lipid organization. Biochemistry **33**, 4291–4299 (1994).
24. Mobashery, N., Nielsen, C., and Andersen, O.S. The conformational preference of gramicidin channels is a function of lipid bilayer thickness. FEBS Lett. **412**, 15–20 (1997).
25. Salom, D., Pérez-Payá, E., Pascal, J., and Abad, C. Environment- and sequence-dependent modulation of the double-stranded to single-stranded conformational transition of gramicidin A in membranes. Biochemistry **37**, 14279–14291 (1998).
26. Veatch, W.R., Fossel, E.T., and Blout, E.R. Conformation of gramicidin A. Biochemistry **13**, 5249–5256 (1974).
27. Urry, D.W. The gramicidin A transmembrane channel: a proposed π(L,D) helix. Proc. Natl. Acad. Sci. U.S.A. **68**, 672–676 (1971).
28. Andersen, O.S., Apell, H.-J., Bamberg, E., Busath, D.D., Koeppe, R.E., II, Sigworth, F.J., Szabo, G., Urry, D.W., and Woolley, A. Gramicidin channel controversy — the structure in a lipid environment. Nat. Struct. Biol. **6**, 609 (1999).
29. Koeppe, R.E., II, Greathouse, D.V., Providence, L.L., and Andersen, O.S. [L-Leu9-D-Trp10-L-Leu11-D-Trp12-L-Leu13-D-Trp14-L-Leu15]-gramicidin forms both single- and double-helical channels. Biophys. J. **59**, 319a(abstract) (1991).
30. Durkin, J.T., Providence, L.L., Koeppe, R.E., II, and Andersen, O.S. Formation of non-$\beta^{6.3}$-helical gramicidin channels between sequence-substituted gramicidin analogues. Biophys. J. **62**, 145–159 (1992).
31. O'Connell, A.M., Koeppe, R.E., II, and Andersen, O.S. Kinetics of gramicidin channel formation in lipid bilayers: transmembrane monomer association. Science **250**, 1256–1259 (1990).
32. Rosenberg, P.A. and Finkelstein, A. Interaction of ions and water in gramicidin A channels. Streaming potentials across lipid bilayer membranes. J. Gen. Physiol. **72**, 327–340 (1978).
33. Busath, D.D. The use of physical methods in determining gramicidin channel structure and function. Ann. Rev. Physiol. **55**, 473–501 (1993).
34. Neher, E., Sandblom, J., and Eisenman, G. Ionic selectivity, saturation, and block in gramicidin A channels. II. Saturation behavior of single channel conductances and evidence for the existence of multiple binding sites in the channel. J. Membr. Biol. **40**, 97–116 (1978).
35. Andersen, O.S. Ion movement through gramicidin A channels. Single-channel measurements at very high potentials. Biophys. J. **41**, 119–133 (1983).
36. Bañó, M.C., Braco, L., and Abad, C. Conformational transitions of gramicidin A in phospholipid model membranes. A high-performance liquid chromatography assessment. Biochemistry **4**, 886–894 (1991).
37. Scarlata, S.F. The effects of viscosity on gramicidin tryptophan rotational motion. Biophys. J. **54**, 1149–1157 (1988).
38. Mukherjee, S. and Chattopadhyay, A. Motionally restricted tryptophan environments at the peptide-lipid interface of gramicidin channels. Biochemistry **33**, 5089–5097 (1994).
39. Kelkar, D.A. and Chattopadhyay, A. The gramicidin ion channel: a model membrane protein. Biochim. Biophys. Acta **1768**, 2011–2025 (2007).
40. Arseniev, A.S., Barsukov, I.L., Bystrov, V.F., Lomize, A.L., and Ovchinnikov, Yu.A. Proton NMR study of gramicidin A transmembrane ion channel. Head-to-head right-handed, single-stranded helixes. FEBS Lett. **186**, 168–174 (1985).
41. Arseniev, A.S., Lomize, A.L., Barsukov, I.L., and Bystrov, V.F. Gramicidin A transmembrane ion-channel. Three-dimensional structure reconstruction based on NMR spectroscopy and energy refinement. Biol. Membr. **3**, 1077–1104 (1986).
42. Ketchem, R.R., Hu, W., and Cross, T.A. High-resolution of gramicidin A in a lipid bilayer by solid-state NMR. Science **261**, 1457–1460 (1993).

43. Ketchem, R.R., Roux, B., and Cross, T.A. High resolution polypeptide structure in a lamellar phase lipid environment from solid state NMR-derived orientational constraints. Structure **5**, 1655–1669 (1997).
44. Cross, T.A., Arseniev, A., Cornell, B.A., Davis, J.H., Killian, J.A., Koeppe, R.E., II, Nicholson, L.K., Separovic, F., and Wallace, B.A. Gramicidin channel controversy — revisited. Nat. Struct. Biol. **6**, 610–611 (1999).
45. He, K., Ludtke, S.J., Wu, Y., Huang, H.W., Andersen, O.S., Greathouse, D., and Koeppe, R.E., II. Closed state of gramicidin channel detected by X-ray in-plane scattering. Biophys. Chem. **49**, 83–89 (1994).
46. Koeppe, R.E., II and Andersen, O.S. Engineering the gramicidin channel. Annu. Rev. Biophys. Biomol. Struct. **25**, 231–258 (1996).
47. Oiki, S., Koeppe, R.E., II, and Andersen, O.S. Asymmetric gramicidin channels: heterodimeric channels with a single F6-Val-1 residue. Biophys. J. **66**, 1823–1832 (1994).
48. Oiki, S., Koeppe, R.E., II, and Andersen, O.S. Voltage-dependent gating of an asymmetric gramicidin channel. Proc. Natl. Acad. Sci. U.S.A. **92**, 2121–2125 (1995).
49. Mattice, G.L., Koeppe, R.E., II, Providence, L.L., and Andersen, O.S. Stabilizing effect of D-alanine-2 in gramicidin channels. Biochemistry **34**, 6827–6837 (1995).
50. Smith, R., Thomas, D.E., Separovic, F., Atkins, A.R., and Cornell, B.A. Determination of the structure of a membrane-incorporated ion channel. Solid-state nuclear magnetic resonance studies of gramicidin A. Biophys. J. **56**, 307–314 (1989).
51. Nicholson, L.K. and Cross, T.A. Gramicidin cation channel: an experimental determination of the right-handed helix sense and verification of á-type hydrogen bonding. Biochemistry **28**, 9379–9385 (1989).
52. Levitt, D.G., Elias, S.R., and Hautman, J.M. Number of water molecules coupled to the transport of sodium, potassium and hydrogen ions via gramicidin, nonactin or valinomycin. Biochim. Biophys. Acta **512**, 436–451 (1978).
53. Andersen, O.S., Koeppe, R.E., II, and Roux, B. Gramicidin channels: versatile tools. In Chung, S.H., Andersen, O.S., and Krishnamurthy, V. (eds.) Biological Membrane Ion Channels. Springer, New York (2007).
54. Myers, V.B. and Haydon, D.A. Ion transfer across lipid membranes in the presence of gramicidin. II. The ion selectivity. Biochim. Biophys. Acta **274**, 313–322 (1972).
55. Russell, E.W.B., Weiss, L.B., Navetta, F.I., Koeppe, R.E., II, and Andersen, O.S. Single-channel studies on linear gramicidins with altered amino acid side chains. Effects of altering the polarity of the side chain at position 1 in gramicidin A. Biophys. J. **49**, 673–686 (1986).
56. Becker, M.D., Greathouse, D.V., Koeppe, R.E., II, and Andersen, O.S. Amino acid sequence modulation of gramicidin channel function: effects of tryptophan-to-phenylalanine substitutions on the single-channel conductance and duration. Biochemistry **30**, 8830–8839 (1991).
57. Andersen, O.S., Greathouse, D.V., Providence, L.L., Becker, M.D., and Koeppe, R.E., II. Importance of tryptophan dipoles for protein function: 5-fluorination of tryptophans in gramicidin A channels. J. Am. Chem. Soc. **120**, 5142–5146 (1998).
58. Cotten, M., Tian, C., Busath, D.D., Shirts, R.B., and Cross, T.A. Modulating dipoles for structure-function correlations in the gramicidin A channel. Biochemistry **38**, 9185–9197 (1999).
59. Jude, A.R., Greathouse, D.V., Koeppe, R.E., II, Providence, L.L., and Andersen, O.S. Modulation of gramicidin channel structure and function by the aliphatic "spacer" residues 10, 12 and 14 between the tryptophans. Biochemistry **38**, 1030–1039 (1999).
60. Weiss, L.B. and Koeppe, R.E., II. Semisynthesis of linear gramicidins using diphenyl phosphorazidate (DPPA). Int. J. Pept. Protein Res. **26**, 305–310 (1985).
61. Mouritsen, O.G. and Bloom, M. Mattress model of lipid–protein interactions in membranes. Biophys. J. **46**, 141–153 (1984).
62. Killian, J.A., Salemink, I., De Planque, M.R., Lindblom, G., Koeppe, R.E., II, and Greathouse, D.V. Induction of non-bilayer structures in diacylphosphatidylcholine model membranes by

transmembrane α-helical peptides. Importance of hydrophobic mismatch and proposed role of tryptophans. Biochemistry **35**, 1037–1045 (1996).
63. Andersen, O.S. and Koeppe, R.E., II. Bilayer thickness and membrane protein function: an energetic perspective. Annu. Rev. Biophys. Biomol. Struct. **36**, 107–130 (2007).
64. Jordan, J.B., Shobana, S., Andersen, O.S., and Hinton, J.F. Effects of glycine substitutions on the structure and function of gramicidin A channels. Biochemistry **45**, 14012–14020 (2006).
65. Durkin, J.T., Koeppe, R.E., II, and Andersen, O.S. Energetics of gramicidin hybrid channel formation as a test for structural equivalence. Side-chain substitutions in the native sequence. J. Mol. Biol. **211**, 221–234 (1990).
66. Durkin, J.T., Providence, L.L., Koeppe, R.E., II, and Andersen, O.S. Energetics of heterodimer formation among gramicidin analogues with an NH_2-terminal addition or deletion: consequences of missing a residue at the join in the channel. J. Mol. Biol. **231**, 1102–1121 (1993).
67. Sun, H. Applications of Gramicidin Channels: I. Function of Tryptophan at the Membrane/Water Interface. II. Molecular Design of Membrane-Spanning Force Transducers. Ph.D. Thesis, University of Arkansas (2003).
68. Fonseca, V., Daumas, P., Ranjalahy Rasoloarijao, L., Heitz, F., Lazaro, R., Trudelle, Y., and Andersen, O.S. Gramicidin channels that have no tryptophan residues. Biochemistry **31**, 5340–5350 (1992).
69. Lundbæk, J.A. and Andersen, O.S. Spring constants for channel-induced lipid bilayer deformations. Estimates using gramicidin channels. Biophys. J. **76**, 889–895 (1999).
70. Miloshevsky, G. and Jordan, P. The open state gating mechanism of gramicidin a requires relative opposed monomer rotation and simultaneous lateral displacement. Structure **14**, 1241–1249 (2006).
71. Goforth, R.L., Chi, A.K., Greathouse, D.V., Providence, L.L., Koeppe, R.E., II, and Andersen, O.S. Hydrophobic coupling of lipid bilayer energetics to channel function. J. Gen. Physiol. **121**, 477–493 (2003).
72. Huang, H.W. Deformation free energy of bilayer membrane and its effect on gramicidin channel lifetime. Biophys. J. **50**, 1061–1070 (1986).
73. Nielsen, C., Goulian, M., and Andersen, O.S. Energetics of inclusion-induced bilayer deformations. Biophys. J. **74**, 1966–1983 (1998).
74. Lundbæk, J.A., Maer, A.M., and Andersen, O.S. Lipid bilayer electrostatic energy, curvature stress, and assembly of gramicidin channels. Biochemistry **36**, 5695–5701 (1997).
75. Andersen, O.S., Bruno, M.J., Sun, H., and Koeppe, R.E., II. Single-molecule methods for monitoring changes in bilayer elastic properties. Methods Mol. Biol. **400**, 543–570 (2007).
76. Helfrich, P. and Jakobsson, E. Calculation of deformation energies and conformations in lipid membranes containing gramicidin channels. Biophys. J. **57**, 1075–1084 (1990).
77. Nielsen, C. and Andersen, O.S. Inclusion-induced bilayer deformations: effects of monolayer equilibrium curvature. Biophys. J. **79**, 2583–2604 (2000).
78. Elliott, J.R., Needham, D., Dilger, J.P., and Haydon, D.A. The effects of bilayer thickness and tension on gramicidin single-channel lifetime. Biochim. Biophys. Acta. **735**, 95–103 (1983).
79. Andersen, O.S., Nielsen, C., Maer, A.M., Lundbaek, J.A., Goulian, M., and Koeppe, R.E., II. Ion channels as tools to monitor lipid bilayer-membrane protein interactions: gramicidin channels as molecular force transducers. Methods Enzymol. **294**, 208–224 (1999).
80. Hille, B. Ionic Channels of Excitable Membranes. Sinauer, Sunderland, MA (2001).

Chapter 3
Self-Assembled Organic Nanotubes and Their Applications in Nano-Bio Fields

Toshimi Shimizu

Abstract We focus on the template-independent self-assembly of organic nanotubes, which are recently breaking new ground every day under the influence of the bloom of carbon nanotube (CNT) researches. In particular, the spontaneous self-assembly of amphiphilic molecules (lipids) into hollow cylindrical tubular architectures is discussed in terms of molecular design, possible formation scheme, dimension control, and potential applications in nano-bio fields. The advanced researches on a variety of self-assembled lipid nanotubes (LNTs), including diacetylenic phospholipid nanotubes (PLNTs), peptide nanotubes (PNTs), glycolipid nanotubes (GLNTs), and molecular graphene-based nanotubes (graphitic nanotubes, GNTs), have been detailed, focusing on current progress and topics on and after 2000. The attempts to use the LNT as a nanocontainer, nanochannel, and nanopipette have also been described to feature the concept of "attoliter chemistry", which will occur only in a confined liquid nanospace shaped by the LNT hollow cylinder.

Abbreviations

CMC	Critical micelle concentration
CNT	Carbon nanotube
DLS	Dynamic light scattering
DNA	Deoxyribonucleic acid
Dps	DNA-binding protein from starved cells
FRET	Fluorescence resonance energy transfer
GFP	Green fluorescence protein
GLNT	Glycolipid nanotube
GNT	Graphitic nanotube
HBC	Hexa-*peri*-hexabenzocoronene
LNT	Lipid nanotube
MEMS	Micro-electro mechanical system
MLM	Monolayer membrane
PLNT	Phospholipid nanotube

T. Shimizu (✉)
Nanotube Research Center (NTRC), National Institute of Advanced Industrial Science and Technology (AIST), Tsukuba, Ibaraki 305-8565, Japan
e-mail: tshmz-shimizu@aist.go.jp

O. Hayden, K. Nielsch (eds.), *Molecular- and Nano-Tubes*,
DOI 10.1007/978-1-4419-9443-1_3, © Springer Science+Business Media, LLC 2011

PNT	Peptide nanotube
RNA	Ribonucleic acid
SEM	Scanning electron microscopy
TAS	Total analysis system
TEM	Transmission electron microscopy
T_{g-l}	Gel-to-liquid crystalline phase transition temperature
TMV	Tobacco mosaic virus
P	Critical packing parameter

3.1 Introduction

The biological structure shaping viruses gives a typical example of well-defined, self-organized products made of naturally-occurring molecular building blocks. In particular, Tobacco Mosaic Virus (TMV), a pathogenic organism of tobacco mosaic diseases, is known to form a hollow cylindrical architecture, in which 2,130 pieces of identical proteins assemble helically around a single chain polymer RNA as a template with a period of 49 proteins/3-turns (Fig. 3.1, left) [1, 2]. The dimensions of TMV, including inner and outer diameters, thickness, and length, can characterize the homogeneous and perfect nanotubular structure of TMV. It should be pointed out that mixing purely isolated TMV proteins with ribonucleic acid (RNA) as a

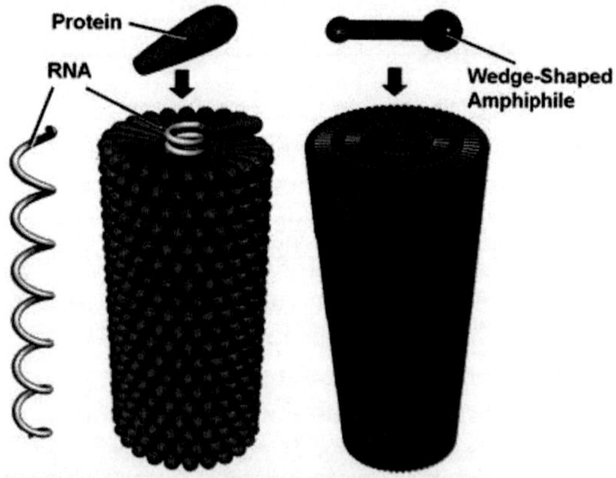

	TMV	Organic NT
Length	300 nm	~10 μm
Outer Diameter	18 nm	30~43 nm
Inner Diameter	4 nm	14~29 nm
Building Block	Protein	Amphiphile
Unit Number	2,130	A large number
Template	RNA	None

Fig. 3.1 Comparison of the size dimensions and tubular architectures for TMV and the self-assembled organic nanotubes from a wedge-shaped amphiphile **22**(n) [see Fig. 3.4 as to the molecular structure of **22**(n)]

template under appropriate conditions can allow the re-self-organization into the intrinsic tubular structure of TMV endowed with an original biological activity [1]. Every structural information to assemble should be installed in advance to each protein molecule as a molecular building block. Besides such self-assembling proteins, certain amphiphilic molecules that were rationally designed have been found to produce tubular structures with well-defined dimensions and morphologies through molecular self-assembly [3–6]. There is no need in this case to use the core materials for exo-templating [6–11], such as hollow-forming fibrous species or RNA, for the preparation of hollow cylindrical organic nanotubes or TMV, respectively. The amphiphilic molecules, once rationally designed as tube-forming compounds, can self-assemble into nanotube architectures with 7–500 nm inner diameters, 10–1,000 nm outer diameters, and few to hundreds μm length, with maximum accuracy and minimum energy [6]. Nowadays, only self-assembly system, independent on the nanoporous template synthesis pioneered by Martin et al. [12–16], enables us to obtain sophisticated organic nanotubes with unsymmetrical inner and outer surfaces covered with different chemical functionalities (Fig. 3.1, right) [17–20]. Here we focus on the spontaneous self-assembly of hollow cylindrical tubular architectures from amphiphilic molecules (abbreviated as lipids hereafter unless specified) in liquid media. In particular, a close relationship between the molecular structures and the resultant self-assembled morphologies of the lipids is discussed as well as current situation for the dimension control of the self-assembled lipid nanotubes (LNTs). We also touch recent topics on intriguing encapsulation properties of the organic nanotubes toward biomacromolecules and nanomaterials more than 10 nm wide [21]. Finally, possible unprecedented applications of the LNTs are described in currently emerging nano-bio fields, benefited from their encapsulation, diffusion, separation, and adsorption capabilities.

3.2 Molecular Building Blocks for Nanotube Formation

Here we consider the template-independent formation of discrete organic nanotubes characterized by well-defined dimensions as well as two open ends. Those nanotubes self-assemble from a variety of molecules and produce no bundles or no further organized two-dimensional structures [6, 22]. Molecular building blocks for the self-assembly may be categorized into eight groups in rough view of the formation mechanism and molecular shape of each molecule (Fig. 3.2). The molecular shape, topological arrangement of diverse functionalities, molecular recognition site, and the local environment of solvophilic and solvophobic moieties are responsible for diverse morphologies of nanotube-like assemblies, while involving different types of self-assembling mechanism. In case of cyclic molecules, for example, the molecular architecture itself strictly determines the inner or outer diameters of the resultant nanotubes since the molecular size is less than few nm (Fig. 3.2e) [23–25]. If the cyclic molecules assemble to stack in one-dimensional fashion, the intensity and dimensionality of intermolecular interactions will direct the nanotube length.

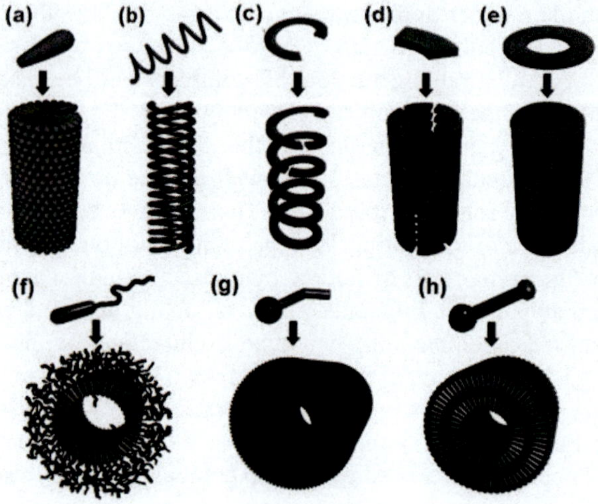

Fig. 3.2 Classification of molecular building blocks that self-assemble into discrete organic nanotubes with well-defined dimensions as well as two open ends. (**a**) Proteins, (**b**) helical chains, (**c**) rigid helices, (**d**) dendrons, (**e**) rings, (**f**) block copolymers, (**g**) general amphiphiles, and (**h**) wedge-shaped amphiphiles

Main driving forces for the molecular self-assembly in Fig. 3.2 are van der Waals force, hydrogen bonds, coordination bonds, and π–π stacking.

Besides natural proteins associated with viruses self-assembly, chemically-derived synthetic proteins were recently shown to self-assemble into nanometer-sized tubular structures (Fig. 3.2a) [26, 27]. Partial hydrolysis of the milk protein α-lactalbumin by a protease from *Bacillus Licheniformis* resulted into protein nanotubes that have potential for use in nanotechnology with both food and non-food applications. Schizophyllan, β-1,3-glucan polysaccharide produces in aqueous solutions a tube-like triple helix structure in induced-fit-type fashion from a random coil state in dimethylsulfoxide (Fig. 3.2b) [28–30]. If one-dimensional guest species, such as carbon nanotube (CNT), a single-chain polymer, or deoxyribonucleic acid (DNA), co-exists on self-assembly of schizophyllan, the size for the inner cavity of the triple helix is adjustable to fit the dimension along the short axis of the guest [31–37].

Rigid helical oligomers and polymers can also provide hollow cylindrical structures in their interior regions [38–55]. The rigidity and directionality associated with aromatic rings and/or hydrogen-bond functionalities like an amide group drive the formation of the tube-like architectures. However, the number of repeating units or the degree of polymerization is still not enough to shape the tubes with high-aspect ratios (Fig. 3.2c). On the other hand, cyclic molecules [23–25, 56–61] and panel- [62, 63] or fan-like molecules [64–72] form at first cyclic architectures through elaborate complementary hydrogen bonding and the resultant cyclic molecular motifs assemble to grow into tubular structures (Fig. 3.2d, e). Although this ring-to-tube formation scheme appears simple and clear cut, one has to pay detailed and careful attention to molecular design for the self-assembly of organic nanotubes. It is interesting that nature has never adopted the ring-to-tube formation scheme to shape, for example, ion channels [22]. Although each molecular modeling apparently shows a definite hollow (Fig. 3.2d, e), there have been few examples that demonstrated the direct evidence for the hollow cylinder structures in a discrete nanotube observed using, for example, high-resolution transmission electron microscopy (TEM) or scanning electron microscopy (SEM) [73, 74].

Self-assembled polymer nanotubes or LNTs, stabilized by lamellar-bilayer, -monolayer membranes, or hard-core–soft-shell structures, often give flexible and fibrillar nanotubes with the aspect ratios of more than 10^4 (Fig. 3.2f–h). They can be well dispersed in liquid medium, depending on their surface properties. Hereafter we focus on the LNTs that are recently breaking new ground every day [6, 75–77].

3.3 Formation Scheme of LNTs

A variety of formation scheme to produce LNT structures using tube-forming amphiphilic molecules (lipids) as molecular building blocks has been reported so far [6], including chiral molecular [78–81] and packing directed self-assembly [17, 19, 20], polymer self-assembly [82–84], molecular sculpting [60, 85], and templated assembly using a nanopore [12, 13, 15, 16, 86, 87]. The most popular way to yield the LNTs is chiral molecular self-assembly, which starts with the formation of fluid spherical vesicles (or micelles) by a chiral amphiphile and eventually results into solid tubular architectures via solid bilayer helices (Fig. 3.3a) [5, 78, 88–92]. At temperatures above the gel-to-liquid crystalline phase transition temperature ($T_{g\text{–}l}$)

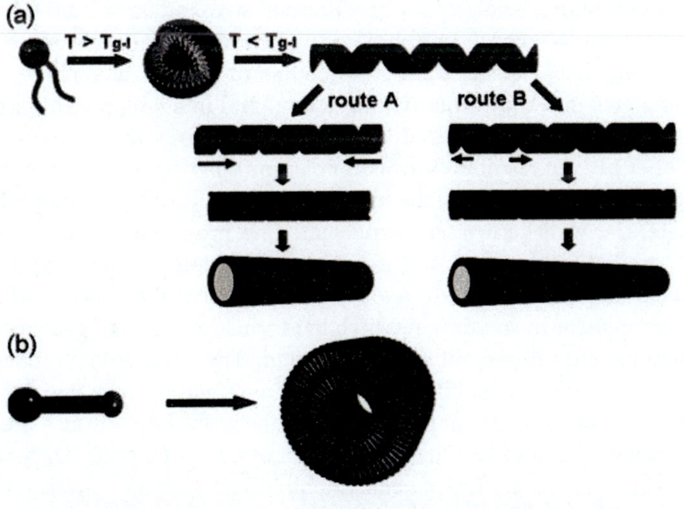

Fig. 3.3 Possible formation scheme of amphiphiles into self-assembled nanotubes. (**a**) Chiral molecular self-assembly and (**b**) packing-directed self-assembly. $T_{g\text{–}l}$ means a gel-to-liquid crystalline phase transition temperature of each amphiphile

of the chiral amphiphile, it forms vesicular assemblies. When the hot aqueous solution of the vesicles is allowed to gradually cool to room temperature (generally, $T < T_{g\text{-}l}$), the resultant spherical morphologies are converted into helical ribbons consisting of bilayer membranes. Two different types of routes follow the formation of helical ribbons, leading to the nanotube formation [6, 93–96]. One route involves the reduction in the helical pitch length of the ribbon (route A) [90, 97, 98], whereas another route proceeds with widening of the tape width of the ribbon (route B) [89, 99–103]. The latter route has been reported more commonly than the former one. Interestingly, it has also been reported that the addition of cyclodextrin to the vesicles, self-assembled from a dendritic building block with a focal pyrene unit, induces reversible vesicle–nanotube conversion [104]. Worth noticing that both the helical ribbon and nanotube architectures are stable in a *solid* state only at temperatures below $T_{g\text{-}l}$ of the corresponding amphiphile. Heating the aqueous dispersion of the nanotubes to temperatures above the $T_{g\text{-}l}$ instantly causes morphological conversion from the nanotubes to spheres in a *fluid* state. Although the mutual conversion between the nanotubes and spheres is reversible, the growth of the nanotubes from the vesicles takes relatively longer time (1 day ~ few weeks) than the instant conversion from the nanotubes to vesicles. Currently, advanced micromanipulation methodologies, which bridge several spherical vesicles with lipid nanotube networks in a *fluid* state at temperatures above the $T_{g\text{-}l}$, have been developed by Tirrel [105] and Orwar [106]. The integrated nanotube–vesicle networks are now gaining much interest in analytical chemistry fields toward single molecule detection [107–110]. The network also contributes to direct transport of membrane vesicles between different, relatively larger mother vesicles [111].

Unlike the conventional chiral self-assembly mentioned above, a new type of self-assembly scheme has recently been found for wedge-shaped amphiphiles possessing two hydrophilic headgroups of different sizes at each end (Fig. 3.3b) [4, 17–20, 112–114]. No chiral morphologies, such as helically twisted or coiled ribbons as intermediates, appear during the course of the self-assembly. The self-assembly completes in a short time (within a day) and in a single step. In general, if the hydrocarbon chain of a desired amphiphile packs in a crystalline state, one cannot predict easily the resultant self-assembled morphologies as well as packing profile of the molecule. Only when the hydrophobic chains of the amphiphiles pack in a fluid state, the well-known structural guideline proposed by Israelachivili is applicable to predict the resultant self-assembled morphologies [6, 115]. In chemistry, the critical micelle concentration (CMC) is defined as the concentration of an amphiphilic component in solution at which the formation of aggregates (micelles, lamellar structures etc.) in the solution is initiated. The geometrical model of the guideline defines the critical packing parameter $P = v/(a_0 l_c)$, where v is the volume of the hydrophobic chain of an amphiphile, a_0 is the polar head surface area at the CMC of the amphiphile, and l_c is the chain length of the amphiphile. Depending on the critical packing parameter (P) of each amphiphilic molecule, spherical micelles ($P < 0.33$), cylindrical micelles ($0.33 < P < 0.5$), spherical vesicles ($0.5 < P < 1$), planar bilayers ($P \sim 1$), and reversed micelles ($P > 1$) will be favored to produce through self-assembly (see the illustration in the reference paper [6]). As described later (see the Section 3.7), both the rational molecular design of wedge-shaped amphiphiles and optimized experimental conditions result in the unification of the polymorphism of molecular packing, giving single and suitable molecular packing with parallel arrangement. Those situations can stabilize nanometer-sized tubular architectures [17–19].

3.4 A Variety of LNTs

3.4.1 Diacetylenic Phospholipid Nanotubes (PLNTs)

The history of LNTs started with accidental findings during the self-assembly experiment using amphiphilic molecules. At almost same time in 1984, 7 years before the first discovery of CNTs [116], three research groups in Japan and United States encountered independently the formation of tubular assemblies from amphiphilic molecules [88, 89, 91, 92]. In particular, Schnur's research groups have been studying intensively and extensively the LNTs self-assembled from synthetic diacetylenic phospholipids **1**(m,n) (Fig. 3.4) [5, 79, 81, 117]. The most commonly studied lipid 1,2-bis(tricosa-10,12-diynoyl)-*sn*-glycero-3-phosphocoline **1**(8,9) was reported first by Yager and Schoen [91], and is now commercially available. The introduction of a diacetylenic triple bond causes a bent structure into the acyl chain of **1**(m,n), imposing a steric hindrance upon the molecules that are packing parallel to each other. Thus, the diacetylenic phospholipids tend to pack at nonzero angle with respect to the nearest neighbors, inducing chiral molecular packing (Fig. 3.5) [79, 118, 119]. Eventually, the generated bilayer membranes start to twist or coil and in some cases they convert into tubular architectures. Polymerizable glycolipids also corroborate the view that the introduction of diacetylenic bonds into *n*-alkyl chains is effective to form tubular structures [120–124]. In addition to these findings with triple bonds, many examples support that the introduction of unsaturated bonds, including a double bond as well, in the middle of *n*-alkyl chains is critical for the nanotube formation based on chiral molecular self-assembly [98, 125–127].

3.4.2 Peptide Nanotubes (PNTs)

Since the pioneering work for cyclic peptide nanotubes (PNTs) by Ghadiri et al. [23–25, 59, 128, 129], many reports that address the nanotube formation from peptide itself or peptide derivatives are currently arousing considerable interest in the non-CNT world [130–135]. Interestingly, the self-assembly of the Alzheimer's β-amyloid diphenylalanine motif **2** was found to produce discrete and stiff nanotubes in 1,1,1,3,3,3-hexafluoro-2-propanol at high concentrations [136]. The biocompatible and water-soluble nanotubes that were developed in this way by Gazit et al. are producible under mild conditions and are inexpensive and easy to manufacture. These benefits allow them to direct the research on PNTs toward the

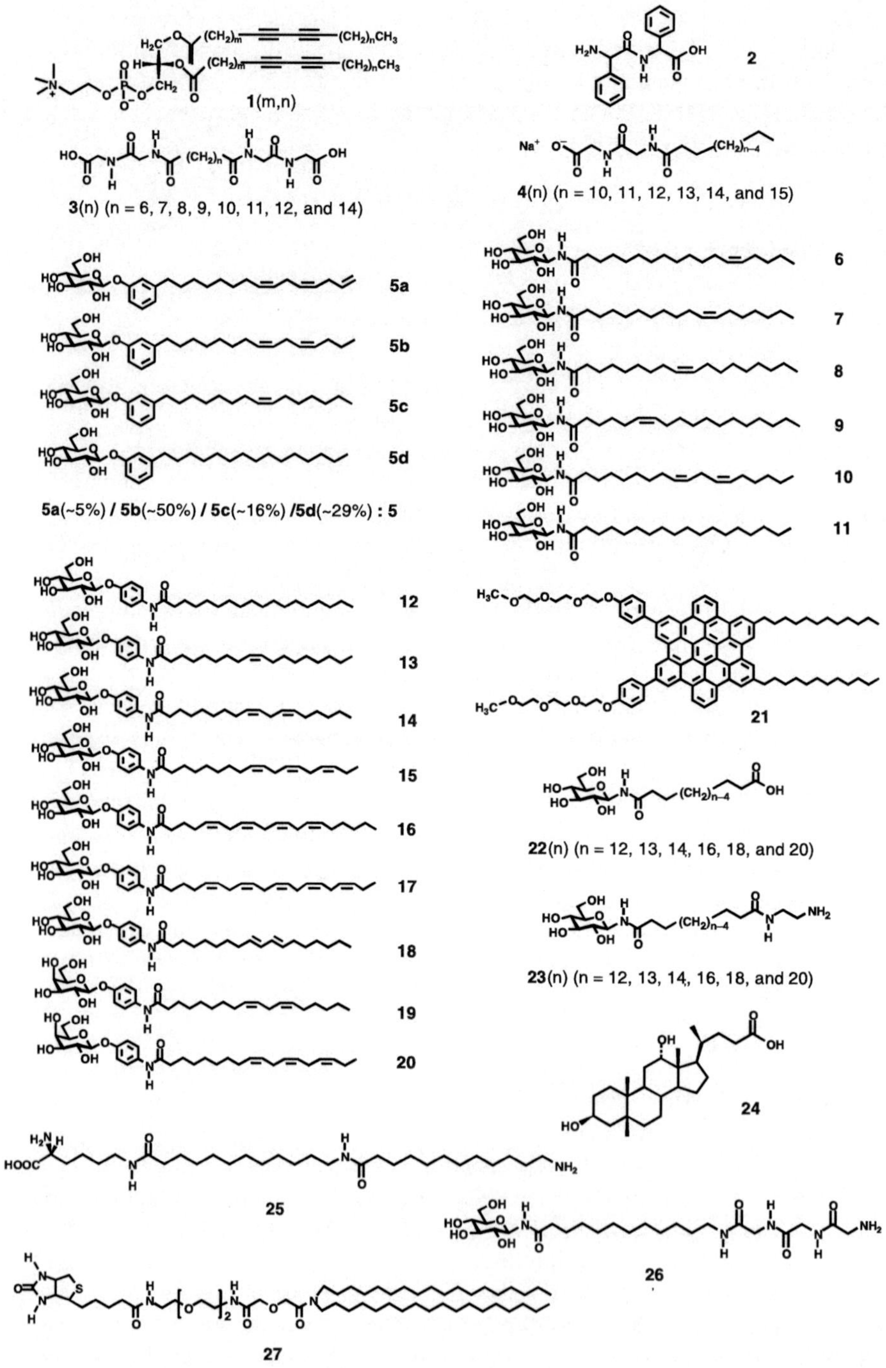

Fig. 3.4 Molecular structures of amphiphiles that self-assemble into organic nanotubes

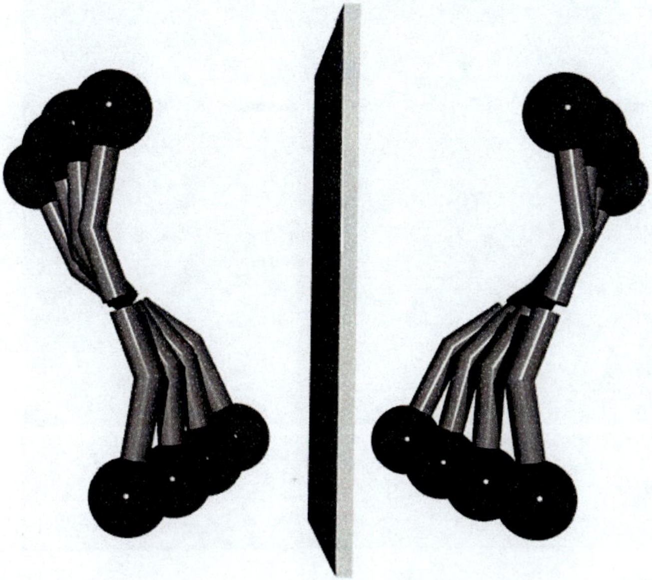

Fig. 3.5 Schematic illustration for chiral molecular packing of a molecule, in which the molecules pack at a nonzero angle with respect to the nearest neighbors. The image indicates a mirror-imaged relationship that is nonsuperimposable. The diacetylene group introduces a kink into the acyl chain of **1**(8,9), which imposes a steric hindrance to the molecules packing parallel to each other

practical use in nanotechnological and biosensor applications [137–144]. An interesting suggestion by Perutz et al. that amyloid fibers are water-filled nanotubes excites our curiosity more and more [145]. It is well-known for approximately two decades that even the peptide derivatives associated with no biological activities can also form nanotubes by self-assembly [90, 146, 147]. Among them, dicarboxylic salts (for example, the Na^+ salt) of dumbbell-shaped peptidic amphiphiles **3**(n), in which two oligoglycine moieties are connected to the both ends of an oligomethylene spacer, can provide typical examples. They proved to self-assemble in weakly alkaline aqueous solutions into microtube architectures [148–151]. The obtained tubes encapsulate vesicular assemblies in their hollow cylinders, possessing 1–3 μm outer diameters and several-hundreds μm length (Fig. 3.6a, b). The unique morphology of the microtube is chemically, physically, or thermodynamically stable, exhibiting no remarkable morphological changes on heating to boiling conditions or on sonication treatment [148]. The vesicle-encapsulated microtubes that were discovered in 1996 remarkably contrast to fullerene-encapsulated CNTs (so-called peapods) [152] and precede 2 years to appear. Notably, the spherical molecular self-assemblies, so-called liposomes by Bangham et al. [153] and vesicles by Kunitake and Okahata [154] emerged in 1965 and 1977, respectively. Their findings precede more than 8 years before the discovery of carbon spheres fullerene [155]. Infinitely possible molecular design together with intrinsic aggregation properties in a solid state will have enabled one to encounter such epoch-making findings on molecular self-assembly.

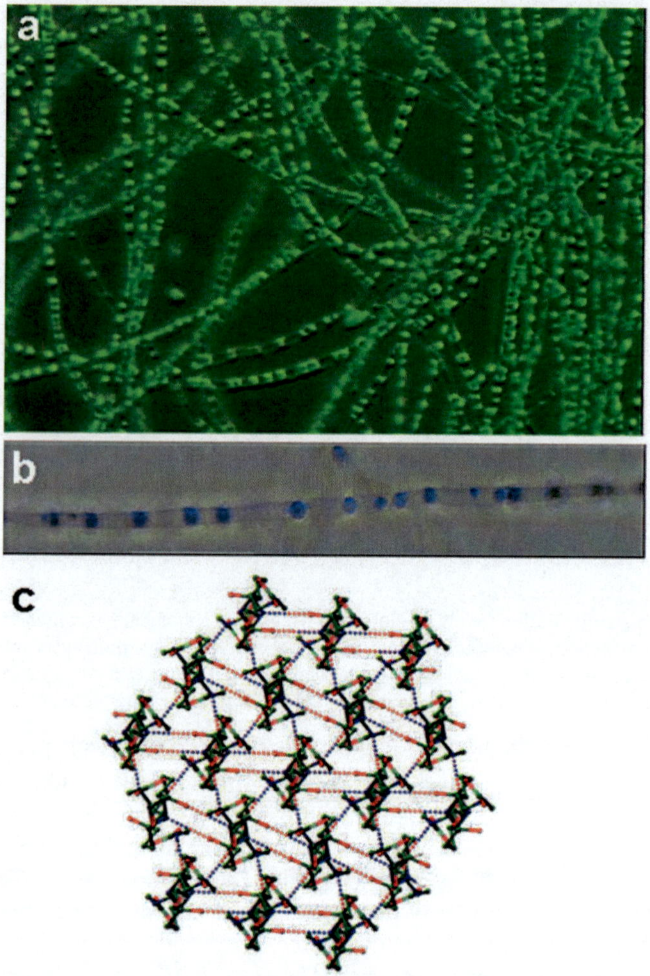

Fig. 3.6 Vesicle-encapsulated microtubes self-assembled from **3**(10) in aqueous solutions. (**a**) Dark-field and (**b**) phase-contrast optical micrographs. (**c**) A single crystal structure of polyglycine II-type hydrogen-bond network stabilized by hexagonal self-assembly of linear polymolecular chains of **3**(10)

Atomic force microscopy revealed that the headgroups of oligoglycine moiety in **3**(n) are aligned perpendicularly to the tube membrane surface and eventually form distorted hexagonal packing [151]. This result is well compatible with the view that glycine residues hydrogen bond to the neighboring six glycine residues in a pseudo polyglycine-II-type hydrogen bond network (Fig. 3.6c) [150, 156–159]. In contrast, the dumbbell-shaped amphiphiles appending valylvaline residues at each end produce no tubular architectures, but only fiber structures [160].

Similarly, multiple hydrogen bond networks between amino-acid residues as well as hydrophobic interaction between saturated hydrocarbon chains can allow another type of oligoglycine-appended lipids **4**(n) to easily organize into tubular

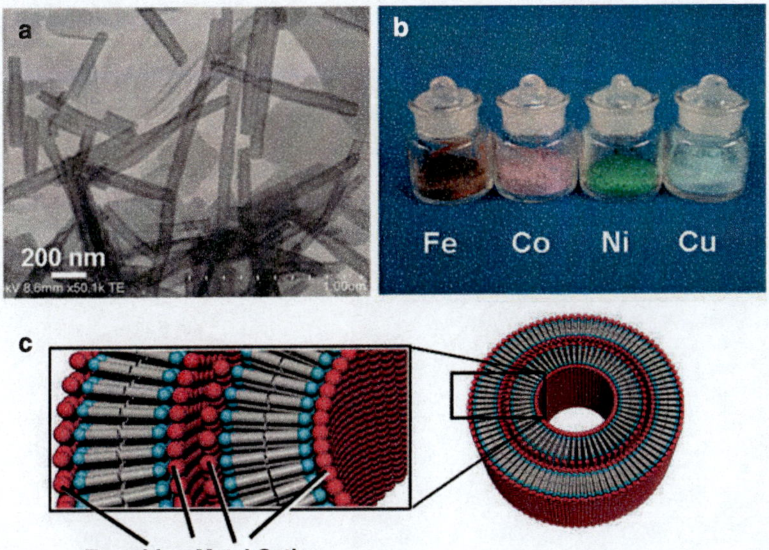

Fig. 3.7 (**a**) TEM image, (**b**) powder solids (10 g in amount), and (**c**) schematic illustration of metal cation-coordinated organic nanotubes from **4**(n)

structures. For example, by just mixing the aqueous solution of **4**(n) with that containing diluted acetic acid (H^+), copper (Cu^{2+})-, mangane (Mn^{2+})-, ferri (Fe^{2+})-, or cobalt (Co^{2+})-acetate under ambient conditions, one can instantly obtain LNTs almost quantitatively (Fig. 3.7) [161]. The length of the hydrocarbon chains and the kind of transition metals including proton strongly affect the inner and outer diameters of the resultant LNTs, regulating the outer diameters from 60 to 500 nm and the inner diameters from 20 to 150 nm. Transition metal cations and lipid bilayers alternatively stack in this LNT to form nanotube membrane structures, in which the terminal carboxylate anions are coordinating to the metal cations. By using the advantage of metal-coordinated PNTs, our own research group has recently developed three kinds of metal nanoparticle-hybridized glycolipid nanotubes (GLNTs) and PNTs, in which the nanoparticles are localized in a specific space. Namely, surface deposition [162], capillary action [163], and in-membrane crystallization [164] allowed GLNTs and PNTs to transform into CdS-deposited GLNTs on the outer surfaces (Fig. 3.8a), CdS-encapsulated PNTs in the hollow cylinder (Fig. 3.8b), and fluorescent PNTs involving CdS nanoparticles only in the wall membranes, respectively (Fig. 3.8c).

3.4.3 Glycolipid Nanotubes (GLNTs)

As compared with the extensive studies on the diacetylenic PLNTs (see the Section 3.4.1), only a few reports were addressed to GLNTs till 2000 [127, 165,

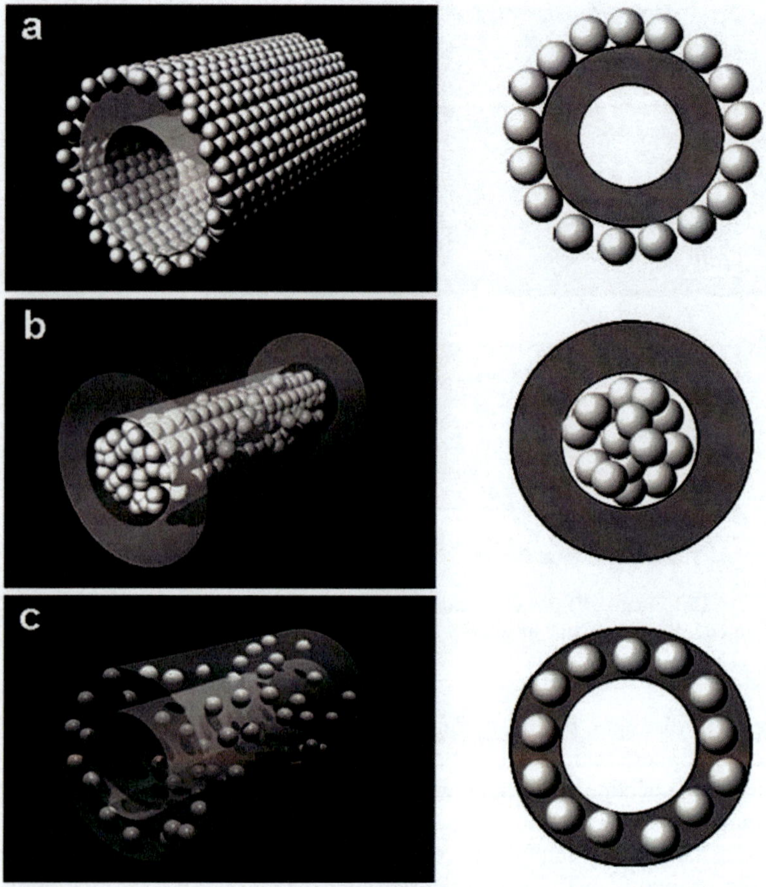

Fig. 3.8 Schematic illustration for the metal nanoparticle-hybridized nanotubes, in which CdS nanoparticles are (**a**) deposited on the outer surface, (**b**) encapsulated in the hollow cylinder, and (**c**) embedded in the membrane wall

166] in spite of great number of physicochemical and biological studies of glycolipids themselves [167–172]. Somewhat difficulties in the chemical synthesis of glycolipids may have prevented the evolution of the study related to glycolipid-forming nanotubes. In this background, Cardanol, well-known long-chain phenol mixtures derived from cashew nut shell liquid (CNSL) as renewable plant-derived resources [173], gave a chance to become an important material for nanotube formation. By covalently linking the cardanol moiety via *O*-glycoside bond with a hydrophilic glucose headgroup, John et al. synthesized glycoside mixtures **5** carrying triene-, diene-, monoene-, and saturated-type hydrocarbon chains. The mixtures were shown to efficiently self-assemble into fibrous nanotubes when dispersed in aqueous solutions under boiling conditions and allowed to gradually cool [94, 174]. The dimensions of the resultant nanotube, which possess 10 and 40–50 nm inner and

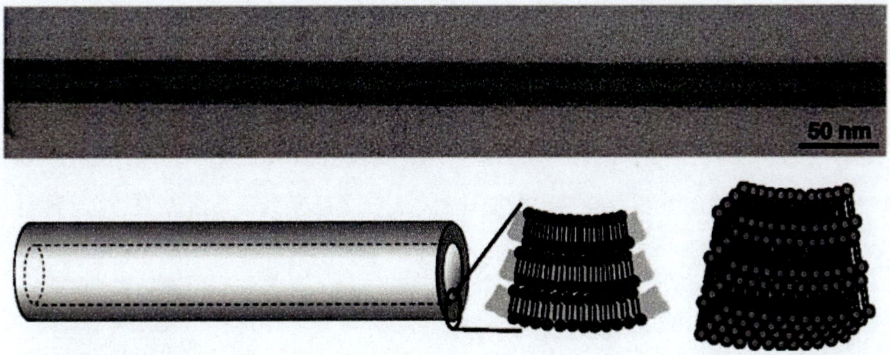

Fig. 3.9 TEM image and plausible molecular packing for the self-assembled lipid nanotube from **5**

outer diameters, respectively, and several tens to hundred μm length, were shown to be much smaller among the LNTs so far studied (Fig. 3.9) [6].

To overcome the thermal instability of the nanotubes from **5** ($T_{g\text{-}l}$ = 41°C) for the practical use, our own research group newly designed a series of glucopyranosylamide lipids **6–11**. Unfatty acids with different introduction position of a *cis*-double bond in the C18 hydrocarbon chains were connected via amide linkage to the glucose headgroup. Consequently, uniformity of the dimension and yields for the self-assembled LNTs strongly depended on the location of a *cis*-double bond (see the details shown in the Section 3.5.1) [126]. This finding means that the involvement of an unsaturated hydrocarbon chain as a hydrophobic moiety in the glycolipids is also crucial to form nanotube structures exclusively.

In light of this guideline for molecular design of tube-forming glycolipids, Jung et al. have synthesized seven glucoside **12–18** and two galactoside **19** and **20** amphiphiles, which possess long chain *p*-alkanoylamino phenol groups as hydrophobic moieties. They systematically investigated the relationship between the degree of unsaturation and the formation capability of self-assembled LNTs [98, 125]. The introduced numbers of a *cis*-double bond in the hydrophobic chains were varied from one to five. Fully hydrated samples of **13**, **14**, and **15** gave relatively higher $T_{g\text{-}l}$ by 40–80°C as compared to that of *m*-phenol glucoside derivatives **5**. Therefore, all the glucoside compounds **13**, **14**, and **15** are found to exist in a solid state at room temperature. Depending on the degree of unsaturation, the glycolipids **13**, **14**, and **15** self-assembled into twisted nanofibers, helical ribbons, and nanotubes with 70 nm in inner diameters, respectively.

3.4.4 Graphitic Nanotubes (GNTs)

Aida's research group has recently succeeded in the formation of nanotube architectures by the self-assembly of amphiphilic hexa-*peri*-hexabenzocoronene (HBC) derivatives that involve the 13 species of benzene units carved out from graphene

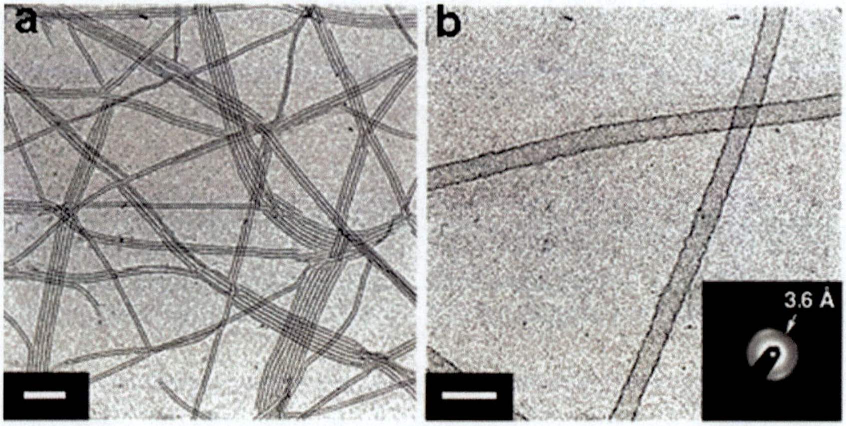

Fig. 3.10 TEM images for the self-assembled nanotubes from **21**. The *inset* in (**b**) shows an electron diffraction pattern of the tubular assembly. Bars: (**a**) 200 and (**b**) 50 nm. Reproduced with permission from Hill et al., 1481 [175]. ©2004, AAAS

structure [175]. CNT is well-known to be the ultimate one-dimensional materials with π-electron system, in which the graphene sheets condensedly packed with sp^2 carbons are rolling up. However, it is too difficult not only to prepare them only by conventional organic synthesis, but also to chemically modify the CNT surface with a definite functionality. They have developed an optimized structure **21**, in which the hydrophobic side of the HBC carries two dodecyl chains and the hydrophilic side consists of triethylene glycol moieties, for the nanotube formation in tetrahydrofuran. The resultant nanotubes with 20 nm in outer diameters are in a monodispersed state, having two open ends and a single bilayer wall 3 nm thick (Fig. 3.10). A single piece of the HBC nanotube was randomly positioned across two-probe Pt electrodes with a 180 nm gap on a SiO_2 substrate. The nanotube after oxidation with $NOBF_4$ showed a conducting *I–V* profile that is compatible with the resistivity of 2.5 M at 285 K [175]. It should be pointed out that a variety of functionalization on the surfaces enabled them to provide diverse graphitic nanotubes (GNTs) including the polymerization-stabilized [176–178], selectively-constructed chiral [179], photo-conductive [180], anion-capturing nanotubes [181], and microscopic fibers [182].

3.5 Recent Progress in Dimension Control

3.5.1 Outer Diameters

The diameter control for the self-assembled LNTs from the diacetylenic phospholipids **1**(m,n) has been extensively investigated by optimizing the mixing ratio of alcohol and water as a dispersion media, the type of alcohol used [183, 184], and cooling rate [185] for self-assembly. As far as we know, the dimensions of outer and inner diameters of self-assembled LNTs strongly depend on the intrinsic structures of molecule itself as well as the functionalities appended [6]. Furthermore, fine tuning of the molecular structure with keeping the molecular backbone is known to induce slight changes in dimensions experimentally [18, 20, 98, 126]. The replacement of the C–O–P phosphoryl linkage with a C–C–P phosphocholine group [186] in **1**(m,n) or the use of the C–P functionality instead of the C–C–P linkage [187] makes the outer diameters of the resultant PLNTs change remarkably.

On the other hand, the location of an unsaturated *cis*-double bond in the hydrocarbon chains of glucopyanosylamide lipids determines the resultant morphologies of the self-assembled nanotubes and eventually affects the size histogram of the outer diameters [126]. We synthesized a series of glucopyranosylamide lipid **6–11** with different introduction positions of a *cis*-double bond and analyzed carefully the outer diameters of large number pieces (> 250) of self-assembled nanotubes. Consequently, we have recently demonstrated a structurally optimized glycolipid structure **7** to form a uniform nanotube structure with 200 and 61 nm in outer and inner diameters, respectively (Fig. 3.11a, b). The size histogram of outer diameters for this nanotube of **7** showed the narrowest distribution (standard deviation = 23 nm) among those for **6**, **7**, and **8** (Fig. 3.11e, c, and d, respectively). It should be noted here that both the monoene derivative of the cardanyl glycolipid **5c** and the 11-*cis* derivative **7** possess the identical C18 carbon numbers and the same unsaturation position when calculating from the neighbor carbon of the *O*-glycoside or

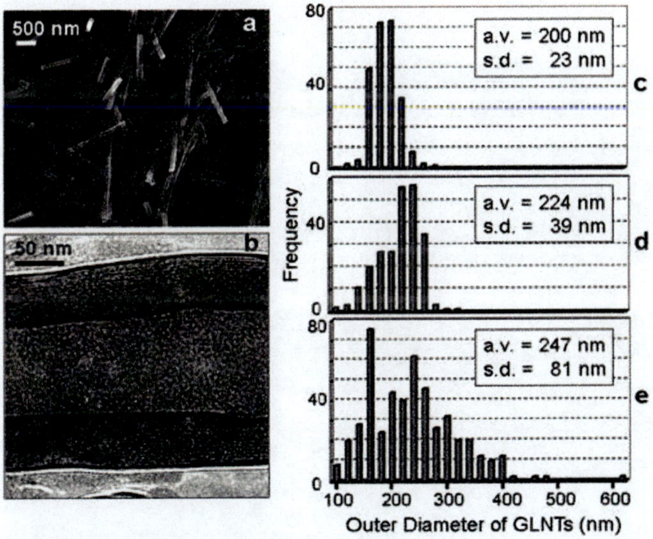

Fig. 3.11 **(a)** SEM and **(b)** high-resolution cryo-TEM images for the self-assembled nanotubes from **7**. The histograms showing the distribution of outer diameters for the self-assembled nanotubes from **(c) 7**, **(d) 8**, and **(e) 6**. The histograms were evaluated for approximately 250 pieces of nanotubes obtained

amide NH linkage. The glycolipids **6** or **8** that possess a *cis*-double bond at the C9 or C13 position widen their size distributions of outer diameters (Fig. 3.11e or 3.11d, respectively). In the similar way to the diacetylenic phospholipids **1**(8,9), the introduction of a *cis*-double bond in the middle of the hydrophobic chain can allow the glycolipids to take a bent structure for optimization of chiral molecular packing [126].

3.5.2 Inner Diameters

The regulation of inner diameters of organic hollow cylinders in the 10–100 nm range is generally difficult to perform without using a nanoporous template like an anodic alumina (AAM) or polycarbonate membrane. By designing wedge-shaped bolaamphiphiles **22**(n), we have recently achieved for the first time the precise control of inner diameters for the self-assembled LNTs (Figs. 3.12 and 3.13) [20]. The bolaamphiphiles consist of two hydrophilic moieties, glucopyranose and carboxylic acid headgroups of different size, which link to an oligomethylene spacer (carbon numbers = 12~20). Except for the bolaamphiphile **22**(12) with the C12 oligomethylene spacer, the evaluated inner diameters of the obtained nanotubes varied from 17.7 to 22.2 nm at intervals of approximately 1.5 nm every increment of two carbons in the spacer chains (Fig. 3.13). The nanotube formation was found to occur in a single step based on the packing-derived self-assembly accompanying with no intermediate morphologies. Detailed X-ray analysis supported that the monolayer

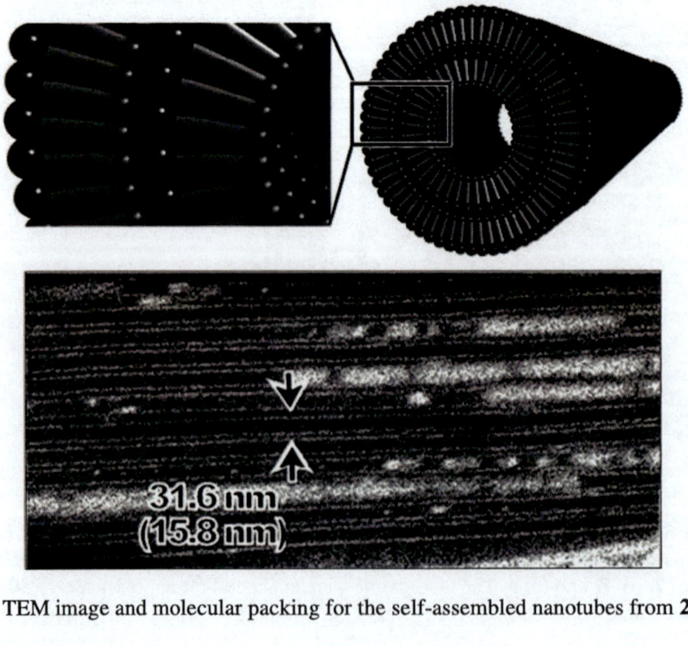

Fig. 3.12 TEM image and molecular packing for the self-assembled nanotubes from **22**(16)

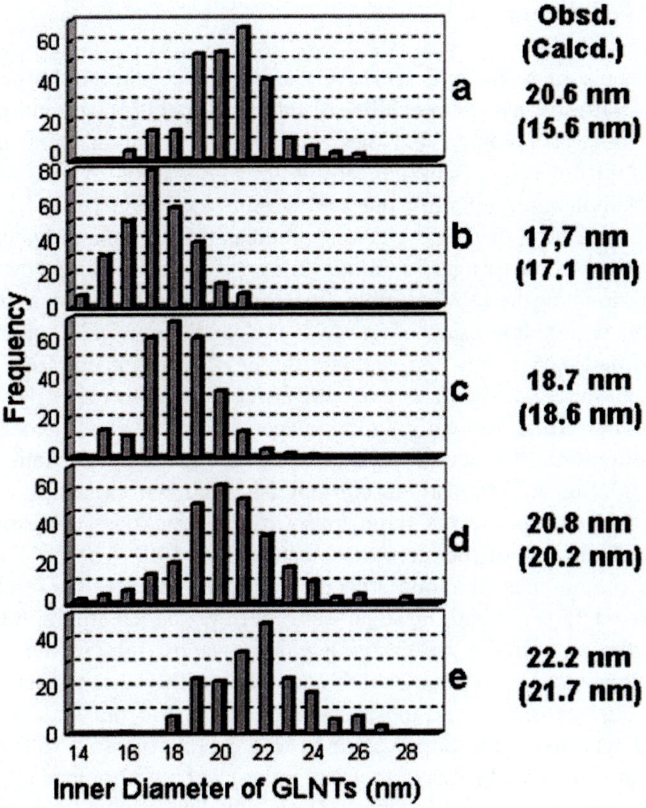

Fig. 3.13 Histgrams showing observed inner diameters of self-assembled nanotubes from (**a**) **22**(12), (**b**) **22** (14), (**c**) **22** (16), (**d**) **22** (18), and (**e**) **22** (20), which were evaluated for approximately 250 pieces of nanotubes obtained. Calculated values based on a simulation are also shown in parentheses (see the details in the Ref. [20])

membranes of the amphiphile pack in a parallel fashion (unsymmetrical MLM), resulting into unsymmetrical inner and outer surfaces covered with carboxylic acid and sugar hydroxyl groups, respectively (Fig. 3.12, upper).

The molecular design similar to this carboxylic acid-based bolaamphiphile **22**(n) have allowed us to form self-assembled LNTs from the amine-based bolaamphiphiles **23**(n) (n = 18 and 20). They also have unsymmetrical inner and outer surfaces covered with amine and sugar hydroxyl headgroups, respectively [17, 18, 114]. The pH values, depending on weakly alkaline or neutral conditions, allowed the inner diameters of the self-assembled LNTs of **23**(18) to have 80 or 20 nm, respectively [17, 18]. Interestingly, the narrower nanotubes with 20 nm inner diameters proved to be driven by chiral molecular self-assembly system. These types of unsymmetrical LNTs from **22**(n) and **23**(n) are of great advantage to chemically modify the carboxylic acid or amine group with any functionalities for the selective encapsulation of 10–50 nm scaled guest substances.

3.5.3 Wall Thickness

The wall thickness of LNTs will sensitively affect the membrane properties such as robustness, stiffness, and permeability of the constituent bilayer and monolayer membranes. Ratna et al. found that the resultant number of bilayers in the self-assembled LNT from **1**(8,9) is influenced by the composition of mixture solvents [183]. In methanol/water mixture, the self-assembly of **1**(8,9) tends to produce the nanotubes consisting of single bilayers, whereas that in ethanol/water or water produces nanotubes consisting of multiple bilayers. By controlling low and high lipid concentrations for the self-assembly, Spector et al. succeeded in preparation of LNTs of **1**(8,9) with single bilayers and two- to four-bilayer walls separately [78, 79]. They claimed that if they can evaluate the crossover concentration, they can produce the nanotubes consisting of two-bilayer membranes exclusively.

Elegant tubular structures with a monodisperse thickness of membrane walls have been also reported for certain tube-forming compounds. For example, a simple aqueous solution of lithocholic acid salt **24** is known to self-assemble into steroid nanotubes that possess a quite uniform wall thickness corresponding to a monomolecular length of the steroid molecule [188–190]. The LNT consisting of monolayer membranes of monomolecular thickness also self-assembles from amino-acid-based **25** [112] and the sugar-based wedge-shaped amphiphile **26** [19]. In particular, the amphiphile **26**, carrying glucose- and triglycine-headgroups at both ends, exclusively self-assembles into nanotubes with 7–9 nm inner diameters. Furthermore, they consist of the molecular monolayer 3–4 nm wide (Fig. 3.14). Polyglycine-II-type-hydrogen-bond networks (Fig. 3.6c) [156, 157, 159] among the triglycine moieties line up the unsymmetrical monolayer membrane to stabilize parallel molecular packing (Fig. 3.14, upper). As already discussed in the Section 3.4.4, Aida's research group has recently reported the self-assembly of nanotubes from **21** with a single bilayer wall 3 nm thick [175]. The LNTs that possess unilamellar with an inner diameter of 16 nm and a constant diameter of 27 nm have been also prepared by biotin-containing dioctadecylamine molecules **27** [191]. At all events, there exists a variety of wall thicknesses from a single to multiple layers in the LNT, like single-wall to multi-wall CNTs.

3.5.4 Length

The control of length dimension for the LNTs is of great importance to perform desired functions of organic nanotubes with high-aspect ratios. For example, the releasing rate of encapsulated drugs from the terminal ends of the LNT should be strongly dependent on the nanotube length [192]. Relatively longer nanotubes with more than 10 μm in length will be applicable to, for example, nanochannel parts in miniaturized systems for chemistry and life sciences (what we call "micro total analysis systems: μ-TAS"). Toward single molecular protein and DNA analyses, medium-sized (1 μm ∼ 10 μm) and relatively shorter nanotubes (less than

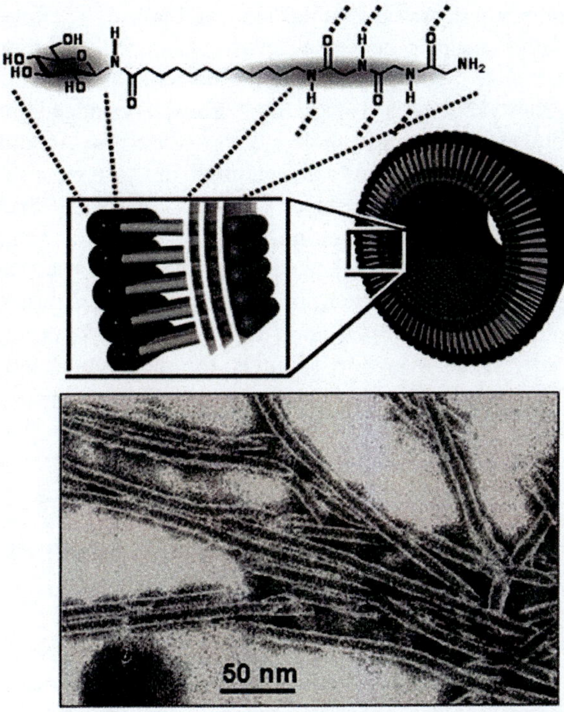

Fig. 3.14 TEM image and molecular packing for the self-assembled nanotubes from 26

1 μm in length) may be usable as microdevice parts including capillaries, connectors, or pipettes, and as supramolecular container- or sensor assemblies, respectively. However, it has been too difficult to achieve the precise length control by rationally designing the internal molecular backbone and the arrangement of diverse functionalities in each tube-forming amphiphile. In other words, the nanotube length cannot be regulated by any intrinsic structural factors in molecular building blocks. Initial attempt has been done more than 10 years ago by tuning the composition of mixture solvents [193] or by addition of alkaline metal salts [147]. Thomas's research group will be the first to achieve the length control by studying carefully the formation kinetics in the self-assembly of multilamellar LNTs from the diacetylenic phospholipids **1**(8,9) [185]. On the basis of variable-temperature X-ray diffraction study, they confirmed the first-order nature of the phase transition at the nanotube formation temperature. This finding suggested that the control of the cooling rate should determine the nanotube morphology. When examining over a wide range of cooling rates from 0.08 °C/h to 105 °C/h, they found that the lengths of the resultant nanotubes are controllable from 1 to 100 μm without affecting the diameter of the LNTs.

A gentle mechanical stirring is also applicable to shorten and regulate the length effectively for the resultant nanotubes of **7** [192]. The stirring time and rate can direct the length distribution. It should be noted that no shorter LNTs less than

1 μm in length are obtainable. This method will be effective to limit the LNT length within its intrinsic persistent length. As described before, Martin et al. developed the template method by wetting porous template with various organic and inorganic materials [12, 13, 15, 16, 86, 87]. This will be indeed promising to generate tubular architectures with uniform diameters and lengths. Actually, polymer nanotubes with diameters ranging from a few tens of nm to μm are obtainable by the template wetting with polymer melts or solutions [16, 194–196]. However, the chemical and physical properties of wetting materials as well as porous membranes sensitively direct the hollowness of the resultant tubes with well-defined open ends. We have developed a rapid and convenient method to prepare LNTs with definite diameters by successive use of vesicle extrusion and AAM porous templates [197]. Even for the self-assembly system of lipids, the mean outer and inner diameters, and wall thickness can be remarkably regulated by the template method. However, the details about the homogeneity of the obtained LNT length still need further investigation.

3.6 Hollow Cylinder as an Encapsulation Field for Biomacromolecular Guests

Yui et al. revealed for the first time the solvent polarity and viscosity of confined water inside the LNT hollow cylinder (10 nm inner diameter) of **5** [198, 199]. Consequently, the water was found to show the physical properties of structured water similar to intracellular water. This finding suggested us the application of LNTs to a nanocontainer, nanochannel, and nanopipette, in which we can stably preserve biomacromolecules sensitive to temperature, pH, and polarity of aqueous environments [21]. Figure 3.15 illustrates the comparison between the sizes of a variety of biomacromolecules and nanoparticles in the 1–100 nm range, and the inner diameters of the LNTs obtained so far [126, 149, 174]. In this size region, one can find, for example, DNA (2 nm wide), Ferritin (12 nm diameter) that is one of the largest spherical proteins [200], and a variety of polymer nanoparticles (5–30 nm diameter) [201–203]. There also exist norovirus (27 nm diameter) that is the smallest spherical virus [204] and magnetosome (50–100 nm wide) that is aligning in magnetotactic bacteria in this region [205, 206]. Attempts to encapsulate such biomacromolecular substances into LNT hollow first started with the utilization of capillary action. As a result, we demonstrated that the LNTs can well encapsulate the aqueous solutions or dispersions containing such a guest substance into the internal hollow cylinders [21, 207, 208]. Before encapsulation procedure, however, one has to remove the confined water in the hollow cylinders by lyophilization. Favorably, no remarkable changes in the hollowness take place during this pre-treatment.

Relationship between the size of the guest substances encapsulated and inner diameters of the LNT hollow cylinders strongly affect their encapsulation ability as well as the one-dimensional confinement patterns of the encapsulated guests. Encapsulation behavior of the LNT from **7** with 30–50 nm inner diameters was

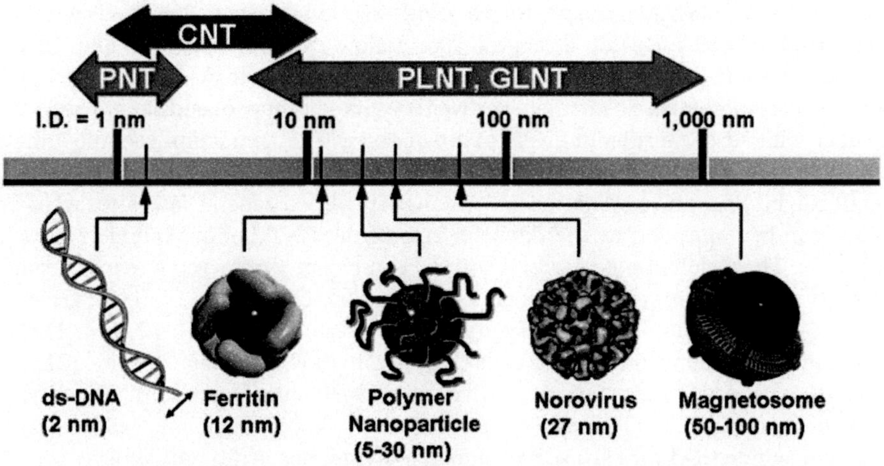

Fig. 3.15 Comparison of size dimensions among carbon- (CNTs), peptide- (PNTs), phospholipid- (PLNTs), glycolipid nanotubes (GLNTs), nanoscale biomacromolecules and biological objects. The images of Ferritin and norovirus are reproduced with permission from Prof. Ichiro Yamashita (CREST, JST) and Voedselinfecties door virussen-virusgevaar onder de loep (Bijkerk [275]. ©2002, RIVM/CIE), respectively

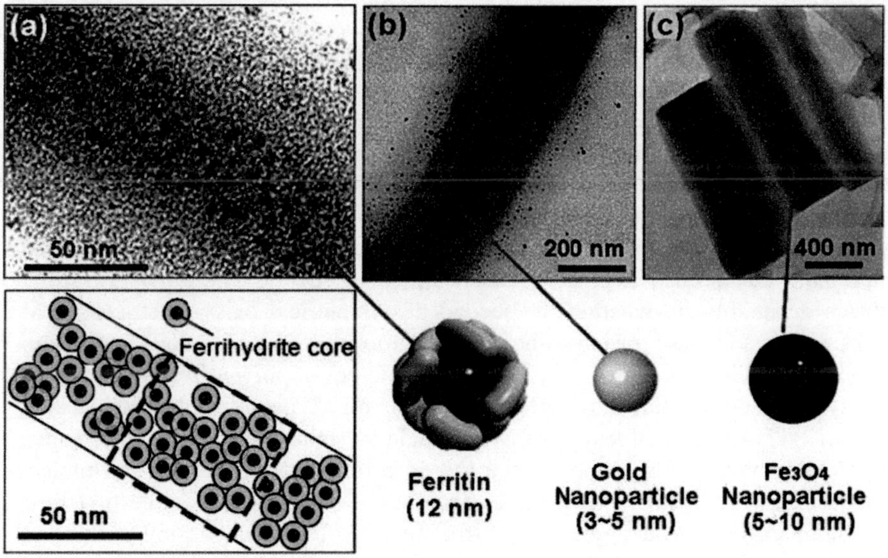

Fig. 3.16 Magnified TEM images for the glycolipid nanotubes from **7** that encapsulate (**a**) Ferritin (12 nm diameter), (**b**) gold nanoparticles (3–5 nm wide), and (**c**) Fe_3O_4 nanoparticles (5–10 nm diameters) in the hollow cylinders. Traced image of (**a**) is also shown in the bottom of (**a**)

investigated by using gold nanoparticles of different diameters as guests. The gold nanoparticles 3–10 nm wide arrange in the hollow cylinder with 4–5 lines in a confined fashion, whereas those 15 nm diameter form a single line [207]. The nanoparticles more than 50 nm diameter were observed to stay outside of the hollow cylinder without encapsulation. Heat treatment or solvent extraction can allow the removal of organic components that shape the outer shell of the obtained organic-metallic hybrid nanotubes. This means that one can fabricate metal nanowires whose widths can be controlled by the diameter size of the LNT hollow cylinders [207, 209, 210]. Thus, supramolecular LNTs have been recently observed to encapsulate a variety of biomacromolecules, such as double-stranded DNA [17, 110], Green Fluorescence Protein (GFP, 3 $\times$ 4.5 nm) [211], Ferritin (Fig. 3.16a) [17, 114, 212], DNA-binding protein from starved cells (Dps, 9 nm) [17], and magnetosome [213], as well as a variety of nanoparticles, such as gold (3–30 nm) (Fig. 3.16b) [207, 208], magnetite nanoparticles (10 nm) (Fig. 3.16c) (Yui & Shimizu, Unpublished results), cadmium sulfide (3–5 nm) [163], and anionic polymer beads (20 nm) [114].

3.7 Control of Polymorphism and Rational Design of Inner Surfaces

The supramolecular nanotubes that possess completely unsymmetrical inner and outer surfaces are promisingly applicable to selective filling of nanomaterials and biomacromolecules into their hollow cylinders. To fabricate them exclusively, one needs to accept only single molecular packing among the possible four types. It is well-known that the molecular packing of wedge-shaped amphiphiles in a solid state includes two types of polymorphism, *polymorph* and *polytype*. The *polymorph* can be classified into unsymmetrical or symmetrical monolayer membranes (MLMs) depending on whether the molecules pack in a parallel or an antiparallel fashion, respectively. Here, relatively larger and smaller headgroups of the wedge-shaped amphiphile are defined as a head and a tail, respectively. The *polytype* further appears depending on whether the second unsymmetrical or symmetrical MLM stacks with head-to-tail (or tail-to-head) or head-to-head (or tail-to-tail) interlamellar interfaces. Therefore, the combination of two types of *polymorph* and two types of *polytype* should result into the occurrence of four types of molecular packing (Fig. 3.17) [214, 215]. All reports of X-ray single crystal analyses for wedge-shaped amphiphiles have so far demonstrated the occurrence of the head-to-tail molecular packing in a *symmetrical* MLM [216, 217]. We reported for the first time the unique example of a single crystal structure, involving head-to-tail packing of *unsymmetrical* MLMs [214, 215].

A variety of amphiphiles are known to self-assemble into spheres, rods, fibers, tapes, helical ribbons, or nanotubes [3, 4, 218–225]. If the molecular interactions between the hydrophilic headgroups of the amphiphiles are weak, it is generally difficult to define how each molecule packs in the assemblies. In case of wedge-shaped amphiphiles, where multiple hydrogen-bonding functionalities are introduced, the

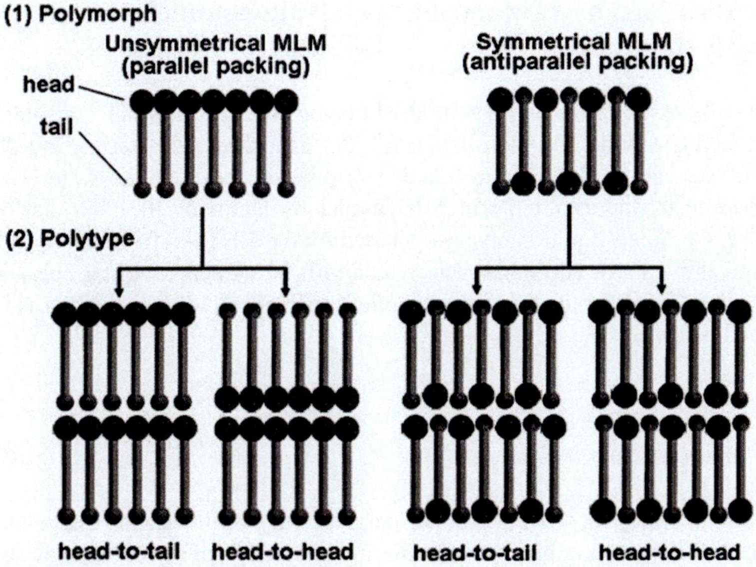

Fig. 3.17 Possible four types of monolayer lipid membranes (MLMs) from wedge-shaped amphiphiles, depending on (1) polymorph and (2) polytype patterns

obtained X-ray diffraction patterns for the resultant assemblies should give useful information about long-range ordering of corresponding monolayer membrane [226]. The frequencies and peak profile of the CH_2 scissoring bands provide valuable information on the sub-cell structure of corresponding oligomethylene spacers [227]. For example, the obtained long-range ordering ($d = 3.65$ nm) and single IR band at 1,464 cm^{-1} for the supramolecular LNT from **23**(18) are exactly compatible with the head-to-tail packing of unsymmetrical MLMs [114]. Nowadays, one can discuss in this way about the detailed molecular orientation and arrangement of the LNTs consisting of solid surfaces [6].

Unlike low-molecular-weight guest substances that are treated in conventional host–guest chemistry [228, 229], meso-scale guests with 10–100 nm dimensions should have a variety of characteristics, including surface hydrophobicity and hydrophilicity, molecular shape, flexibility, and surface charges. When performing the selective and effective encapsulation of such a variety of guest substances into the hollow cylinders, one has to consider how we can design the functionalities, surface charges and their distribution of inner surfaces, the size of inner diameters and length of the LNTs. Those factors that tailor advanced supramolecular hosts have never been addressed when designing conventional low-molecular-weight host compounds. Thus, immobilization of rationally designed functionalities as well as charges onto the inner surfaces of the LNTs will allow for the effective encapsulation of guest substances, leading to the applications as the nanocontainer, nanochannel, and nanopipette.

3.8 Applications as a Nanocontainer, Nanochannel, and Nanopipette

A single cell possesses the water volume of one picoliter inside when considering it as a cube 10 μm wide. On the other hand, the nanotube characterized by 10 nm inner diameters as well as 1 μm length can provide the confined water volume corresponding to one attoliter, which is smaller by factor of 10^6 than that of the single cell. Confined liquid nanospace shaped by the LNTs is favorable to feature the chemical events in attoliter space, which will be named "attoliter chemistry". Typical examples of applications that benefited such characteristics of the LNTs are introduced below.

3.8.1 Nanocontainer

As already mentioned, Kameta et al. has recently synthesized wedge-shaped amphiphiles **23**(n) having glucose and amino headgroups of different sizes at each end [114]. Thin films of **23**(n), obtained from a dimethylformamide solution, were used as self-assembling seeds. The amphiphiles **23**(18) and **23**(20) are found to organize in aqueous solutions at pH ~10 to form the nanotubular structures consisting of unsymmetrical MLMs [18]. When the lyophilized nanotubes from the above dispersions are dispersed again in water, they can possess partially cationic inner surfaces under neutral pH conditions [17]. This new design of supramolecular nanotube hosts not only allow the chemical modification or immobilization of functionalities onto the inner surfaces, but partially implement positive or negative charges. The obtained LNT nanocontainer of **23**(18), which has the inner surfaces partially covered with positively charged amino groups, was mixed in aqueous solutions with spherical protein Ferritin (12 nm diameter) or poly(styrene) nanobeads (20 nm diameter) having anionic outer surfaces. Consequently, they have succeeded in the efficient encapsulation of anionic guests, independent of lyophilization and capillary action [114].

A donor chromophore in its excited state can transfer energy to an acceptor chromophore in close proximity (< 10 nm). If the DNA labeled with a fluorescent donor molecule is certainly confined in a LNT hollow cylinder, the addition of a fluorescent acceptor molecule from outside should induce time-requiring quenching of the fluorescence due to the slow diffusion of the acceptor molecule. On the basis of this fluorescence resonance energy transfer (FRET) experiment, we have very recently found that the LNT of **23**(18) with 80 nm inner diameter can encapsulate the double-stranded DNA (T4GT7-DNA, 166 kbp and 56 μm long) [17]. On the other hand, the LNT of **23**(18) with 20 nm inner diameters cannot encapsulate the DNA. The double-stranded DNA with a persistence length of typically 50 nm long can allow the penetration into the wider nanotube not into the narrower one. Thus, the size and inner surface charge of the nanocontainer strongly affect the effective encapsulation of spherical proteins and double-stranded DNA. Indeed,

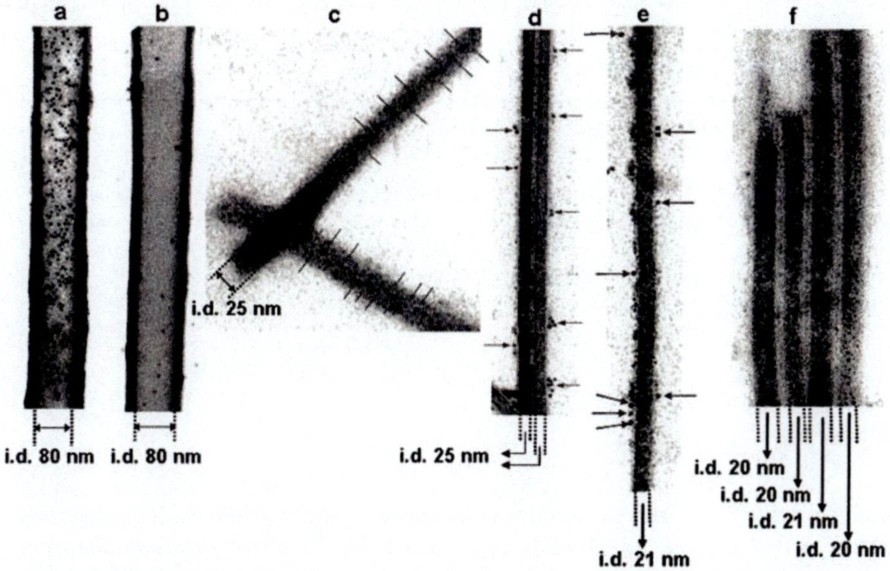

Fig. 3.18 TEM images for the self-assembled nanotubes from a variety of molecular building blocks, which display the encapsulation or no encapsulation abilities toward Ferritin and Dps. (**a**) The nanotube of **23**(18) with 80-nm inner diameters, which encapsulated Ferritin in the hollow cylinder. (**b**) The same nanotube as in (**a**), which showed no encapsulation ability for Dps. (**c**) The nanotube of **23**(18) with 20-nm inner diameters, which was prepared under different pH conditions, also encapsulated Ferritin. (**d**) The same nanotube as in (**c**), which showed no encapsulation ability for Dps. (**e**) The nanotube of **22**(18) with 20-nm inner diameters, which showed no encapsulation ability for Ferritin. (**f**) The same nanotube as in (**e**), which encapsulated Dps. *Arrows* indicate the corresponding biomacromolecules that locate inside or outside of LNTs

electrostatic interaction between the surfaces of the LNT host and the guest protein was found to induce effective encapsulation of the guest into the LNT hollow cylinder (Fig. 3.18, Table 3.1). The LNTs with partially positive charges on the inner surfaces never encapsulated positively charged Dps (Fig. 3.18b, d), although both LNTs can effectively encapsulate the negatively charged Ferritin via electrostatic interaction (Fig. 3.18a, c), irrespective of the inner diameter size. On the other hand, the carboxylate LNTs of **22**(18) displayed the opposite behavior to that of those nanotubes. Namely, the self-assembled LNTs from **22**(18) effectively encapsulated Dps (Fig. 3.18f), whereas they never encapsulated Ferritin because of electrostatic repulsion (Fig. 3.18e).

3.8.2 Nanochannel

To fabricate nanofluidic devices for sensing a single DNA molecule, Yang's research group are currently developing the use of inorganic nanotubes consisting of GaN or silica, which are integrated with microfabrication system [230, 231]. On the

Table 3.1 The effect of the inner diameter size and charges of LNTs on the encapsulation ability of three biomacromolecules

Biomacromolecular guest		Self-assembled nanotube host			
Species (diameter)	Surface charge	**23**(18) i.d. = ~80 nm	**23**(18) i.d. = ~20 nm	**22**(18) i.d. = ~20 nm	**7** i.d. = ~60 nm
		Partly positive	Partly positive	Partly negative	No charge
Ferritin (12 nm)	Negative	Encap.[a]	Encap.[a]	No Encap.[b]	No Encap.[b]
Dps (9 nm)	Positive	No Encap.[b]	No Encap.[b]	Encap.[a]	No Encap.[b]
T4GT7-DNA[c] (2 nm wide)	Negative	Encap.[a]	No Encap.[b]	n.d.[d]	n.d.[d]

[a]Encapsulated.
[b]Not encapsulated.
[c]166 kbp, 56 μm.
[d]Not determined.

other hand, attractive and potential properties of organic nanotubes, including tolerance toward diverse functionalization, possibility of labile manipulation, and dynamic morphologies, have prompted researchers to employ them in nanofluidic applications. The LNTs consisting of fluid-lipid-bilayer membranes were first attempted as a nanochannel system for transport of molecules and nanomaterials. To connect a couple of different fluid vesicles to form branched conduits, Tirrell's research group developed an interesting method by pulling a fluid LNT from a micropipette-held feeder vesicle [105, 232]. The composition of the vesicle involves a mixture of 66 mol.% stearoyl-oleoyl phosphatidylcholine, 33 mol.% cholesterol, and 1 mol.% *N*-([6-(biotinyloyl)amino]hexanoyl)-1,2-dihexadecanoyl-*sn*-glycero-3-phosphoethanolamine, triethylammonium salt. The inner diameters of the nanotubes are controllable in the 20–200 nm range by adjusting the suction pressure in the micropipette. Following this micromanipulation protocols for LNTs, Orwar and co-workers recently developed a new electroinjection technique [233–239]. In this system, the LNTs interconnect fluid-state phospholipid bilayer vesicles, enabling complex two-dimensional microscopic networks (for details, see Chapter 4 by Lobovkina et al.). Long polymer nanotubes have also been created by pulling on the membrane of polymersomes using optical tweezers or a micropipette [240].

On the other hand, the chemical functionalization for the inner surfaces of solid LNTs enabled us to construct an optical recognition system for the diffusion of guest biomacromolecules or metal nanoparticles. Namely, the fluorescence microscopy using FRET system can allow us to visualize the nanofluidic behavior of the guest species in the LNT nanochannels (80 nm inner diameter) of **23**(18) (Fig. 3.19). The FRET occurs only when the fluorescent-acceptor (QSY7)-immobilized guests are approaching to the fluorescent donor dye (NBD-F) within few nm, which is covalently linked to the amino groups on the LNT inner surfaces (Fig. 3.20). Kameta et al. has also recently succeeded in elucidation for the nanofluidic features of QSY7-immobilized Ferritin [14 nm on the basis of dynamic light scattering (DLS)], -gold nanoparticles (1.4 nm), and QSY7 itself in the nanochannel (80 nm inner

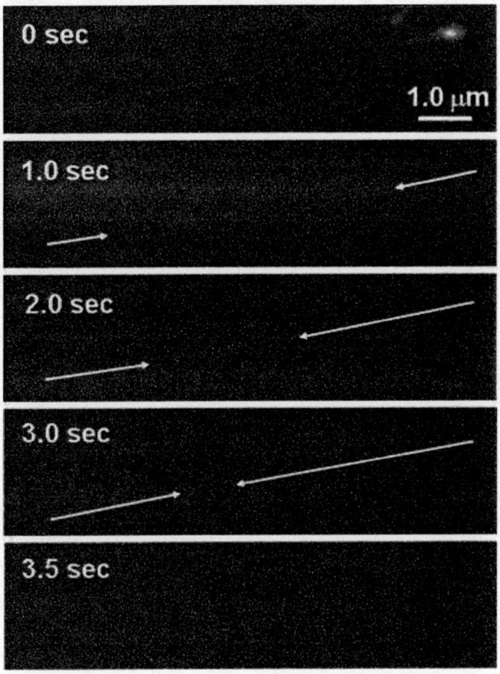

Fig. 3.19 Time-lapse fluorescence microscopic images of the NBD-immobilized LNT (80 nm inner diameter) from **23**(18) upon addition of QSY7-immobilized Ferritin. The time course was indicated in the *top left* of each figure

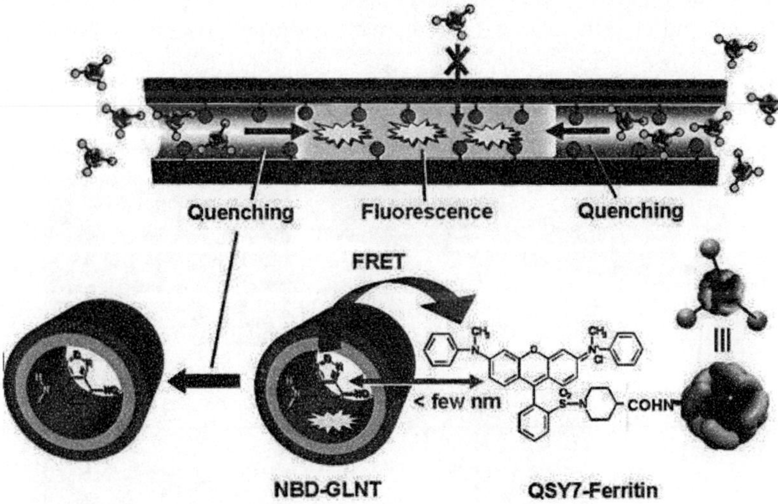

Fig. 3.20 Schematic illustration indicating gradual quenching phenomenon based on FRET from the internal NBD to the QSY7-Ferritin, which was encapsulated in the organic nanochannel of the NBD-immobilized LNT of **23**(18)

diameter) shaped by the LNT hollow cylinder [17]. Interestingly, the encapsulation only occurs under neutral pH conditions. The evaluated diffusion constant ($D = 0.7 \times 10^{-11}$ m^2/s) for the QSY7-immobilized Ferritin is 5 times smaller that that of the QSY7-Ferritin in a bulk aqueous solution ($D = 3.4 \times 10^{-11}$ m^2/s). Orwar et al. suggested that the diffusion constant of the nanoparticle in the fluid-state LNT of ~200 nm inner diameters was similar to the theoretical diffusion constant for a 30 nm particle at 298 K in a 100 nm diameter tube ($D = 0.9 \times 10^{-11}$ m^2/s) [241, 242]. This D-value is well compatible with the experimental evaluation for the opened solid-state LNT.

3.8.3 Nanopipette

Fukuda and Arai's research group, experts of robotics and micro-electro mechanical systems (MEMS), has recently developed a nanopipette device with 50-nm inner diameters. The unique device consists of a single solid LNT of **7**, which was fixed at the tip of a glass micropipette with photo-crosslinkable resin (Fig. 3.21) [243]. Single cell analysis has attracted much attention to clarify the unknown biological and analytical aspects of individual single cells. The probe-type device with nanometer scale dimensions, such as a nanopipette, can be expected as an end-effecter to control the local environment of cells with minimal changes to the environment. Some fabrications of nano pipette-like devices have been developed, including pointed-tip-type carbon whiskers [244], glass micropipette with a stacked nozzle nanostructure [245], and borosilicate glass capillary prepared by a commercially available puller [246]. There still remain many difficult issues to solve in terms

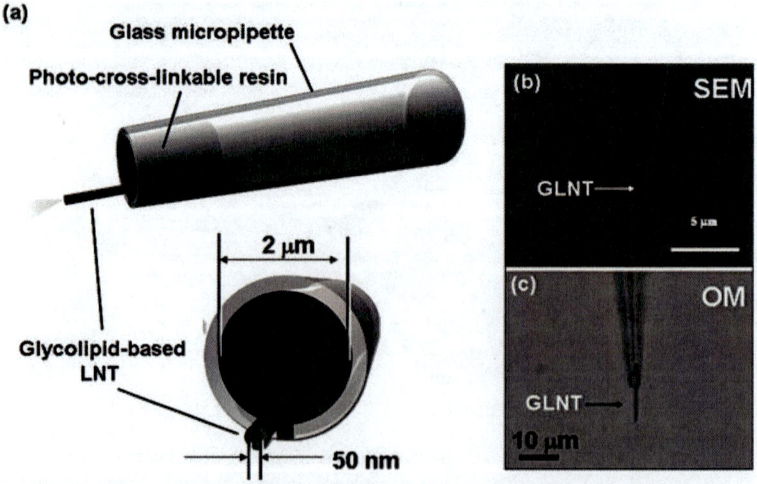

Fig. 3.21 (**a**) Schematic illustration for a nanopipette fixed at the tip of a glass micropipette with a cross-linkable resin. (**b**) SEM and (**b**) optical microscopic images for the nanopipette consisting of the LNT, glass micropipette, and cross-linkable resin

of the maintenance of complete hollowness, the fabrication time, appropriate stiffness, and the cost for the fabrication apparatus. By picking up a LNT and sealing the interspace between the LNT and glass micropipette with photo-crosslinkable resin, they succeeded in the fixation of the LNT. A fluorescent solution of Rhodamine 6G was applied to spout by electroosmotical force. When the applied voltage was 426 V, they confirmed the spout of the solution from the LNT tip using fluorescent microscopy, which remarkably contrasts to that from a glass micropipette in terms of the volume spouted.

3.9 Present and Future Aspects of Self-Assembled Organic Nanotubes

Self-assembly process of amphiphilic molecules in aqueous solutions is well-known as a general preparation method for organic nanotubes formation [6, 89, 91, 92]. Actually, a novel type of LNTs that are of great interest in terms of functions and applications have been formed from diverse amphiphilic compounds, including phosphatidylcholine [247], aminoglycerol [248], *N*-acyl phenylalanine [249], cetyltrimenthylammoniumbromide [250], dumbbell-shaped oligomer consisting of hexa-*para*-phenylene rod and aliphatic polyether dendrons [251], ABA triblock macromonomers [252], and cyclic lipid derivatives [253]. However, complete organization into nanotube morphology generally requires several steps of morphological change as well as long time spanning several days to several weeks. Therefore, it has been impossible in the laboratory to produce more than 1 g of LNTs. Manufacturer will also need water of more than 1,000 L for 1-g production of the LNTs. Synthesis of tube-forming amphiphiles also needs skillful techniques, endurance, and long time of several weeks to complete all chemical treatment. Therefore, aiming at large-scale production with reduced cost, we reconsidered raw materials, consisting of low-cost carbohydrates and peptides used as foods, for the synthesis of the tube-forming compounds. Furthermore, we also attempted the self-assembly in organic solvents like ethanol, instead of aqueous solvents, used as a food. It should be pointed out that the self-assembly of tubular structures in organic solvents is currently not so novel matter since the discovery by Shinkai's research group in 2000 [102, 254–258]. However, white solid powders consisting of supramolecular LNTs, this time, proved to self-assemble 1,000 times more than that by self-assembly in the same amount of water [259]. The yield of more than 100 g and that of more than 10 kg have been becoming possible in a laboratory (Fig. 3.22) and factory, respectively. It should be emphasized that these LNTs are organic nanotubes whose properties and size dimensions differ from well-known CNTs.

Cyclic oligosaccharides named cyclodextrin exhibit advantageous properties [260, 261]. By encapsulating diverse hydrophobic low-molecular weight compounds into the hydrophobic cavity, they can stabilize unstable substances, release drugs and flagrance, and improve the dispersibility of substances insoluble in water.

Fig. 3.22 (**a**) one-hundred gram of organic nanotubes prepared with **8** and (**b**) the SEM images of the solid powders in (**a**)

Therefore, a variety of cyclodextrin are widely used, for example, in food industry, water-based paint, medical application, deodorization materials, and cosmetics fields. By replacing the cyclodextrin with the supramolecular tubular assemblies with 10–100 nm inner diameters, one can capture proteins, metal nanoparticles, and DNA, which the cyclodextrin cannot encapsulate at all. Benefiting from the unique properties of the organic nanotubes in many diverse fields, such as encapsulation, separation, slow release, and adsorption, they can be expected as slow-release fertilizer in agriculture fields, pet and healthy food, skin care materials, targeting drug delivery system, and the removal of metal nanoparticles in environmental fields [262]. In actual, potential medical and biological applications using self-assembled LNTs and polymer nanotubes are in progress, including controlled drug release [263], gene delivery [264], cell adhesion [265], antimicrobial activity [123], helical crystallization of proteins [266, 267], and biomolecule sensing [268]. The alignment and ordered arrays of the solid LNTs on solid substrates [76, 124, 269, 270] are also important issues for their practical use. At the same time, the establishment for the substantial methodologies unveiling the mechanical properties of a single piece of LNT is also developing and gaining a lot of interest [26, 76, 271–273].

The protein named tubulin self-organizes into microtubule with 25 nm inner diameters on demand. The microtubules exhibit a variety of functions to form spindle on cell division, to form cytoskeleton consisting of fibrillar network, and to perform flagella or ciliary movement [274]. It is of great interest that, when unnecessary, depolymerization scheme convert the tubular morphology into intrinsic tubulin protein. Similarly, it is assumed that supramolecular LNTs encapsulating effective drugs are transported to a diseased part and released them slowly. Eventually, the nanotube architectures will decompose into each constituting molecule by the change in the external environment or the stimulating response, in a way similar as that reported by Li's research group [264]. They will certainly gain more benefits if safely adsorbed with biocompatibility through skin. If the molecule itself is safe to the body as well as environment, the bottom-up type fabrication system based on molecular self-assembly will have preference since the exhibition of function occurs only in the state of molecular organization. By assembling and amplifying diverse and intriguing function, with which molecules are naturally equipped, one can make the self-assembled molecules exhibit novel functions. Further progress in the research on supramolecular assembly will certainly result into the appearance of revolutionary supramolecular nanotubes that have big influence in the world.

3.10 Summary

Focusing on discrete, self-assembled organic nanotubes, with well-defined dimensions and morphologies, from amphiphilic compounds, we categorized molecular building blocks for the nanotube formation into eight groups. Among them, the formation scheme for the chiral molecular and packing directed self-assembly has been addressed in detail. The researches on diacetylenic phospholipid-, peptide-, glycolipid-, and molecular graphene-based-nanotubes have been outlined with their characteristics and recent progress. We have also touched on the recent progress in dimension control of self-assembled LNTs, showing a number of achievements of precise control for the inner and outer diameters, wall thickness, and length. To demonstrate the potent functions of the LNT hollow cylinder as an encapsulation field for biomacromolecules, we have introduced several examples for the applications, including nanocontainers, nanochannels, and nanopipettes, which are strongly associated with nano-bio fields. Finally, high expectations for the self-assembled organic nanotubes as new nanomaterials are discussed in comparison with cyclodextrins and microtubules.

Acknowledgment The author thanks his colleagues Dr. Mitsutoshi Masuda, Dr. Hiroyuki Minamikawa, Dr. Masaki Kogiso, Dr. Masumi Asakawa, Dr. Masaru Aoyagi, Dr. Rika Iwaura, Dr. Bo Yang, and Dr. Qingmin Ji for their continuous support at Nanoarchitectonics Research Center (NARC), AIST, during the course of this work on self-assembled LNTs. Dr. Naohiro Kameta, Dr. Yong Zhou, Dr. Nahoko Morii, and Ms. Keiko Sumitomo (SORST, JST); Dr. Shoko Kamiya and Dr. George John (CREST, JST) are acknowledged for collaboration on the synthesis and analysis of self-assembled organic tubular architectures. Dr. Kaname Yoshida and Prof. Seiji Isoda (Kyoto Univ.) are gratefully acknowledged for their carrying out TEM measurements. Prof. Tsuguo Sawada, Prof. Kohzo Ito, Dr. Yasuhiro Sakai, and Dr. Yanli Guo (University of Tokyo); Prof. Hiroharu Yui (Tokyo Univ. of Science), Prof. Yoshinori Yamaguchi (Waseda Univ.), Prof. Hiroshi Frusawa (Kochi Univ. of Tech.), Prof. Ichiro Yamashita (Nara Inst. Sci. Tech. and CREST, JST), Dr. Yumiko Mishima (CREST, JST), Prof. Kuniaki Nagayama (Natl. Inst. Nat. Sci.), Prof. Jong Hwa Jung (Gyeongsang Natl. Univ.), Prof. Toshio Fukuda (Nagoya Univ.), and Prof. Fumito Arai (Tohoku Univ.) are also acknowledged for fruitful collaboration on novel physicochemical and encapsulation properties of LNTs. The Japan Science and Technology Agency (JST) is acknowledged for financial support of the CREST and SORST projects.

References

1. Stubbs, G. in Biological Macromolecules and Assemblies, Vol. 1. (eds. A. McPherson & F. Jurnak) (Wiley, New York; 1984).
2. Stubbs, G. Molecular Structures of Viruses from the Tobacco Mosaic Virus Group. *Semin. Virol.* **1**, 405–412 (1990).

3. Fuhrhop, J.-H. & Koening, J. in Monographs in Supramolecular Chemistry. (ed. J.F. Stoddart) (The Royal Society of Chemistry, Cambridge; 1994).
4. Fuhrhop, J.-H. & Helfrich, W. Fluid and Solid Fibers Made of Lipid Molecular Bilayers. *Chem. Rev.* **93**, 1565–1582 (1993).
5. Schnur, J.M. Lipid Tubules: A Paradigm for Molecularly Engineered Structures. *Science* **262**, 1669–1676 (1993).
6. Shimizu, T., Masuda, M. & Minamikawa, H. Supramolecular Nanotube Architectures Based on Amphiphilic Molecules. *Chem. Rev.* **105**, 1401–1443 (2005).
7. Hou, H. et al. Poly(*p*-xylylene) Nanotubes by Coating and Removal of Ultrathin Polymer Template Fibers. *Macromolecules* **35**, 2429–2431 (2002).
8. Jang, J. & Bae, J. Selective Fabrication of Polymer Nanocapsules and Nanotubes Using Cyclodextrin as a Nanoporogen. *Macromol. Rapid Commun.* **26**, 1320–1324 (2005).
9. Zhu, L. et al. One-Pot Synthesis of Poly(cyclotriphosphazene-co-4,4-sulfonyldiphenol) Nanotubes via an In Situ Template Approach. *Adv. Mater.* **18**, 2997–3000 (2006).
10. Pu, Y.-C. et al. Water-Dissolvable Sodium Sulfate Nanowires as a Versatile Template for the Fabrication of Polyelectrolyte- and Metal-Based Nanotubes. *J. Am. Chem. Soc.* **128**, 11606–11611 (2006).
11. Yang, X., Zhu, Z., Dai, T. & Lu, Y. Facile Fabrication of Functional Polypyrrole Nanotubes via a Reactive Self-Degraded Template. *Macromol. Rapid Commun.* **26**, 1736–1740 (2005).
12. Martin, C.R. Template Synthesis of Polymeric and Metal Microtubules. *Adv. Mater.* **3**, 457–459 (1991).
13. Martin, C.R. Nanomaterials: A Membrane-Based Synthetic Approach. *Science* **266**, 1961–1966 (1994).
14. Martin, C.R. Template Synthesis of Electronically Conductive Polymer Nanostructures. *Acc. Chem. Res.* **28**, 61–68 (1995).
15. Martin, C.R., Dyke, L.S.V., Cai, Z. & Liang, W. Template Synthesis of Organic Microtubules. *J. Am. Chem. Soc.* **112**, 8976–8977 (1990).
16. Steinhart, M., Wehrspohn, R.B., Gösele, U. & Wendorff, J.H. Nanotubes by Template Wetting: A Modular Assembly System. *Angew. Chem., Int. Ed.* **43**, 1334–1344 (2004).
17. Kameta, N. et al. Functionalizable Organic Nanochannels Based on Lipid Nanotubes: Encapsulation and Nanofluidic Behavior of Biomacromolecules. *Chem. Mater.* **19**, 3553–3560 (2007).
18. Kameta, N., Masuda, M., Minamikawa, H. & Shimizu, T. Self-Assembly and Thermal Phase Transition Behavior of Unsymmetrical Bolaamphiphiles Having Glucose- and Amino-Headgroups. *Langmuir* **23**, 4634–4641 (2007).
19. Kameta, N. et al. Molecular Monolayer Nanotubes Having 7–9 nm Inner Diameters Covered with Different Inner and Outer Surfaces. *Chem. Lett.* **36**, 896–897 (2007).
20. Masuda, M. & Shimizu, T. Lipid Nanotubes and Microtubes: Experimental Evidence for Unsymmetrical Monolayer Membrane Formation from Unsymmetrical Bolaamphiphiles. *Langmuir* **20**, 5969–5977 (2004).
21. Shimizu, T. Self-Assembled Lipid Nanotube Hosts: The Dimension Control for Encapsulation of Nanometer-Scale Guest Substances. *J. Poly. Sci.: Part A: Polym Chem.* **44**, 5137–5152 (2006).
22. Block, M.A.B., Kaiser, C., Khan, A. & Hecht, S. in Functional Molecular Nanostructures (Top. Curr. Chem.) Vol. 245. (ed. A.D. Schlueter) 89–150 (Springer, Berlin; 2005).
23. Bong, D.T., Clark, T.D., Granja, J.R. & Ghadiri, M.R. Self-Assembling Organic Nanotubes. *Angew. Chem., Int. Ed.* **40**, 988–1011 (2001).
24. Ghadiri, M.R., Granja, J.R. & Buehler, L.K. Artificial Transmembrane Ion Channels from Self-Assembling Peptide Nanotubes. *Nature* **369**, 301–304 (1994).
25. Ghadiri, M.R., Granja, J.R., Milligan, R.A., McRee, D.E. & Khazanovich, N. Seif-Assembling Organic Nanotubes Based on a Cyclic Peptide Architecture. *Nature* **366**, 324–327 (1993).

26. Graveland-Bikker, J.F. & Kruif, C.G. Unique Milk Protein Based Nanotubes: Food and Nanotechnology Meet. *Trends in Food Science & Technology* **17**, 196–203 (2006).
27. Graveland-Bikker, J.F., Schaap, I.A.T., Schmidt, C.F. & de Kruif, C.G. Structural and Mechanical Study of a Self-Assembling Protein Nanotube. *Nano Lett.* **6**, 616–621 (2006).
28. Hasegawa, T. et al. Schizophyllan-Folate Conjugate as a New Non-cytotoxic and Cancer-Targeted Antisense Carrier. *Bioorg. Med. Chem. Lett.* **15**, 327–330 (2005).
29. Hasegawa, T. et al. Schizophyllan Acts as a One-dimensional Host to Accommodate 5, 10, 15, 20-Tetrakis(4-carboxyphenyl)porphyrinatozinc Acetate to Produce Its Fibrous Superstructure. *Chem. Lett.* **34**, 1118–1119 (2005).
30. Hasegawa, T. et al. Schizophyllan Can Act as a One-dimensional Host to Construct poly(diacetylene) Nanofibers. *Chem. Lett.* **34**, 40–41 (2005).
31. Matsumoto, T. et al. Inclusion of Cut and As-Grown Single-Walled Carbon Nanotubes in the Helical Superstructure of Schizophyllan and Curdlan (β-1,3-Glucans). *J. Am. Chem. Soc.* **127**, 5875–5884 (2005).
32. Matsumoto, T. et al. Poly(*N*-substituted-acrylamide)-Branched Schizophyllans Are Useful for Selective Recovery of Homopolynucleotides Through Convenient and Quick Precipitation Procedures. *Polymer J.* **37**, 177–185 (2005).
33. Numata, M. et al. Curdlan and Schizophyllan (β-1,3-Glucans) Can Entrap Single-Wall Carbon Nanotubes in Their Helical Superstructure. *Chem. Lett.*, 232–233 (2004).
34. Numata, M. et al. β-1,3-Glucan (schizophyllan) Can Act as a One-Dimensional Host for Creating Chirally Twisted Poly(*p*-phenylene Ethynylene). *Supramol. Chem.* **19**, 107–113 (2007).
35. Numata, M., Hasegawa, T., Fujisawa, T., Sakurai, K. & Shinkai, S. β-1,3-Glucan (schizophyllan) Can Act as a One-Dimensional Host for Creation of Novel Poly(aniline) Nanofiber Structures. *Org. Lett.* **6**, 4447–4450 (2004).
36. Numata, M. et al. β-1.3-Glucan Polysaccharide Can Act as a One-Dimensional Host to Create Novel Silica Nanofiber Structures. *Chem. Commun.* **2005**, 4655–4657 (2005).
37. Numata, M. et al. β-1,3-Glucan Polysaccharide (Schizophyllan) Acts as a One-Dimensional Host for Creating Supramolecular Dye Assemblies. *Org. Lett.* **8**, 5533–5536 (2006).
38. Meijere, A.D.de et al. Syntheses and Properties of Enantiomerically Pure Higher ($n \geq 7$) [n–2]Triangulanedimethanols and σ-[n]Helicenes. *Chem. Eur. J.* **12**, 5697–5721 (2006).
39. Berl, V., Huc, I., Khoury, R.G., Krische, M.J. & Lehn, J.-M. Interconversion of Single and Double Helices Formed from Synthetic Molecular Strands. *Nature* **407**, 720–723 (2000).
40. Berl, V., Huc, I., Khoury, R.G. & Lehn, J.-M. Helical Molecular Programming: Folding of Oligopyridine-dicarboxamides into Molecular Single Helices. *Chem. Eur. J.* **7**, 2798–2809 (2001).
41. Berl, V., Huc, I., Khoury, R.G. & Lehn, J.-M. Helical Molecular Programming: Supramolecular Double Helices by Dimerization of Helical Oligopyridine-dicarboxamide Strands. *Chem. Eur. J.* **7**, 2810–2820 (2001).
42. Cary, J.M. & Moore, J.S. Hydrogen Bond-Stabilized Helix Formation of a *m*-Phenylene Ethynylene Oligomer. *Org. Lett.* **4**, 4663–4666 (2002).
43. Cuccia, L.A., Lehn, J.-M., Homo, J.-C. & Schmutz, M. Encoded Helical Self-Organization and Self-Assembly into Helical Fibers of an Oligoheterocyclic Pyridine-Pyridazine Molecular Strand. *Angew. Chem., Int. Ed.* **39**, 233–237 (2000).
44. Panto, G.D., Pengo, P. & Sanders, J.K.M. Hydrogen-Bonded Helical Organic Nanotubes. *Angew. Chem., Int. Ed.* **46**, 194–197 (2007).
45. Gong, B. Crescent Oligoamides: From Acyclic Macrocycles to Folding Nanotubes. *Chem. Eur. J.* **7**, 4336–4342 (2001).
46. Gong, B. et al. Creating Nanocavities of Tunable Sizes: Hollow Helices. *Proc. Natl. Acad. Sci. USA.* **99**, 11583–11588 (2002).
47. Gorp, J.J.van, Vekemans, J.A.J.M. & Meijer, E.W. Facile Synthesis of Chiral Polymeric Helix; Folding by Intramolecular Hydrogen Bonding. *Chem. Commun.*, 60–61 (2004).

48. Han, S. et al. Total Syntheses of Angular [7]-, [8]-, and [9]Phenylene by Triple Cobalt-Catalyzed Cycloisomerization: Remarkably Flexible Heliphenes. *Angew. Chem., Int. Ed.* **41**, 3227–3230 (2002).
49. Han, S. et al. Total Syntheses and Structures of Angular [6]- and [7]Phenylene: The First Helical Phenylenes (Heliphenes). *Angew. Chem., Int. Ed.* **41**, 3223–3227 (2002).
50. Hecht, S. & Khan, A. Intramolecular Cross-Linking of Helical Folds: An Approach to Organic Nanotubes. *Angew. Chem., Int. Ed.* **42**, 6021–6024 (2003).
51. Khan, A. & Hecht, S. Functional Organic Nanotubes from Hollow Helical Scaffolds. *Synthetic Metals* **147**, 37–42 (2004).
52. Nelson, J.C., Saven, J.G., Moore, J.S. & Wolynes, P.G. Solvophobically Driven Folding of Nonbiological Oligomers. *Science* **277**, 1793–1796 (1997).
53. Petitjean, A., Cuccia, L.A., Lehn, J.-M., Nierengarten, H. & Schmutz, M. Cation-Promoted Hierarchical Formation of Supramolecular Assemblies of Self-Organized Helical Molecular Components. *Angew. Chem., Int. Ed.* **41**, 1195–1198 (2002).
54. Yang, X. et al. A New Strategy for Folding Oligo(*m*-phenylene ethynylenes). *Chem. Commun.*, 56–57 (2003).
55. Zhu, J. et al. A New Class of Folding Oligomers: Crescent Oligoamides. *J. Am. Chem. Soc.* **122**, 4219–4220 (2000).
56. Ghadiri, M.R., Kobayashi, K., Granja, J.R., Chadha, R.K. & McRee, D.E. The Structural and Thermodynamic Basis for the Formation of Self-Assembled Peptide Nanotubes. *Angew. Chem., Int. Ed. Engl.* **34**, 93–95 (1995).
57. Harada, A., Li, J. & Kamachi, M. Synthesis of a Tubular Polymer from Threaded Cyclodextrins. *Nature* **364**, 516–518 (1993).
58. Harada, A., Li, J. & Kamachi, M. Preparation and Characterization of a Polyrotaxane Consisting of Monodisperse Poly(ethylene glycol) and α-Cyclodextrines. *J. Am. Chem. Soc.* **116**, 3192–3196 (1994).
59. Hartgerink, J.D., Granja, J.R., Milligan, R.A. & Ghadiri, M.R. Self-Assembling Peptide Nanotubes. *J. Am. Chem. Soc.* **118**, 43–50 (1996).
60. Kim, Y., Mayer, M.F. & Zimmerman, S.C. A New Route to Organic Nanotubes from Porphyrin Dendrimers. *Angew. Chem., Int. Ed.* **42**, 1121–1126 (2003).
61. Organo, V.G., Leontiev, A.V., Sgarlata, V., Dias, H.V.R. & Rudkevich, D.M. Supramolecular Features of Calixarene-Based Synthetic Nanotubes. *Angew. Chem., Int. Ed.* **44**, 3043–3047 (2005).
62. Aoyagi, M., Biradha, K. & Fujita, M. Quantitative Formation of Coordination Nanotubes by Templated Rodlike Guests. *J. Am. Chem. Soc.* **121**, 7457–7458 (1999).
63. Yamaguchi, T. et al. Self-Assembly and Host–Guest Chemistry of a 3.5-nm Coordination Nanotube. *Chem. Asian J.* **2**, 468–476 (2007).
64. Bonazzi, S. et al. Four-Stranded Aggregates of Oligodeoxyguanylates Forming Lyotropic Liquid Crystals: A Study by Circular Dichroism, Optical Microscopy, and X-ray Diffraction. *J. Am. Chem. Soc.* **113**, 5809–5816 (1991).
65. Choi, I.S., Li, X., Simanek, E.E., Akaba, R. & Whitesides, G.M. Self-Assembly of Hydrogen-Bonded Polymeric Rods Based on the Cyanuric Acid•Melamine Lattice. *Chem. Mater.* **11**, 684–690 (1999).
66. Fenniri, H. et al. Entropically Driven Self-Assembly of Multichannel Rosette Nanotubes. *Proc. Natl. Acad. Sci. USA.* **99**, 6487–6492 (2002).
67. Fenniri, H. et al. Helical Rosette Nanotubes: Design, Self-Assembly, and Characterization. *J. Am. Chem. Soc.* **123**, 3854–3855 (2001).
68. Jonkheijm, P. et al. π-Conjugated Oligo-(ρ-phenylenevinylene) Rosettes and Their Tubular Self-Assembly. *Angew. Chem., Int. Ed.* **43**, 74–78 (2004).
69. Kimizuka, N. et al. Mesoscopic Supramolecular Assembly of a 'Janus' Molecule and a Melamine Derivative via Complementary Hydrogen Bonds. *J. Chem. Soc., Chem. Commun.*, 2103–2104 (1995).

70. Kimizuka, N., Kawasaki, T., Hirata, K. & Kunitake, K. Tube-like Nanostructures Composed of Networks of Complementary Hydrogen Bonds. *J. Am. Chem. Soc.* **117**, 6360–6361 (1995).
71. Mezzina, E. et al. The Self-Assembly of a Lipophilic Guanosine Nucleoside into Polymeric Columnar Aggregates: The Nucleoside Structure Contains Sufficient Information to Drive the Process towards a Strikingly Regular Polymer. *Chem. Eur. J.* **7**, 388–395 (2001).
72. Percec, V. et al. Controlling Polymer Shape Through the Self-Assembly of Dendritic Side-Groups. *Nature* **391**, 161–164 (1998).
73. Fenniri, H., Deng, B.-L. & Ribbe, A.E. Helical Rosette Nanotubes with Tunable Chiroptical Properties. *J. Am. Chem. Soc.* **124**, 11064–11072 (2002).
74. Lu, N., Wang, G., Yang, W., Du, Z. & Li, T. Supramolecular Self-Assembly of Nanotubes and Supercoils from Long-Chain Derivatives of Pyrimindinetrione and Triazine. *J. Colloid Interface Sci.* **252**, 222–225 (2002).
75. Gao, X. et al. Peptide Nanotubes: Simple Separation Using Size-Exclusion Columns and Use as Templates for Fabricating One-Dimensional Single Chains of Au Nanoparticles. *Adv. Mater.* **242**, 1492–1496 (2005).
76. Zhao, Y. & Fang, J. Positioning and Alignment of Lipid Tubules on Patterned Au Substrates. *Langmuir* **22**, 1891–1895 (2006).
77. Zhao, Y., Mahajan, N. & Fang, J. Bending and Radial Deformation of Lipid Tubules on Self-Assembled Thiol Monolayers. *J. Phys. Chem. B* **110**, 22060–22063 (2006).
78. Spector, M.S. et al. Chiral Molecular Self-Assembly of Phospholipid Tubules: A Circular Dichroism Study. *Proc. Natl. Acad. Sci. USA.* **93**, 12943–12946 (1996).
79. Spector, M.S. et al. Controlling the Morphology of Chiral Lipid Tubules. *Langmuir* **14**, 3493–3500 (1998).
80. Spector, M.S., Price, R.R. & Schnur, J.M. Chiral Lipid Tubules. *Adv. Mater.* **11**, 337–340 (1999).
81. Spector, M.S., Singh, A., Messersmith, P.B. & Schnur, J.M. Chiral Self-Assembly of Nanotubules and Ribbons from Phospholipid Mixtures. *Nano Lett.*, **1**, 375–378 (2001).
82. Jenekhe, S.A. & Chen, X.L. Self-Assembled Aggregates of Rod-Coil Block Copolymers and Their Solubilization and Encapsulation of Fullerenes. *Science* **279**, 1903–1907 (1998).
83. Raez, J., Barjovanu, R., Massey, J.A., Winnik, M.A. & Manners, I. Self-Assembled Organometallic Block Copolymer Nanotubes. *Angew. Chem., Int. Ed.* **39**, 3862–3865 (2000).
84. Yu, K. & Eisenberg, A. Bilayer Morphologies of Self-Assembled Crew-Cut Aggregations of Amphiphilic PS-*b*-PEO Diblock Copolymers in Solution. *Macromolecules* **31**, 3509–3518 (1998).
85. Stewart, S. & Liu, G. Block Copolymer Nanotubes. *Angew. Chem., Int. Ed.* **39**, 340–344 (2000).
86. Steinhart, M. et al. Palladium Nanotubes with Tailored Wall Morphologies. *Adv. Mater.* **15**, 706–709 (2003).
87. Steinhart, M. et al. Polymer Nanotubes by Wetting of Ordered Porous Templates. *Science* **296**, 1997–1997 (2002).
88. Nakashima, N., Asakuma, S., Kim, J.-M. & Kunitake, T. Helical Superstructures Are Formed from Chiral Ammonium Bilayers. *Chem. Lett.*, 1709–1712 (1984).
89. Nakashima, N., Asakuma, S. & Kunitake, T. Optical Microscopic Study of Helical Superstructures of Chiral Bilayer Membranes. *J. Am. Chem. Soc.* **107**, 509–510 (1985).
90. Shimizu, T. & Hato, M. Self-Assembling Properties of Synthetic Peptidic Lipids. *Biochim. Biophys. Acta* **1147**, 50–58 (1993).
91. Yager, P. & Schoen, P.E. Formation of Tubules by Polymerizable Surfactant. *Mol. Cryst. Liq. Cryst.* **106**, 371–381 (1984).
92. Yamada, K., Ihara, H., Ide, T., Fukumoto, T. & Hirayama, C. Formation of Helical Super Structure from Single-Walled Bilayers by Amphiphiles with Oligo-L-glutamic Acid-Head Group. *Chem. Lett.*, 1713–1716 (1984).
93. John, G., Jung, J.H., Masuda, M. & Shimizu, T. Unsaturation Effect on Gelation Behavior of Aryl Glycolipids. *Langmuir* **20**, 2060–2065 (2004).

94. John, G., Jung, J.H., Minamikawa, H., Yoshida, K. & Shimizu, T. Morphological Control of Helical Solid Bilayers in High-Axial-Ratio Nanostructures Through Binary Self-Assembly. *Chem. Eur. J.* **8**, 5494–5500 (2002).
95. Jung, J.H. & Shinkai, S. Gels as Templates for Nanotubes. *Top. Curr. Chem.* **58**, 223–260 (2004).
96. Jung, J.H., Shinkai, S. & Shimizu, T. Organic Supramolecular Architectures and Their Sol-Gel Transcription to Silica Nanotubes. *Chem. Rec.* **3**, 212–224 (2003).
97. Chung, D.S., Benedek, G.B., Konikoff, F.M. & Donovan, J.M. Elastic free energy of anisotropic helical ribbons as metastable intermediates in the crystallization of cholesterol. *Proc. Natl. Acad. Sci. USA.* **90**, 11341–11345 (1993).
98. Jung, J.H., John, G., Yoshida, K. & Shimizu, T. Self-Assembling Structures of Long-Chain Phenyl Glucoside Influenced by the Introduction of Double Bonds. *J. Am. Chem. Soc.* **124**, 10674–10675 (2002).
99. Jung, J.H., Kobayashi, H., Masuda, M., Shimizu, T. & Shinkai, S. Helical Ribbon Aggregate Composed of a Crown-Appended Cholesterol Derivative Which Acts as an Amphiphilic Gelator of Organic Solvents and as a Template for Chiral Silica Transcription. *J. Am. Chem. Soc.* **123**, 8785–8789 (2001).
100. Jung, J.H., Ono, Y., Sakurai, K., Sano, M. & Shinkai, S. Novel Vesicular Aggregates of Crown-Appended Cholesterol Derivatives Which Act as Gelators of Organic Solvents and as Templates for Silica Transcription. *J. Am. Chem. Soc.* **122**, 8648–8653 (2000).
101. Jung, J.H., Ono, Y. & Shinkai, S. Novel Preparation Method for Multi-Layered, Tubular Silica Using an Azacrown-Appended Cholesterol as Template and Metal-Deposition into the Interlayer Space. *J. Chem. Soc., Perkin Trans* **2**, 1289–1291 (1999).
102. Jung, J.H., Ono, Y. & Shinkai, S. Novel Silica Structures Which Are Prepared by Transcription of Various Superstructures Formed in Organogels. *Langmuir* **16**, 1643–1649 (2000).
103. Jung, J.H., Ono, Y. & Shinkai, S. Sol-Gel Polycondensation of Tetraethoxysilane in a Cholesterol-Based Organogel System in Chiral Spiral Silica. *Angew. Chem., Int. Ed.* **39**, 1862–1865 (2000).
104. Park, C. et al. Cyclodextrin-Covered Organic Nanotubes Derived from Self-Assembly of Dendrons and Their Supramolecular Transformation. *Proc. Natl. Acad. Sci. USA.* **103**, 1199–1203 (2006).
105. Evans, E., Bowman, H., Leung, A., Needham, D. & Tirrel, D. Biomembrane Templates for Nanoscale Conduits and Networks. *Science* **273**, 933–935 (1996).
106. Karlsson, M. et al. Electroinjection of Colloid Particles and Biopolymers into Single Unilamellar Liposomes and Cells for Bioanalytical Applications. *Anal. Chem.* **72**, 5857–5862 (2000).
107. Hurtig, J., Gustafsson, B., Tokarz, M. & Orwar, O. Electrophoretic Transport in Surfactant Nanotube Networks Wired on Microfabricated Substrates. *Anal. Chem.* **78**, 5281–5288 (2006).
108. Hurtig, J., Karlsson, M. & Orwar, O. Topographic SU-8 Substrates for Immobilization of Three-Dimensional Nanotube-Vesicle Networks. *Langmuir* **20**, 5637–5641 (2004).
109. Jofre, A., Hutchison, J.B., Kishore, R., Locascio, L.E. & Helmerson, K. Amphiphilic Block Copolymer Nanotubes and Vesicles Stabilized by Photopolymerization. *J. Phys. Chem. B.* **111**, 5162–5166 (2007).
110. Tokarz, M. et al. Single-File Electrophoretic Transport and Counting of Individual DNA Molecules in Surfactant Nanotubes. *Proc. Natl. Acad. Sci. USA.* **102**, 9127–9132 (2007).
111. Iglic, A., Hagerstrand, H., Bobrowska-Hagerstrand, M., Arrigler, V. & Kralj-Iglic, V. Possible Role of Phospholipid Nanotubes in Directed Transport of Membrane Vesicles. *Phys. Lett. A* **310**, 493–497 (2003).
112. Fuhrhop, J.-H., Spiroski, D. & Boettcher, C. Molecular Monolayer Rods and Tubules Made of α-(L-Lysine), ω-(Amino) Bolaamphiphiles. *J. Am. Chem. Soc.* **115**, 1600–1601 (1993).

113. Jung, J.H. et al. Stabilization of an Asymmetric Bolaamphiphilic Sugar-Based Hydrogel by Hydrogen Bonding Interaction and Its Sol-Gel Transcription. *Tetrahedron* **63**, 7449–7456 (2007).
114. Kameta, N. et al. Selective Construction of Supramolecular Nanotube Hosts with Cationic Inner Surfaces. *Adv. Mater.* **17**, 2736–2741 (2005).
115. Israelachivili, J.N. in Intermolecular and Surface Forces. (Academic Press, Inc., New York; 1985).
116. Iijima, S. Helical Microtubules of Graphitic Carbon. *Nature* **354**, 56–58 (1991).
117. Singh, A., Wong, E.M. & Schnur, J.M. Toward the Rational Control of Nanoscale Structures Using Chiral Self-Assembly: Diacetylenic Phosphocholines. *Langmuir* **19**, 1888–1898 (2003).
118. Selinger, J.V., MacKintosh, F.C. & Schnur, J.M. Theory of Cylindrical Tubules and Helical Ribbons of Chiral Lipid Membranes. *Phys. Rev. E* **53**, 3804–3818 (1996).
119. Spector, M.S., Selinger, J.V. & Schnur, J.M. in Materials Chirality (Topics in Stereochemistry), Vol. 24. (eds. M.M. Green, R.J.M. Nolte & E.W. Meijer) 281–372 (Wiley-Interscience, New York; 2003).
120. Frankel, D.A. & O'Brien, D.F. Supramolecular Assemblies of Diacetylenic Aldonamides. *J. Am. Chem. Soc.* **113**, 7436–7437 (1991).
121. Frankel, D.A. & O'Brien, D.F. Supramolecular Assemblies of Diacetylenic Aldonamides. *J. Am. Chem. Soc.* **116**, 10057–10069 (1994).
122. Fuhrhop, J.-H., Blumtritt, P., Lehmann, C. & Luger, P. Supramolecular Assemblies, a Crystal Structure, and a Polymer of *N*-Diacetykenic Gluconamide. *J. Am. Chem. Soc.* **113**, 7437–7439 (1991).
123. Lee, S.B., Koepsel, R., Stolz, D.B., Warriner, H.E. & Russell, A.J. Self-Assembly of Biocidal Nanotubes from a Single-Chain Diacetylene Amine Salt. *J. Am. Chem. Soc.* **126**, 13400–13405 (2004).
124. Lee, S.B., Koepsel, R.R. & Russell, A.J. Surface Dispersion and Hardening of Self-Assembled Diacetylene Nanotubes. *Nano Lett.* **5**, 2202–2206 (2005).
125. Jung, J.H. & Shimizu, T. Self-Assembling Structures of Long-Chain Sugar-Based Amphiphiles Influenced by the Introduction of Double Bonds. *Chem. Eur. J.* **11**, 5538–5544 (2005).
126. Kamiya, S. et al. Molecular Structure of Glucopyranosylamide Lipid and Nanotube Morphology. *Langmuir* **21**, 743–750 (2005).
127. Kulkarni, V.S., Anderson, W.H. & Brown, R.E. Bilayer Nanotubes and Helical Ribbons Formed by Hydrated Galactosylceramides: Acyl Chain and Headgroup Effects. *Biophys. J.* **69**, 1976–1986 (1995).
128. Ghadiri, M.R. Self-Assembled Nanoscale Tubular Ensembles. *Adv. Mater.* **7**, 675–677 (1995).
129. Hartgerink, J.D., Clark, T.D. & Ghadiri, M.R. Peptide Nanotubes and Beyond. *Chem. Eur. J.* **4**, 1367–1372 (1998).
130. Couet, J. & Biesalski, M. Surface-Initiated ATRP of *N*-Isopropylacrylamide from Initiator-Modified Self-Assembled Peptide Nanotubes. *Macromolecules* **39**, 7258–7268 (2006).
131. Couet, J., Samuel, J.D.J.S., Kopyshev, A., Santer, S. & Biesalski, M. Peptide-Polymer Hybrid Nanotubes. *Angew. Chem., Int. Ed.* **44**, 3297–3301 (2005).
132. Lu, K., Jacob, J., Thiyagarajan, P., Conticello, V.P. & Lynn, D.G. Exploiting Amyloid Fibril Lamination for Nanotube Self-Assembly. *J. Am. Chem. Soc.* **125**, 6391–6393 (2003).
133. Elgersma, R.C., Meijneke, T., Posthuma, G., Rijkers, D.T.S. & Liskamp, R.M.J. Self-Assembly of Amylin(20-29) Amide-Bond Derivatives into Helical Ribbons and Peptide Nanotubes Rather than Fibrils. *Chem. Eur. J.* **12**, 3714–3725 (2006).
134. Santoso, S., Hwang, W., Hartman, H. & Zhang, S. Self-Assembly of Surfactant-Like Peptides with Variable Glycine Tails to Form Nanotubes and Nanovesicles. *Nano Lett.* **2**, 687–691 (2002).

135. Vauthey, S., Santoso, S., Gong, H., Watson, N. & Zhang, S. Molecular Self-Assembly of Surfactant-Like Peptides to Form Nanotubes and Nanovesicles. *Proc. Natl. Acad. Sci. USA.* **99**, 5355–5360 (2002).
136. Reches, M. & Gazit, E. Casting Metal Nanowires Within Discrete Self-Assembled Peptide Nanotubes. *Science* **300**, 625–627 (2003).
137. Adler-Abramovich, L. et al. Thermal and Chemical Stability of Diphenylalanine Peptide Nanotubes: Implications for Nanotechnological Applications. *Langmuir* **22**, 1313–1320 (2006).
138. Kol, N. et al. Self-Assembled Peptide Nanotubes Are Uniquely Rigid Bioinspired Supramolecular Structures. *Nano Lett.* **5**, 1343–1346 (2005).
139. Hendler, N. et al. Formation of Well-Organized Self-Assembled Films from Peptide Nanotubes. *Adv. Mater.* **19**, 1485–1488 (2007).
140. Sedman, V.L., Adler-Abramovich, L., Allen, S., Gazit, E. & Tendler, S.J.B. Direct Observation of the Release of Phenylalanine from Diphenylalanine Nanotubes. *J. Am. Chem. Soc.* **128**, 6903–6908 (2006).
141. Song, Y. et al. Synthesis of Peptide-Nanotube Platinum-Nanoparticle Composites. *Chem. Commun.* **9**, 1044–1045 (2004).
142. Gao, X. & Matsui, H. Peptide-Based Nanotubes and Their Applications in Bionanotechnology. *Adv. Mater.* **17**, 2037–2050 (2005).
143. Yemini, M., Reches, M., Rishpon, J. & Gazit, E. Novel Electrochemical Biosensing Platform Using Self-Assembled Peptide Nanotubes. *Nano Lett.* **5**, 183–186 (2005).
144. Yemini, M., Reches, M., Gazit, E. & Rishpon, J. Peptide Nanotube-Modified Electrodes for Enzyme-Biosensor Applications. *Anal. Chem.* **77**, 5155–5159 (2005).
145. Perutz, M.F., Finch, J.T., Berrinman, J. & Lesk, A. Amyloid Fibers Are Water-Filled Nanotubes. *Proc. Natl. Acad. Sci. USA.* **99**, 5591–5595 (2002).
146. Shimizu, T. & Hato, M. Aggregation Morphology of Synthetic Peptidic Amphiphiles. *Thin Solid Films* **180**, 179–183 (1989).
147. Shimizu, T., Mori, M., Minamikawa, H. & Hato, M. Synthetic Peptidic Amphiphile: Reduction in Length of a Helical Bilayer Assembly due to Interaction with a Metal Cation. *J. Chem. Soc., Chem. Commun.*, 183–185 (1990).
148. Kogiso, M., Ohnishi, S., Yase, K., Masuda, M. & Shimizu, T. Dicarboxylic Oligopeptide Bolaamphiphiles: Proton-Triggered Self-Assembly of Microtubes with Loose Solid Surfaces. *Langmuir* **14**, 4978–4986 (1998).
149. Shimizu, T., Kogiso, M. & Masuda, M. Vesicle Assembly in Microtubes. *Nature* **383**, 487–488 (1996).
150. Shimizu, T., Kogiso, M. & Masuda, M. Noncovalent Formation of Polyglycine II-Type Structure by Hexagonal Self-Assembly of Linear Polymolecular Chains. *J. Am. Chem. Soc.* **119**, 6209–6210 (1997).
151. Shimizu, T., Ohnishi, S. & Kogiso, M. Cross Section Molecular Imaging of Supramolecular Microtube Using Contact Atomic Force Microscopy. *Angew. Chem., Int. Ed. Engl.* **37**, 3260–3262 (1998).
152. Smith, B.W., Monthioux, M. & Luzzi, D.E. Encapsulated C_{60} in Carbon Nanotubes. *Nature* **396**, 323–324 (1998).
153. Bangham, A.D., Standish, M.M. & Watkins, J.C. Diffusion of Univalent Ions Across the Lamellae of Swollen Phospholipids. *J. Mol. Biol.* **13**, 238–252 (1965).
154. Kunitake, T. & Okahata, Y. A Totally Synthetic Bilayer Membrane. *J. Am. Chem. Soc.* **99**, 3860–3861 (1977).
155. Kroto, H.W., Heath, J.R., O'Brien, S.C., Curl, R.F. & Smalley, R.E. C_{60}: Buckminsterfullerene. *Nature* **318**, 162–163 (1985).
156. Bella, J., Puiggali, J. & Subirana, J.A. Glycine Residues Induce a Helical Structure in Polyamides. *Polymer* **35**, 1291–1297 (1994).
157. Crick, F.H.C. & Rich, A. Structure of Polyglycine II. *Nature* **176**, 780–781 (1955).

158. Kogiso, M., Masuda, M. & Shimizu, T. Supramolecular Polyglycine II-Type Structure of Glycylglycine Bolaamphiphile. *Supramol. Chem.* **9**, 183–189 (1998).
159. Navarro, E., Tereshko, V., Subirana, J.A. & Puiggali, J. Incorporation of Diacids into the Polyglycine II Structure: Nodel Studies. *Biopolymers* **36**, 711–722 (1995).
160. Kogiso, M., Okada, Y., Hanada, T., Yase, K. & Shimizu, T. Self-Assembled Peptide Fibers from Valylvaline Bolaamphiphiles by a Parallel β-Sheet Network. *Biochim. Biophys. Acta* **1475**, 346–352 (2000).
161. Kogiso, M., Zhou, Y. & Shimizu, T. Instant Preparation of Self-Assembled Metal-Complexed Lipid Nanotubes That Act as Templates to Produce Metal-Oxide Nanotubes. *Adv. Mater.* **19**, 242–246 (2007).
162. Zhou, Y., Ji, Q., Masuda, M., Kamiya, S. & Shimizu, T. Helical Arrays of CdS Nanoparticle Tracing on a Functionalized Chiral Template of Glycolipid Nanotubes. *Chem. Mater.* **18**, 403–406 (2006).
163. Zhou, Y. & Shimizu, T. Lipid Nanotubes: A Unique Template to Create Diverse One-Dimensional Nanostructures. *Chem. Mater.* **20**, 625–633 (2008).
164. Zhou, Y. et al. Fluorescent Nanotubes Consisting of CdS-Embedded Bilayer Membranes of a Peptide Lipid. *Adv. Mater.* **19**, 1055–1058 (2007).
165. Imae, T. et al. Small-Angle Scattering and Electron Microscopy Investigation of Nanotubules Made from a Perfluoroalkylated Glucophospholipid. *J. Colloid Interface Sci.* **212**, 330–337 (1999).
166. Kulkarni, V.S., Boggs, J.M. & Brown, R.E. Modulation of Nanotube Formation by Structural Modifications of Sphigolipids. *Biophys. J.* **77**, 319–330 (1999).
167. Curatolo, W. Glycolipid Function. *Biochim. Biophys. Acta* **906**, 137–160 (1987).
168. Curatolo, W. The Physical Properties of Glycolipids. *Biochim. Biophys. Acta* **906**, 111–136 (1987).
169. Hinz, H.-J., Kuttenreich, H., Meyer, R., Renner, M. & Frund, R. Stereochemistry and Size of Sugar Head Groups Determine Structure and Phase Behavior of Glycolipid Membranes: Densitometric, Calorimetric, and X-ray. *Biochemistry* **30**, 5125–5138 (1991).
170. Hinz, H.-J., Six, L., Ruess, K.-P. & Lieflander, M. Head-Group Contributions to Bilayer Stability: Monolayer and Calorimetric Studies on Synthetic, Stereochemically Uniform Glucolipids. *Biochemistry* **24**, 806–813 (1985).
171. Mannock, D.A., Brain, A.P.R. & Williams, W.P. The Phase Behaviour of 1,2-diacyl-3-monogalactosyl-*sn*-glycerol Derivatives. *Biochim. Biophys. Acta* **817**, 289–298 (1985).
172. Mannock, D.A., Lewis, R.N.A.H., Sen, A. & McElhaney, R.N. The Physical Properties of Glycosyldiacylglycerols. Calorinetric Studies of a Homologous Series of 1,2-Di-*O*-acyl-3-*O*-(β-D-glucopyranosyl)-*sn*-glycerols. *Biochemistry* **27**, 6852–6859 (1988).
173. Tyman, J.H.P. Non-isoprenoid Long Chain Phenols. *Chem. Soc. Rev.* **8**, 499–537 (1979).
174. John, G., Masuda, M., Okada, Y., Yase, K. & Shimizu, T. Nanotube Formation from Renewable Resources via Coiled Nanofibers. *Adv. Mater.* **13**, 715–718 (2001).
175. Hill, J.P. et al. Self-Assembled Hexa-*peri*-hexabenzocoronene Graphitic Nanotube. *Science* **304**, 1481–1483 (2004).
176. Jin, W. et al. Controlled Self-Assembly Triggered by Olefin Metathesis: Cross-Linked Graphitic Nanotubes from an Amphiphilic Hexa-*peri*-hexabenzocoronene. *J. Am. Chem. Soc.* **127**, 8284–8285 (2005).
177. Motoyanagi, J., Fukushima, T., Kosaka, A., Ishii, N. & Aida, T. Self-Assembled Graphitic Nanotubes from an Amphiphilic Hexabenzocoronene Bearing Thiol Functionalities: Redox-Mediated Polymerization and Depolymerization. *J. Polym. Sci. Part A: Polym. Chem.* **44**, 5120–5127 (2006).
178. Yamamoto, T. et al. Stabilization of a Kinetically Favored Nanostructure: Surface ROMP of Self-Assembled Conductive Nanocoils from a Norbornene-Appended Hexa-*peri*-hexabenzocoronene. *J. Am. Chem. Soc.* **128**, 14337–14340 (2006).

179. Jin, W. et al. Self-Assembled Graphitic Nanotubes with One-Handed Helical Arrays of a Chiral Amphiphilic Molecular Graphene. *Proc. Natl. Acad. Sci. USA.* **102**, 10801–10806 (2005).
180. Yamamoto, Y. et al. Photoconductive Coaxial Nanotubes of Molecularly Connected Electron Donor and Acceptor Layers. *Science* **314**, 1761–1764 (2006).
181. Zhang, G. et al. Formation of Water-Dispersible Nanotubular Graphitic Assembly Decorated with Isothiouronium Ion Groups and Its Supramolecular Functionalization. *J. Am. Chem. Soc.* **129**, 719–722 (2007).
182. Yamamoto, Y. et al. A Glass Hook Allows Fishing of Hexa-*peri*-hexabenzocoronene Graphitic Nanotubes: Fabrication of a Macroscopic Fiber with Anisotropic Electrical Conduction. *Adv. Mater.* **18**, 1297–1300 (2006).
183. Ratna, B.R., Baral-Tosh, S., Kahn, B., Schnur, J.M. & Rudolph, A.S. Effect of Alcohol Chain Length on Tubule Formation in 1,2-bis(10,12-tricosadiynoyl)-*sn*-glycero-3-phosphocholine. *Chem. Phys. Lipids* **63**, 47–53 (1992).
184. Singh, A. et al. Lateral Phase Separation Based on Chirality in a Polymerizable Lipid and Its Influence on Formation of Tubular Microstructures. *Chem. Phys. Lipids* **47**, 135–148 (1988).
185. Thomas, B.N., Safinya, C.R., Plano, R.J. & Clark, N.A. Lipid Tubule Self-Assembly Length: Dependence on Cooling Rate Through a First-Order Phase Transition. *Science* **267**, 1635–1638 (1995).
186. Thomas, B.N. et al. Phosphonate Lipid Tubules. I. *J. Am. Chem. Soc.* **120**, 12178–12186 (1998).
187. Thomas, B.N. et al. Phosphonate Lipid Tubules. II. *J. Am. Chem. Soc.* **124**, 1227–1233 (2002).
188. Jean, B., Oss-Ronen, L., Terech, P. & Talmon, Y. Monodisperse Bile-Salt Nanotubes in Water: Kinetics of Formation. *Adv. Mater.* **17**, 728–731 (2005).
189. Terech, P., Geyer, A.D., Struth, B. & Talmon, Y. Self-Assembled Monodisperse Steroid Nanotubes in Water. *Adv. Mater.* **14**, 495–498 (2002).
190. Terech, P., Jean, B. & Ne, F. Hexagonally Ordered Ammonium Lithocholate Self-Assembled Nanotubes with Highly Monodisperse Sections. *Adv. Mater.* **18**, 1571–1574 (2006).
191. Lepault, J., Pattus, F. & Martin, N. Cryo-Electron Microscopy of Artificial Biological Membranes. *Biochim. Biophys. Acta* **820**, 315–318 (1985).
192. Yang, B., Kamiya, S., Yui, H., Masuda, M. & Shimizu, T. Effective Shortening in Length of Glycolipid Nanotubes with High Axial Ratios. *Chem. Lett.* **32**, 1146–1147 (2003).
193. Georger, J.H. et al. Helical and Tubular Microstructures Formed by Polymerizable Phosphatidylchholines. *J. Am. Chem. Soc.* **109**, 6169–6175 (1987).
194. Jin, M. et al. Superhydrophobic Aligned Polystyrene Nanotube Film with High Adhesive Force. *Adv. Mater.* **17**, 1977–1981 (2005).
195. Liang, Z., Susha, A.S., Yu, A. & Caruso, F. Nanotubes Prepared by Layer-by-Layer Coating of Porous Membrane Templates. *Adv. Mater.* **15**, 1849–1853 (2003).
196. Cui, Y. et al. Synthesis of Thermosensitive PNIPAM-co-MBAA Nanotubes by Atom Transfer Radical Polymerization within a Porous Membrane. *Macromol. Rapid Commun.* **26**, 1552–1556 (2005).
197. Guo, Y. et al. Dimension Control of Glycolipid Nanotubes by Successive Use of Vesicle Extrusion and Porous Template. *Chem. Mater.* **18**, 1577–1580 (2006).
198. Guo, Y. et al. FT-IR Study of the Interlamellar Water Confined in Glycolipid Nanotube Walls. *Langmuir* **21**, 4610–4614 (2005).
199. Yui, H. et al. Local Environment and Property of Water Inside the Hollow Cylinder of a Lipid Nanotube. *Langmuir* **21**, 721–727 (2005).
200. Proulx-Curry, P.M. & Chasteen, N.D. Molecular Aspects of Iron Uptake and Storage in Ferritin. *Coord. Chem. Rev.* **144**, 347–368 (1995).
201. Bosman, A.W., Janssen, H.M. & Meijer, E.W. About Dendrimers: Structures, Physical Properties, and Applications. *Chem. Rev.* **99**, 1665–1688 (1999).

202. Frechet, J.M.J. & Tomalia, D.A. Dendrimers and Other Dendritic Polymers. (Wiley, West Sussex; 2001).
203. Tomalia, D.A. in Supramolecular Polymers. (ed. A. Ciferri) 187–256 (CRC Press, Boca Raton; 2005).
204. see, for example: http://www.cdc.gov/ncidod/dvrd/revb/gastro/norovirus.htm).
205. Matsunaga, T., Kawasaki, M., Yu, X., Tsujimura, N. & Nakamura, N. Chemiluminescence Enzyme Immunoassay Using Bacterial Magnetic Particle. *Anal. Chem.* **68**, 3551–3554 (1996).
206. Yang, C.D., Takeyama, H., Tanaka, T. & Matsunaga, T. Effects of Growth Medium Composition, Iron Sources and Atmospheric Oxygen Concentrations on Production of Luciferase-Bacterial Magnetic Particle Complex by a Recombinant Magnetospirillum Magneticum AMB-1. *Enzyme Microb. Technol.* **29**, 13–19 (2001).
207. Yang, B., Kamiya, S., Shimizu, Y., Koshizaki, N. & Shimizu, T. Glycolipid Nanotube Hollow Cylinders as Substrates: Fabrication of One-Dimensional Metallic-Organic Nanocomposites and Metal Nanowires. *Chem. Mater.* **16**, 2826–2831 (2004).
208. Yang, B., Kamiya, S., Yoshida, K. & Shimizu, T. Confined Organization of Au Nanocrystals in Glycolipid Nanotube Hollow Cylinders. *Chem. Commun.* **5**, 500–501 (2004).
209. Takahashi, R. & Ishiwatari, T. Preparation of Helical Gold Nanowires on Surfactant Tubules. *Chem. Commun.* **12**, 1406–1407 (2004).
210. Zhu, H., John, G. & Wei, B. Synthesis of Assembled Copper Nanoparticles from Copper-Chelating Glycolipid Nanotubes. *Chem. Phys. Lett.* **405**, 49–52 (2005).
211. Masuda, M. et al. Room-Temperature Synthesis of Lipid Nanotubes and Simultaneous Encapsulation of Biopolymers. *Polym. Prepr.* **56**, 602 (2007).
212. Yui, H. et al. Encapsulation of Ferritin Within a Hollow Cylinder of Glycolipid Nanotubes. *Chem. Lett.* **34**, 232–233 (2005).
213. Banerjee, I.A. et al. Magnetic Nanotube Fabrication by Using Bacterial Magnetic Nanocrystals. *Adv. Mater.* **17**, 1128–1131 (2005).
214. Masuda, M. & Shimizu, T. Multilayer Structure of Unsymmetrical Monolayer Lipid Membrane with 'Head-to-Tail' Interface. *Chem. Commun.* **2001**, 2442–2443 (2001).
215. Masuda, M. & Shimizu, T. Polymorphism of Monolayer Lipid Membrane Structures Made from Unsymmetrical Bolaamphiphiles. *Carbohydr. Res.* **340**, 2502–2509 (2005).
216. Aigouy, P.T., Costeseque, P., Sempere, R. & Senac, T. Caractérisation Structurale et Évolution Thermiquue de l'Acide Amino-11-undécanoïque($C_{11}H_{23}NO_2{\cdot}1.5H_2O$). *Acta Cystallogr.* **B51**, 55–61 (1995).
217. Szafran, M. et al. Electrostatic Interactions and Conformations of Zwitterionic Pyridinium Alkanoates. *J. Org. Chem.* **63**, 2898–2908 (1998).
218. Iwaura, R. & Shimizu, T. Reversible Photochemical Conversion of Helicity in Self-Assembled Nanofibers from a 1,ω-Thymidylic Acid Appended Bolaamphiphile. *Angew. Chem., Int. Ed.* **45**, 4601–4604 (2006).
219. Iwaura, R. et al. Molecular-Level Helical Stack of a Nucleotide-Appended Oligo(*p*-phenylenevinylene) Directed by Supramolecular Self-Assembly with a Complementary Oligonucleotide as a Template. *J. Am. Chem. Soc.* **128**, 13298–13304 (2006).
220. Iwaura, R., Yoshida, K., Masuda, M., Yase, K. & Shimizu, T. Spontaneous Fiber Formation and Hydrogelation of Nucleotide Bolaamphiphile. *Chem. Mater.* **14**, 3047–3053 (2002).
221. Kunitake, T. Synthetic Bilayer Membrane: Molecular Design, Self-Organization, and Application. *Angew. Chem., Int. Ed. Engl.* **31**, 709–726 (1992).
222. Shimizu, T. Bottom-Up Synthesis and Structural Properties of Self-Assembled High-Axial-Ratio Nanostructures. *Macromol. Rapid Commun.* **23**, 311–331 (2002).
223. Shimizu, T. Bottom-Up Synthesis and Morphological Control of High-Axial-Ratio Nanostructures through Molecular Self-Assembly. *Polymer J.* **35**, 1–22 (2003).
224. Shimizu, T., Iwaura, R., Masuda, M., Hanada, T. & Yase, K. Internucleobase-Interaction-Directed Self-Assembly of Nanofibers from Homo- and Heteroditopic 1,ω-Nucleobase Bolaamphiphiles. *J. Am. Chem. Soc.* **123**, 5947–5955 (2001).

225. Shimizu, T. & Masuda, M. Stereochemical Effect of Even-Odd Connecting Links on Supramolecular Assemblies Made of 1-Glucosamide Bolaamphiphiles. *J. Am. Chem. Soc.* **119**, 2812–2818 (1997).
226. Garti, N. & Sato, K. Crystallization and Polymorphism of Fats and Fatty Acids. (Marcel Dekker, New York; 1988).
227. Masuda, M., Vill, V. & Shimizu, T. Conformational and Thermal Phase Behavior of Oligomethylene Chains Constrained by Carbohydrate Hydrogen-Bond Networks. *J. Am. Chem. Soc.* **122**, 12327–12333 (2000).
228. Biradha, K. & Fujita, M. in Advanced in Supramolecular Chemistry, Vol. 6. (ed. G.W. Gokel) 1–39 (JAI Press Inc., Stamford, CT; 2000).
229. MacGilivray, L.R. & Atwood, J.L. in Advances in Supramolecular Chemistry, Vol. 6. (ed. G.W. Gokel) (JAI Press Inc., Stamford, CT; 2000).
230. Fan, R. et al. DNA Translocation in Inorganic Nanotubes. *Nano Lett.* **5**, 1633–1637 (2005).
231. Goldberger, J., Fan, R. & Yang, P. Inorganic Nanotubes: A Novel Platform for Nanofluidics. *Acc. Chem. Res.* **39**, 239–248 (2006).
232. Waugh, R.E. & Hochmuth, R.M. Mechanical Equilibrium of Thick, Hollow, Liquid Membrane Cylinders. *Biophys. J.* **1987**, 391–400 (1987).
233. Davidson, M., Karlsson, M., Sinclair, J., Sott, K. & Orwar, O. Nanotube-Vesicle Networks with Functionalized Membranes and Interiors. *J. Am. Chem. Soc.* **125**, 374–378 (2003).
234. Karlsson, A. et al. Nanofluidic Networks Based on Surfactant Membrane Technology. *Anal. Chem.* **75**, 2529–2537 (2003).
235. Karlsson, R., Karlsson, A. & Orwar, O. A Nanofluidic Switching Device. *J. Am. Chem. Soc.* **125**, 8442–8443 (2003).
236. Karlsson, A. et al. Networks of Nanotubes and Containers. *Nature* **409**, 150–152 (2001).
237. Karlsson, M. et al. Micropipet-Assisted Formation of Microscopic Networks of Unilamellar Lipid Bilayer Nanotubes and Containers. *Langmuir* **17**, 6754–6758 (2001).
238. Karlsson, M. et al. Formation of Geometrically Complex Lipid Nanotube-Vesicle Networks of Higher-Order Topologies. *Proc. Natl. Acad. Sci. USA.* **99**, 11573–11578 (2002).
239. Sott, K. et al. Micropipet Writing Technique for Production of Two-Dimensional Lipid Bilayer Nanotube-Vesicle Networks on Functionalized and Patterned Surfaces. *Langmuir* **19**, 3904–3910 (2003).
240. Reiner, J.E., Wells, J.M., Kishore, R.B., Phefferkorn, C. & Helmerson, K. Stable and Robust Polymer Nanotubes Stretched from Polymersomes. *Proc. Natl. Acad. Sci. USA.* **103**, 1173–1177 (2006).
241. Brenner, H. & Gaydos, L.J. The Constrained Brownian Movement of Spherical Particles in Cylindrical Pores of Comparable Radius: Models of the Diffusive and Convective Transport of Solute Molecules in Membranes and Porous Media. *J. Colloid Interface Sci.* **58**, 312–356 (1977).
242. Karlsson, R. et al. Moving-Wall-Driven Flows in Nanofluidic Systems. *Langmuir* **18**, 4186–4190 (2002).
243. Nogawa, K. et al. Development of Novel Nanopipette with a Lipid Nanotube as Nanochannel. *J. Robotics Mechatronics* **19**, 528–534 (2007).
244. Mani, R.C., Li, X., Sunkara, M.K. & Rajan, K. Carbon Nanopipettes. *Nano. Lett.* **3**, 671–673 (2003).
245. Kometani, R. et al. Nozzle-Nanostructure Fabrication on Glass Capillary by Focused-Ion-Beam Chemical Vapor Deposition and Etching. *Jpn. J. Appl. Phys.* **42**, 4107–4110 (2003).
246. Clarke, R.W., White, S.S., Zhou, D., Ying, L. & Klenerman, D. Trapping of Proteins Under Physiological Conditions in a Nanopipette. *Angew. Chem., Int. Ed.* **44**, 3747–3750 (2005).
247. Lauf, U., Fahr, A., Westesen, K. & Ulrich, A.S. Novel Lipid Nanotubes in Dispersions of DMPC. *Chem. Phys. Chem.* **5**, 1246–1249 (2004).
248. Polidori, A., N. Michel, Fabiano, A.S. & Pucci, B. Exotic Aqueous Behavior of Synthetic Lipids: Formation of Vesicular Nanotubes. *Chem. Phys. Lipids* **136**, 23–46 (2005).

249. Ohta, A. et al. Transition from Nanotubes to Micelles with Increasing Concentration in Dilute Aqueous Solution of Potassium *N*-Acyl Phenylalaninate. *Langmuir* **22**, 8472–8477 (2006).
250. Zhai, L., Herzog, B., Drechsler, M. & Hoffmann, H. Novel Nanotubes from a Cationic Surfactant and an Anionic Stiff Aromatic Counter-Ion. *J. Phys. Chem. B.* **110**, 17697–17701 (2006).
251. Ryu, J.-H., Kim, H.-J., Huang, Z., Lee, E. & Lee, M. Self-Assembling Molecular Dumbbells: From Nanohelices to Nanocapsules Triggered by Guest Intercalation. *Angew. Chem., Int. Ed.* **45**, 5304–5307 (2006).
252. Grumelard, J., Taubert, A. & Meier, W. Soft Nanotubes from Amphiphilic ABA Triblock Macromonomers. *Chem. Commun.* **2004**, 1462–1463 (2004).
253. Nakamura, M., Goto, R. & Shibakami, M. Closed- and Open-End Tubules Composed of Cyclic Lipids: Effect of Chain Length on Morphological Change. *Colloids Surf A Physicochem Eng Asp* **304**, 54–59 (2007).
254. Ajayaghosh, A., Varghese, R., Mahesh, S. & Praveen, V.K. From Vesicles to Helical Nanotubes: A Sergeant-and-Soldiers Effect in the Self-Assembly of Oligo(*p*-phenyleneethynylene)s. *Angew. Chem., Int. Ed.* **45**, 7729–7732 (2006).
255. Boettcher, C., Schade, B. & Fuhrhop, J.-H. Comparative Cryo-Electron Microscopy of Noncovalent *N*-Dodecanoyl-(D- and L-) Serine Assemblies in Vitreous Toluene and Water. *Langmuir* **17**, 873–877 (2001).
256. Díaz, N. et al. Self-Assembled Diamide Nanotubes in Organic Solvents. *Angew. Chem., Int. Ed.* **44**, 3260–3264 (2005).
257. Jung, J.H., Lee, S.M., Han, S.-H., Shinkai, S. & Shimizu, T. Ultrastable Steroidal Nanotube Formed in Organic Solvents. *Chem. Lett.* **34**, 532–533 (2005).
258. Diáz, N., Simon, F.-X., Schmutz, M. & Mésini, P. Self-Assembled Nanotubes in Organic Solvents. *Macromol. Symp.* **241**, 68–74 (2006).
259. Shimizu, T. & Asakawa, M. Latest Researches in AIST, http://www.aist.go.jp/aist_e/latest_research/2006/20060807/20060807.html. (2006).
260. Dodziuk, H. Cyclodextrins and Their Complexes: Chemistry, Analytical Methods, Applications. (Wiley-VCH, Weinheim; 2006).
261. Szejtli, J. Cyclodextrin Technology. (Kluwer Academic Publishers, Boston, MA; 1988).
262. Martin, C.R. & Kohli, P. The Emerging Field of Nanotube Biotechnology. *Nat. Rev. Drug Discov.* **2**, 29–37 (2003).
263. M. R. Abidian, D.-H. Kim & Martin, D.C. Conducting-Polymer Nanotubes for Controlled Drug Release. *Adv. Mater.* **18**, 405–409 (2006).
264. Yan, X. et al. Transition of Cationic Dipeptide Nanotubes into Vesicles and Oligonucleotide Delivery. *Angew. Chem., Int. Ed.* **88**, 2431–2434 (2007).
265. Chun, A.L., Moralez, J.G., Webster, T.J. & Fenniri, H. Helical Rosette Nanotubes: A Biomimetic Coating for Orthopedics? *Biomaterials* **26**, 7304–7309 (2005).
266. Dang, T.X., Milligan, R.A., Tweten, R.K. & Wilson-Kubalek, E.M. Helical Crystallization on Nickel–Lipid Nanotubes: Perfringolysin *O* as a Model Protein. *J. Struct. Biol.* **152**, 129–139 (2005).
267. Dang, T.X. et al. Helical Crystallization on Lipid Nanotubes: Streptavidin as a Model Protein. *J. Struct. Biol.* **150**, 90–99 (2005).
268. John, G., Mason, M., Ajayan, P.M. & Dordick, J.S. Lipid-Based Nanotubes as Functional Architectures with Embedded Fluorescence and Recognition Capabilities. *J. Am. Chem. Soc.* **126**, 15012–15013 (2004).
269. Guo, Y. et al. Alignment of Glycolipid Nanotubes on a Planar Glass Substrate Using a Two-Step Microextrusion Technique. *J. Nanosci. Nanotech.* **6**, 1464–1466 (2006).
270. Mahajan, N. & Fang, J. Two-Dimensional Ordered Arrays of Aligned Lipid Tubules on Substrates with Microfluidic Networks. *Langmuir* **21**, 3153–3157 (2005).
271. Frusawa, H. et al. Aligning a Single-Lipid Nanotube with Moderate Stiffness *Angew. Chem., Int. Ed.* **42**, 72–74 (2003).

272. Fujima, T., Frusawa, H., Minamikawa, H., Ito, K. & Shimizu, T. Elastic Precursor of the Transformation from Glycolipid Nanotube to Vesicle. *J. Phys.: Condens. Matter* **18**, 1577–1580 (2006).
273. Zhao, Y., An, L. & Fang, J. Buckling of Lipid Tubules in Shrinking Liquid Droplets. *Nano. Lett.* **7**, 1360–1363 (2007).
274. Avila, J. Microtubule Proteins. (CRC Press, Boca Raton, FL; 1989).
275. Bijkerk, P. *Infectieziekten Bull.* **13**, 271 (2002).

Chapter 4
Soft-Matter Nanotubes

Tatsiana Lobovkina, Aldo Jesorka, Björn Önfelt, Jan Lagerwall, Paul Dommersnes, and Owe Orwar

Abstract This chapter provides an overview over the extended area of surfactant nanotubes research, covering theoretical as well as experimental approaches. Fabrication strategies for nanotube assemblies include surfactant or amphiphile self-assembly, forced shape transformations and, in case of appropriately functionalized building blocks, polymerization. The main body of this review is dedicated to the dynamic properties of lipid nanotubes, their role in vesicle-nanotube networks and derivatives consisting of branched, knotted, or circular nanotubes. Transport modes and enzymatic reactions in nanotube-interconnected vesicles are discussed, in particular their application as unique research tools to mimic sub-cellular conditions, and to explore concepts of unconventional, miniaturized chemical reactors in a biocompatible environment. Lipid nanotubes can be expected to further gain in importance as model systems for nanotube-based (transport) processes in biological cells. They facilitate the investigation of chemical reaction kinetics in complex structured geometries as well as studies of transport phenomena involving ultra small volumes and single molecules, and provide new insights into fundamental aspects of the biophysics of membranes. This review is particularly committed to highlight the rich opportunities to engineer and utilize surfactant nanotube assemblies, leading the way to future biological and technological applications.

Symbols

a_o	Interface area occupied by the polar head group of a surfactant molecule.
v_{ch}	Volume of the hydrocarbon region of a surfactant molecule.
l_{ch}	Length of the hydrocarbon region of a surfactant molecule.
D	Diffusion constant.
$F_{nanotube}$	Free energy of lipid nanotube.
κ	Bending rigidity.
C_1, C_2	Two principal curvatures.
C_0	Spontaneous curvature.
C	Curvature of the string.

O. Orwar (✉)
Department of Chemical and Biological Engineering, Chalmers University of Technology, SE-41296 Göteborg, Sweden
e-mail: orwar@chalmers.se

O. Hayden, K. Nielsch (eds.), *Molecular- and Nano-Tubes*,
DOI 10.1007/978-1-4419-9443-1_4,

A	Area.
dA	Surface area element.
σ	Membrane tension.
f	Pulling force.
f_0	Equilibrium pulling force.
l	Length.
p	Pressure.
V	Volume.
g	Genus number.
r	Lipid nanotube radius.
r_0	Equilibrium radius of lipid nanotube.
k_B	Boltzmann constant.
T	Temperature.
$\zeta_{\|\|}, \zeta_{\perp}$	Friction coefficients (per unit length) for parallel and perpendicular directions, respectively.
η	Solvent viscosity.
$v_{\|\|}, v_{\perp}$	String velocities in the parallel and perpendicular directions.
v_Y	Velocity of the Y*-junction.
s	Arch length.
τ	String tension.
τ_0	Equilibrium string tension.
$\delta\tau$	Deviation from the equilibrium string tension.
t	Time.
x	Coordinate.
h	Parameter, describing shape of lipid nanotubes.
h_{eq}	Equilibrium shape of lipid nanotube
u	Deviation from the equilibrium shape.
Z	Height of the Y-junction network.
X	Half-width of the Y-junction network.
u_x, u_{xx}	First and second derivatives of u in x-direction.
$\frac{\partial u}{\partial t}$	Time-derivative of u.
t_1	Relaxation time.
L_p	Persistence length.
R	Vesicle radius.
$t_{mix}, t_{mix}^{sphere}, t_{mix}^{tube}$	Mixing time.
$t_{traffic}$	Traffic time.
τ_{relax}	Relaxation time.
J	Resulting flow in the lipid nanotube interior.
J_P	Poiseuille flow.
J_M	Marangoni flow.
Φ	Volume fraction.
$\Delta\sigma$	Change in membrane tension.
i, j	Container number.
$c_{q_j}(t)$	Concentration of substance q at time moment t in the container j.

$\partial_t c_{q_j}(t)$	Change in concentration of substance q as a function of time.
$k_{ij}^{(q)}$	Rate of the diffusive transport of substance q from container i to container j.
D_q	Diffusion coefficient of substance q.
ℓ_{ij}	Length of the tube connecting containers i and j.
V_j	Volume of vesicle j.
k_1, k_{-1}, k_{-cat}	Reaction constants.
K_M	Michaelis constant.
$k_{dissip}^{(q)}$	Phenomenological loss term.
$\mathbf{k}$	Bilayer normal.
$\mathbf{n}$	Average chain direction.
θ	Tilt angle.
φ	Tilting direction.

4.1 Introduction

The control of chemical reactions combined with manipulation and transport of ultra-small amounts of reactants down to the limit of single molecules is a considerable scientific and engineering challenge.

Stimulated by the rapidly increasing knowledge of biological systems, where nanoscale reaction containers and interconnecting tubular assemblies are the foundation of many sub-cellular processes, man-made soft-matter nanotube structures and associated systems have already emerged to provide in-depth insight into reaction and transport phenomena on the ultra-small scale. Self-assembled as well as mechanically formed nanosized conduits and tubular arrangements made from soft-matter materials such as surfactant membranes offer interesting opportunities in nanotechnology research. Unlike hard material nanodevices, which are typically produced by top-down fabrication using e-beam lithography or a variety of other methods, soft-matter nanostructures are accessible through unconventional fabrication routes, such as self-assembly, self-organization, and forced shape transitions. Surfactant membranes are flexible, offering unprecedented control over device dimensions, geometries, fluidity, and functionality. These properties can often be changed on demand resulting in responsive and dynamic devices. Furthermore, such devices can incorporate and interact with a variety of biological materials, components, and systems, thus providing a level of complexity that is difficult to attain using conventional cleanroom technologies.

Surfactant soft-matter nanotube assemblies, and associated devices find use in biophysics, biology, biomedicine, and bioanalysis where requirements for similarity, compatibility, and compliance with living matter systems are particularly high. Applications may range from basic research such as investigations of reaction-diffusion phenomena in nanoscale networks to transport and identification of single DNA molecules confined in nanotubes. Soft matter surfactant structures combine a variety of attractive features and properties, such as compatibility with aqueous environments, chemical stability, ease of preparation, and accessibility to different modes of material transport through their fluid boundary or interior. Furthermore, the tubular structures can be prepared with an unusually large length-to-diameter ratio, as well as in different complex geometric and topological shapes. Soft-matter nanotubes, predominantly consisting of non-covalently interacting surfactants such as phospholipids, serve, on one hand, important roles in biological systems, mainly in maintaining inter- and intracellular transport and communication, and, on the other hand, gain more and more importance in artificial biomimetic devices.

In summary, this chapter highlights recent advances in soft-matter (surfactant) nanotube technology, largely based on biomimetic phospholipid membrane assemblies. The material in this section should provide biomedical researchers as well as professionals in the biophysical sciences with state-of-the-art material required to understand and evaluate the opportunities and technological challenges of soft matter nanotubular assemblies. In the first section, physico-chemical properties of surfactants and surfactant membranes are discussed, the second section is dedicated to the fabrication, properties, and applications of phospholipid bilayer nanotubes, the third section is dedicated to nanotubular structures in biology. Here, several new functions for nanotubes have been found recently, in particular in cell-to-cell communication and transport. The fourth section introduces self-assembled lipid nanotubes and the final section covers aspects of polymer nanotubes, largely based on application examples.

4.2 Self-Assembly of Surfactants

Central to almost all biological and biomimetic tubular structures is the fluid boundary, a self-assembled bilayer membrane composed of phosphorylated fatty acid amphiphiles, i.e. lipids (Fig. 4.1). In general, amphiphiles combine a spatially separated hydrophobic and hydrophilic region in the same molecule, enabling spontaneous alignment of the respective regions to form extended supramolecular membrane structures. The morphology of these structures, or aggregates, depends on the shape of the surfactant molecule, which is characterized by a shape factor $v_{ch}/a_o l_{ch}$, where a_o is the interface area occupied by the polar head group, and v_{ch} and l_{ch} are the volume and length of the hydrocarbon region, respectively. The value of the shape factor determines whether surfactants will form micelles ($v_{ch}/a_o l_{ch} < 1/3$), non-spherical (globular or cylindrical) micelles ($1/3 < v_{ch}/a_o l_{ch} < 1/2$), bilayers ($1/2 < v_{ch}/a_o l_{ch} < 1$), or inverted micelles ($v_{ch}/a_o l_{ch} > 1$) [1]. In particular, cylindrical micelles are composed of single-chained lipids with small head group areas, while bilayers are formed by double-chained lipids with large head group areas (Fig. 4.1a–c).

The most common biological membrane lipids are the phospholipids. Phospholipids consist of a hydrophobic head group which is linked to two hydrophobic hydrocarbon tails though a phosphate group (Fig. 4.1a–d). In most biological membranes, the hydrocarbon tails contain from 10 to 18 carbons per chain,

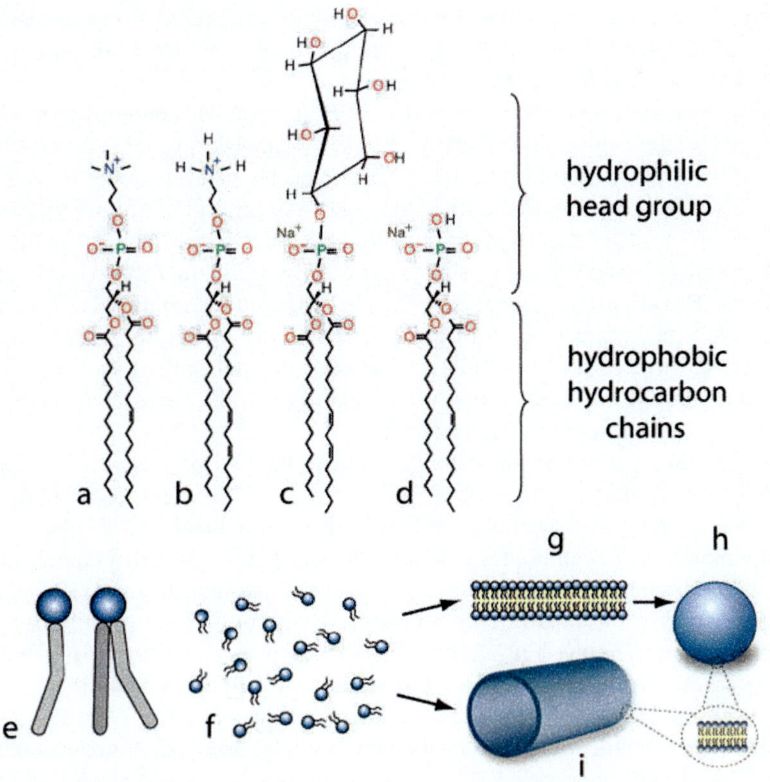

Fig. 4.1 Self-assembly of surfactant molecules. (**a–d**) Molecular structures of some common phospholipids (**a**) L-α-phosphatidylcholine (1-palmitoyl-2-oleoyl; egg/chicken), (**b**) L-α-phosphatidylethanolamine (1-palmitoyl-2-linoleoyl; plant/soy), (**c**) L-α-phosphatidylinositol (1-palmitoyl-2-linoleoyl, sodium salt; plant/soy), (**d**) L-α-phosphatidic acid (1-palmitoyl-2-oleoyl, sodium salt; egg/chicken). (**e**) Schematic drawing of surfactant molecules, consisting of a hydrophilic head group, and one or two (as it is in the case of phospholipids) hydrophobic tails, i.e. hydrocarbon chains. The kink in the hydrophobic tail represents a *cis*-double bond in the hydrocarbon chain. Depending on the conditions, surfactant molecules can, for example, self-assemble into bilayers (**g**), which in turn close up and form vesicles (**f**), or hollow tubular structures (**i**) depending on the chemical nature of the aggregate lipids

one of which is unsaturated or branched. Examples of the dominating phospholipids in the plasma membrane of mammalian cells include phosphatidylcholine, phosphatidylethanolamine, and phosphatidylserine. Due to their cylindrical shape and amphiphilic nature, phospholipids spontaneously form bilayers (Fig. 4.1g). The bilayer is a relatively flat structure, where the hydrophilic head groups are facing the aqueous medium, whereas the hydrophobic tails of the molecules are shielded from the aqueous medium, sandwiched between the hydrophilic head groups. The thickness of phospholipid bilayers is only ~5 nm. Often, lipid bilayers are considered as molecularly thin 2D sheets, sometimes referred to us as 2D liquids. The term "fluidity" is normally used when describing the diffusion of molecules within a bilayer.

The lateral diffusion of lipid molecules is rather rapid, with a diffusion constant D in the range of 10^{-8}–10^{-10} cm^2/s [2]. In general, the fluidity of a lipid bilayer depends on its composition and on the temperature.

Lipid molecules have a strong preference for the lamellar configuration, which is evidenced by the negligible solubility of lipid molecules in water. The extremely low solubility of phospholipids implies almost no exchange of material between the membrane and the solution. The critical aggregate concentration of lipid bilayers is only 10^{-12}–10^{-10} M. In this condensed state, lipid membranes have a limited surface compressibility, since compression is opposed by steric interaction between the amphiphiles. For almost all phenomena, the phospholipid membrane can be considered as locally incompressible with great resistance to change in surface density [3, 4]. Several properties of lipid bilayers account for their unique characteristics: fluidity, low solubility and compressibility, flexibility, high resistance to stretching deformation, but no in-plane shear resistance.

The chemical structure of the individual lipids, mainly the presence of unsaturated elements within the hydrophobic tails as well as the presence of lipophilic additive compounds, such as cholesterol, tailor the mechanical and chemical properties of a membrane. Membranes assembled through self-association mechanisms can under certain conditions separate portions of the medium they were suspended in, forming spherical compartments (liposomes) with an interior volume isolated from the external medium (Fig. 4.1h). They can also spontaneously form tubular structures (nanotubes) depending on the structure of the lipids (Fig. 4.1i).

Numerous other synthetic or semi synthetic amphiphiles with the ability to self-assemble to membrane structures are known, and stabilization strategies such as cross-linking and polymerization of one or more molecular layer(s) also exist. In the following sections of this chapter, mainly phospholipid-based membrane structures are discussed from different points of view, and in the final section, polymerized amphiphile-based nanotubes are covered.

4.3 Mechanically Formed Lipid Nanotubes

Lipid nanotubes can be drawn from lipid bilayer membranes by the action of a highly localized mechanical force. The force pulls out a thin membrane cylinder which is generally referred to as a membrane tether or a membrane nanotube. Lipid nanotubes were initially observed when red blood cells, anchored to a glass surface, were exposed to a fluid flow which caused detachment of the cells. However, as the cells were still anchored to the surface, a long membrane tether could be observed extending from the cell body to the substrate surface [5]. In the following years, formation of nanotubes from cells and lipid vesicles has been studied by applying a point-force to their membrane. For example, a nanotube can be pulled using a micrometer-size bead attached to a membrane, manipulated with magnetic [6] or optical tweezers [7, 8], or, alternatively, by hydrodynamic extrusion [9]. Nanotubes can also be formed through adhesion to the tip of an atomic force microscope (AFM) cantilever [10], by using a micromanipulation technique [11, 12], or by using motor proteins [13]. These techniques allow for studying properties of bilayer membranes, such as measurements of the membrane bending rigidity [6, 14, 15], and the adhesion energy of the membrane to the cytoskeleton [16, 17]. By pulling membrane tethers with the help of motor proteins, such as kinesin, insights into the range of forces acting during formation and modulation of biological compartments are given [13].

In the same way as artificially formed lipid vesicles are used as a model for studying complex biological membranes, lipid nanotubes can be used to explore properties of tubulo-vesicular cell compartments. Due to their small size (5–300 nm in radius [14, 18]), lipid nanotubes allow for routing of ultra small amounts of materials, down to the level of single molecules, which can be detected and interrogated by means of confocal fluorescence microscopy. Various transportation modes can be employed in lipid nanotubes, such as diffusion, tension-driven transport, and electrophoresis. By using micromanipulation methods, not only membrane nanotubes can be formed, but also rather complex nanotubular networks, including tubulo-vesicular networks [19, 20]. These networks allow for studying dynamic properties of nanotubes, such as their merging, shape transitions, and self-organization behavior. Investigations of shape transformations in lipid nanotubes are especially interesting, since cellular tubulo-vesicular compartments (such as the endoplasmatic reticulum or Golgi apparatus) are dynamic, and continuously undergo shape transformations. All these aspect and properties of lipid nanotubes will be discussed in further details below.

4.3.1 Mechanical Properties of Lipid Nanotubes

When a highly localized load is applied to a lipid bilayer membrane vesicle, a membrane nanotube is formed (Fig. 4.2). In experiments, lipid vesicles (or cells) are usually held by aspirating them into a micropipette [14, 21] or by adhesion to a surface [22]. The membrane can be considered as a thin elastic sheet with elastic

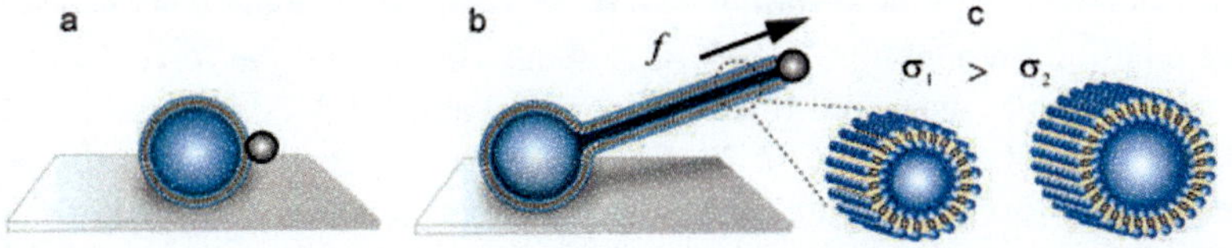

Fig. 4.2 Schematic drawings showing formation of a lipid nanotube due to action of a mechanical force. (**a**) A micrometer-sized bead is attached to a surface of an adhered lipid vesicle. (**b**) When a pulling force (*f*) is applied to a bead, a membrane nanotube is formed. The radius of the nanotube is exaggerated for clarity of presentation; the radius of a lipid nanotube is typically more than thousand times smaller than the radius of a vesicle. The equilibrium radius of a nanotube is set by membrane tension and bending rigidity. For the same lipid composition (i.e. the same bending rigidity), the higher the tension ($\sigma_1 > \sigma_2$), the smaller the nanotube radius (**c**). Cross sections of lipid nanotubes under different tension regimes

bending as well as surface tension energy. The external pulling force adds a negative mechanical work term to the free energy [23–25]:

$$F_{nanotube} = \frac{\kappa}{2} \oint (C_1 + C_2 - C_0)^2 dA + \sigma A - fl, \quad (4.1)$$

where κ is the bending rigidity, C_1 and C_2 are the two principal curvatures of the membrane, C_0 is the spontaneous curvature, which is a parameter accounting for a possible asymmetry between the two monolayers (for example, due to different lipid composition in the two monolayers), dA is a surface area element, A is the membrane area, σ is the membrane tension (which is kept constant), f is the force pulling on the tether, and l is the length of the membrane tether. In general, the energy functional also includes a Gaussian curvature term, and a pressure term $-pV$ (where p is the pressure difference between the inside and the outside of the membrane, and V is the volume of the vesicle). Since the topology of the membrane is fixed, the Gaussian curvature term can be omitted in the energy minimization. The topology of an object is defined by the genus number (g), which is a measure of the number of holes (or handles) in an object. The pressure term is negligible for lipid nanotubes, and can also be omitted [24, 26]. If both of the leaflets of the bilayer have identical lipid composition, the spontaneous curvature C_0 is equal to zero. Under these assumptions, the free energy of a nanotube is

$$F_{nanotube} = \frac{\kappa \pi l}{2r} + \sigma 2\pi rl - fl \quad (4.2)$$

where $\sigma A = \sigma 2\pi rl$ is the stretching energy of the nanotube, $\frac{\kappa}{2} \oint \left(\frac{1}{r}\right)^2 dA = \frac{\kappa \pi l}{r}$ is the bending energy (one of the principal curvatures is equal to zero and the other is equal to $\frac{1}{r}$), and $-fl$ represents the effect of the external force f, that pulls on the tether. Membrane tension exerts a Laplace pressure on the nanotube, which tends to reduce the nanotube radius, while the bending rigidity counteracts this. This results in a finite equilibrium radius. From Eq. (4.2), the equilibrium radius (at $\frac{\partial F_{nanotube}}{\partial r} = 0$) is found to be $r_0 = \sqrt{\kappa/2\sigma}$, and the equilibrium force (at $\frac{\partial F_{nanotube}}{\partial l} = 0$) is $f_0 = 2\pi\sqrt{2\kappa\sigma}$, which can also be expressed as $f_0 = \frac{2\pi\kappa}{r_0}$ or $f_0 = 4\pi\sigma r_0$. By taking a typical value of the bending rigidity $\kappa = 10^{-19}$ J, and membrane tension $\sigma = 10^{-5}$ N/m, the equilibrium nanotube radius r_0 is about 70 nm, and the force f_0 is about 9 pN. Experimentally measured forces falls in the range of 3–50 pN, and nanotube radii in the range of 5–300 nm [6, 14, 21]. It has also been shown that the point-force required to form a nanotube is somewhat larger (about 13%), than the force required to continue pulling a nanotube (with equilibrium radius r_0) or hold it once it has been formed [25]. In practice, the pulling force is applied not to a single point, but to the membrane patch with a finite area. In this case, a force barrier grows linearly with the size of the membrane area on which the pulling force is exerted [27].

From the membrane tension σ, equilibrium radius r_0, and force f_0, the bending rigidity can be calculated as $\kappa = 2\sigma r_0^2$ or $\kappa = \frac{f_0^2}{8\pi^2\sigma}$. Depending on the composition of the membrane, the bending modulus varies in the range of 0.2–1 $\times$ 10^{-19} J [14, 28–30], which corresponds to a few $k_B T$, where k_B is the Boltzmann constant and T is room temperature. For example, the bilayer bending rigidity is about 11 $k_B T$ for diarachidonyl-phosphatidylcholine (diAPC) and digalactosyl-diacylglycerol (diGDG) [29], 11–14 $k_B T$ for dimyristoyl-phosphatidylcholine (diMPC) [29, 30], 21 $k_B T$ for dioleoyl-phosphatidylcholine (diOPC) [31], and for the red blood cell plasma membrane the values of bending rigidity are in the range of 8–35 $k_B T$ [28, 30, 32]. By doping bilayer membranes with sterols (such as cholesterol, lanosterol, and ergosterol) the membrane rigidity increases, which in turn lead to an increase of the equilibrium radius of nanotubes formed from such membranes [15, 29, 33].

4.3.2 Lipid Nanotubes and Nanotube-Vesicle Networks

Micromanipulation methods in combination with microelectroinjection allows not only for pulling membrane nanotubes from lipid vesicles, but also for constructing nanotube-vesicle networks (NVNs), and controlling the interior solution composition of the network [12, 34]. In this method, a micropipette filled with a buffer solution and containing an electrode, is inserted into a lipid vesicle. The insertion is done by applying electric pulses, and simultaneously piercing the vesicles with the micropipette (Fig. 4.3a). When the micropipette tip is inserted into the vesicle, the membrane seals around the micropipette tip. By pulling the micropipette away from the vesicle, a lipid nanotube which connects the pipette tip and the vesicle is created (Fig. 4.3b). If the nanotube is released from the pipette tip, it retracts to the vesicle in order to minimize the total surface area, and therefore the membrane tension. Thus, means for supporting and maintaining lipid nanotubes are required. Such a problem can be resolved by building a NVN, where the nanotubes are maintained by being suspended between the surface-adhered vesicles. Briefly, formation of NVNs can be described as follows.

First, giant unilamellar vesicles (a vesicle composed of a single bilayer) attached to multilamellar vesicles are prepared. The lipid mixture commonly used is a soybean lecithin (a polar lipid extract), which is a mixture of phosphatidylcholine (45.7%), phosphatidylethanolamine (22.1%), phosphatidylinositol (18.4%), phosphatidic acid (6.9%), and other lipids (6.9%). The multilamellar vesicle is required as a source for lipid material used for building the network.

After formation of a lipid nanotube (Fig. 4.3b), a positive pressure is applied through the micropipette orifice. This leads to injection of a buffer solution, and the formation of a small (daughter) vesicle at the pipette tip (Fig. 4.3c). The newly formed vesicle is then positioned on the surface (Fig. 4.3d). The pipette can be detached from the vesicle by pulling the pipette away from the vesicle and by applying electric pulses at the same time. By repeating this procedure, a nanotube-vesicle network can be formed (Fig. 4.3f). The nanotubes in the network (with typical radius of 100 nm) can consequently conjugate vesicles, or form a multilayer system, and

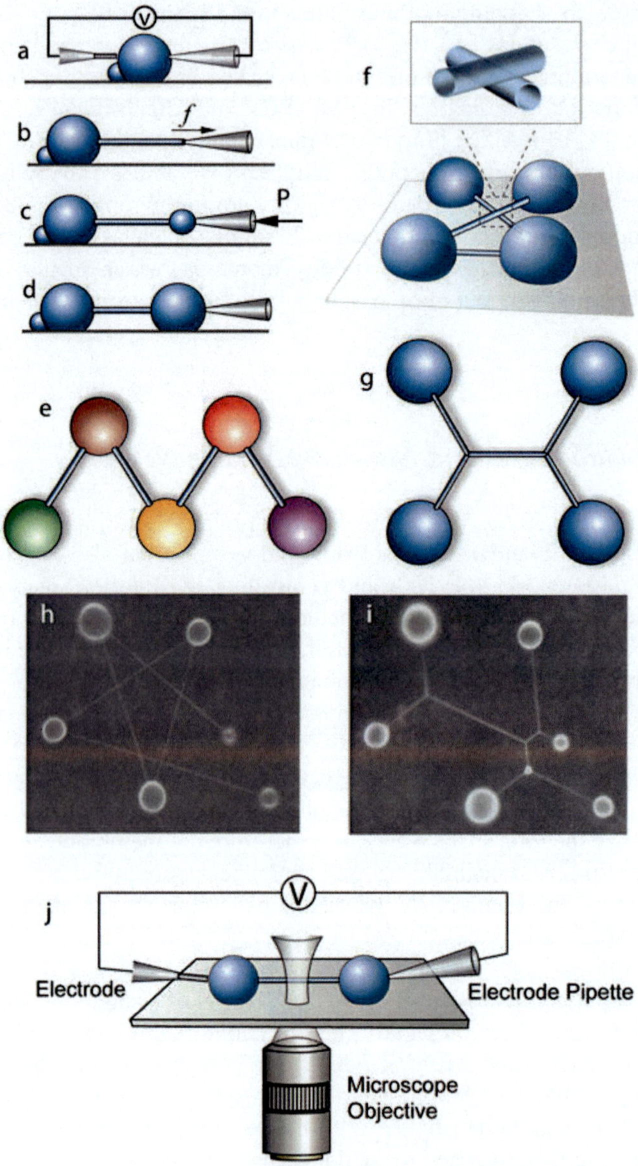

Fig. 4.3 Formation of lipid nanotubes, and nanotube-vesicle networks. (**a**) By applying an electric pulse (10–40 V/cm during ~1 ms [20]) and piercing the unilamellar liposome with a micropipette, it is possible to penetrate the membrane so that the pipette tip enters the vesicle. The membrane then seals around the micropipette tip. Large and small vesicles correspond to giant unilamellar and multilamellar vesicles, respectively. (**b**) Pulling of the micropipette away from a vesicle with a force f, creates a membrane nanotube. (**c**) A positive pressure P is applied through the micropipette. This leads to the injection of buffer solution, and formation of a small (daughter) vesicle at the pipette tip. The size of the daughter vesicle is controlled by the amount of injected liquid. (**d**) The newly

be spatially separated in the z-direction (Fig. 4.3f–g). The composition of the buffer solution contained in the micropipette sets the interior solution of the vesicles. Thus, different nodes within a network can initially have different contents. Note, that a NVN is made of a continuous lipid bilayer, and represents a topological sphere. A nanotube-vesicle network is stable for 4–5 h depending on the experimental conditions. However, by careful control of solution and surface conditions, such as addition of water to counteract evaporation and build up of osmotic pressure, and by a proper pre-treatment of surfaces, for example, by coating with a lipid-friendly polymer, such as SU-8 [35], much longer lifetimes can be achieved.

A NVN can be forced to change its connectivity, so that the lipid nanotubes are connected by a series of branching points, i.e. three-way junctions. For example, the four-vesicle network in Fig. 4.3f can be forced to rearrange the nanotubes into a network containing two branching points (Fig. 4.3g). Figure 4.3h shows a NVN initially containing six vesicles, and five nanotubes. After tube rearrangement, this network contains four branching points (Fig. 4.3i). These shape transitions are accompanied by formation of unstable nanotube junctions. A detailed explanation for formation and propagation of such junctions is given in the next section.

Giant lipid vesicles and lipid nanotubes can be directly observed and manipulated under the optical microscope, using e.g. differential interference contrast (DIC) for contrast enhancement Fig. 4.3j. Fluorescence microscopy in combination with fluorescent labels or membrane stains is also widely used for monitoring and studying NVNs and encapsulated compounds. To be able to track single molecules, a more sensitive fluorescence detection method, such as photon counting, has to be employed.

4.3.3 Self-Organization in Lipid Nanotubes: Zipper Dynamics of Merging Nanotubes

Membrane tubulo-vesicular complexes in biological cells are organized into networks that continuously undergo shape transitions, and function as highways for intra-and inter-cellular transport [36, 37]. In such networks, tubular elements are often connected through dynamic three-way junctions [38].

Fig. 4.3 (continued) formed vesicle is then positioned on the surface. (**e**) The content of the individual vesicles is determined by the composition of the injected solution, which is shown by the different colors of the vesicles. (**f**) Four-vesicle network with three nanotubes. The *inset* shows the spatial arrangement of the nanotubes; the nanotubes cross each other at different planes. (**g**) Forced transition from the initial configuration (**f**) results in a network where the nanotubes are connected through Y-junctions. (**h**) A fluorescence micrograph of a six-vesicle network connected by five lipid nanotubes. The nanotubes cross each other at different levels. (**i**) The same network after a forced shape transition, where the lipid vesicles are connected by a network of lipid nanotubes, containing three-way junctions. Scale bar 15 μm. (**j**) A schematic drawing showing a microscopy/electroporation system setup for the study of nanotube-vesicle networks

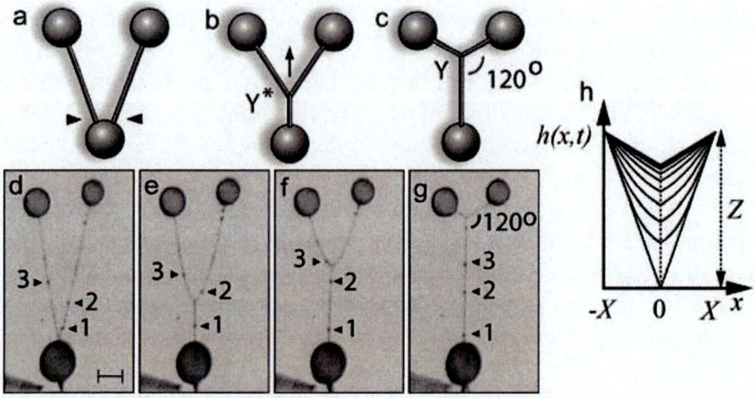

Fig. 4.4 Three-way nanotube junctions. (**a**) A schematic drawing of a three-liposome network. The *arrows* are pointing at the area where the nanotubes are triggered to merge. (**b**) Merging of the nanotubes results in the formation of an unstable Y*-junction; *arrow* shows the direction of the Y*-junction movement. (**c**) The final, stable Y-state of the network. The nanotubes are connected by a Y-junction, and intersect at angles of 120°. (**d**) Inverted fluorescence micrograph of 3 vesicles connected in a V-shape. The *arrows* show the location of the tracer particles on the nanotubes. Reprinted with permission from Lobovkina et al. [41]. Copyright 2007 by the American Physical Society. (**e**) After the junction point is formed, it moves towards a new stable configuration, minimizing the free energy of the system. Here, the first tracer particle has passed the junction point and stays on the lower nanotube, while the second and third particles are still positioned above the Y*-junction. (**f**) As the junction point propagates upwards, the second tracer particle flows through the junction point and gets deposited on the lower nanotube. The distance between the particles 1 and 2 has increased after passing the junction point. (**g**) The third tracer particle has also passed the junction point. The network reaches the geometry with a minimum of free energy. The nanotubes are straight with angles of 120° between them. (**h**) Numerically calculated shapes of the nanotubes during 3-vesicle network evolution. Z is the height and $2X$ is the width is of the network. Scale bar 30 μm

To get an insight into the mechanism of propagation of nanotube three-way junction, let us consider a 3-vesicle network (Fig. 4.4a–g) with two nanotubes connected to one of the vesicles in a V-shape. When electric pulses in combination with mechanical force are applied at the surface of the lower vesicle (shown by the arrows in Fig. 4.4a), the two nanotubes come close together. At a critical distance, the nanotubes will merge. Initially, a 3-way junction is created, which connects three nanotubes (Fig. 4.4b, e–f). Such a junction is unstable, and spontaneously moves towards a state with a minimum free energy. A transitional state of the network with an unstable nanotube junction is referred to as a Y*-state, and the junction itself is called a Y*-junction. Eventually, the network reaches its stable configuration, called the Y-state (Fig. 4.4c, g). The network energy in the Y-state is minimal, which according to Eq. (4.2) is equivalent to the state with minimum total nanotube length. In the Y-state the nanotubes are straight and meet at angles of 120° at the Y-junction. The property of the networks to self-generate Y-junctions is one of the main principles for constructing complex nanotubular networks.

In order to reveal the mechanism of Y*-junction propagation, i.e. to understand the flow of lipids during network evolution, the movement of fluorescent heterogeneities or tracer particles on the nanotubes were studied. From such an analysis, it is possible to draw conclusions about the flow of the lipids during network evolution.

In Fig. 4.4d–g, a 3-vesicle network containing three tracer particles is shown: the nanotube on the right contains two (*1* and *2*) tracer particles, and the nanotube on the left contains one (*3*) tracer particle. The nanotubes are triggered to coalesce by mechanical force, and the movement of the Y*-junction together with the tracer particles is monitored. During propagation of the Y*-junction, the tracer particles pass the junction and reside on the third (lower) nanotube (Fig. 4.4e, f). It is also observed that the tracer particles preserve their relative order in the y-direction before, and after network evolution (Fig. 4.4d, g). However, the distance between tracer particles *1* and *2* originating from the same nanotube increases after passing the Y*-junction. Therefore, it is concluded that the third (lower) nanotube consists of a mixture of lipids coming from the two merging (upper) nanotubes. In other words, the Y*-junction is zipping the two upper nanotubes, and thereby builds up the third, lower, nanotube.

In order to describe the zipper dynamics of nanotubes, the nanotubes are considered to be analogous to semi-flexible polymers and modeled as strings under tension that are able to merge into one single string at the Y*-junction. Using the standard approach for semi-flexible polymers [39], the friction coefficients (per unit length) are assigned for parallel ($\zeta_{||}$) and perpendicular ($\zeta_{\perp}$) displacement of a point on the string. A first estimate of these friction coefficients is obtained from those of a rigid cylinder: $\zeta_{||} = \frac{2\pi\eta}{\ln(l/r)}$ and $\zeta_{\perp} = 2\zeta_{||}$, where l is the length of the tube and η the solvent viscosity. The dynamic equations for the string follow from force balance in the parallel and perpendicular directions:

$$\zeta_{||}v_{||} = \frac{\partial\tau}{\partial s}, \tag{4.3}$$

$$\zeta_{\perp}v_{\perp} = C\tau, \tag{4.4}$$

where C is the curvature of the string, τ is the string tension, which has the dimensions of energy per unit length, $v_{||} = v_{||}(x,t)$ and $v_{\perp} = v_{\perp}(x,t)$ are the string velocities in the parallel, and perpendicular directions at the coordinate x and time t, and s is the arch length of the string. Variations in nanotube radii are assumed to be sufficiently small to be neglected. In the following, the network is considered to be close to the equilibrium state, i.e. when the network is approaching the final geometry, such that $\tau = \tau_0 + \delta\tau \approx \tau_0$ where τ_0 is the equilibrium string tension of the Y*-junction, and $\delta\tau$ is the deviation from the equilibrium tension. The shape of the upper nanotubes is parameterized with coordinates $[x, h = h(x,t)]$ with the origin at the lower vesicle-nanotube junction (Fig. 4.4h). The shape of the upper nanotube in terms of h is $h = h_{eq} + u$, where $u = u(x,t)$ is the deviation from the equilibrium shape, $h_{eq} = Z + \frac{1}{\sqrt{3}}(|x| - X)$. Here the height of the Y-junction network is denoted by Z and the width by $2X$ (Fig. 4.4h). For $u << h_{eq}$, Eq. (4.4) can be linearized, such that one obtains a one-dimensional diffusion equation for u such that:

$$\frac{\partial u}{\partial t} = \frac{3\tau_0}{4\zeta_\perp} u_{xx} \tag{4.5}$$

The boundary conditions for Eq. (4.5) are:

$$u(X,t) = 0, \tag{4.6}$$

$$u_x(0,t) = 0. \tag{4.7}$$

Boundary condition (4.6) implies that the position of the nanotubes is fixed at the nanotube-liposome junction, and condition (4.7) implies a constant angle of 120° between the nanotubes at the Y*-junction. This can be explained as follows. The fluidity of the nanotubes implies continuity of the tension over the Y*-junction. Therefore, in the proximity of the junction point, the nanotubes have the same tension and are locally in equilibrium. The force balance on the junction point in the equilibrium requires that the angles between the nanotubes are 120°.

The solution of Eq. (4.5) is a sum of exponentially relaxing Fourier modes. The largest relaxation time is $t_1 = \frac{16\zeta_\perp X^2}{3\pi^2\tau_0}$. The shape relaxation time t_1 increases as the squared size of the system (i.e. $\sim X^2$). For the half width of the network $X \sim 50$ μm, the relaxation time is on the order of $t_1 \sim 4$ s, which is in good agreement with the experimental data. Taking the length of the nanotube $l \sim 100$ μm, $r \sim 100$ nm, and using $\zeta_{||} = \frac{2\pi\eta}{\ln(l/r)}$ and $\zeta_\perp = 2\zeta_{||}$ with the water viscosity $\eta = 10^{-3}$ Pa·s, the string tension is estimated to be $\tau_0 \sim 0.6$ pN, which corresponds to a membrane tension $\sigma = \frac{\tau_0}{4\pi r} \sim 0.5 \times 10^{-6}$ N/m. The literature values for σ are between 10^{-7} and 10^{-4} N/m [3, 8, 29, 40]. Note that here, τ_0 is essentially of the same magnitude as the pulling force f_0 needed to form a nanotube.

In order to confirm the zipper dynamics of the Y*-junction, the ratio $\frac{v_{||}(0,t)}{v_Y(t)}$ is examined, where $v_Y(t) = \frac{\partial u}{\partial t}(0,t)$ is the velocity of the Y*-junction, and $v_{||}(0,t)$ is the velocity of the lipids along the upper nanotubes at the Y*-junction. It is possible to show that $\frac{v_{||}(0,t)}{v_Y(t)}$ has a negative sign. This negative ratio supports the requirement of opposite directions of $v_Y(t)$ and $v_{||}(0,t)$. The Y*-junction propagates upwards, so v_Y is positive. Thus, the velocity $v_{||}(0,t)$ of the upper nanotubes at the Y*-junction must be negative. Indeed, it is observed experimentally that lipids from the upper nanotubes flow downwards and pass the junction point, i.e. the Y*-junction is zipped. At the junction point, lipids mix and reside on the lower nanotube.

The analysis described above is valid when the shape of the network is close to its equilibrium shape. In order to find the nanotube shape and tension during the whole process of network evolution, Eqs. (4.3) and (4.4) need to be integrated numerically. Integration gives the shape of the nanotubes and the velocity of the lipid flow. Numerically calculated shapes of nanotubes are shown in Fig. 4.4h. The calculated ratio $\frac{v_{||}(0,t)}{v_Y(t)}$ is always negative, and the zipper mechanism can therefore act at any moment of Y*-junction propagation [41].

4.3.4 Formation of Branched, Knotted and Circular Nanotubes

In complex networks, where more than two nanotubes interact, the final configuration of a network depends on the initial arrangement of the nanotubes as well as the relative position of the vesicles. As an example, let us consider two 5-liposome networks having initial arrangements of the nanotubes as shown in Figs. 4.5a, f and 4.6a, f, respectively. Self-organization in these networks leads to different final configurations: in the case of the network in Fig. 4.5, the final configuration is a branched nanotube network, while the nanotubes in Fig. 4.6 form a tight knot.

Initially, the nanotubes in both networks are spatially separated in the z-direction and form a multilayer system, as shown in Figs. 4.5a, f and 4.6a, f, respectively. Self-organization is triggered by merging two adjacent nanotubes at the surface of the lower vesicle. A three-way junction (Y_1*) is formed immediately in both networks, and it starts to propagate towards a new stable configuration, Figs. 4.5b, g and 4.6b. During the propagation of the Y_1*-junction, nanotubes *I*, *II* and *III*, *IV* move towards each other at the surface of the two upper vesicles *3* and *4*, respectively, and eventually start to merge. As a result, two new junction points (Y_2* and Y_3*) are created (Figs. 4.5c, h and 4.6c, g). Note that formation of junctions Y_2* and Y_3* do not require any energy input; they form and evolve spontaneously in the direction of the first Y_1*-junction. Until this point, both networks in Figs. 4.5 and 4.6 evolve identically. However, when junctions Y_1*, Y_2* and Y_3* meet, a different behavior is observed. In the case of the network in Fig. 4.5, an transient 5-way

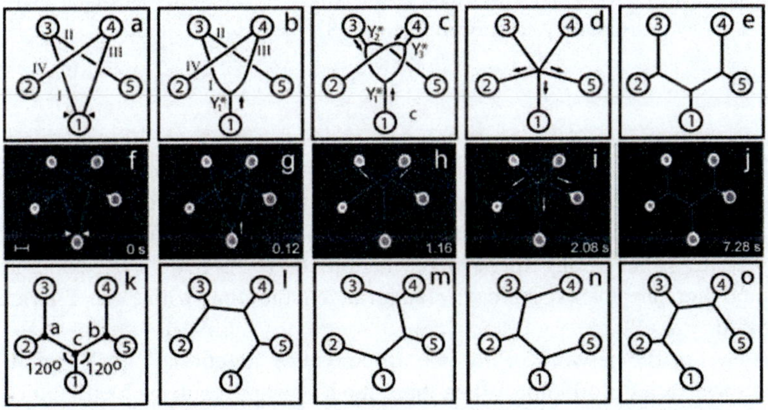

Fig. 4.5 Formation of tree-like structures in lipid nanotube networks. Figures **(a–e)** represent schematic drawings of a 5-vesicle network, and figures **(f–j)** are the corresponding fluorescence micrographs obtained from experiments. **(a)** and **(f)** show the initial network configuration. Network evolution begins with the formation of a Y^*_1 junction, **(b)** and **(g)**, leading to subsequent formation of Y^*_2 and Y^*_3 junctions, **(c)**, **(h)**. Evolution of the network involves formation of an intermediate 5-way junction, **(d)**, **(i)**. This intermediate junction splits into the final tree-like structure **(e)**, **(j)**. Figures **(k)**–**(o)** show five final configurations that correspond to the local optimal solution of an ESTP. Y-junctions *a*, *b*, and *c* in figure **(k)** represent Steiner points. Scale bar, 15 μm

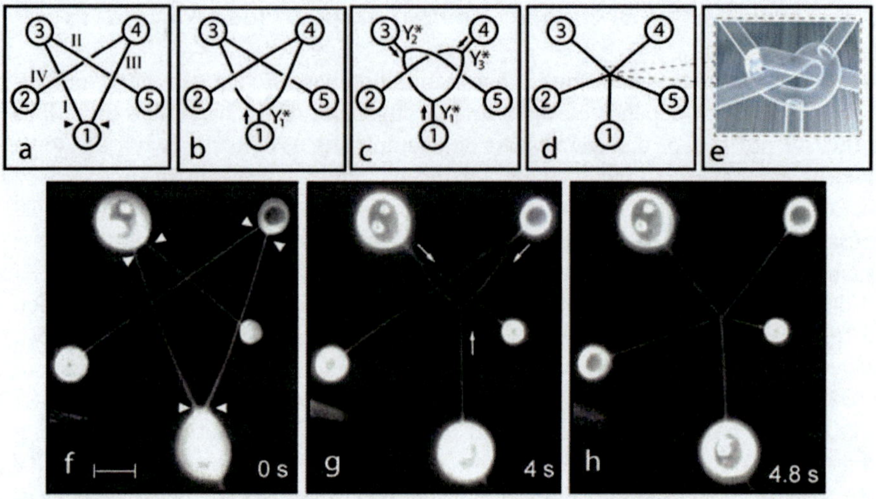

Fig. 4.6 Formation of a trefoil knot in lipid nanotubes. Figures (**a–e**) show schematic drawings of the networks, and figures (**f–h**) are fluorescence micrographs of knot-formation in a NVN. Initially, the NVN contain five vesicles connected with four nanotubes (**a**) and (**f**), similar to the initial configuration shown in Fig. 4.5**a**, **f**. However, comparing to Fig. 4.5**a**, **f**, the nanotubes are built in different order, and their relative arrangement in z-direction is different. Network evolution begins with formation of a $Y^*{}_1$ junction, (**b**). Similar to the network evolution shown in Fig. 4.5, two traveling junctions $Y^*{}_2$ and $Y^*{}_3$ are spontaneously formed, (**c**), (**g**). Junctions $Y^*{}_1$, $Y^*{}_2$ and $Y^*{}_3$ represent the loops of a trefoil knot. When these junctions meet, the knot is tightened (**d**), (**h**), which is the final stable configuration of the network. *Inset* (**e**) gives a schematic representation of the knotted region. Scale bar, 15 μm. Reprinted with permission from Lobovkina et al. [42]. Copyright 2007 by the National Academy of Science, USA

junction is created (Fig. 4.5d, i). In the next moment (less than 40 ms), this junction splits into three new interconnected 3-way junctions, which evolve towards the final configuration of the network (Fig. 4.5e, j). Finally, three Y-junctions are formed, each created by nanotubes intersecting at an angle of 120°. It has to be said that the final arrangement of the nanotubes for 5-vesicle networks is not unique, and there are five possible ways to arrange nanotubes with three Y-junctions (Fig. 4.5k–o).

In the case of the network in Fig. 4.6, the traveling junctions $Y_1{}^*$, $Y_2{}^*$ and $Y_3{}^*$ converge towards a trefoil knot. When these junctions approach each other, the knot is getting tight (Fig. 4.6c–e, g, h). In the final stable configuration, the knot remains tight, and has an equilibrium size of about $1.3r_0$, where r_0 is the equilibrium radius of a nanotube [42].

When a network self-organizes into a tree-like structure (Fig. 4.5), it shows a striking similarity with a well-known optimization problem, which is called the Euclidian Steiner Tree Problem (ESTP). ESTP is the problem of finding a network connecting a given set of terminal points on a plane, allowing addition of auxiliary points, also called Steiner points, with the objective to minimize the total network length. ESTP, and related problems such as the Traveling Salesman problem [43] are Nondeterministic Polynomial time hard (i.e. NP-hard) problems which are computationally hard to solve. It has been shown that lipid nanotube networks may self-organize into geometries that are locally optimal solutions to such problems [44]. As an example, a problem with 5 terminal points is taken, since smaller problems are fairly straightforward to solve. The lipid nanotube network is a scaled model where nanotube junctions give the Steiner points, and the final bifurcated arrangement of nanotubes represents a solution to the specific ESTP. The initial network configuration is shown in Fig. 4.5a. Network self-organization is triggered mechanically by forming a single Y_1*-junction (Fig. 4.5b), which immediately starts to move towards a lower energy level. Subsequently, new Y_2*and Y_3* junctions are spontaneously created which in turn evolve, merge, and split (Fig. 4.5c, d), until a stable final configuration is reached after a few seconds (Fig. 4.5e). This final configuration represents a locally optimal solution to the present ESTP. Such a solution must be a tree with up to 3 Steiner points (points a, b, c in Fig. 4.5k), having exactly three edges intersecting at each Steiner point at angles 120° [43]. The solution is locally optimal with respect to selection of Steiner points, since alternative solutions exist as shown above (Fig. 4.5k–o).

Similarities to the Steiner Tree problem are also found in biology. For example, the amoeboid organism *Physarum polycephalum* forms optimized networks of tubular elements through which chemical nutrients and intracellular signals are transported, Fig. 4.7a [45]. Furthermore, when endoplasmatic reticulum and Golgi are extracted from cells they form similar branched networks with membrane nanotubes connected through Y-junctions, Fig. 4.7b [46]. It is tempting to believe that biological organisms have the capability to solve pathway minimization problems.

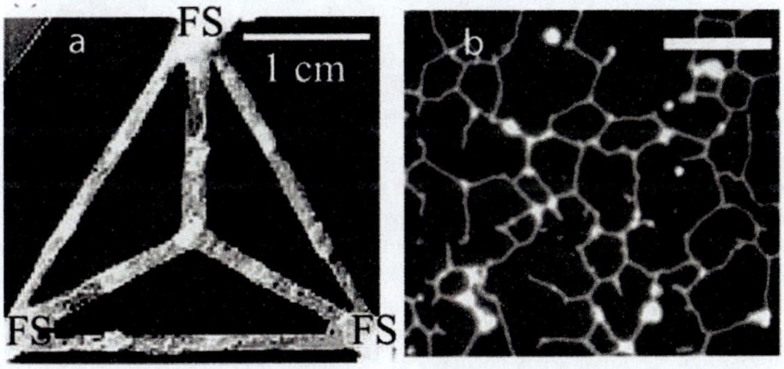

Fig. 4.7 Images of nanotubular networks of biological origin. (**a**) The *Physarum polycephalum* plasmodium forms an optimized network of tubular elements through which chemical nutrient and intracellular signals are transported when the food sources (FS) are presented at different locations. Reprinted with permission from Nakagaki et al. [45]. Copyright 2008 by the Royal Society of London. (**b**) Fluorescence microscopy image of endoplasmic reticulum formed in vitro. Scale bar 10 μm. Reprinted with permission from Dreier and Rapoport [46]. Copyright 2008 by the Rockefeller University Press

Given the logistical challenge of transporting thousands of building blocks and reactants to diverse locations within the cell, such a capability would result in a much more efficient organism.

So far, only networks with spherical topology have been considered. However, inside biological cells, the membrane structures undergo not only shape transformations, but also topological changes. Several techniques exist for building networks with a given topology [20] as well as to form circular nanotubes [47]. These techniques require fusion of the membrane, and therefore addition of energy – for example, by applying an electric field in order to fuse two lipid vesicles. Topological changes with subsequent shape transitions in the nanotube networks does, in a very simplified way, mimic topological transformations of cell membranes. In order to show transformation from spherical topology, $g=0$, to $g=1$, let us consider formation of circular lipid nanotubes. Formation of circular lipid nanotubes requires two nanotube-vesicle networks: (i) a principal network, which is used for constructing a circular nanotube (marked with green color in Fig. 4.8a), and (ii) a supporting network (marked with red color in Fig. 4.8a) which is used for restraining the

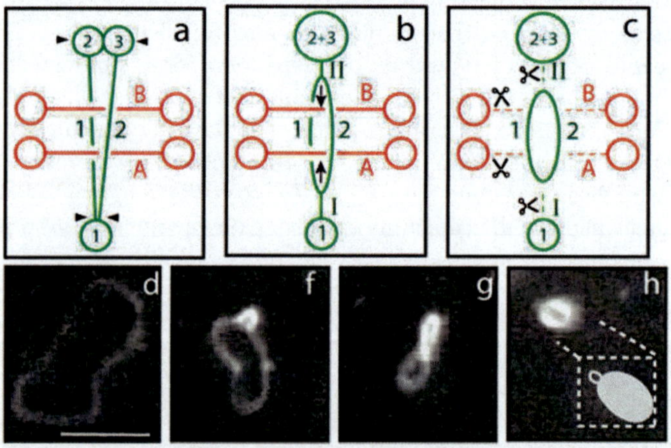

Fig. 4.8 Circular lipid nanotubes. Figures (**a–c**) represent schematic drawings showing formation of a circular nanotube, and figures (**d–h**) are fluorescence images showing shape transformations of unconstrained circular nanotube. (**a**) Two NVN networks are built: the principal network (shown in *green*), used to construct a circular nanotube, and the supporting network (*red*) used to restrain the nanotubes from the principal network. Initially, the principal network has genus 0 topology. *Arrows* indicate the areas where the transformation of the principal network is triggered: (i) the nanotube junction is formed by merging nanotubes 1, and 2 at the surface of vesicle 1; (ii) vesicles 2 and 3 are fused and the topology is changed. The distance between the nanotubes 1 and 2 at the surface of the newly formed vesicle decreases so that nanotubes 1 and 2 merge and produce a nanotube junction connecting nanotubes 1, 2 and II. (**b**) Nanotubes 1, and 2 form a circular nanotube. (**c**) After cutting nanotubes I and II from the principal network and nanotubes A and B from the supporting networks, a circular nanotube is released into the surrounding solution. (**d**) The circular nanotube is curved due to thermal undulations. (**f**) Formation of a small bud on the *upper right area* of the circular nanotube. (**g**) The bud grows into the vesicle by collecting lipids from the circular part. (**h**) The circular nanotube transforms into a small vesicle with a handle. The *inset* shows a schematic drawing of the final state. Scale bar 20 μm. Reprinted with permission from Lobovkina et al. [47]

circular nanotube before releasing it into the surrounding solution. Figure 4.8a–c schematically show how a circular lipid nanotube can be formed.

Free circular nanotubes undergo strong thermal fluctuations (Fig. 4.8d), and from the fluorescence images, the persistence length of the nanotube can be estimated, $L_p = \kappa\pi r/k_B T \sim 5 - 10$ µm. Next, a small bud is formed on the nanotube (Fig. 4.8f). The bud grows into a vesicle by collecting lipids from the circular part of the nanotube (Fig 4.8g). As the vesicle grows, the size of the circular part decreases, and in the final state, a vesicle a few micrometers large, with a small fluorescent spot on the side, is observed (Fig. 4.8h). Presumably, this fluorescent spot is a submicron handle attached to the vesicle. It is reasonable to assume such a final arrangement since it is energetically unfavorable to change the topology of the system.

The described technique can be extended to obtain, and study objects with more complex topologies (g>1). The results may also be important in applications where templates are required for giving complex shapes to polymer gels by solidifying the interior of the nanotubes.

4.3.5 Diffusion in Nanotube-Vesicle Networks

Hydrodynamics in micrometer-sized liquid systems is over damped, i.e. inertia effects can safely be neglected as viscous forces are dominant. Thus, on the micrometer-scale and below, an object comes to rest right after the driving force stops [48]. Under most circumstances, the flow of liquid surrounding membrane nanotubes, and the fluid flow of the membrane itself, both represent laminar flows. In the absence of convective mixing and other forces that may induce transport of particles trapped in the nanotubes (such as tension-driven transport or electrophoresis), diffusion is the only mode of transporting, and mixing particles in the solution contained inside NVNs.

There are three characteristic time scales governing the behavior of diffusing particles in nanotube-vesicle networks. These time scales depend on the geometry of the network and are discussed below.

First, let us consider a single lipid vesicle with radius R, containing particles (or molecules). The time it takes for a particle to explore all volume elements of a container is given by the mixing time, $t_{mix}^{sphere} = \frac{R^2}{D}$, where D is the diffusion coefficient. Released at some point, the particle can be found with equal probability anywhere in the volume after the time lapse t_{mix}^{sphere} [49].

The second characteristic time scale is the time needed for a particle to cross a nanotube with length l, $t_{mix}^{tube} = \frac{l^2}{D}$. In NVNs, the nanotube length is typically longer than the radius of a vesicle. However, their values are often of the same order of magnitude (typically 50 µm vs. 10 µm, respectively). Therefore, the corresponding mixing times are comparable to each other $t_{mix}^{sphere} \sim t_{mix}^{tube}$ and will be denoted by t_{mix}.

The third important time parameter is the traffic time, which is the characteristic time needed for a particle to encounter an immobilized target. For a particle diffusing inside a vesicle, this target is represented by the nanotube entrance. The traffic time can be estimated as $t_{traffic} = \frac{R^3}{Dr}$, where r is the radius of the nanotube

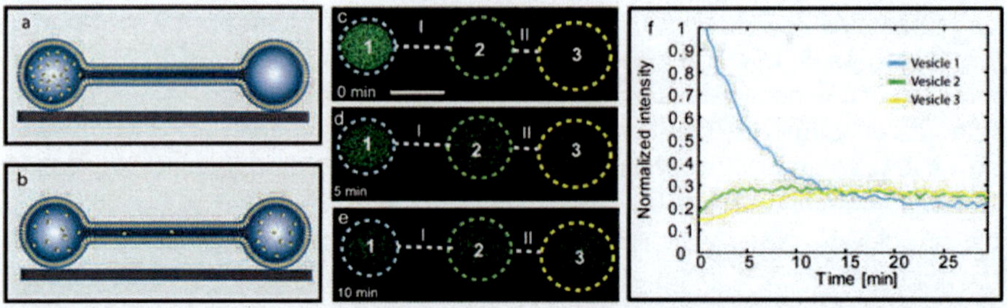

Fig. 4.9 Diffusion relaxation of particles inside a two-vesicle network. (**a**) Schematic drawing showing the initial state of the network, where one of the vesicles is filled with particles. (**b**) Diffusion of the particles through the nanotube smears out the concentration gradient. (**c**) Fluorescence micrograph of three interconnected vesicles filled with fluorescein. Vesicles 2 and 3 are photobleached by laser illumination. *Dashed lines* show the contours of the vesicles and the nanotubes. (**d**)–(**e**) Diffusion of fluorescein from vesicle 1 through the network at 5 and 10 min after photobleaching. The poor fluorescence recovery is due to bleaching and leakage of fluorescein. (**f**) Graph showing normalized fluorescence intensity plotted versus time. Scale bar, 10 μm. Reprinted with permission from Sott et al. [52], Copyright 2007 by the American Chemical Society

(i.e. target) [49]. The traffic time is related to the mixing time as $t_{traffic} \sim \frac{R}{r} t_{mix}$ [49]. For a vesicle with radius $R \sim 10$ μm, and a nanotube radius $r \sim 100$ nm, the ratio $\frac{R}{r} >> 1$, yielding $t_{mix} << t_{traffic}$. Thus, in a NVN, the traffic time ($t_{traffic}$) is much longer than the mixing time (t_{mix}), and the particles inside a vesicle are assumed to be ideally mixed.

Now, let us consider a two-vesicle network, where one of the vesicles is filled with particles (Fig. 4.9a). The particles diffuse along their concentration gradient through the nanotube towards the empty vesicle. After a while, the difference in the number of particles in the two vesicles will vanish, and the concentration gradient will disappear (Fig. 4.9b). The concentration difference between the compartments decays exponentially $\exp(-t/\tau_{relax})$ [50, 51], where τ_{relax} is the relaxation time. For a simple network consisting of two vesicles (Fig. 4.9b), the relaxation time is $\tau_{relax} = \frac{V_1 V_2}{V_1 + V_2} \frac{l}{\pi r^2 D}$, where l is the length of the tube and V_1 and V_2 are the volumes of the containers [50, 51].

Let us compare the relaxation times for a substrate with a diffusion coefficient $D \sim 10^{-6}$ cm^2/s, in a microfluidic system and in a NVN. A microfluidic system with characteristic dimensions: $l \sim 0.1 \times 10^{-3}$ m, $r \sim 10^{-6}$ m, and volume $V \sim 10^{-12}$ m^3 will have relaxation times (τ_{relax}) on the order of hours. Miniaturizing the system to the nanoscale, where $l \sim 10 \times 10^{-6}$ m, $r \sim 100 \times 10^{-9}$ m and $V \sim 10^{-15}$ m^3, leads to relaxation times (τ_{relax}) on the order of minutes. Thus, in contrast to microfluidic systems, diffusion is an efficient way of transporting materials in nanochannels [52, 53].

Figure 4.9c–f show rapid diffusion of fluorescein (molecular weight: 332.3, diffusion coefficient: 4.25×10^{-6} cm^2/s [54]) in a three-vesicle network. Initially, all the vesicles are filled with fluorescein. Then, vesicles 2, and 3 were bleached with laser light, and diffusion of fluorescein from vesicle 1 to vesicles 2, and 3 was monitored over time by confocal fluorescence microscopy. After ~15 min, the fluorescence intensity in each vesicle is approximately the same, indicating an even distribution of fluorescein in the network.

Technological applications of complex nanofluidic networks, as well as studies of transport phenomena in biological systems require theoretical models of diffusive transport. In principle, one could try to obtain the distribution of particles throughout a given network by solving the diffusion equation numerically. However, for large networks, such an approach amounts to dealing with a large number of variables, which can be highly impractical. Instead, a simplified model that describes the most important aspects of particle transport in such complex networks has been developed [51]. This model provides a set of rate equations that govern the time dependence of the number of particles in each vesicle. In such a way, the complicated transport problem is reduced to a set of first order integro-differential equations in time. These equations can be solved efficiently by specifically developed algorithms. The rate equations are valid for any network topology under assumptions that the nanotubes are thin in comparison to their length, and to the radii of the containers, and that particles move solely by diffusion and do not interact with each other. Some simple examples of diffusional relaxation in quite simple networks already a show number of interesting properties, such as sensitivity to the structure (geometry) of the network or wave-like behavior in generated particle number (with one or two extreme points) [51]. Experimental results support these predictions, and will be discussed in further detail in the section below which describes chemical reactions in nanotube-vesicle networks.

4.3.6 Tension-Driven Transport in Lipid Nanotubes

If there is a tension gradient across a surfactant membrane, it will cause a movement of lipid molecules from the region of lower tension towards the region of higher membrane tension. For example, in a network with two vesicles, a tension gradient can be induced by gently deforming one of the vesicles with a micropipette (Fig. 4.10a). The tension of the deformed vesicle increases, and a tension gradient develops along the nanotube. The tension difference between the vesicles gives rise to two oppositely directed flows (see insets of Fig. 4.10a). One is the Poiseuille flow (J_P) which is due to the hydrostatic pressure difference between the vesicles. The second flow is due to the membrane tension difference across the network, and can be explained as follows. The tension gradient between the vesicles results in a flow of lipids from the undisturbed region towards the region where the additional stress is applied. This tension-driven flow is also known as Marangoni flow (J_M), where the liquid (in this case membrane) is pulled out from the area of lower tension towards the region of higher tension due to the surface tension gradient. Marangoni flow is accompanied by the movement of the solution inside the nanotube together with trapped particles.

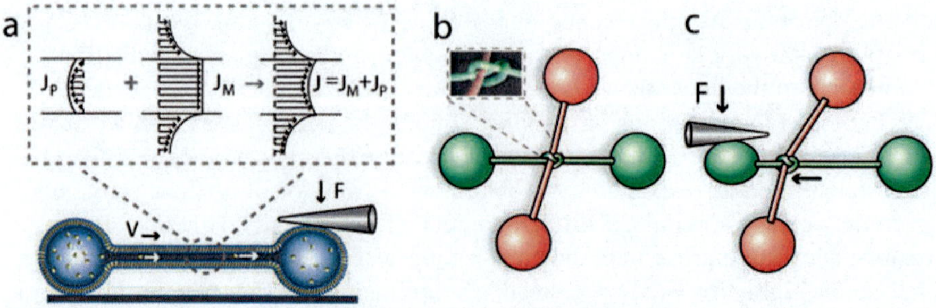

Fig. 4.10 Schematic drawings showing tension-driven transport also known as Marangoni flow. (**a**) Movement of the membrane together with the nanotube interior due to a Marangoni flow. The *inset* explains the effect on flow profile caused by the interplay between the Poiseuille (J_P), and the Marangoni flow (J_M). (**b**) Schematic drawing of a nanotube mechanical tweezer. A nanotube (marked with *red color*) is captured inside the nanotube knot (marked with *green color*). The *inset* shows the structure of the knot. (**c**) One of the vesicles connected to the knotted nanotube is deformed by applying mechanical force F. Deformation of the vesicle causes displacement of the knotted nanotube together with the captured (*red*) nanotube

The resulting flow (J), in the nanotube interior (liquid with trapped particles) is a combination of Poiseuille flow (J_P) and Marangoni flow (J_M) (Fig. 4.10a). These two flows are oppositely directed; however, for lipid nanotubes with a length much larger than the nanotube radius, the Marangoni flow is large compared to the Poiseuille flow [55]. The effective flow profile is shown in the inset of Fig. 4.10a. Note that in the performed experiments, variations in membrane tension are relatively small, and are not sufficient to cause e.g. Rayleigh pearling instabilities [56].

The time needed to transport a fraction (Φ) of the volume of one vesicle with radius R to another vesicle can be estimated as $t \approx \left(\frac{R}{r}\right)^3 \frac{l\eta}{\Delta\sigma}\Phi$, where r is the radius, and l is the length of the nanotube, η is the viscosity of a film with thickness equal to the radius of the tube, and $\Delta\sigma$ is the change in membrane tension [55]. For example, it takes 0.75 s to transport a volume fraction $\Phi = 6 \times 10^{-3}$ of a vesicle with a radius of 5 μm. Such a fraction is equivalent to ten nanotube volumes for a nanotube with radius $r = 100$ nm and length $l = 10$ μm.

Note that the Marangoni flow can be exploited not only to transport material inside the nanotube, but also to transport materials bound to the wall. Furthermore, Marangoni flow can be used to move small high-aspect ratio objects in aqueous solution by winding nanotubes around them. A nanotube knot can be tied around micrometer-sized objects, and it has been shown that a nanotube knot (marked with green color in Fig. 4.10b) can be tied around another nanotube (marked with red color in Fig. 4.10b) [42]. Both ends of the knotted nanotube are connected to vesicles. Deformation of one of the vesicles (Fig. 4.10c) causes Marangoni flow in the nanotube containing the knot, causing movement of the knot itself along the tension gradient. As the knot moves, the nanotube captured inside the knot is also translated towards the region where the force is applied. For such a mechanical torus tweezer, the force required to translocate objects is in the range of several pN [42].

4.3.7 Electrophoresis in Lipid Nanotubes

Electric fields, applied along a lipid nanotube, can be used to transport charged particles, and single molecules, such as DNA [57, 58]. To achieve electrophoretic transport in lipid nanotubes, two micropipettes containing Ag/AgCl electrodes are inserted into a lipid vesicle which contains charged particles or molecules by using electroporation. The surface of the micropipettes is modified with a hydrophobic agent, and the micropipettes are gel-capped in order to suppress electroosmotic flow through the pipettes. When the micropipettes are inserted into a lipid vesicle, the membrane adheres around the micropipettes forming a high electric resistance seal. After that, one of the micropipettes is drawn away from the lipid vesicle, and a nanotube is formed, Fig. 4.11a, b. A positive potential (100 mV) is applied to the pipette holding the nanotube, and an electric field is created inside the tube. Outside the nanotube, the electric field is assumed to be close to zero due to the low conductivity of the membrane. Figure 4.11d shows a finite element method simulation of the electric field strength distribution inside, and outside the nanotube.

Application of an electric field causes movement of the negatively charged lipid membrane (velocities up to ~50 μm/s), and a counter-directional electroosmotic flow. The movement of the nanotube wall causes liquid inside the nanotube to move as well. However, such a wall-driven movement of the liquid column inside the

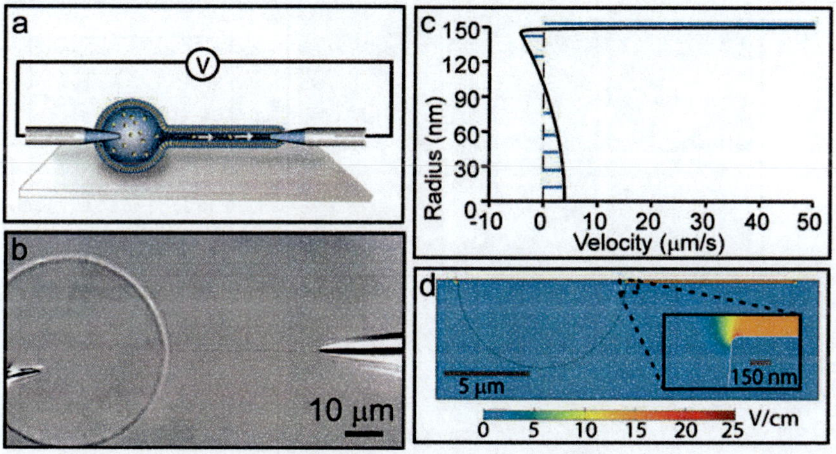

Fig. 4.11 Schematic drawings showing electrophoretic transport in lipid nanotubes. (**a**) The experimental setup. The membrane is sealed around gel-filled micropipettes equipped with electrodes. The nanotube is held by the pipette on the right, to which a positive voltage is applied (100 mV). (**b**) Differential interference contrast microscopy image of a vesicle and nanotube together with the two electrodes. (**c**) Velocity flow profile of the liquid inside the nanotube, drawn from the nanotube center line to the membrane surface. (**d**) Electric field strength distribution inside and outside of lipid nanotube, drawn from the nanotube center line to the membrane surface. Reprinted with permission from M. Tokarz, B. Akerman, J. Olofsson et al., *Proceedings of the National Academy of Sciences of the United States of America* **102** (26), 9127 (2005). Copyright 2007 by the National Academy of Science, USA

nanotubes is counteracted by the electroosmotic flow which acts in the other direction. Thus, the internal liquid is essentially stationary, and the resulting flow profile of solvent inside the nanotube and away from the walls is essentially flat, Fig. 4.11c. Experiments have shown that latex beads of different sizes (30, 100, and 200 nm in diameter), as well as double stranded DNA (5.4–166 kbp) can be transported by electrophoresis inside surfactant nanotubes [57].

4.3.8 Chemical Reactions in Nanotube-Vesicle Networks

The rate of chemical reactions is, dependent on the effective reactant concentration, and temperature, but also dependent on geometrical factors such as size, shape, and dimensionality of the chemical reactor [49, 59]. For example, in some enzymatic reaction systems, oscillatory behavior in product formation can be induced by reducing the size of the reactor [49]. In structured volumes, such as NVNs, geometrical parameters of the network influence the interplay between transport properties of reacting molecules and the observed chemical reaction rates. In such systems, transitions from compact to structured geometries may lead to unusual chemical behavior (Fig. 4.12a), and by manipulating the geometry of the reactor network, the dynamics of confined chemical reactions can be controlled [52].

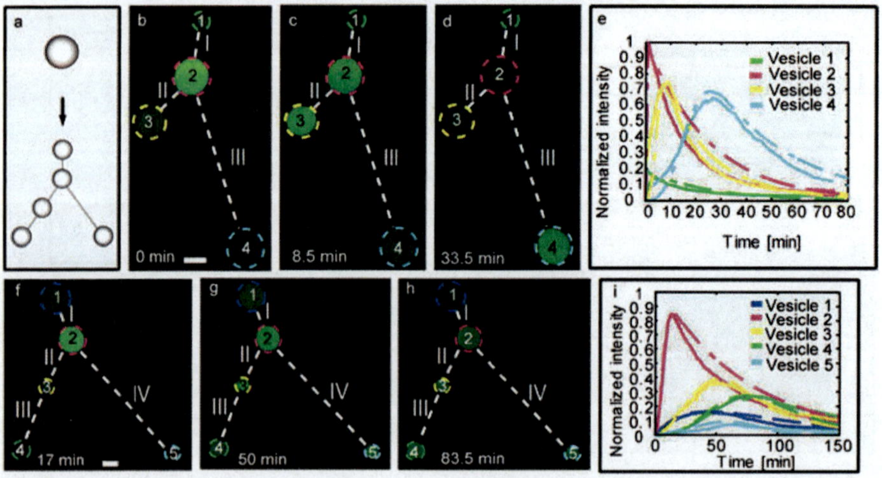

Fig. 4.12 Enzymatic reaction in NVNs. (**a**) Schematic drawing of the transition from a compact (spherical) to a structured geometry (conjugated spheres). Spheres outline lipid vesicles, *lines* represent the interconnecting nanotubes. (**b–d**) and (**f–h**) fluorescence microscopy images showing formation of a product (flourescein) in four- and five-vesicle networks respectively. Alkaline phosphatase was used to convert fluorescein diphosphate (FDP) to fluorescein. Initially, vesicle 1 (in figures **b** and **f**) is filled with enzyme while all other vesicles contain substrate molecules. Product formation is monitored by measuring fluorescence intensity of the product in the vesicles. Graphs (**e**) and (**i**) show normalized intensity plots of product formation (*solid lines*) versus time, and theoretical fits to the experimental data (*dash-dotted lines*). Scale bar: 10 μm. Reprinted with permission from Sott et al. [52], Copyright 2007 by the American Chemical Society

During the formation of NVNs, transitions from a compact (spherical) geometry to a structured geometry (i.e. spheres connected by nanoconduits) occurs (Fig. 4.12a). Such geometric transitions essentially change the spatial framework for the contained reaction systems. Normally, if e.g. enzyme molecules are locally injected into a substrate-filled spherical vesicle with a radius of 20–30 μm, an even distribution of all molecules in the reaction volume will be achieved within a few seconds. Thus, product formation will be homogeneous throughout the reaction volume which is small enough for rapid diffusive mixing of the reactants, but it is still not small enough for unusual reaction dynamics to be observed [60]. However, in the case of structured geometries, product formation may occur as a cascade through a series of vesicles connected by lipid nanotubes, and displays wave-like propagation behavior (Fig. 4.12b–i). Such behavior is caused by the, on the time scale of chemical reactions themselves, long time period required for enzyme molecules to explore the reaction volume, which can take from several minutes up to several hours depending on the geometry of the system.

As an example of a structured geometry, let us consider a 4-vesicle network (Fig. 4.12b–d), where initially vesicle 1 contains enzyme molecules, and vesicles 2–4 are filled with substrate. Starting from vesicle 1, enzyme molecules first diffuse into vesicle 2, and initiate product formation there. From vesicle 2, enzyme molecules diffuse towards vesicle 3 or 4 (or return to vesicle 1). Since nanotube II is shorter than nanotube III, the enzyme molecules first reach vesicle 3, and than vesicle 4. Product formation is observed in the respective order (Fig. 4.12b–d).

The rate of product formation in a NVN depends not only on the length of the nanotubes, but also on the number of vesicles through which enzyme molecules are diffusing. For example, in Fig. 4.12f–h, vesicle 1 is filled with enzyme, and vesicles 2–5 are filled with substrate. Such a network is similar to the network in Fig. 4.12b–d, however, vesicle 3 is introduced between vesicle 2 and 4. Although the distance between vesicles 2–4 is shorter than the distance between vesicles 2–5, the enzyme molecules first reach vesicle 5, and then vesicle 4. Vesicle 3 locally dilutes the enzyme concentration and functions as a diffusion barrier. The rate limiting step for an enzymatic reaction in NVNs is the low probability of finding the entrance to the nanotube by random motion, rather than diffusion inside the nanotubes, which is fast once the enzyme molecules have entered the nanotube. Constraining the diffusive transport in the network, for example, by introducing additional vesicles between the reaction containers, leads to direct control over the reaction rate.

A model that describes the dynamics of the reaction-diffusion system has been developed [51]. The rate equations (4.8a), (4.8b), and (4.8c) allow for mapping the concentration of enzyme (E), substrate (S) and product (P) (respectively) in the different reaction containers as a function of time. Reactions in the nanotubes are neglected due to their very small volume. The rate equations are:

$$\partial_t c_{P_j}(t) = \sum_i k_{ij}^{(P)} \lfloor c_{P_i}(t) - c_{P_j}(t) \rfloor + k_{cat}/K_M c_{E_j}(t) c_{S_j}(t) - k_{dissip.}^{(P)} c_{P_i}(t) \tag{4.8a}$$

$$\partial_t c_{S_j}(t) = \sum_i k_{ij}^{(S)} \lfloor c_{S_i}(t) - c_{S_j}(t) \rfloor - k_{cat}/K_M c_{E_j}(t) c_{S_j}(t) - k_{dissip.}^{(S)} c_{S_i}(t) \tag{4.8b}$$

$$\partial_t c_{E_j}(t) = \sum_i k_{ij}^{(E)} \lfloor c_{E_i}(t) - c_{E_j}(t) \rfloor \tag{4.8c}$$

The three terms on the right hand side represent transport (diffusion), reaction and dissipation, respectively. In the model, the rate of the diffusive transport between two containers, i.e. from container i to container j, is given by $k_{ij}^{(q)} = D_q \pi r^2 / V_j \ell_{ij}$ where D_q is the diffusion coefficient of substance q, r is the tube radius, ℓ_{ij} is the length of the tube connecting containers i and j (please note that $\ell_{ij} = \ell_{ji}$) and V_j is the volume of vesicle j. The terms containing $c_{E_i}(t)c_{S_i}(t)$ arise from a steady state approximation on the intermediate enzyme-substrate complex (ES), under the assumption that the reaction $E + S \underset{k_{-1}}{\overset{k_1}{\rightleftarrows}} ES \xrightarrow{k_{cat}} E + P$ follows Michaelis-Menten kinetics: $K_M = (k_{cat} + k_{-1})/k_1$. The negative dissipation factor in the model is introduced to compensate for photobleaching, and leakage of molecules through the vesicle walls. Graphs e and i in Fig. 4.12 show fluorescence intensity measurements of product formation versus time (solid lines) together with theoretical predictions of product formation (dash-dotted lines). In the graphs, the peaks for product formation are well separated in time, and display wave-like characteristics which can be well controlled by network geometry.

In summary, experimental and theoretical results strongly suggest that the spatial arrangement of reaction containers in a given network can directly modulate the dynamics of chemical reactions. Such a conclusion can both be applied to reactions taking place in nanofluidic systems, and in compartments of biological cells. Many reactions that show periodic, time-varying, or other peculiar behavior might thus sometimes be explained by a particular spatial structuring of the reaction volume rather than being inherently autocatalytic or feed-back modulated.

In the considered examples, the geometry of the networks was preserved during the course of the experiments. However, the unique softness and fluidity of the membrane also allows for altering the connectivity of the network during the course of a chemical reaction, which can be utilized to further differentiate local chemical environments. Moreover, an environment simulating macromolecular crowding conditions inside a NVN can be achieved by internalizing water-soluble stimuli-responsive polymer materials [61, 62]. Different membrane compositions, including membrane proteins, are well-suited to condition a reactor for a certain pattern in product formation [63].

A number of theoretical studies have recently been employed for investigating the properties of reaction-diffusion networks. For example, a numerical study shows filtering capabilities of a nanotube-vesicle network [64]. In a two-container filter, two vesicles (e.g. vesicle *1* and vesicle *2*) are connected by a nanotube. Vesicle *1* is initially filled with enzyme molecules. In vesicle *2*, substrate molecules are introduced by a series of injections which represent the input signals to the system. Substrate molecules diffuse into the enzyme-filled vesicle *1*, and react with enzyme molecules, under the assumption that the enzymes can not leave vesicle *1*, and that reactions in the nanotube can be neglected. Although hindering of the diffusive motion represents a technical challenge, it can be overcome, for example, by trapping the enzyme molecules in polymer gel aggregates [61]. The output of this system is obtained by monitoring the substrate concentration in vesicle *1* over time. A selection criterion is imposed on the output signal, e.g. such that the substrate concentration in vesicle *1* must be located in a certain concentration interval. If the maximum substrate concentration lies within the given interval, the signal is recorded as accepted. The filter characteristic is obtained by plotting the number of accepted signals as a function of the lower threshold level. The study shows that the characteristics of the filter are tunable by simple changes in geometry, i.e. the nanotube length. Such a filter can be viewed as a simple chemistry-based computational element, which makes use of the principles of chemistry to perform certain computational tasks.

The results from these diverse studies support the notion that NVNs are a highly suitable system for studying dynamics of (bio-) chemical reactions under conditions which are close to the biological cell environment, as well as for developing artificial or biomimetic nanofluidic reaction devices, chemical wave-form synthesizers, and tools for unconventional computing.

4.4 Biological Lipid Nanotubes

The biological role of membrane nanotubes is still a sparsely explored field, but it is emerging that membrane nanotubes have important functions in various processes throughout cell biology, e.g. inter- and intracellular communication, leukocyte rolling, and transendothelial migration [65, 66]. The section below overviews a few of the roles and properties found for membrane nanotubes in cell biology this far.

4.4.1 Intercellular Communication via Tunneling Nanotubes

Neighboring cells can be connected by membrane tubules that serve as intercellular communication channels that has been referred to as tunneling nanotubes (TNT) [67–69]. Intercellular membrane nanotubes, or TNTs, have been observed for various cell types, including neuronal cells [68] and immune cells [67, 69].

Membrane nanotubes tend to form as cells come apart after transient contacts, e.g. disassembly of immune synapses [67], or, as has been reported for rat pheochromocytoma PC12 cells, through protrusion of philopodia-like cellular extensions that subsequently connect with an adjacent cell [68]. Membrane nanotubes are fragile and sensitive to mechanic stress but can connect cells at a distance of several cell diameters (nanotube lengths above 100 μm have been reported for several cell types [67, 70, 71]). Nanotube diameters are generally estimated to be 50–200 nm but also larger nanotubes (diameter $> 0.7\ \mu$m) have been reported [37, 70]. Typical signatures for membrane nanotubes are that they are raised above the surface where cells are resting and they are stretched, generally minimizing the distance between connected cells. Occasionally, and in resemblance with artificial vesicle nanotube networks, Y-junctions can be observed for membrane nanotubes connecting three cells [67, 68]. Thus, membrane nanotubes between cells share many characteristics with artificial membrane nanotubes in vesicle networks [72].

One major difference between artificially constructed lipid nanotubes and membrane nanotubes bridging cells is that the latter generally, but not always, contain filamentous (f-) actin [37, 68, 73]. Thicker nanotubes (diameter $> \sim 0.7\ \mu m$) connecting primary macrophages were found to contain f-actin as well as microtubules [37]. Microtubules were also found in intercellular bridges between prostate cancer cells [70]. Presence of structural proteins inside intercellular membrane nanotubes most likely make them more rigid than nanotubes mechanically formed from model membranes. Interestingly, Y-junctions has been shown to contain f-actin [37] and, thus, to get a mechanistic understanding of membrane nanotubes between cells, one would have to consider the driving forces behind both the lipid flow and that of the remodeling of f-actin.

Various objects have been observed to move along membrane nanotubes, including bulges (Fig. 4.13), lipid vesicles, such as endosomes and lyosomes, and cell organelles, such as mitochondria, membrane proteins, virus particles and bacteria [37, 67, 68, 70, 71, 73]. Rustom reported that small organelles belonging to the endosomal/lysosomal system could be transported unidirectionally inside TNTs connecting PC12 cells. Remarkably, these vesicles were observed to move in a manner indicating that TNTs could allow passage of organelles from one cell to the next. Furthermore, in the same study, green fluorescent protein (GFP) anchored to the plasma membrane through fusion with the farnesylation signal of c-HA-Ras was observed to move along the surface of a nanotube from a donor cell onto a connected target cell [68]. Thus, these observations suggest that TNTs could under some circumstances serve as conduits for intercellular transport of whole organelles and membrane components by forming seamless bridges between connected cells. However, this is not undisputed since it was recently shown that membrane nanotubes connecting T cells instead have a distinct junction, across which membrane molecules did not transfer efficiently Sowinski [73].

The fact that membrane nanotubes between cells contain structural proteins makes these nanotubes particular in another aspect. F-actin as well as microtubules

Fig. 4.13 ***Formation and transport of bulges along membrane nanotubes.*** (**a**) Microscope images showing that two HEK 293T cells, stained overnight with separate fluorescent lipids, DiD (white) or DiO (not shown) are connected by a membrane nanotube. Bright-field and fluorescence images indicate that this nanotube originated from the cell on the right. As the nanotube was stretched, a large bulge, 4 μm across, formed along the nanotube (**b**). Bright-field and DiD fluorescence time-lapse images show that the bulge (arrowhead) moved towards, and fused with the cell on the right. (**c**) Graph of the distance between the bulge and the cell on the right as a function of time. The sloped lines drawn in the graph indicates that the bulge slows down as it approaches the cell it eventually fuses with

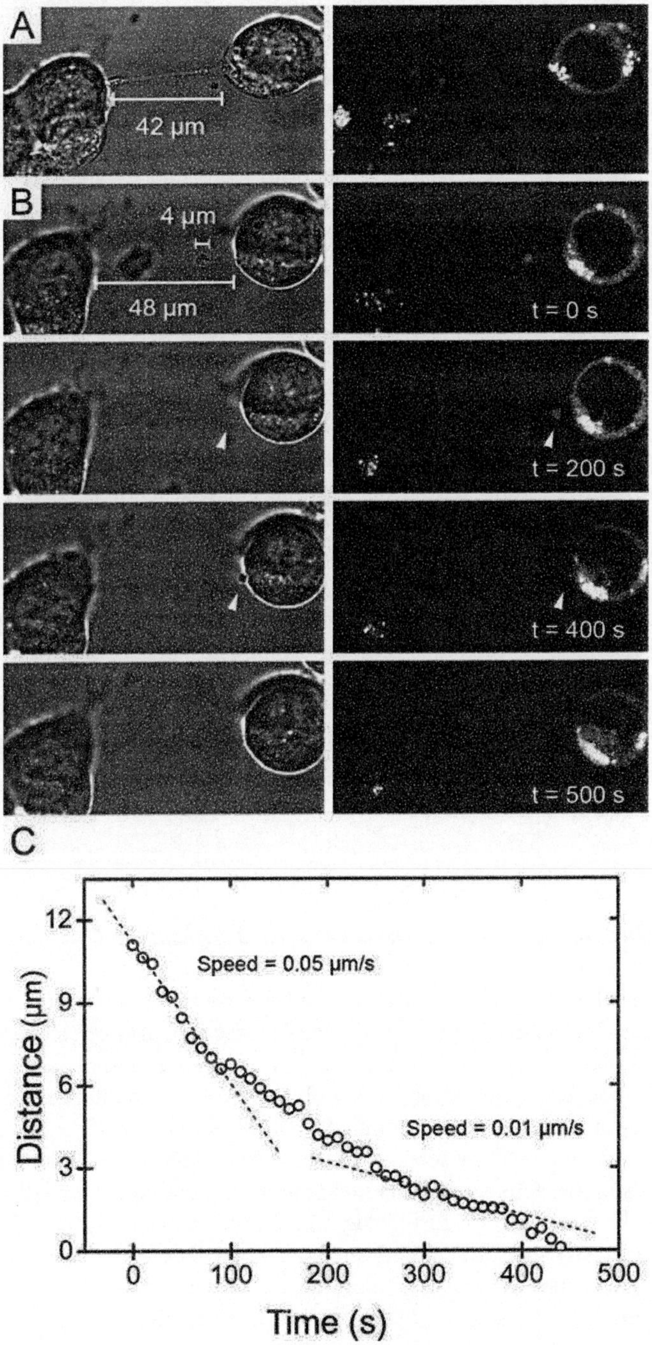

Fig. 4.13 (continued)

can serve as tracks for transport of molecular cargo carried by motor proteins such as myosin (moving along f-actin) or kinesin, and dynein (both moving along microtubules). Indeed, myosin VI were found to be present inside TNTs, and in close proximity to lipid vesicles, indicating that transport of lipid vesicles is driven by motor proteins [68]. Furthermore, calcium signals have been observed to propagate through groups of myeloid-lineage dendritic cells and monocytes networked by membrane nanotubes [69]. These results indicate that signaling through nanotubes could function to activate cells that are distal to a triggering antigen, which could facilitate and enhance an efficient immune response.

It remains an outstanding goal in biology to understand how cells exchange materials and communicate with each other to collectively perform functions that give the organism its properties. Transport through membrane nanotubes represent a novel mechanism that could allow for formation of supra cellular structures where nutrients, genetic material and signaling molecules could be efficiently shared between connected cells [65]. Furthermore, similar to filopodial bridges [74], membrane nanotubes may play a general role in infection since a nanotube network could be exploited for efficient cell-to-cell spread of pathogens. Indeed, it was recently reported that HIV-1 could be transmitted via membrane nanotubes formed between infected and previously uninfected T cells [73]. Such a pathway of transmission may in part sidestep current therapeutic strategies, e.g. treatment with monoclonal antibodies or by vaccination- mediated antibody response. Future challenges involves finding the role of membrane nanotubes in vivo and solving the structure of membrane nanotube – cell body junctions to understand under what conditions the membranes of different cells are allowed to fuse and if there is a "gate keeping" mechanism selecting what cargo can be exchanged between cells.

4.4.2 Membrane Tethers Formed During Leukocyte Rolling

Leukocytes, or white blood cells, circulate and defend the body against infections and foreign material. As leukocytes reach a site of infection, they can leave the blood flow and traverse the vessel walls in a process called transendothelial migration or diapedesis. Leukocytes adhere transiently with activated endothelial cells and "roll" along the walls of blood vessels continuously making and breaking contact. This initial adhesion is supported by selectin proteins expressed by the endothelium and cognate ligands on the leukocyte. This interaction with the vessel walls is believed to yield the leukocytes time for further signal integration that can lead to migration arrest and diapedesis when inflamed tissue is reached. The lifetime of the selectin bond is sensitive to the shear stress caused by the blood flow and in theories only taking the strength of the selectin bond into account, predict that increased blood flow rapidly should increase the rolling speed, and eventually cause detachment of the leukocyte [75]. However, in vivo leukocytes are observed to roll in a fairly steady speed over a wide range of shear stresses at the blood vessel walls [76], suggesting the existence of a mechanism that counterbalance the force generated by increased blood flow.

When leukocytes are exposed to an external flow and allowed to roll along a layer of endothelial cells, they deform and extend thin membrane tethers or membrane nanotubes that connect the cells together [77]. Such tethers also form when leukocytes are rolling on a surface coated with platelets or P-selectin [78], and the number of tethers increase with increased shear stress [79]. Interestingly, formation of these tethers act to decrease the pulling force imposed on adhesive bonds [80], and cells forming nanotubes exhibit slower and more uniform rolling velocities [79]. Thus, extension of nanotubes connecting the leukocyte with the endothelium contributes to the mechanism that dampens the response to variations in blood flow.

Recent data also suggest that membrane tethers, expressing high levels of the integrins ICAM-1, and VCAM-1, extending from the endothelial cells are important for diapedesis of leukocytes [77]. In addition, micropipette experiments have characterized membrane tethers extracted from endothelial cells with the conclusion that if the endothelium extends tethers in parallel with those extended from the leukocyte it could contribute significantly to the stabilizing effect observed for the rolling process [81, 82]. Thus, it is becoming increasingly recognized that intercellular membrane nanotubes are important structures involved in several steps of leukocyte recruitment to sites of inflammation.

4.4.3 Intracellular Membrane Nanotubes

Membranous tubular structures are important for intracellular transport of proteins between the Golgi network and endosomes [83], the endoplasmatic reticulum (ER) [84, 85], as well as the plasma membrane (PM) [86]. Shuttling of membrane proteins from the Golgi network to the PM is mediated by transport of elongated membranous containers called Golgi-to-plasma-membrane-carriers (GPCs) along microtubules. Recent results have shown that formation of GPCs is preceded by extension of membrane nanotubes from the trans-Golgi network, which are subsequently pinched off to generate one or several GPCs [86]. Similarly, extensive tubular networks connecting the ER with the Golgi have been observed to form in a microtubule-dependent manner [84, 85]. These networks could serve as transport channels of both soluble and membrane-bound material, predominantly in the direction from the ER to the Golgi. Thus, it is emerging that membrane nanotubular structures are important for transport between different intracellular membrane compartments.

It has been hypothesized that the initial extrusion process of tubular structures could be caused by microtubule-based motor proteins mechanically pulling a tether from the donor membrane, a process that has been shown to be possible in experiments with giant liposomes [13]. However, to accomplish selectivity in what is transported by these structures the mechanism of formation is likely to be rather complex and regulated by several parameters [84, 86].

4.4.4 Membrane Nanotubes Mechanically Drawn from the Cell Membrane

In order to gain further insight into the mechanical properties of membrane nanotubes, experiments have been performed where tethers are mechanically pulled from the cell surface. Such experiments have been done using a range of different cell types including endothelial cells [81], neurons [17], fibroblasts [87], and red blood cells [88]. These experiments are important since they allow for measuring various biophysical properties of the plasma membrane, e.g. adhesion forces at the interface between the cytoskeleton and the plasma membrane [17, 87], and how tether formation depend on the nature, and specificity of the mediating interaction [10, 81]. Further knowledge about these parameters will increase the understanding of processes such as leukocyte rolling and TNTs mediated intercellular communication.

Not only membrane tethers can be pulled out of biological cells, but also nanotube-vesicle networks can be formed by using plasma membrane of cultured cells. [89]. In these experiments, a combination of dithiothreitol and formaldehyde was applied to produce micron-sized plasma membrane vesicles, i.e. blebs. From these blebs, nanotubes can be pulled and NVNs can be created, Fig. 4.14. These experiments allow for preserving the transmembrane proteins in their natural environment. Under such conditions, membrane proteins maintain their correct orientations in the membrane matrix, which is important when studying, e.g., transport phenomena across cell membranes.

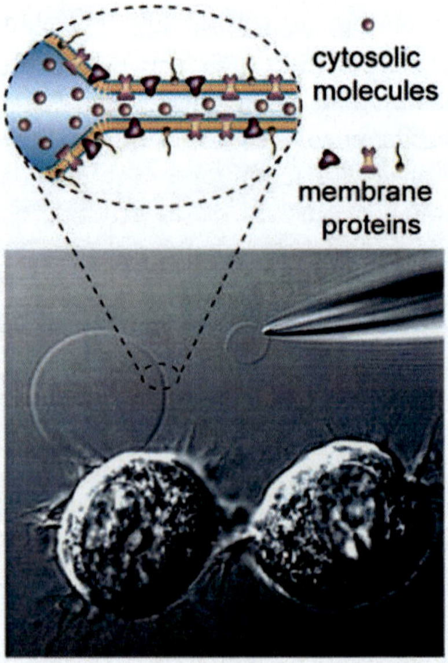

Fig. 4.14 Differential interference contrast microscopy of adherent cells (NG108-15 cells), displaying membrane blebs, from which a nanotube-vesicle connection is formed. The *inset* is a schematic drawing of a bleb-nanotube junction. The membrane proteins are imbedded in the membrane. Reprinted with permission from Bauer et al. [89]. Copyright 2007 by the American Chemical Society

4.5 Self-Assembled Lipid Nanotubes

The nanotubes discussed so far are either a result of careful manipulation of fluid lipid bilayers in the laboratory or a result of biological activity, i.e., formation of TNTs. In both instances, however, nanotubes are formed from essentially planar bilayers which prefer to form spherical, not tubular, aggregates. The tubular morphology must be stabilized, for example, by immobilizing the vesicles on surfaces as discussed above or by chemical crosslinking. Interestingly, there are a few lipid classes that can form tubular membrane aggregates *spontaneously*, without external interaction. Alternatively, and often as a transitional morphology between sphere and tube, helical or twisted membrane ribbons appear. In this section we first summarize the empirical knowledge of how and in which systems this lipid nanotube self-assembly process occurs, then we will give a condensed picture of our (still incomplete) theoretical understanding of why it takes place and, finally, we will give some examples of how the spontaneously formed nanotubes can be put to use technologically, and how they might play a role in biology. Our aim is to give an easy-to-read introduction to the phenomenon and its implications. More detailed accounts with broader scopes can be found in a few comprehensive reviews [90–92].

4.5.1 The Empirical Knowledge on Lipid Nanotube Self-Assembly

In the mid 1980s, an American [93] and a Japanese [94, 95] group independently reported the unexpected observation of lipid bilayer aggregates changing morphology from spherical vesicle to tubules or helicoidally curling or twisting ribbons (Fig. 4.15) when cooling the systems past the temperature where the bilayers change liquid crystal phase from the high-symmetry L_α phase to the low-symmetry $L_{\beta'}$ phase (see below). The phenomenon was found to be fully reversible on heating back into L_α (Fig. 4.15b). The membranes of most phospholipids do not exhibit such a morphology change as a result of the $L_\alpha - L_{\beta'}$ transition. Among those that do (some of which are depicted in Fig. 4.15) we can identify two structural characteristics which seem to be of particular importance for the phenomenon.

First, as in the case of the diacetylene-containing phospholipid $DC_{8,9}PC$ (molecule **1** in Fig. 4.16, the single most studied compound in this context), a kink in the hydrophobic chain is often present. Sometimes the molecule may be bent rather than kinked, as in molecule **3** in Fig. 4.16. Such a kink or bend has important consequences for the way molecules pack, certain angles between adjacent molecules often being highly preferable [90, 96, 97]. This tendency is particularly enhanced in the $L_{\beta'}$ phase where the molecules are tilted with respect to the bilayer normal. The tilt together with the common kink/bend direction gives the membrane a two-fold tangential director field in the plane of the membrane and this turns out to have an important impact on which aggregate morphologies that can be expected, as discussed below. Moreover, the three axes defined by layer normal, tilt direction and kink/bend direction give this type of phase a full coordinate system, thus a handedness (right- or left-handed). In other words, the appearance of tilt and a preferred

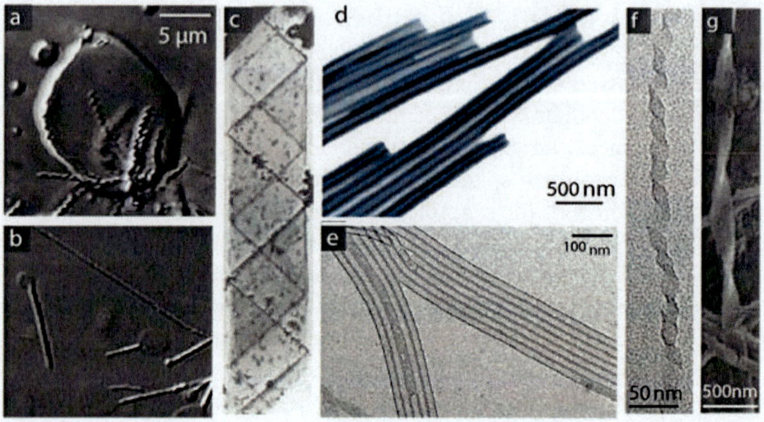

Fig. 4.15 Examples of structures encountered in lipid nanotube self-assembly. **a**, **b** (Reprinted with permission from Thomas et al. [120]. Copyright 2007 by the American Physical Society.): On cooling an aqueous $DC_{8,9}PC$ (see Fig. 4.16, subfigure **1**) dispersion from the L_α to the $L_{\beta'}$ phase, transitional helical ribbons sprout from the initially present vesicles (**a**). Tens of minutes later the helical ribbons have disappeared and been replaced by tubules, which upon heating back into the L_α phase change into spherical vesicles (**b**). **c** (Reprinted with permission from Selinger and Schnur [133]. Copyright 2007 by the American Physical Society.): The surface of lipid nanotubules is often decorated with a helical ribbon pattern, which here is enhanced by the adsorption of colloidal particles at ribbon boundaries. (**d**) (Reprinted with permission from Kamiya et al. [107]. Copyright 2007 by the American Chemical Society.): Thin nanotubes without helical decoration formed from an aqueous dispersion of molecules shown in Fig. 4.16, subfigure **3**. (**e**), (**f**) (Reprinted with permission from Jin et al. [102]. Copyright 2007 by the National Academy of Science, USA): Thin nanotubes (**e**) and helical ribbon (**f**) formed by the hexabenzocoronene-based amphiphile shown in Fig. 4.16, subfigure **5**. (**g**) (Reprinted with permission from John et al. [108]. Copyright 2007 by the Wiley-VCH Verlag GmbH & Co. KGaA.): Twisted ribbon formed by the saturated single-chain amphiphile, shown in Fig. 4.16, subfigure **4**

direction of the kink/bend constitutes a spontaneous breaking of mirror symmetry, rendering the membrane chiral, regardless of if the constituent molecules are chiral or not [98, 99].

The second important recurring characteristic is exceptionally strong intermolecular interactions within the membrane plane, due to hydrogen bonding as in **2** [94, 100, 101] or aromatic π-π-stacking as in **4** and **5** [102–105] in Fig. 4.16. This gives the membrane a particularly high translational order along the bilayer normal, i.e. the out-of-layer fluctuations of molecules are suppressed. Actually, a kink or bend in the molecule structure has exactly the same effect, as evidenced in the field of bent-shaped thermotropic liquid crystals [106], hence we can expect also the membranes of molecules **1** and **3** in Fig. 4.16 to be highly ordered in this sense. Summarizing, it seems an amphiphile optimized for nanotube self-assembly should at least be designed for strongly correlated membrane bilayers, suppressing out-of-layer molecular fluctuations either by strong lateral intermolecular interactions or via a bent or kinked chain shape.

1

2

3

4

5

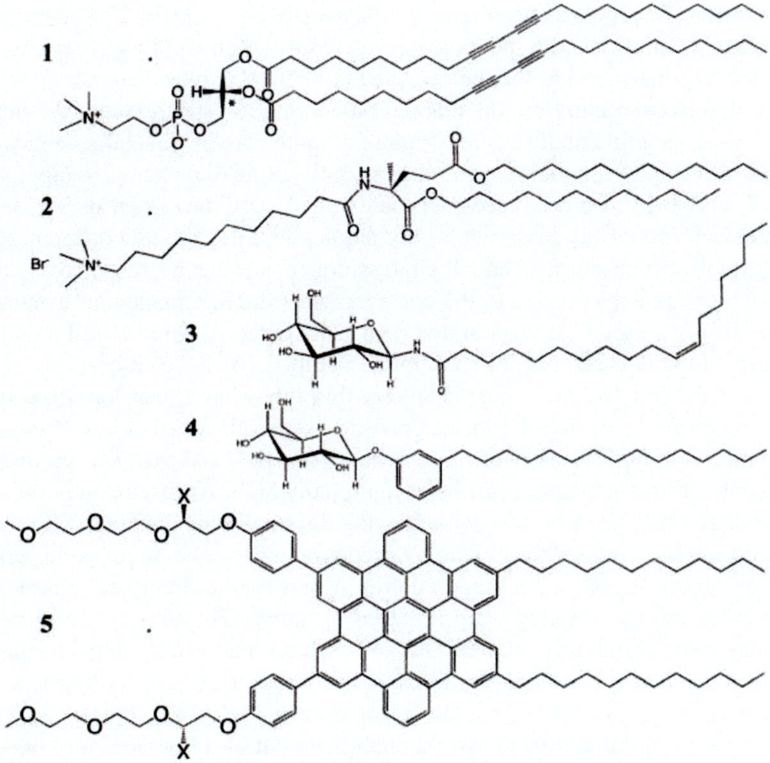

Fig. 4.16 Two of the oldest nanotube-forming lipids (**1** $DC_{8,9}PC$, 1,2-bis(tricosa-10,12-diynoyl)-*sn*-glycero-3-phosphocholine [93], and **2** $2C_{12}$-L-Glu-$C_{11}N^+$ [94]) as well as three more recent examples (**3** *N*-(11-*cis*-octadecenoyl)-β-D-glucopyranosylamine [107], **4** 1-*O*-3′-*n*-(pentadecyl)phenyl-glucopyranoside [108], **5**; *X* is either H for the achiral version or CH_3 for the chiral one [102])

The tubules are generally quite monodisperse in terms of diameter – ranging from less than 50 nm [92, 94, 95, 100, 102–105, 107, 109–111] to a few micrometers [93, 101, 112–115] depending on the system – but the length can vary relatively much in one and the same sample, from a few to hundreds of micrometers. With time, a few methods have been developed to influence the length, and its distribution width [96, 116]. Reported wall thicknesses range from a single bilayer to tens of layers, depending on the initial lipid concentration [96], and the tubes are generally open-ended [93, 107, 117]. Sometimes liposomes are, however, found encapsulated inside the tubes [118]. The yield in the morphology conversion was early on found to depend on the size of the initial aggregates: vesicles smaller than a certain limiting size remained even when the system was in the $L_{\beta'}$ phase, whereas larger vesicles were all consumed by tubules [118].

One of the most important geometrical parameters to control is the tube diameter, the initially found tubules with diameters of 0.5–1 μm being too large for many purposes. Success has been reached only relatively recently. The many signs of relevance of chirality in the spontaneous tubule formation rapidly gave rise to theoretical models (discussed further below) predicting that the tube diameter is a direct function of molecular chirality, the diameter diverging in case of racemic samples. This was soon experimentally demonstrated not to be correct, several racemic (and even achiral) samples producing nanotubes with the same diameter as related chiral lipids [97, 117, 119]. A more successful route turned out to be to vary the structures of the molecules involved, in particular mixing amphiphiles of quite different character, e.g. with stiff unsaturated and flexible saturated acyl chains, respectively [103, 108–110]. This strategy has led to the considerable reduction in nanotube diameter seen the last few years. The unsaturated bent glucopyranosylamide lipid **3** and the hexabenzocoronene-based gemini amphiphile **5** (Fig. 4.16) and its modifications are examples of the relatively rare case that very thin tubes have been formed using a single amphiphile type, Fig. 4.15d, e. Looking at the bilayer thickness compared to the molecule length in each of these cases [102, 104, 105, 107], it seems that the molecule tilt in the bilayer must be exceptionally high. As discussed in the next subsection, this may be important for achieving the small tube diameter.

Initially, all experiments were done by preparing a vesicular aqueous dispersion of the lipid in the L_α phase and then cooling it past the liquid crystal phase transition, whereupon the vesicles were replaced by tubes. The time required varied greatly, the most commonly studied $DC_{8,9}PC$ system being fast (helix formation time scale of seconds and tube equilibrium after tens of minutes [120]) while in other cases the process could take hours, days or even weeks to complete [94, 121]. As it was discovered that tubules would precipitate out of a low-concentration (on the order of mg/mL) solution of $DC_{8,9}PC$ in alcohol if water was added, an alternative production route was established, allowing tubule production on a time scale of seconds or minutes [118]. This procedure, which avoids the initial ordinary vesicle morphology, was highly optimized over time, to the extent that a protocol for achieving quite long high-quality tubes with exactly two bilayers of $DC_{8,9}PC$ in the walls could be identified [96]. While the precipitation method is attractive due to its speed it seems to generally produce thicker nanotubes. Most reports of high-quality nanotubes of small diameter instead follow the initial method [103, 107, 108, 110], the hexabenzocoronene-based **5** (Fig. 4.16) again constituting an exception. Here, precipitation from solutions in tetrahydrofuran (THF) or similar solvents turned out to be the method of choice, yielding nanotubes with ~15–20 nm diameter [102, 105]. In these cases, the presence of water in the initial solution should actually be avoided, as this produces helical ribbons along with the nanotubes [104].

The recurring observation of chiral superstructures among the self-assembled products gives a strong hint that chirality plays an important role in spontaneous nanotube formation. Helical/twisted ribbon structures (Fig. 4.15f, g) are found mainly in freshly formed samples, whereas seamless tubes tend to dominate after long time, but also these generally have visible helical markings on them (Fig. 4.15c), the exception being tubes with particularly small diameter (<100 nm) [102, 103, 107, 109, 110], cf. Fig. 4.15d, e. All initial experiments demonstrated a perfect correlation between handedness of the products and the enantiomeric form of the molecule used [94, 101, 118, 122]. Racemic samples produced either no helices [94, 101] or an equal amount of left- and right-handed helices [96, 117, 119]. Even more convincing evidence for the relation between molecular and superstructure chirality has been given by circular dichroism (CD) data. In repeated experiments investigating helical ribbons and tubules, produced with various amphiphiles and including thin tubes without visible helical markings, a pronounced CD signal was found with clear correlation between the sign of the signal and the enantiomer of the molecule used [95, 96, 102, 123, 124]. In the vesicle-forming L_α phase as well as in solution, no comparable CD signal was produced [95, 102, 107, 108, 123], giving strong evidence that the molecules pack in a chiral manner in the tubules or the related helicoidal/twisted structures, but not in ordinary vesicles.

It therefore came as a great surprise when a careful real-time study of the vesicle-helix-tube transition revealed that *both* left- and right-handed transitional helices were formed, at equal frequency, from an enantiomerically pure sample [114, 115, 120]. Although the handedness of helical patterns on the equilibrium tubule structures did show a 100% handedness correlation with the molecule chirality, this observation constituted quite a challenge to the understanding of the time. Some years later the observation of chiral tubular structures (right- as well as left-handed) formed by a *non-chiral* analog to $DC_{8,9}PC$ cast further doubts on the relevance of molecular chirality [97]. These findings do not rule out the importance of chirality, but they indicate a much more subtle role than initially anticipated, as discussed in the following.

4.5.2 Theoretical Models for Explaining Lipid Nanotube Self-Assembly

There is a relatively rich body of theoretical models developed for describing the spontaneous formation of lipid nanotubes. Reflecting the empirical knowledge about the phenomenon, chirality in most cases plays a central role, as does the nature of the L_α–$L_{\beta'}$ liquid crystal phase transition. To understand the reasoning, we thus first need to briefly describe the two liquid crystal phases that are relevant in the context.

The L_α and $L_{\beta'}$ phases are the main two *lamellar lyotropic liquid crystal* phases of the bilayers that amphiphilic lipids form in water, whether in aggregates such as vesicles or in bulk samples. While both phases exhibit 1D translational order along the bilayer normal **k** (Fig. 4.17), they differ strongly in the degree and type of in-plane order. In the L_α phase the hydrophobic acyl chains are in a melted state, with an average orientation (indicated by the *director*, **n**) along **k**, and there is no long-range in-plane translational order. The $L_{\beta'}$ phase occurs below L_α in temperature[1] and is consequently characterized by a higher degree of order. The acyl chains

[1] In strongly hydrated lipid systems, like the systems used for self-assembled nanotubes, the transition often occurs in the vicinity of room temperature.

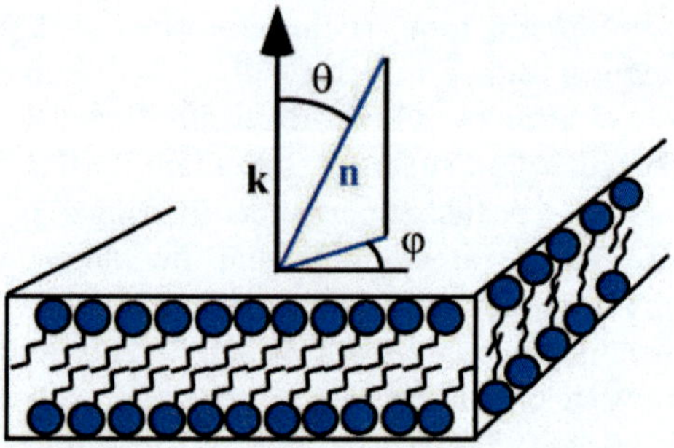

Fig. 4.17 Schematic figure of a tilted amphiphile bilayer (*circles* = hydrophilic head, *zig zag chain* = hydrophobic tail), with the tilt angle θ and the tilting direction φ defined. The average chain direction is indicated with **n** and the bilayer normal with **k**

are normally frozen in an all-*trans* configuration and tilted with respect to **k**, and some degree of in-plane translational order exists as well. The tilt is well-defined only if both its magnitude θ, and its direction φ, in the plane of the bilayer is defined, cf. Fig. 4.17. The L_α–$L_{\beta'}$ phase transition that induces the morphology change from spherical to tubular aggregates is thus connected to the appearance of long-range tangential orientational order in φ. This turns out to be fundamental for understanding spontaneous formation of nanotubes.

Among the first to propose a theoretical framework for describing nanotube self-assembly was de Gennes [125]. He pointed out that the director tilt creates a problem for membrane aggregates with the topology of a sphere, since the tilt direction constitutes a tangential director field. Such a field cannot be spatially uniform on a spherical surface[2] since the spherical topology exhibits a total defect strength (or vorticity) of 2. Defects costing energy, a membrane with tilted molecules would thus prefer a tubular morphology, since a uniform, defect-free director field is allowed on the surface of a cylinder. He also presented a possible driving force for the nanotube formation of electrostatic origin, by considering the symmetrical equivalence of the chiral $L_{\beta'}$ phase and the corresponding thermotropic liquid crystal phase smectic-C*, suggesting the presence of a spontaneous electric polarization [126]. While physically perfectly correct, the argument was soon shown not to be relevant in this case in a simple experiment where nanotube formation was found to persist even after electrolytes had been added to the aqueous lipid solutions, effectively screening out electrostatic interactions [127].

In contrast to the electrostatic argument, de Gennes' point concerning the tangential tilt director field could not be rejected, and this was instead the basis of several later theoretical approaches [128, 129]. Also in the models of Helfrich [130, 131], tilt and chirality were pointed out as important parameters but the relevant energy considered was here elastic rather than electrostatic. This approach now became the paradigm, and basically all further models (the development being driven forward mainly by Lubensky and Prost [129, 131], Zhong-Can and Ji-Xing [128], Nelson

[2]The same phenomenon renders the directions north, south, east and west undefined if you are standing on the North or the South pole of the earth.

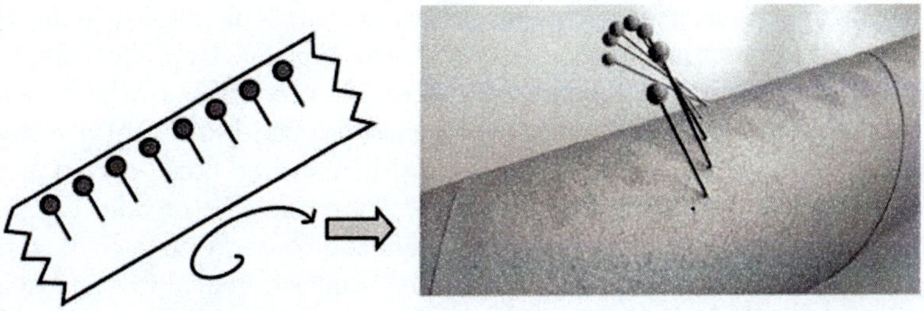

Fig. 4.18 When rolling up a ribbon of "molecules" tilting uniformly (same tilt angle and same tilt direction) into a tube, a twist between adjacent molecules appears, as seen in the photo of a very simple model, yielding a chiral structure

and Powers [132] and Selinger and Schnur [90, 133, 134]) used as their starting point the continuum elasticity theory available for describing a layered liquid crystal phase with non-zero molecule tilt (smectic-C-type phase).

Chirality is present in all recent models (as a chiral term in the elastic free energy), reflecting the fact that chiral molecules in general prefer to pack not parallel, but at a certain angle, to their neighbors. In other words, chirality favors a twist in the molecular orientation. In a tilted bilayer membrane, such a twist can easiest be accomplished if the whole membrane curves, as illustrated with a very simple model in Fig. 4.18. This is the basic concept for explaining the tubular morphology as a result of chiral packing.

The chiral packing argument has been particularly strongly promoted by Selinger, Schnur, and Spector, who addressed a number of subtle aspects of lipid nanotube self-assembly. For instance, by pointing out that the chiral packing argument also favors an in-plane tilt direction modulation, the team provided an explanation for the frequent observations of coiling ribbons [123, 133]. The modulation in the plane of the bilayer can only extend over a certain distance. A line defect then appears after which the pattern repeats itself. Thus, the appearance of chiral tilt order below the $L_\alpha - L_{\beta'}$ phase transition would give the membrane of a vesicle a striped pattern of repeated tilt direction modulation. When the membrane breaks up in order to accommodate the morphology change to the defect-minimizing tubular shape, it will do so along the mechanically weak line defects (figure 12 in [123]). This would on the one hand explain the transitional helical ribbon coils, on the other it shows that the helical ribbon can actually be an equilibrium structure, hence the permanent helical ribbon markings on the surfaces of the tubular end products. Finally, this scenario for the morphology change explained why small vesicles did not produce tubules [118]: the vesicle diameter must be at least large enough to accommodate the ribbon formation [123].

The Selinger and Schnur team pointed out that the experiments with right- as well as left-handed helices forming from enantiomerically pure lipids [114, 115, 120] and of non-chiral lipids self-assembling into nanotubes [97], did not rule out the importance of chirality [112], since the issue need not be *molecular* chirality. It could instead be the chirality of the supramolecular structure of lipid molecules that are tilted in the membrane plane, with one further tangential direction, e.g. molecule kink or bend, also exhibiting long-range order.[3] This type of supramolecular chirality from achiral molecules with bent core has been a very active research topic in the field of thermotropic liquid crystals over the last 15 years [98, 99, 135], entailing the development of completely new physics for tilted layered liquid crystal phases. It is very likely that the results are applicable also to the case of lipid nanotubes [97]. Indeed, the most famous bent-molecule smectic phase exhibits characteristic twisted and helical features [99, 135, 136], very similar to the helical and twisted ribbon shapes found in the phospholipid systems.

Spector et al., pointed out that the kink in the $DC_{8,9}PC$ molecule imposes a steric hindrance for parallel molecule packing [96], irrespective of the molecular chirality (the stereogenic center of this molecule, and actually all chiral lipids found to self-assemble into nanotubes, is located at the head group). Instead, there are two symmetric optimum arrangements where each molecule is rotated by a certain angle with respect to its nearest neighbor to the left or to the right. Supramolecular chirality in terms of twist would thus be promoted just by the shape of the molecule. Without a molecular stereogenic center there would be a degeneracy in twist handedness, but this would be lifted in case of an enantiomerically enriched sample of a chiral lipid. Although non-zero, the energy difference between the two oppositely twisted molecule organizations in such a case of chiral molecules is however expected to be much less than that between either twisted supramolecular arrangement and a non-twisted one. The role of molecular chirality would thus be more of a biasing than of a driving nature.

The exact nature of this biasing becomes particularly interesting when analyzing the circular dichroism results discussed above. In a careful investigation of mixtures of opposite enantiomers of $DC_{8,9}PC$ at different ratios, Spector et al., showed that the CD signal is a linear function of enantiomeric excess [96]. They concluded from this that a spontaneous separation of enantiomers must take place during the nanotube formation process, yielding left- and right-handed tubes at the same ratio as the (*R*) to (*S*) enantiomer ratio of the mixture, resulting in the observed linear relationship between CD signal and molecular enantiomeric enrichment. If no spontaneous resolution had occurred and every tube contained both enantiomers, one would rather expect that the majority enantiomer determines the handedness of the whole tube, which would produce only left-handed (or only right-handed) tubes as soon as either enantiomer is present at larger concentration than the other, and this would obviously give a completely different relationship between CD signal and enantiomer excess of the lipid sample. Interestingly, exactly this type of "seargants and soldiers" type of biasing was later found by Aida and co-workers when studying nanotube formation in enantiomeric mixtures of the chiral hexabenzocoronene amphiphile **5** (Fig. 4.16) [102]. CD measurements in that case showed that the chiral

[3]This aspect was somewhat later developed by Seifert et al. in an original direction, removing the requirement for a molecule kink or bend by considering the possibility that the molecule tilt must not be identical in the two monolayers of each membrane bilayer [154].

superstructure was essentially independent of enantiomeric excess over a very large range (20–100%), the handedness of all tubes dictated by the majority enantiomer.

The chiral elasticity theories all predict that the tubule radius should depend inversely on some kind of chiral "strength" parameter. The question is what parameter this would be, the most obvious one – the enantiomeric excess – having been experimentally proven irrelevant for the tubule size. Rather, the parameter must quantify the degree of spontaneous supramolecular mirror symmetry breaking at the L_α – $L_{\beta'}$ tilting transition, but it is far from obvious how – if at all – such a parameter can be determined and related to the characteristics of the amphiphiles, solvent or any other component of the system. Such a relation might be necessary in order to fully control the self-assembly process.

A recent work by Douliez and co-workers [137] hints at a route forwards. Using a surprisingly simple set of surfactants, they achieved tubules with diameters ranging from about half a micron to several microns, the control parameter being temperature. The authors propose an explanation based on the mirror symmetry breaking model, suggesting that the effective quantitative chirality parameter decreases on heating as a result of increased chain disorder. Another variable that is likely to play a role is the magnitude of the molecule tilt, which typically decreases on heating, at least in the case of the analog thermotropic phase transition. This would explain why the nanotubes formed by **3** and the various derivatives of **5** (Fig. 4.16) all have very small radii: as mentioned earlier the molecules must be very strongly tilted with respect to the bilayer normal in order to fit within the slim membranes [102, 104, 105, 107].

An important unresolved issue regards the true thermodynamic state of the membrane bilayers in the tubules. The periodic tilt direction modulation proposed by Selinger and co-workers requires the membrane to be in a fluid, liquid crystalline state, like the thermotropic SmC phase. Generally, however, the $L_{\beta'}$ phase is said to be in a gel state, where the fluidity is considered to be restricted mainly to the water separating consecutive bilayers, leaving the actual bilayers in a more or less crystalline state. This would be incompatible with modulated tilt direction, since the tilting direction in a crystalline state is locked to fit the crystal lattice. Considering the evidence for tilt modulation, the nature of the $L_{\beta'}$ phase in the tubular aggregates might thus be quite different from a bulk $L_{\beta'}$ phase with multiple flat bilayers. This question has an interesting analogy in current solid state research on graphene sheets. In a recent paper, Meyer et al., showed that freely suspended graphene sheets exhibit considerable deformations in the direction perpendicular to the sheet plane [138]. This is a consequence of the Landau-Peierls instability, stating that thermal fluctuations will destroy long-range order in 2D systems, thereby melting a 2D crystal lattice at any temperature. The same line of reasoning was used by Nelson, Peliti and Seung in the end of the 1980s to show that a flexible lipid bilayer membrane cannot exhibit defect-free crystalline order [139].

Experimental tests of the actual condition are difficult but the first report of an effort to verify the tilting direction on the tube surface was recently published [140], suggesting that modulated as well as uniform tubules are present. Moreover, the fact that the thinnest and most high-quality nanotubes have been produced using mixtures of lipids, some saturated and flexible, others unsaturated and thus more stiff and kinked, is a strong argument for the liquid crystalline state of the membrane. In fact, exactly this type of multi-component mixing is a fundamental strategy in the thermotropic liquid crystal industry for producing substances that do *not* crystallize even at low temperatures.

4.5.3 Applications in Technology and Possible Biological Implications

As fascinating as the lipid nanotube self-assemply phenomenon is in itself, research on the topic is not only curiosity-driven. The potential for technological applications of the tubules have since early on been a major motivation, and there are indications that the physics behind the phenomenon plays a role in some biological processes, giving it medical significance as well. We end this overview of lipid nanotube self-assembly by briefly discussing these issues.

By the mid 1990s, application-oriented research on self-assembled lipid nanotubes had reached a certain level of maturity mainly within two fields [112]. On the one hand, the tubes could be used as templates for metallization to produce conducting tubes with the geometry of the lipid ones, with potential applications e.g. for conducting composites [141] or field-emitting cathodes. On the other hand, they are of interest as convenient carrier vehicles of active substances for release into the environment at controlled rate. Early experiments with copper-coated lipid nanotubes filled with antifouling agents showed promising results when applied to the problem of marine fouling [142]. Since then numerous applications of the nanotubules in agriculture, medicine and drug delivery, filtration and purification, as well as industrial encapsulation, have been proposed.

A recent *tour de force* in terms of tailoring the nanotubes for specific purposes is the achievement by Aida and co-workers of self-assembled photoconducting nanotubes by functionalizing the electron-rich hexabenzocoronene-containing gemini amphiphile **5** (Fig. 4.16) with an electron-accepting trinitrofluorenone unit [105]. These results open opportunities for application of such functional self-assembled nanotubes in the field of nanoscale/molecular electronics and photovoltaics [143], recent experiments achieving external quantum efficiencies in the range of 0.1% and an irradiation-enhancement of the current by more than four orders of magnitude [143]. The low measured quantum efficiency is, however, expected to be largely due to artifacts such as nanotubes not actually spanning the full distance between the electrodes due to insufficient control of tube orientation.

In the field of materials science, several routes are currently being pursued to apply self-assembled lipid nanotubes. Shimizu, and co-workers prepared self-assembled metal-complexed lipid nanotubes that could then act as templates for metal oxide tubes [111]. The same team earlier used the self-assembly process to produce hybrid nanotubes with concentric organic and inorganic composition [144] as well as metallic or metallo-organic nanowires [145]. These are just a few examples of current activities with technological relevance. The interested reader is referred to some of the more recent reviews mentioned earlier.

When helical ribbons and nanotubes of the type discussed here were identified in bile (natural as well as synthetic analogues) [113, 146], it became clear that the phenomenon of lipid nanotube self-assembly might well have biological relevance, in this particular case possibly being involved in the formation of gallstones. Very recently, it was suggested that the membrane budding preceding the formation of transport carriers in cells, e.g. vesicles and tubules, is driven by chirality and molecule tilt in the bilayer membranes [147]. It thus seems that the physical chemistry underlying the spontaneous lipid nanotube self-assembly process has considerable relevance for the routine processes carried out by our healthy cells as well as for our common diseases. Although the study of self-assembled lipid nanotubes is by now more than 20 years old, it is clear that a great deal of exciting research on the topic still lies ahead of us.

4.6 Soft-Matter Nanotubes Composed of Crosslinked Amphiphiles

Manipulations of lipid nanotubes are often limited by the fragility of the fluid membrane. Stabilization of the membrane is therefore an important consideration in order to open pathways to applications of such structures. In principle, cross-linkeable phospholipids can be utilized to create chemically interlinked membrane structures, but limited chemical reactivity and membrane fragility under mechanical, thermal and osmotic stress accompany typical cross-linking conditions in locally confined spaces [148]. Other strategies such as adding cross-linking components after assembly [149], or templating non-amphiphilic monomers [150], have similar shortcomings.

More advantageous are synthetically obtained block copolymers with amphiphile properties. Such materials can be designed in a way that the concentration of polymerizable groups is considerably higher and structures such as tubes and vesicles are, typically, more robust than self-assembled bioamphiphile structures such as phospholipid bilayers. Synthetic amphiphiles have an amphiphilicity similar to lipids, but they have much larger molecular weights, accordingly increasing the efficiency of cross-linking.

Directly following advances in the preparation of polymerizeable self-assemblies, such as worm-like micelles of diblock copolymers [151], both spontaneous formation and subsequent cross-linking of nanosized tubular structures obtained by self-assembly of amphiphilic triblock copolymers [152] and direct formation of nanotubes by forced shape transformations were reported. In the latter case the procedure is based on amphiphilic diblock copolymers, with the nanotubes stretched from polymersome membranes by means of an attached microbead and optical tweezers [153]. The study showed that due to larger intermonolayer friction, the force required to pull a nanotube from a polymersome membrane was a factor of ten higher than that required for typical phospholipid membrane. Subsequently, improved methods using either optical tweezers or a micropipette setup have become available. Through incorporation of a triblock copolymer in the formation of the polymersomes, the membranes could be rendered sufficiently fluid to form nanotubes by pulling directly on the membranes with optical tweezers or a micropipet. A free radical polymerization reaction was employed to cross-link the membrane material and form stable and robust nanotubes that can be removed from the solution. The formation of polymer nanotube-vesicle networks was demonstrated as well [148].

Figure 4.19 shows an example of a polymerized vesicle-nanotube-network fabricated based on this method. In the inset in the figure, the nanotubes appear corrugated, which was attributed to effects of the crosslinking conditions. Evidence has been obtained that the polymer nanotubes created by this process possess an aqueous core, which indicated that transport of materials through the tubes or, due to the pronounced corrugations, even sorting of appropriately sized molecules such as DNA may be possible [148].

However, the aggressive conditions of radical polymerization, found to be responsible for the shape distortion of the resulting nanotubes, have been overcome and a comparatively simple technique to photopolymerize block copolymers, which spontaneously self-assemble into membrane structures, has been recently

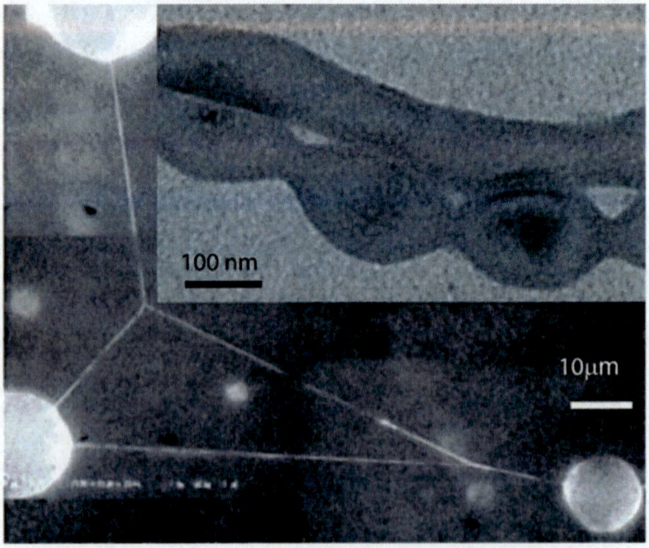

Fig. 4.19 Composite image obtained from video fluorescence microscopy of a network of polymer nanotubes and polymersomes, labeled with the membrane dye DiO-C16. *Inset*: high-resolution TEM image of washed, dried, and uranyl acetate-stained polymer nanotubes after cross-linking. Reprinted with permission from Reiner et al. [148]. Copyright 2007 by the National Academy of Science, USA

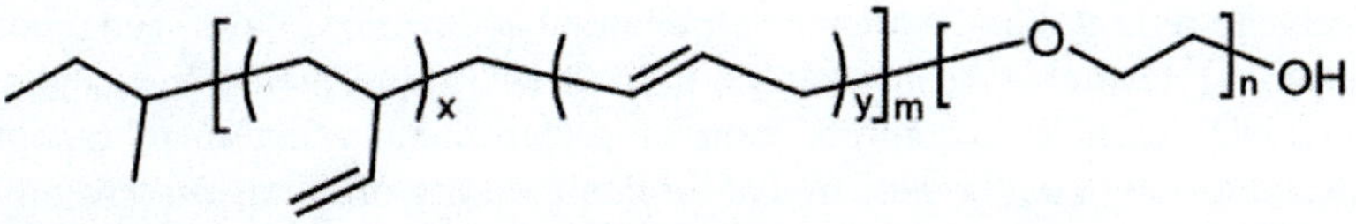

Fig. 4.20 Polybutadiene(1,800 Da)-block-poly(ethylene oxide) (900 Da) with 94% 1,2- (i.e., vinyl) groups (PB-b-PEO or B_{35}-b-EO_{22}) used as polymerizable nanotube precursor in Ref. [150]

demonstrated [150]. Figure 4.20 shows exemplarily the structure of the commercially available diblock copolymer utilized in the study.

Photocrosslinking with ultraviolet light has the advantage that nanotubes and vesicles are minimally disturbed and the rate of polymerization is easily controlled by adjusting the light intensity. This methodology makes access to polymer nanotubes easier, since precursor amphiphiles do not need to be obtained by typically elaborate syntheses. Additionally, photopolymerization facilitates microscopic "spot-curing" to stabilize selected regions of the nanotube, enabling creation of nanotubes with a curved or bent morphology.

This improved method yields nanotubular structures that are stable enough to be subjected to material transport protocols. Electrophoresis of DNA through a polymer nanotube integrated with a spherical vesicle reservoir was demonstrated (Fig. 4.21), showing that practically relevant applications are coming within reach.

Nanotube assemblies composed of chemically crosslinked amphiphiles have the potential to become important materials for nanofluidic applications. Improved fabrication strategies ensure minimal disturbances of the membranes and internalized

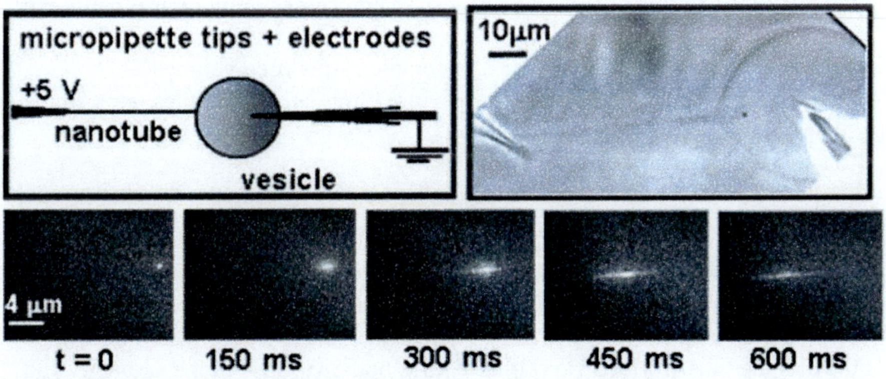

Fig. 4.21 DNA transport in a polymerized block copolymer nanotube-vesicle assembly, suspended between two glass micropipettes with internal electrodes [150]. Once the nanotube and vesicle were photocross-linked, an electric field was applied between the tips to electrophoretically move fluorescently labeled DNA through the nanotube. The images display a schematic view of the setup, a brightfield micrograph of the nanotube assembly and a time series of the transport process (position vs. time). Reprinted with permission from Jofre et al. [150]. Copyright 2007 by the American Chemical Society

additional materials and first examples of potential applications have been reported. Although many favorable biocompatible properties of lipid membranes and associated benefits, such as accommodation of proteins and other membranophilic materials, controlled membrane permeability and fluidity of the membrane are lost in these polymerized structures, the greatly enhanced stability and portability of the structures open a wide window of opportunity for technological advances.

4.7 Conclusions

This chapter provides an overview of the recent activities within the extended area of lipid nanotubes research, covering a broad range of experimental and theoretical studies as well as a directions leading to future biological and technological applications. We have initially reviewed fabrication strategies based on surfactant or amphiphile self-assembly, forced shape transformations and, in case of appropriately functionalized building blocks, polymerization. A major part of this review is dedicated to the dynamic properties of lipid nanotubes, their role in vesicle-nanotube networks and varieties of branched, knotted. and circular nanotubes derived thereof. Transport modes and enzymatic reactions in nanotubes-interconnected vesicles were presented as unique research tools to mimic sub-cellular conditions, and explore concepts of unconventional, miniaturized chemical reactors in a biocompatible environment.

Lipid nanotubes can be expected to further gain in importance in five major scientific research and technology areas:

1. Model systems for nanotube-based (transport) processes in biological cells, and biophysics of membranes,
2. Studies of chemical reaction kinetics in complex structured geometries,
3. Studies of transport phenomena involving ultra small volumes and single molecules,
4. Bionanotechnological and sensor devices,
5. Templates for hard-material nanotubes and wires.

Obviously, strong competition exists from a variety of other nanoscale materials and objects, such as carbon nanotubes. The future success of lipid nanotubes in various technological applications will depend greatly on the extent to which we will be able to tailor the tubes formed via self-assembly, both in terms of tube geometry and functionalization. Improved chemical stability has been achieved by polymerization of the tube walls, and new amphiphilic polymer building blocks will lead to further progress.

In conclusion, there are rich opportunities to engineer and utilize surfactant nanotube assemblies in order to obtain specific functionalities, even in areas that are not yet connected to ultra-miniaturized biomimetic structures.

Acknowledgements BÖ was supported by a fellowship from the Wenner-Gren Foundations and JL by a fellowship from the Knut & Alice Wallenberg Foundation. The data for Fig. 4.12 was collected by BÖ in the lab of Daniel Davis, Imperial College London. OO acknowledges the Royal Swedish Academy of Sciences, the Swedish Research Council (VR), and the Swedish Foundation for Strategic Research (SSF, Nano-X, INGVAR), the Wallenberg Foundation, and the Göran Gustafsson Foundation.

References

1. J. N. Israelachvili, *Intermolecular and Surface Forces, Second Edition: With Applications to Colloidal and Biological Systems.* (Academic Press Inc., London, 1985).
2. R. Lipowsky, *Nature* **349** (6309), 475 (1991); T. Schmidt, G. J. Schutz, W. Baumgartner et al., *Journal of Physical Chemistry* **99** (49), 17662 (1995); W. L. C. Vaz, J. Stumpel, D. Hallmann et al., *European Biophysics Journal with Biophysics Letters* **15** (2), 111 (1987).
3. P. L. (Editor) Luisi and P. (Editor) Walde, *Giant Vesicles.* (Wiley, New York, 1988).
4. E. Evans and D. Needham, *Journal of Physical Chemistry* **91** (16), 4219 (1987).
5. R. M. Hochmuth, N. Mohandas, and P. I. Blackshe, *Biophysical Journal* **13** (8), 747 (1973).
6. V. Heinrich and R. E. Waugh, *Annals of Biomedical Engineering* **24** (5), 595 (1996).
7. J. W. Dai and M. P. Sheetz, *Biophysical Journal* **68** (3), 988 (1995).
8. D. Cuvelier, I. Derenyi, P. Bassereau et al., *Biophysical Journal* **88** (4), 2714 (2005).
9. O. Rossier, D. Cuvelier, N. Borghi et al., *Langmuir* **19** (3), 575 (2003).
10. B. G. Hosu, M. Sun, F. Marga et al., *Physical Biology* **4** (2), 67 (2007).
11. E. Evans and A. Yeung, *Chemistry and Physics of Lipids* **73** (1–2), 39 (1994).
12. M. Karlsson, K. Sott, A. S. Cans et al., *Langmuir* **17** (22), 6754 (2001).
13. A. Roux, G. Cappello, J. Cartaud et al., *Proceedings of the National Academy of Sciences of the United States of America* **99** (8), 5394 (2002).
14. L. Bo and R. E. Waugh, *Biophysical Journal* **55** (3), 509 (1989).
15. J. B. Song and R. E. Waugh, *Biophysical Journal* **64** (6), 1967 (1993).
16. R. E. Waugh and R. G. Bauserman, *Annals of Biomedical Engineering* **23** (3), 308 (1995); J. Y. Shao and R. M. Hochmuth, *Biophysical Journal* **71** (5), 2892 (1996); M. P. Sheetz, *Nature Reviews Molecular Cell Biology* **2** (5), 392 (2001).
17. F. M. Hochmuth, J. Y. Shao, J. Dai et al., *Biophysical Journal* **70** (1), 358 (1996).
18. R. E. Waugh, J. Song, S. Svetina et al., *Biophysical Journal* **61** (4), 974 (1992); E. Evans, H. Bowman, A. Leung et al., *Science* **273** (5277), 933 (1996).
19. A. Karlsson, R. Karlsson, M. Karlsson et al., *Nature* **409** (6817), 150 (2001).
20. M. Karlsson, K. Sott, M. Davidson et al., *Proceedings of the National Academy of Sciences of the United States of America* **99** (18), 11573 (2002).
21. R. E. Waugh and R. M. Hochmuth, *Biophysical Journal* **52** (3), 391 (1987); R. M. Hochmuth, H. C. Wiles, E. A. Evans et al., *Biophysical Journal* **39** (1), 83 (1982).
22. D. Cuvelier, N. Chiaruttini, P. Bassereau et al., *Europhysics Letters* **71** (6), 1015 (2005).
23. W. Helfrich, *Zeitschrift Fur Naturforschung C-a Journal of Biosciences* **C 28** (11-1), 693 (1973); H. J. Deuling and W. Helfrich, *Journal De Physique* **37** (11), 1335 (1976).
24. D. J. Bukman, J. H. Yao, and M. Wortis, *Physical Review E* **54** (5), 5463 (1996).
25. I. Derenyi, F. Julicher, and J. Prost, *Physical Review Letters* **88** (23), 238101/1 (2002).
26. T. R. Powers, G. Huber, and R. E. Goldstein, *Physical Review E* **65** (4) (2002); V. Heinrich, B. Bozic, S. Svetina et al., *Biophysical Journal* **76** (4), 2056 (1999); I. Derenyi, G. Koster, M. M. van Duijn et al., *Membrane Nanotubes* (Springer, Berlin, Heidelberg, 2007).
27. G. Koster, A. Cacciuto, I. Derenyi et al., *Physical Review Letters* **94** (6) (2005).
28. F. Brochard and J. F. Lennon, *Journal De Physique* **36** (11), 1035 (1975).
29. E. Evans and W. Rawicz, *Physical Review Letters* **64** (17), 2094 (1990).
30. A. Zilker, M. Ziegler, and E. Sackmann, *Physical Review A* **46** (12), 7998 (1992).

31. W. Rawicz, K. C. Olbrich, T. McIntosh et al., *Biophysical Journal* **79** (1), 328 (2000).
32. E. A. Evans, *Biophysical Journal* **43** (1), 27 (1983).
33. J. Henriksen, A. C. Rowat, and J. H. Ipsen, *European Biophysics Journal with Biophysics Letters* **33** (8), 732 (2004).
34. M. Karlsson, K. Nolkrantz, M. J. Davidson et al., *Analytical Chemistry* **72** (23), 5857 (2000).
35. J. Hurtig and O. Orwar, *Biophysical Journal* **88** (1), 6A (2005).
36. J. F. Presley, N. B. Cole, T. A. Schroer et al., *Nature* **389** (6646), 81 (1997); C. Appenzeller-Herzog and H. P. Hauri, *Journal of Cell Science* **119** (11), 2173 (2006).
37. B. Onfelt, S. Nedvetzki, R. K. Benninger et al., *Journal of Immunology* **177** (12), 8476 (2006).
38. C. Hawes and B. A. Satiat-Jeunemaitre, *BBA Molecular Cell Research* **1744** (3), 465 (2005).
39. M. Doi and S. F. Edwards, *The Theory of Polymer Dynamics*. (Oxford University Press, Oxford, 1988).
40. E. A. Evans and R. Skalak, *Mechanics and Thermodynamics of Biomembranes*. (CRC Press, Boca Raton, FL, 1980).
41. T. Lobovkina, P. Dommersnes, J. Hurtig et al., *Physical Review Letters* **97** (18), 188105/1 (2006).
42. T. Lobovkina, P. Dommersnes, J. F. Joanny et al., *Proceedings of the National Academy of Sciences of the United States of America* **101** (21), 7949 (2004).
43. M. R. Garey, R. L. Graham, and D. S. Johnson, *Siam Journal on Applied Mathematics* **32** (4), 835 (1977).
44. T. Lobovkina, P. G. Dommersnes, S. Tiourine, F. F. Joanny, and O. Orwar, *Eur. Phys. J. E* **26** (3), 295 (2008).
45. T. Nakagaki, R. Kobayashi, Y. Nishiura et al., *Proceedings of the Royal Society of London Series B-Biological Sciences* **271** (1554), 2305 (2004).
46. L. Dreier and T. A. Rapoport, *Journal of Cell Biology* **148** (5), 883 (2000); A. Upadhyaya and M. P. Sheetz, *Biophysical Journal* **86** (5), 2923 (2004).
47. T. Lobovkina, P. G. Dommersnes, J. F. Joanny, and O. Orwar, *Soft Matter* **4** (3), 467 (2008).
48. J. Happel and H. Brenner, *Low Reynolds Number Hydrodynamics*. (Martunus Nihhoff Publishers, Dordrecht, 1983).
49. P. Stange, D. Zanette, A. Mikhailov et al., *Biophysical Chemistry* **72** (1–2), 73 (1998).
50. L. Dagdug, A. M. Berezhkovskii, S. Y. Shvartsman et al., *Journal of Chemical Physics* **119** (23), 12473 (2003).
51. L. Lizana and Z. Konkoli, *Physical Review E* **72** (2) (2005).
52. K. Sott, T. Lobovkina, L. Lizana et al., *Nano Letters* **6** (2), 209 (2006).
53. R. Karlsson, A. Karlsson, A. Ewing et al., *Analytical Chemistry* **78** (17), 5960 (2006).
54. C. T. Culbertson, S. C. Jacobson, and J. M. Ramsey, *Talanta* **56** (2), 365 (2002).
55. P. G. Dommersnes, O. Orwar, F. Brochard-Wyart et al., *Europhysics Letters* **70** (2), 271 (2005).
56. R. Barziv and E. Moses, *Physical Review Letters* **73** (10), 1392 (1994).
57. M. Tokarz, B. Akerman, J. Olofsson et al., *Proceedings of the National Academy of Sciences of the United States of America* **102** (26), 9127 (2005); M. Tokarz, B. Hakonen, P. Dommersnes et al., *Langmuir* **23** (14), 7652 (2007).
58. J. Hurtig, B. Gustafsson, M. Tokarz et al., *Analytical Chemistry* **78** (15), 5281 (2006).
59. R. F. Khairutdinov and N. Serpone, *Progress in Reaction Kinetics* **21** (1), 1 (1996).
60. A. Karlsson, K. Sott, M. Markstrom et al., *Journal of Physical Chemistry B* **109** (4), 1609 (2005).
61. A. Jesorka, M. Markstrom, and O. Orwar, *Langmuir* **21** (4), 1230 (2005).
62. A. Jesorka, M. Markstrom, M. Karlsson et al., *Journal of Physical Chemistry B* **109** (31), 14759 (2005).
63. M. Davidson, M. Karlsson, J. Sinclair et al., *Journal of the American Chemical Society* **125** (2), 374 (2003).
64. L. Lizana, Z. Konkoli, and O. Orwar, *Journal of Physical Chemistry B* **111** (22), 6214 (2007).

65. F. Baluska, A. Hlavacka, D. Volkmann et al., *Trends Cell Biol* **14** (8), 404 (2004).
66. B. Önfelt, M. A. Purbhoo, S. Nedvetzki et al., *Sci STKE* **2005** (313), pe55 (2005); H. H. Gerdes, N. V. Bukoreshtliev, and J. F. Barroso, *FEBS Lett* **581** (11), 2194 (2007).
67. B. Önfelt and D. M. Davis, *Biochemical Society Transactions* **32** (Pt 5), 676 (2004).
68. A. Rustom, R. Saffrich, I. Markovic et al., *Science* **303** (5660), 1007 (2004).
69. S. C. Watkins and R. D. Salter, *Immunity* **23** (3), 309 (2005).
70. C. Vidulescu, S. Clejan, and K. C. O'Connor, *Journal of Cellular and Molecular Medicine* **8** (3), 388 (2004).
71. M. Koyanagi, R. P. Brandes, J. Haendeler et al., *Circulation Research* **96** (10), 1039 (2005).
72. M. Karlsson, M. Davidson, R. Karlsson et al., *Annual Review of Physical Chemistry* **55**, 613 (2004).
73. S. Sowinski, C. Jolly, O. Berninghausen et al., *Nature Cell Biology* **10** (2), 211 (2008).
74. N. M. Sherer, M. J. Lehmann, L. F. Jimenez-Soto et al., *Nature Cell Biology* **9** (3), 310 (2007).
75. S. Chen and T. A. Springer, *Journal of Cell Biology* **144** (1), 185 (1999).
76. A. Atherton and G. V. Born, *Journal of Physiology* **233** (1), 157 (1973); J. C. Firrell and H. H. Lipowsky, *American Journal of Physiology* **256** (6 Pt 2), H1667 (1989).
77. C. V. Carman and T. A. Springer, *Journal of Cell Biology* **167** (2), 377 (2004).
78. D. W. Schmidtke and S. L. Diamond, *Journal of Cell Biology* **149** (3), 719 (2000); E. Y. Park, M. J. Smith, E. S. Stropp et al., *Biophysical Journal* **82** (4), 1835 (2002).
79. V. Ramachandran, M. Williams, T. Yago et al., *Proceedings of the National Academy of Sciences of the United States of America* **101** (37), 13519 (2004).
80. J. Y. Shao, H. P. Ting-Beall, and R. M. Hochmuth, *Proceedings of the National Academy of Sciences of the United States of America* **95** (12), 6797 (1998).
81. G. Girdhar and J. Y. Shao, *Biophysical Journal* **87** (5), 3561 (2004).
82. G. Girdhar and J. Y. Shao, *Biophysical Journal* **93** (11), 4041 (2007).
83. J. S. Bonifacino and R. Rojas, *Nature Reviews Molecular Cell Biology* **7** (8), 568 (2006).
84. J. C. Simpson, T. Nilsson, and R. Pepperkok, *Molecular Biology of the Cell* **17** (2), 723 (2006).
85. R. Blum, D. J. Stephens, and I. Schulz, *Journal of Cell Science* **113** (Pt 18), 3151 (2000).
86. E. V. Polishchuk, A. Di Pentima, A. Luini et al., *Molecular Biology of the Cell* **14** (11), 4470 (2003).
87. D. Raucher, T. Stauffer, W. Chen et al., *Cell* **100** (2), 221 (2000).
88. R. M. Hochmuth, C. A. Evans, H. C. Wiles et al., *Science* **220** (4592), 101 (1983).
89. B. Bauer, M. Davidson, and O. Orwar, *Langmuir* **22** (22), 9329 (2006).
90. J. V. Selinger, M. S. Spector, and J. M. Schnur, *Journal of Physical Chemistry B* **105** (30), 7157 (2001).
91. T. Shimizu, M. Masuda, and H. Minamikawa, *Chemical Reviews* **105** (4), 1401 (2005); D. T. Bong, T. D. Clark, J. R. Granja et al., *Angewandte Chemie-International Edition* **40** (6), 988 (2001); J. H. Fuhrhop and T. Wang, *Chemical Reviews* **104** (6), 2901 (2004); C. R. Martin and P. Kohli, *Nature Reviews Drug Discovery* **2** (1), 29 (2003); T. Kunitake, *Angewandte Chemie-International Edition in English* **31** (6), 709 (1992).
92. J. H. Fuhrhop, D. Spiroski, and C. Boettcher, *Journal of the American Chemical Society* **115** (4), 1600 (1993).
93. P. Yager and P. E. Schoen, *Molecular Crystals and Liquid Crystals* **106** (3–4), 371 (1984).
94. N. Nakashima, S. Asakuma, J. M. Kim et al., *Chemistry Letters* (10), 1709 (1984).
95. K. Yamada, H. Ihara, T. Ide et al., *Chemistry Letters* (10), 1713 (1984).
96. M. S. Spector, J. V. Selinger, A. Singh et al., *Langmuir* **14** (13), 3493 (1998).
97. S. Pakhomov, R. P. Hammer, B. K. Mishra et al., *Proceedings of the National Academy of Sciences of the United States of America* **100** (6), 3040 (2003).
98. G. Heppke and D. Moro, *Science* **279** (5358), 1872 (1998).
99. D. R. Link, G. Natale, R. Shao et al., *Science* **278** (5345), 1924 (1997).

100. T. Imae, Y. Takahashi, and H. Muramatsu, *Journal of the American Chemical Society* **114** (9), 3414 (1992).
101. D. A. Frankel and D. F. Obrien, *Journal of the American Chemical Society* **116** (22), 10057 (1994).
102. W. Jin, T. Fukushima, M. Niki et al., *Proceedings of the National Academy of Sciences of the United States of America* **102** (31), 10801 (2005).
103. G. John, M. Masuda, Y. Okada et al., *Advanced Materials* **13** (10), 715 (2001).
104. J. P. Hill, W. Jin, A. Kosaka et al., *Science* **304** (5676), 1481 (2004).
105. Y. Yamamoto, T. Fukushima, Y. Suna et al., *Science* **314** (5806), 1761 (2006).
106. J. P. F. Lagerwall, F. Giesselmann, M. D. Wand et al., *Chemistry of Materials* **16** (19), 3606 (2004); D. M. Walba, in *Topics in Stereochemistry, Materials-Chirality*, Vol. 24, edited by M. M. Green, R. J. M. Nolte, E. W. Meijer et al. (Wiley-VCH, New York, 2003), pp. 457.
107. S. Kamiya, H. Minamikawa, J. H. Jung et al., *Langmuir* **21** (2), 743 (2005).
108. G. John, J. H. Jung, H. Minamikawa et al., *Chemistry* **8** (23), 5494 (2002).
109. S. Svenson and P. B. Messersmith, *Langmuir* **15** (13), 4464 (1999).
110. M. S. Spector, A. Singh, P. B. Messersmith et al., *Nano Letters* **1** (7), 375 (2001).
111. M. Kogiso, Y. Zhou, and T. Shimizu, *Advanced Materials* **19** (2), 242 (2007).
112. J. M. Schnur, *Science* **262** (5140), 1669 (1993).
113. D. S. Chung, G. B. Benedek, F. M. Konikoff et al., *Proceedings of the National Academy of Sciences of the United States of America* **90** (23), 11341 (1993).
114. B. N. Thomas, R. C. Corcoran, C. L. Cotant et al., *Journal of the American Chemical Society* **120** (47), 12178 (1998).
115. B. N. Thomas, C. M. Lindemann, R. C. Corcoran et al., *Journal of the American Chemical Society* **124** (7), 1227 (2002).
116. B. N. Thomas, C. R. Safinya, R. J. Plano et al., *Science* **267** (5204), 1635 (1995).
117. A. Singh, T. G. Burke, J. M. Calvert et al., *Chemistry and Physics of Lipids* **47** (2), 135 (1988).
118. J. H. Georger, A. Singh, R. R. Price et al., *Journal of the American Chemical Society* **109** (20), 6169 (1987).
119. N. Yamada, T. Sasaki, H. Murata et al., *Chemistry Letters* (2), 205 (1989).
120. B. N. Thomas, C. M. Lindemann, and N. A. Clark, *Physical Review E* **59** (3), 3040 (1999).
121. N. Nakashima, S. Asakuma, and T. Kunitake, *Journal of the American Chemical Society* **107** (2), 509 (1985).
122. P. Yager, P. E. Schoen, C. Davies et al., *Biophysical Journal* **48** (6), 899 (1985).
123. J. M. Schnur, B. R. Ratna, J. V. Selinger et al., *Science* **264** (5161), 945 (1994).
124. M. S. Spector, K. R. Easwaran, G. Jyothi et al., *Proceedings of the National Academy of Sciences of the United States of America* **93** (23), 12943 (1996).
125. P. G. de Gennes, *Comptes Rendus de l'Académie des Sciences* **304**, 259 (1987).
126. R. B. Meyer, L. Liebert, L. Strzelecki et al., *Journal De Physique Letters* **36** (3), L69 (1975).
127. J. S. Chappell and P. Yager, *Biophysical Journal* **60** (4), 952 (1991).
128. O. Zhong-Can and L. Ji-Xing, *Physical Review Letters* **65** (13), 1679 (1990); O. Zhong-Can and L. Ji-Xing, *Physical Review A* **43** (12), 6826 (1991).
129. F. C. MacKintosh and T. C. Lubensky, *Physical Review Letters* **67** (9), 1169 (1991); T. C. Lubensky and J. Prost, *Journal De Physique Ii* **2** (3), 371 (1992).
130. W. Helfrich, *Journal of Chemical Physics* **85** (2), 1085 (1986).
131. W. Helfrich and J. Prost, *Physical Review A* **38** (6), 3065 (1988).
132. P. Nelson and T. Powers, *Journal De Physique Ii* **3** (10), 1535 (1993).
133. J. V. Selinger and J. M. Schnur, *Physical Review Letters* **71** (24), 4091 (1993); J. V. Selinger, E. C. Mackintosh, and J. M. Schnur, *Physical Review E* **53** (4), 3804 (1996).
134. R. L. B. Selinger, J. V. Selinger, A. P. Malanoski et al., *Physical Review Letters* **93** (15), 158103 (2004).
135. G. Pelzl, S. Diele, and W. Weissflog, *Advanced Materials* **11** (9), 707 (1999); D. A. Coleman, J. Fernsler, N. Chattham et al., *Science* **301** (5637), 1204 (2003).

136. Y. A. Nastishin, M. F. Achard, H. T. Nguyen et al., *European Physical Journal E* **12** (4), 581 (2003); M. F. Achard, M. Kleman, Y. A. Nastishin et al., *European Physical Journal E* **16** (1), 37 (2005).
137. J. P. Douliez, B. Pontoire, and C. Gaillard, *Chemphyschem* **7** (10), 2071 (2006).
138. J. C. Meyer, A. K. Geim, M. I. Katsnelson et al., *Nature* **446** (7131), 60 (2007).
139. D. R. Nelson and L. Peliti, *Journal de Physique* **48** (7), 1085 (1987); H. S. Seung and D. R. Nelson, *Physical Review A* **38** (2), 1005 (1988).
140. Y. Zhao, N. Mahajan, R. B. Lu et al., *Proceedings of the National Academy of Sciences of the United States of America* **102** (21), 7438 (2005).
141. S. L. Browning, J. Lodge, R. R. Price et al., *Journal of Applied Physics* **84** (11), 6109 (1998).
142. R. Price and M. Patchan, *Journal of Microencapsulation* **8** (3), 301 (1991); R. R. Price and M. Patchan, *Journal of Microencapsulation* **10** (2), 215 (1993).
143. F. Wurthner, *Science* **314** (5806), 1693 (2006); W. F. Smith, *Nature Nanotechnology* **2** (2), 77 (2007); J. Ebert, *Chemistry World* **4** (2), 20 (2007).
144. Q. M. Ji, S. Kamiya, J. H. Jung et al., *Journal of Materials Chemistry* **15** (7), 743 (2005).
145. B. Yang, S. Kamiya, Y. Shimizu et al., *Chemistry of Materials* **16** (14), 2826 (2004).
146. Y. V. Zastavker, N. Asherie, A. Lomakin et al., *Proceedings of the National Academy of Sciences of the United States of America* **96** (14), 7883 (1999); B. Khaykovich, C. Hossain, J. J. McManus et al., *Proceedings of the National Academy of Sciences of the United States of America* **104** (23), 9656 (2007).
147. R. C. Sarasij, S. Mayor, and M. Rao, *Biophysical Journal* **92** (9), 3140 (2007).
148. J. E. Reiner, J. M. Wells, R. B. Kishore et al., *Proceedings of the National Academy of Sciences of the United States of America* **103** (5), 1173 (2006).
149. H. Zhu, Q. C. Liu, and Y. M. Chen, *Langmuir* **23** (2), 790 (2007).
150. A. Jofre, J. B. Hutchison, R. Kishore et al., *Journal of Physical Chemistry B* **111** (19), 5162 (2007).
151. Y. Y. Won, H. T. Davis, and F. S. Bates, *Science* **283** (5404), 960 (1999).
152. J. Grumelard, A. Taubert, and W. Meier, *Chemical Communications* (13), 1462 (2004).
153. R. Dimova, U. Seifert, B. Pouligny et al., *European Physical Journal E* **7** (3), 241 (2002).
154. U. Seifert, J. Shillcock, and P. Nelson, *Physical Review Letters* **77** (26), 5237 (1996).

Chapter 5
Mesoscopic Structure Formation in the Walls of Nanotubes Confined to Nanoporous Hard Templates

Martin Steinhart

Abstract Nanoporous hard templates containing arrays of parallel, cylindrical nanopores functionalized with nanotubes as well as arrays of released, aligned nanotubes are a versatile material platform for miniaturized device components. While nanotubes of virtually any soft material can be produced by means of shape-defining nanoporous hard templates, it remains a challenge to tailor their properties by controlling the internal mesoscopic fine structure of their walls. Their supramolecular organization largely determines their mechanical, chemical, optical, and electronic properties as well as their specific surface. Hence, the rational design of the internal morphology of nanotubes will pave the way for their use as functional device components. In this chapter, recent efforts to control mesoscopic structure formation processes in the course of the preparation of nanotubes consisting of soft matter or prepared by exploiting self-assembly of soft matter will be reviewed, including the mechanisms governing the deposition of precursors and target materials into the nanopores. Specific attention will be paid to structure formation processes such as crystallization, phase separation and mesophase formation under the influence of the two-dimensional confinement imposed by the pore geometry and the interfacial interactions with the pore walls. Nanoporous hard templates allow rationally generating mesocopic fine structures in nanotubes because equilibrium and non-equilibrium states as well as unprecedented confinement-induced morphologies with new and exciting properties can be realized.

Abbreviations and Symbols

AAO	Anodic aluminum oxide
BCP	Block copolymer
DMF	dimethylformamide
D_p	Pore diameter
DSC	Differential scanning calorimetry
HA	Hard anodization
HRTEM	High-resolution transmission electron microscopy
IR	Infrared

M. Steinhart (✉)
Institute for Chemistry, University of Osnabrück, 49069 Osnabrück, Germany
e-mail: martin.steinhart@uos.de

O. Hayden, K. Nielsch (eds.), *Molecular- and Nano-Tubes*,
DOI 10.1007/978-1-4419-9443-1_5, © Springer Science+Business Media, LLC 2011

L_0	Bulk period of a BCP
LC	Liquid crystal
MA	Mild anodization
M_W	Weight-average molecular weight
NMR	Nuclear magnetic resonance
PAN	Poly(acrylonitrile)
PC	Polycarbonate
Pd	Palladium
PDMS	Poly(dimethylsiloxane)
PE	Poly(ethylene)
PEO	Poly(ethylene oxide)
PL	Photoluminescence
PLA	Poly(lactide)
Pluronic F-127	$Ethyleneoxide_{106}$-$propyleneoxide_{70}$-$ethyleneoxide_{106}$
PMMA	Poly(methyl methacrylate)
PPO	Poly(propylene oxide)
PS	Poly(styrene)
PS-*b*-PBD	Poly(styrene-*block*-butadiene)
PS-*b*-PMMA	Poly(styrene-*block*-methyl methacrylate)
Pt	Platinum
PVDF	Poly(vinylidene difluoride)
SAED	Selected-area electron diffraction
SAXS	Small-angle X-ray scattering
SEM	Scanning electron microscopy
TEM	Transmission electron microscopy
TEOS	Tetraethoxysilane
T_c	Crystallization temperature
THF	Tetrahydrofuran
T_M	Melting temperature
T_p	Pore depth
WAXS	Wide angle X-ray scattering

5.1 Introduction

Tailoring the internal fine structure of nanotube walls by controlling structure formation processes such as crystallization, mesophase formation and phase separation is still challenging, even though a broad range of materials can easily formed into tubular nanostructures. The mesoscopic organization of the material the nanotube walls consist of largely determines both the chemical and the physical properties of the nanotubes and therefore their potential as device components. For example, size and orientation of crystals in the walls of nanotubes will largely influence their optical, electronic, mechanical and ferroelectric properties. Mesoporous nanofibers that contain arrays of aligned cylindrical pores or hollow spaces with non-conventional geometries, such as helical pores, can be produced by the self-assembly of molecular block copolymer (BCP) soft templates under varying degrees of geometric confinement. It is therefore indispensable to control the formation of the internal morphology during the preparation of the nanotubes as far as possible.

Direct self-assembly of molecular and supramolecular building blocks or the use of soft templates that direct the formation of tubular structures is a simple experimental configuration, because all components required for the synthesis are contained in one and the same solution and self-assembly resulting in the formation of nanotubes often takes place under mild conditions. The nanotubes can be produced in large quantities, and their separation as well as their purification is possible with common methods such as filtration and centrifugation. However, the target materials the nanotubes consist of (or the corresponding precursors) must exhibit the intrinsic ability to self-assemble or they must interact with a structure-directing soft template in a very specific way. The range of materials that show these properties is limited. Moreover, it is difficult to arrange the nanotubes thus obtained into ordered arrays, as required for their integration into functional device architectures.

The use of nanoporous hard templates as shape-defining molds consisting of arrays of aligned cylindrical nanopores has turned out to be promising alternative strategy. Specific target materials or precursors thereof are infiltrated into the nanopores, and arrays of aligned nanotubes with precisely adjustable aspect ratios are obtained. Several kinds of nanoporous membranes [1] have been used for this purpose, predominantly track-etch membranes and nanoporous anodic aluminum oxide (AAO). The arrangement of the tubular structures thus fabricated is determined by that of the pores in the hard template. The material the nanotubes consist of can directly be deposited onto the pore walls. It is also possible to infiltrate precursors or monomers into the nanoporous hard templates and to convert them into the target materials. A comprehensive body of literature, including many excellent review articles, deals with this topic [2]. The preparation of nanotubes inside the pores of nanoporous hard templates, which was pioneered by Martin and co-workers [3–9], automatically yields porous hybrid membranes whose channels are functionalized with the nanotubes in their interior. Examples for this are DNA-functionalized nanotube membranes with single-base mismatch selectivity [10] or membranes for enantioselective separations [11]. The release of the nanotubes is commonly achieved by a wet-chemical etching step destroying the hard template, which is a drawback for the up-scaling of template-based approaches to the fabrication of nanotubes. If they are attached to a support, they form more or less ordered arrays. Such arrays are of interest since they may exhibit specific wetting and adhesive properties [12]. Recently reported approaches to the mechanical extraction of fiber arrays from porous templates [13] need to be optimized and require that the nanotubes are tightly connected with an underlying substrate.

Despite these still-challenging issues, the fabrication of nanotubes using nanoporous hard templates is associated with several advantages beyond the possibility to align them. Readily available porous hosts such as self-ordered porous AAO have narrow pore diameter (D_p) distributions and pores with D_p-values ranging from about 20 nm up to a few 100 nm. The pore depths (T_p) can be adjusted to values between about 1 μm and several 100 μm. Therefore, it is easily possible to tailor the diameter and the aspect ratio of the nanotubes. However, the most important advantage is the possibility to control mesoscopic structure formation processes inside the pores. There are relatively few limitations regarding the materials that can be formed into nanotubes via hard templates. Mixtures, sols, semicrystalline and liquid-crystalline polymers, other thermoplastics, as well as BCPs, are eligible for this approach so that the mesoscopic structure formation processes these materials undergo, such as phase separation, crystallization and mesophase formation, can be exploited to rationally generate specific, self-assembled supramolecular architectures in the tube walls. The degree of geometric confinement can be adjusted by the D_p-value. The chemical properties of the rigid pore walls of the hard template can by modified too, for example, by silanization or atomic layer deposition [14]. The high surface-to-volume ratio of the nanoporous hard templates makes it possible to control self-assembly processes by adjusting the properties of the pore walls. This is particularly the case for self-assembly processes based on phase separation, as discussed below. Moreover, the walls of the template pores can be functionalized in such a way that they are electrically charged, a prerequisite for the fabrication of nanotubes by layer-by-layer deposition of polyelectrolytes. As common hard templates consist of inorganic materials such as alumina, they are stable at temperatures at which soft matter is typically processed. There are no limitations regarding the temperature profile applied to the samples, that is, molten polymers can be crystallized either isothermally or non-isothermally. Hence, using hard templates in the synthesis of nanotubes is associated with various handles to tailor the internal fine structure of their walls.

This chapter is organized as follows. Commonly used hard template systems will be described in Section 5.2. Section 5.3 deals with the infiltration of the target materials the nanotubes consist of or corresponding precursors into the pores. Crystallization and the formation of mesophases from liquid-crystalline molecular building will be discussed as examples of self-organization processes of single-component materials that can be exploited to generate textured nanotubes in Section 5.4. Self-organization based on phase separation, leading to mesoporous nanofibers, will be covered in Section 5.5. This includes the generation of fine structures by spinodal decomposition of mixtures as well as the formation of "complex" nanotubes with non-conventional morphologies by self-assembling BCPs inside the template pores. Section 5.6 deals with layer-by-layer deposition of polyelectrolytes and other polymeric materials into nanoporous hard templates, resulting in the formation of nanotubes having walls with a well-defined multilayer structure. This methodology also allows fabricating nanotubes composed of complex macromolecules such as dendrimeric polyelectrolytes and the controlled incorporation of nanoparticles such as luminescent quantum dots into the nanotube walls.

5.2 Nanoporous Materials as Hard Templates

It is beyond the scope of this contribution to exhaustively review the fabrication and properties of nanporous hard templates. Instead, a brief overview of commonly used molds containing arrays of aligned cylindrical channels will be given. Two types of hard templates are employed in the production of nanotubes and nanorods: track-etch polymer membranes and nanoporous AAO. Track-etch membranes [15, 16] most commonly consist of polycarbonate and polyethylene terephthalate. They are fabricated by irradiating polymeric films with a thickness ranging from a few microns to a few tens of microns with ion beams, thus producing latent tracks penetrating through the bombarded films, and the subsequent generation of pores at the positions of the latent tracks by wet-chemical etching. The pore walls are often hydrophilized by plasma treatment or by adsorbing or grafting hydrophilic polymers, such as polyvinyl pyrrolidone, onto the pore walls. Pore size, shape and density can be varied in a controllable manner by the proper selection of the conditions under which irradiation and post-treatment procedures are carried out. The D_p-value and the pore density can be adjusted independently from 10 nm to the micron range and from 1 to 10^{10} pores per centimeter, respectively. The limitations of track-etch membranes are their limited stability at elevated temperatures and their poor resistance to organic solvents, which poses problems for many of the self-assembly processes discussed below. Since the random arrangement of the ion tracks is an inherent feature of ion bombardment, the arrangement of the pores in track-etch membranes is random too. Moreover, because of their poor rigidity and their lack of chemical resistance to organic solvents, it is difficult to remove residual material from their surface after their infiltration, a process step that is crucial to the template-based fabrication of nanotubes and nanorods. The pronounced roughness of the pore walls in track-etch membranes revealed by SEM and adsorption experiments prevents uniform orientation of anisotropic species infiltrated into the pores [16]. However, because of their commercial availability and versatility, track-etch membranes are being routinely used for the production of one-dimensional nanostructures.

The second common hard template is nanoporous AAO produced by the electrochemical anodization of aluminum substrates. The pore walls in native AAO are hydroxyl-terminated high-energy surfaces [17]. AAO is stable at temperatures at which the target materials or precursors thereof are commonly processed. Moreover, AAO is resistant to organic solvents but can be selectively etched with aqueous acids and bases to release the nanotubes. In AAO prepared under so-called mild anodization (MA) conditions, the amorphous pore walls consist of an outer layer containing water, electrolyte anions and positively charged defects, and an inner layer consisting of pure alumina [18]. The concentration profile of these contaminations across the pore walls is inhomogeneous (see, for example, [19]). Disordered, about 60 μm thick AAO membranes with a mean D_p-value of the order of 200 nm featuring a broad pore size distribution as well as irregular pore shapes (Fig. 5.1a) are commercially available (Whatman Anopore) [16]. The dispersity of the pore diameter distribution, calculated by dividing the standard deviation by the mean pore

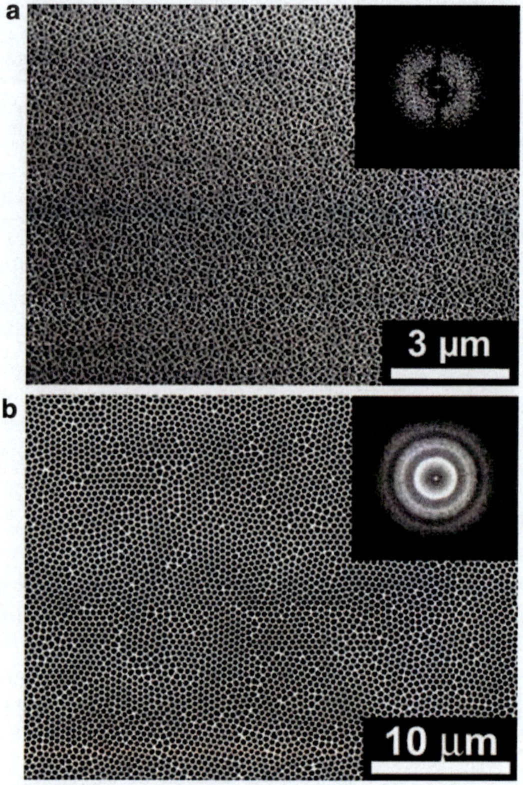

Fig. 5.1 Anodic aluminum oxide. (**a**) Example of disordered AAO with a mean D_p-value of 200 nm; (**b**) self-ordered AAO anodized with a phosphoric acid electrolyte solution under MA conditions. *Insets*: Fourier transforms

diameter, is typically larger than 20%. The membrane surface of such disordered AAOs exhibits pronounced roughness on different length scales that complicates the removal of residual material after infiltration and the purification of the nanostructures thus obtained. The two-step MA process reported by Masuda and Fukuda involving self-ordered pore growth can be considered as a major breakthrough in AAO-based nanoprocessing [20]. The pores are arranged in hexagonal lattices characterized by a polycrystalline degree of order that consist of grains extending 10–20 lattice constants. Three self-ordered MA regimes have been identified. Using sulfuric acid as an electrolyte solution at an anodization voltage of about 25 V yields self-ordered AAO with a lattice constant of 65 nm and a D_p-value of about 25 nm [21]. Anodization with oxalic acid solutions at anodization voltages of about 40 V yields AAO with a lattice constant of 100 nm and a D_p-value of 35 nm [20], and anodization in phosphoric acid solutions at 195 V yields AAO with a lattice constant of 500 nm and a pore diameter of about 180 nm [22]. The pore diameter distributions of self-ordered AAO have a dispersity of less than 8% and are therefore significantly sharper than those of disordered AAO. It was noted that the pore arrays produced in the self-ordering MA regimes exhibit a porosity of 10% [23]. Porosities up to

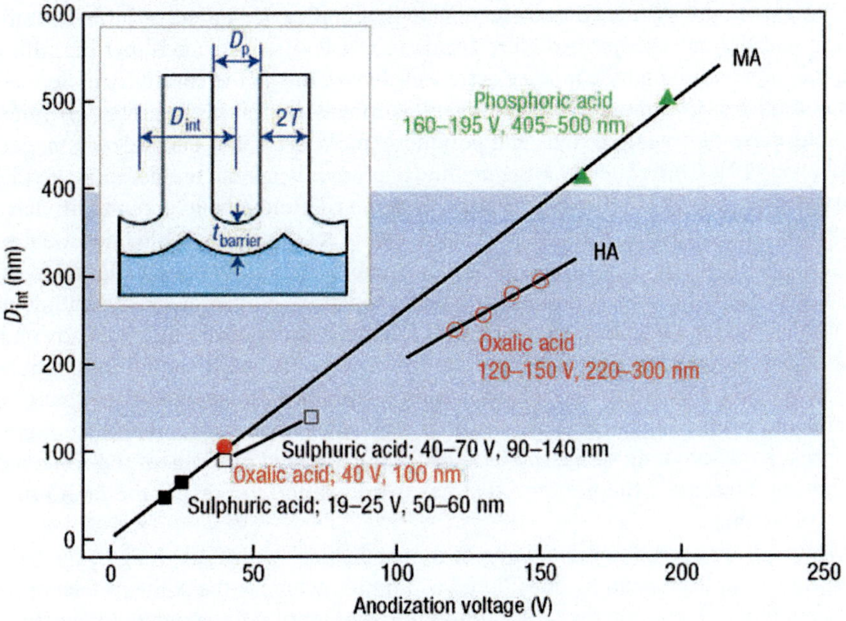

Fig. 5.2 Overview of currently identified self-ordering MA and HA regimes for the production of AAO. Reproduced from [24]. Copyright (2006): Nature Publishing Group

50% can be achieved if the pores are widened by isotropic wet-chemical etching (Fig. 5.1b). The surface of self-ordered AAO is significantly smoother than that of disordered AAO so that fabrication, purification and characterization of nanoobjects prepared inside the pores, as well as the fabrication of functional membranes, is substantially facilitated.

Hard anodization (HA), which is performed at higher anodization voltages than MA, is associated with growth rates of the alumina layers orders of magnitude faster than MA. Lee et al. reported that HA with oxalic acid solutions at anodization voltages between 120 and 150 V yields AAO with D_p-values of 49–59 nm, lattice constants of 220–300 nm and initial porosities from 3.3 to 3.4%, a range not covered by the MA self-ordering regimes [24]. Consequently, HA membranes could enable the production of mechanically stable nanotube arrays in which large distances between the nanofibers are required to prevent them from condensation. An overview of currently identified self-ordering MA and HA regimes is given in Fig. 5.2.

5.3 Infiltration of Nanoporous Hard Templates

The fabrication of nanotubes using nanoporous hard templates starts with the infiltration of the target materials, of precursors thereof, or of monomers, into the pores. The underlying mechanisms are only partially understood. It is reasonable to assume that interfacial interactions dominate, or at least significantly influence both the infiltration and the mesoscopic structure formation. It is often assumed that the filling of nanopores is driven by capillary action. Actually, this is the case only under certain conditions. Commonly, polar inorganic surfaces exhibit high surface energies, whereas those of organic liquids and polymeric melts are about one order of magnitude lower [17]. Consequently, organic fluids commonly spread on inorganic, oxidic substrates such as AAO. Even for viscous polymeric fluids, which commonly have low surface energies, the formation of large-area precursor films could be evidenced on smooth substrates having a high surface energy [25]. At the spreading front, the film is thinner than a monolayer, as determined by ellipsometry, indicating an incomplete surface coverage. Even though a finite amount of the fluid spreads on a surface large enough to be considered as "infinite", a "pancake" equilibrium structure, that is, a liquid film covering a finite area with a thickness exceeding that of a monolayer, forms. This is because the interactions between the solid and the liquid comprise attractive long-range interactions [26] expressed in terms of the so-called "disjoining pressure", the pressure that has to be exerted to prevent the liquid film from thickening.

When a fluid spreads on the walls of a nanochannel with finite length, the bulk reservoir of the liquid can be considered as infinite, whereas the solid surface to be wetted is finite. The infiltration of a liquid into an empty pore is qualitatively similar to the replacement of a liquid phase filling a cylindrical channel by another one having higher affinity to the walls of the channel. This process, the complex physics of which depends on the dimensions of the channels, hydrodynamic phenomena and interfacial phenomena, was theoretically and experimentally investigated [27]. At first, a wetting film consisting of the intruding liquid covers the pore walls (Fig. 5.3a). Instabilities in this film may occur (Fig. 5.3b), and as more and more liquid moves into the pores, these instabilities begin to grow until a "snap off" or "pinch off" takes place, that is, a meniscus forms (Fig. 5.3c). The interfaces of the menicus move in opposite directions, and the pore volume is eventually completely

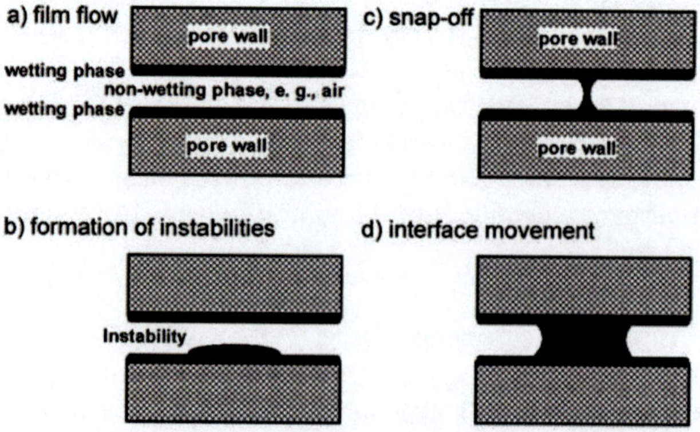

Fig. 5.3 Infiltration of a low molecular weight liquid into a cylindrical channel

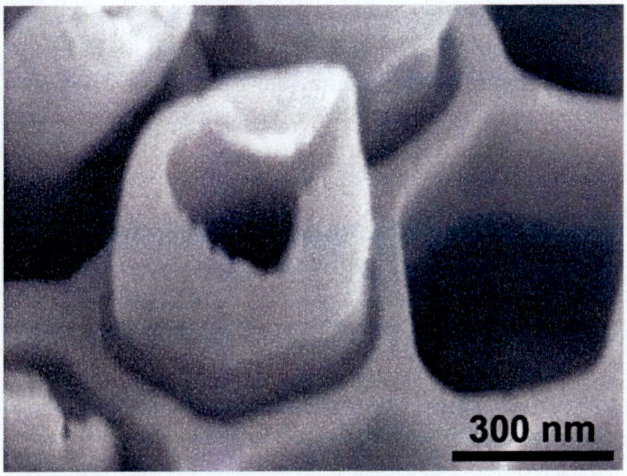

Fig. 5.4 PS tube prepared by precursor wetting protruding from an AAO hard template. Reproduced from [28]. Copyright (2002): AAAS

filled with the liquid (Fig. 5.3d). It is reasonable to assume that this so-called "snap-off" mechanism generally guides the infiltration of nanoporous and microporous materials with fluids.

If liquid, disordered polymers are brought into contact with porous hard templates exhibiting high surface energy and D_p-values significantly exceeding twice the radius of gyration of the infiltrated polymer, a polymeric precursor film with a thickness of a few tens of nanometers rapidly covers the entire area of the pore walls on a time scale of seconds to minutes even if the pores have a depth of the order of 100 μm [28, 29]. This behavior is commonly observed for AAO hard templates. Despite the fact that the pores should be completely filled in equilibrium, the liquid polymer layer is stable at least for several days (Fig. 5.4). Apparently, the macromolecules have to be mobile enough to be removed from the bulk and to diffuse into the pores. Both entropic relaxation of the polymer chains and the disjoining pressure may lead to the generation of a polymer layer having a mesoscopic thickness. When the diameter of the pores in the hard template is reduced below about twice the radius of gyration of the infiltrated polymeric species, the hollow space in the tubes disappears and solid rods are obtained [30–32]. In general, precursor wetting takes place if the pore walls of the hard template exhibit high surface energy and if the polymeric melts are heated to temperatures well above their glass transition temperature [33, 34]. However, little is known about the related relaxation processes and about the parameters influencing the thickness of the nanotube walls.

If polymeric melts are infiltrated into nanoporous hard templates under conditions where the formation of a precursor film is suppressed, the filling of the pore volume is governed by classical capillarity, a mechanism that was intensively investigated in the context of capillary molding [35]. The strong adhesive forces between the polymer and the pore walls are still effective but are not strong enough to drive single molecules out of the polymeric bulk reservoir on top of the hard template. Instead, a solid cylinder of the liquid but viscous polymer, preceded by a meniscus, slowly moves into the pores of the hard template [36, 37] until the entire pore volume is filled. The length of the fibers is proportional to the square root of the infiltration time [33, 36–38]. The time scale on which the pores are filled conveniently allows adjusting the length of the polymeric fibers by quenching the infiltrated polymer below its glass transition temperature or its crystallization temperature, respectively. For instance, Moon and McCarthy adjusted the lengths of PS fibers prepared by melting PS ($M_W = 280{,}000$ g/mol) at 200°C in contact with an AAO membrane having a pore diameter of 200 nm to 0.6, 0.9, 1.2 and 1.6 μm by heating the polymer for 5, 10, 15 and 20 min, respectively [36]. Kriha et al. reported that loading a BCP melt with weight accelerated the infiltration, and that template pores ($D_p = 400$ nm) with a T_p-value of 100 μm were completely filled after 6 h [37].

Microphase-separated BCPs commonly fill the pores of hard templates via capillary wetting [37–39] since the removal of single molecules from the bulk would disturb the ordered structure in the BCP. Moreover, the blocks had to diffuse (or to drift) through domains consisting of the other component. Thus, they had to overcome repulsive enthalpic interactions. In the case of disordered homopolymer melts, apparently a transition from precursor wetting to capillary wetting occurs that appears to be related to an increase in the viscosity of the polymeric melt, if hard templates having pore walls with a high surface energy are used. For example, PS ($M_n = 30{,}500$ g/mol) forms solid rods in AAO membranes with a D_p-value of 200 nm after annealing for 2 h at 130°C. However, increasing the infiltration temperature to 205°C resulted in the instantaneous formation of nanotubes with lengths of 60 μm, corresponding to the T_p-value of the template pores [33]. Correspondingly, at a given infiltration temperature and for a given polymer, precursor wetting will occur if the molecular weight of the polymer is relatively low, whereas capillary wetting will occur in the case of relatively high molecular weights. For example, PS with a M_n-value of 30,500 g/mol forms nanotubes when infiltrated into AAO with a D_p-Value of 200 nm at 205°C, as discussed above. However, if the PS has a molecular weight of about 760,000 g/mol, again short nanorods where obtained [33]. It was suggested to exploit the dependence of the infiltration mechanism on the molecular weight for fractionating polymers with different molecular weights [33]. It should be noted that precursor wetting and capillary wetting represent only to different kinetic routes to equilibrium that is characterized by complete filling of the pores with the polymer. In the case of hard templates having pores with D_p-values of a few tens of nm, precursor wetting will yield solid nanorods such as capillary wetting. However, the rates at which the pores are filled with the polymer potentially allow for distinguishing between both mechanisms. Obviously, the parameter that determines the wetting mechanism is the energetic and entropic effort required to remove the polymer chains from the bulk reservoir on top of the hard template and to draw them into the pores.

Combinations of both wetting mechanisms identified so far allow fabricating new and unprecedented one-dimensional nanostructures, for example, tube/rod hybrid fibers. To this end, AAO hard templates were infiltrated with BCPs under conditions of capillary wetting in such a way that pore segments with an adjustable length were filled with solid BCP rods. Then, the hard templates were turned upside down and

Fig. 5.5 Cross-sectional view of an array of tube/rod hybrid nanofibers obtained by infiltrating polymers into hard templates at first under conditions of capillary wetting and subsequently from the reverse side of the hard template under conditions of precursor wetting. Reproduced from [37]. Copyright (2007): Wiley-VCH

a homopolymer was infiltrated from the reverse side under conditions of precursor wetting. The composite fibers thus obtained consisted of a stiffer, solid BCP segment and a more flexible, tubular homopolymer segment (Fig. 5.5) [37].

The infiltration of solutions consisting of a polymer and a volatile low molecular mass solvent into nanoporous hard templates has been intensively investigated [9, 31, 34, 40–44]. It is far from being trivial to predict whether the properties of such a mixture rather correspond to those of a low-molecular mass liquid or to those of a polymeric melt. In any case, the evaporation of the solvent, a process that can hardly be controlled in a satisfying manner, will influence the morphology of the polymeric nanostructures. Liquid/liquid phase separation [45] and wetting transitions [46] may occur when the composition of the system changes. Structure and density of the absorbed layer will not only strongly depend on the polymer/solvent interactions [47] but also on environmental conditions such as temperature and humidity. Little is known about the conformation of polymer chains covering the pore walls of hard templates infiltrated by polymeric solutions. Primak et al. studied PDMS films (M_w = 10,940 g/mol) deposited from a solution in chloroform into AAO membranes by deuterium nuclear magnetic resonance spectroscopy [43]. They found a high degree of surface-induced ordering inconsistent with the expected loop/tail conformations and suggested that the chains in the proximity of the pore walls were flattened and that particularly strong interactions between the monolayer covering the pore walls and the pore walls were present.

The morphology of polymeric nanostructures can easily be probed by SEM and TEM. Ai et al. reported that nanotubes are obtained if diluted solutions of PS

(M_w = 270,000 g/mol) in cyclohexan infiltrate AAO membranes at 35°C, that is, under θ-conditions, whereas infiltration of concentrated solutions results in the formation of solid rods [44]. This is in line with results reported by Song et al., according to which the wall thickness of PS nanotubes deposited from solutions in dichloromethane increased with increasing concentration of PS [40]. Chen et al. obtained amorphous carbon nanotubes by infiltration of solutions of PAN in DMF, crosslinking of the PAN and subsequent carbonization. Again, the walls of the so-obtained carbonaceous nanotubes were thicker for higher PAN concentrations in the infiltrated stock solutions. In this case, the wall thickness could also be tuned by performing successive infiltration-pyrolysis cycles [48]. However, it should be noted that infiltration of solutions into AAOs often results in the formation of short, defect–rich fiber segments [42].

Only few attempts have been made to rationally design the mesoscopic fine structure of nanotubes fabricated by wetting nanoporous hard templates with polymeric solutions. Chen et al. infiltrated solutions of PS-*b*-PAN in DMF into AAO. As described above, the PAN was at first crosslinked and then carbonized. However, the PS domains were converted into holes, and porous amorphous carbon nanotubes could be fabricated [48]. In a similar approach, Rodriguez et al. used a solution of PS-*b*-PVP and carbohydrates associated with the PVP blocks via hydrogen bonds in DMF into AAO and obtained mesoporous amorphous carbon nanotubes with the positions of the mesopores determined by the positions of the PS domains. Solvent annealing of the BCP/hydrocarbon films in DMF/benzene vapor led to a significantly more uniform distribution of the PS domains and hence of the pores in the amorphous carbon nanotubes [49]. Apparently, in contrast to the infiltration of BCP melts, tubular structures with walls consisting of BCPs can be obtained in this way. In both works, the hard AAO templates had a D_p-value of about 200 nm, and the occurrence of bamboo-like morphologies or ring-like ribs arranged more or less periodically along the nanotubes was reported (Fig. 5.6).

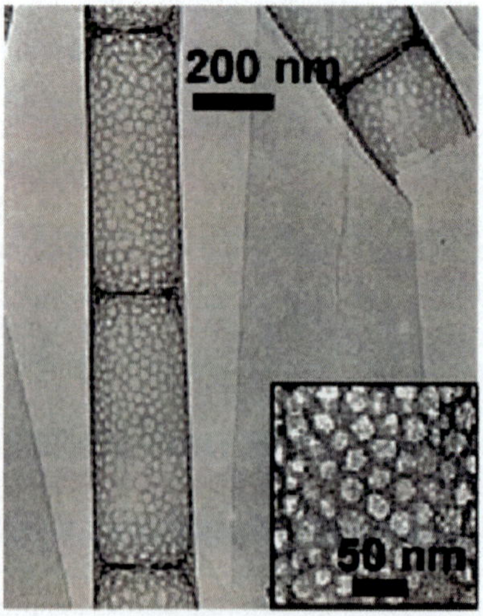

Fig. 5.6 TEM image of nanoporous carbonaceous nanotubes prepared using PS-*b*-P2VP with a bamboo-like structure. *Inset*: Hexagonal arrays of pores on the tube wall. Reproduced from [49]. Copyright (2006): American Chemical Society

An interesting self-ordering phenomenon is the occurrence of Rayleigh-Plateau instabilities. It is well known that annular liquid films are, similar to liquid cylindrical threads, susceptible to the growth of periodic thickness fluctuations [50]. Chen et al. reported that nanotubes prepared by infiltrating AAO hard templates with 5 wt-% solutions of PMMA (M_w = 22,700 g/mol) in chloroform can be converted into nanorods with periodic encapsulated holes driven by the Rayleigh instability [51]. At first, a smooth polymer film covered the pore walls. Thermal annealing of the PMMA/AAO hybrid membrane at temperatures above the glass transition temperature of the PMMA resulted in the growth of thickness undulations in the annular PMMA film and finally in the formation of bridges across the entire nanopore. The wavelength of the periodic structure increases with D_p and amounts to about 1,000 nm for a D_p-value of 200 nm. Figure 5.7a shows a TEM image of a

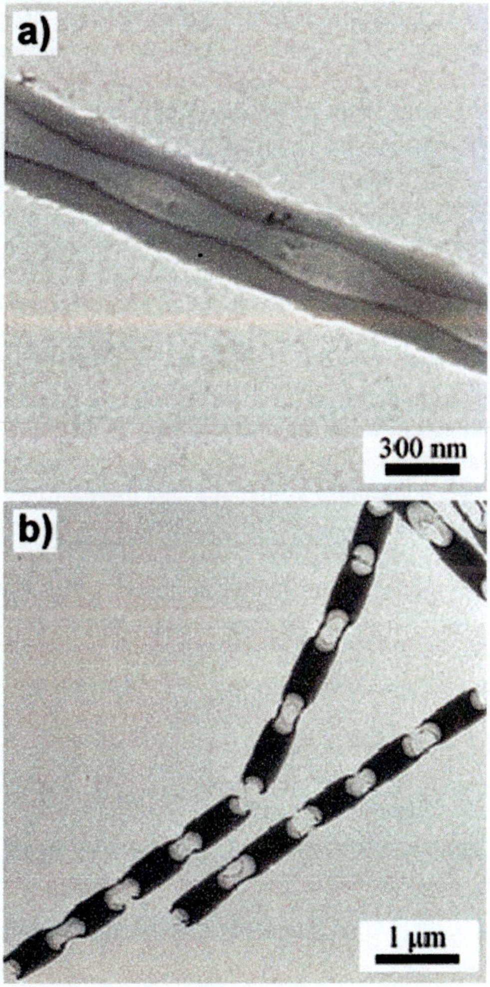

Fig. 5.7 Rayleigh instabilities in PMMA nanofibers. (**a**) TEM image of a PMMA tube with periodically undulated pore walls; (**b**) TEM image of a hole-containing PMMA nanorod. Reproduced from [51]. Copyright (2007): American Chemical Society

PMMA tube with periodically undulated pore walls, and Fig. 5.7b a TEM image of a hole-containing PMMA nanorod.

5.4 Self-Assembly in the Walls of Single-Component Nanotubes

Crystallization and mesophase formation determine the optical, electronic, mechanical, chemical and piezoelectric properties of soft matter to a large extent. Thus, control over crystallization and mesophase formation inside hard templates is crucial to the rational design of one-dimensional nanostructures consisting of such materials. Mesoscopic morphologies characterized by a crystalline or liquid-crystalline degree of order and by pronounced anisotropy can be generated by infiltration of semicrystalline or liquid-crystalline polymers as isotropic liquids and subsequent crystallization or mesophase formation inside the hard template pores. Common experimental techniques, including WAXS [30, 52], SAED [31, 53], HRTEM [54, 55], polarized IR spectroscopy [4, 6, 8, 53] and DSC [30, 52, 56] have been applied to characterize ordered supramolecular architetures in one-dimensional nanostructures. WAXS on ensembles of aligned nanofibers provides valuable information on textures, whereas SAED and HRTEM can be used to locally elucidate crystal orientations and to determine the polymorph formed. Whereas WAXS is easy and straightforward to apply on AAO membranes loaded with soft matter, electron microscopy in general suffers from strong interactions between the nanostructures being probed and the incident electron beam. This drawback may be overcome by short exposure times or by cooling the samples [31]. Up to now, SAXS is no established method for probing textures and mesoscopic features such as long periods in one-dimensional nanostructures confined to hard templates. Background scattering of the matrix material may pose problems for the evaluation of the SAXS patterns, and in the case of released nanofibers the inhomogeneous nature of the powders used for performing SAXS experiments hampers the analysis of the collected data. Much information can be gained by polarized IR spectroscopy [57]. On the one hand, it is often possible to assign specific peaks to amorphous and crystalline material so that the crystallinity can be deduced from the peak areas. On the other hand, anisotropy is obvious from infrared dichroism. DSC yields information on the crystallization kinetics and nucleation mechanisms, as well as on crystal sizes. In principle, NMR and dielectric spectroscopy should also be applicable methods that have, up to now, virtually not been explored for the study of supramolecular architectures in nanoporous hard templates. However, at least for NMR, a proof of concept was reported by Primak et al. [43].

Mesoscopic structure formation processes inside hard templates can be influenced by surface-induced ordering and geometric confinement since their characteristic length scales are on the same order of magnitude than the D_p-values of the nanopores or the thickness of the nanotube walls they contain. Already early works on template syntheses of functional polymers by Martin and co-workers indicated that supramolecular order and properties such as conductivity may be enhanced inside hard templates [7]. Polypyrrole and poly(3-methylthiophene) [3], polyacetylene [4] and polyaniline [6] nanofibers having diameters in the mesoscopic range were obtained by synthesizing the corresponding polymers in the nanopores. Their enhanced conductivity, which was evidenced for nanofibers aligned in the hard templates and for mats consisting of released nanofibers [58] was attributed to enhanced supramolecular order [4, 7, 8, 59]. Polymerization within the pores involves preferential growth of the chains located on the pore walls of the hard template because of their reduced solubility as compared to the monomer. Therefore, after short polymerization times, the nanotubes thus formed predominantly consist of polymer chains in proximity to the pore walls, where they are aligned and the formation of kinks and bends is suppressed. Martin and co-workers therefore concluded that the enhanced supramolecular ordering thus realized is accompanied by an increased conjugation length, which in turn results in higher conductivity. This surface-induced alignment was found to decay when thicker nanotube walls were prepared by longer polymerization times.

One of the most important structure formation processes in polymeric materials having chain architectures that allow, at least to some extent, packing of the chains is crystallization. Polymers usually crystallize as lamellar crystals in which folded chains are oriented approximately perpendicular to the surface of the lamellae [60]. The typical thickness of these crystals lies in the nanometer range, while their lateral dimensions are in the micrometer range, thus by far exceeding typical D_p-values of hard templates. Within the crystals, the growth of the lamellae proceeds in the lateral directions. On a larger scale, the lamellae are organized in spherulites, densely branched, isotropic, polycrystalline superstructures [61]. It is an intriguing question as to how the geometric confinement of the pores in nanoporous hard templates and their large surface-to-volume ratio influences the crystallization of polymers. In the case of melt infiltration of semicrystalline polymers into hard templates, crystallization is an important issue because crystallization may occur upon cooling to room temperature. Even though, up to now, only a limited number of publications deals with this topic [30, 42, 52, 53, 56], it has become clear that crystallization of polymers confined to the pores of hard templates can be influenced, and to some extent engineered, by the presence or absence of a bulk polymer reservoir in contact with the polymer inside the pores, by the D_p-value of the hard template, and by the temperature profile applied.

Generally, the *c*-axis of the polymeric crystals, which is commonly normal to the plane of the lamella crystals, orients perpendicular to the pore axes. This enables the lamellae growing along the pores. Moreover, the crystallinity of the material inside the pores is typically below that of the corresponding bulk material. In the case of non-isothermally crystallized PVDF, the crystallographic direction exhibiting the highest growth rate, that is, the <020> direction, aligns with the pore axis, resulting in uniform crystal orientation inside the pores on a macroscopic scale [42], if crystallization is initiated by heterogeneous nucleation [62] in a bulk PVDF reservoir on top of the hard template. The lamellae in the spherulites in the bulk reservoir are oriented in such a way that their direction of fastest growth points radially outwards. If a growing spherulite hits the surface of a hard template infiltrated with

PVDF, only those lamellae proceed into the pores whose <020> direction is, within a certain tolerance, oriented parallel to the pore axes [30]. If the bulk reservoir is removed from the surface of the hard template, crystallization in each crystallizing entity is predominantly initiated by homogeneous nucleation at high supercooling, because the probability of the occurrence of heterogeneous nuclei in the small volume of the nanopores is negligible. Then, crystals with one of the <*hk*0> directions aligned with the pore axes, a crystal orientation that enables growth of the lamellae along the pores, form with statistical frequency. This was evidenced by WAXS measurements on PVDF nanofibers aligned within the pores of the hard templates, crystallized in the presence or absence of a bulk PVDF reservoir. In the former case, only the (020) peak appears in the XRD pattern (Fig. 5.8a), in the latter case all (*hk*0) peaks show up (Fig. 5.8b) with relative intensities similar to those in the powder pattern of isotropic PVDF (Fig. 5.8c). However, it is striking that (*hkl*) values with non-zero *l*-index are still missing in the pattern of the sample crystallized in the absence of the bulk reservoir. This is because crystals with a corresponding orientation impinge on the pore walls and therefore occupy only a small portion of the pore volume. DSC cooling scans confirmed that homogeneous nucleation significantly contributes to the crystallization of an ensemble of separated PVDF nanotubes inside an AAO template with a D_p-value of 400 nm, whereas exclusively homogeneous nucleation occurs if the D_p-value is reduced to 35 nm. Random PVDF copolymers with trifluoroethylene P(VDF-*ran*-TrFE) infiltrated into AAOs with D_p-values ranging from 55 to 360 nm were also investigated and found to be crystalline. By probing the relative permittivity of arrays of P(VDF-*ran*-TrFE) nanofibers the ferroelectric-to-paraelectric phase transition could be observed [32].

Woo et al. investigated the crystallization kinetics of separated entities of linear PE inside AAO hard templates by DSC. Inside pores with D_p-values below about 50 nm heterogeneous nucleation at the pore walls becomes dominant, whereas for D_p-values of 62 and 110 nm homogeneous nucleation initiates crystallization [56]. Polarized IR spectroscopy was employed by Wu et al. to investigate crystallinity, the formation of different crystal orientations and the crystal texture of syndiotactic

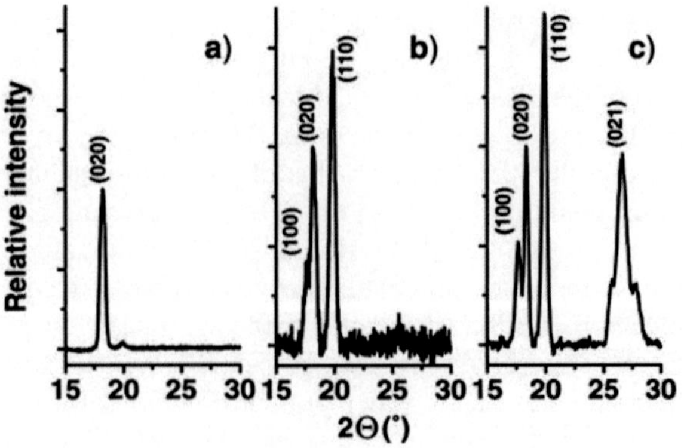

Fig. 5.8 WAXS patterns of PVDF nanofibers aligned in an AAO hard template (D_p = 35 nm) measured in Θ/2Θ geometry. (**a**) Non-isothermal crystallization in the presence of a bulk PVDF surface reservoir; (**b**) non-isothermal crystallization in the absence of a bulk PVDF surface reservoir; (**c**) powder pattern of an isotropic sample

PS (sPS) crystallized inside AAO hard templates while in contact with a bulk reservoir of the same polymer [53]. The β-polymorph was obtained by cooling from the melt to 260°C and crystallizing at this temperature for 2 h, while heating amorphous samples quenched from the molten state to 240°C and heating to this temperature for 2 h resulted in the formation of the α-polymorph. A comparison of the areas of peaks characteristic of amorphous sPS and the β-polymorph revealed again a decreasing crystallinity along with decreasing D_p. For the bulk and D_p-values of 200 and 80 nm, crystallinities of 62.0, 49.8 and 36.2%, respectively, were obtained. The evaluation of polarized IR spectra measured on bulk sPS and sPS confined to AAOs with D_p-values of 200 and 80 nm (Fig. 5.9) revealed that samples crystallized at 240°C consisting of the α-polymorph were isotropic. In samples crystallized at 260°C consisting of the β-polymorph no preferred orientation of bulk sPS was observed, whereas inside the AAO hard templates the *c*-axes in the crystals were

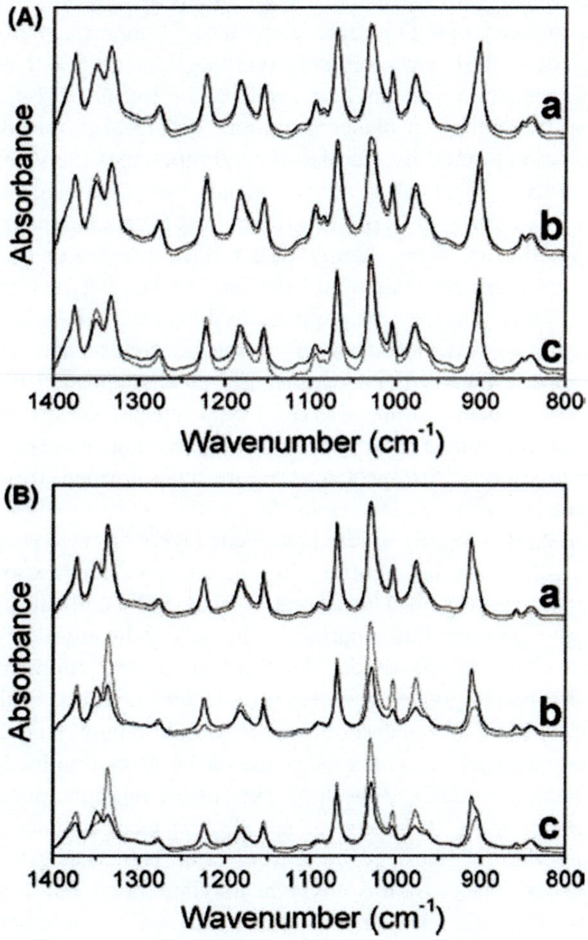

Fig. 5.9 Polarized infrared spectra of sPS crystallized at lower temperatures (A) and at 260 °C (B). (**a**) represents the bulk, (**b**) nanorods prepared inside AAO with $D_p = 200$ nm, and (**c**) nanorods prepared inside AAO with $D_p = 80$ nm. Polarization perpendicular to nanorod axes: *solid lines*; polarization parallel to nanorod axes: *dashed lines*. Reproduced from [53]. Copyright (2007): American Chemical Society

oriented perpendicular to the pore axes. The apparent differences in the degree of crystal orientation in the AAO/sPS hybrid samples were attributed to the different thermal histories. The α-polymorph was obtained by heating quenched amorphous samples from low temperatures to the target crystallization temperature of 240°C. Therefore, a temperature range was passed that is characterized by a high nucleation rate at high supercooling and crystallization was governed by nucleation. Because of the presence of many small growing crystallites no texture could develop. However, the β-polymorph was obtained by cooling from the isotropic melt in contact with a bulk reservoir. Then, it is to be expected that crystallization is initiated by a small number of heterogeneous nuclei so that crystallization is dominated rather by crystal growth than by nucleation. Consequently, crystals having a major growth direction oriented parallel to the pore axes dominate.

Besides the exploitation of melt crystallization in thermoplastics, the formation of ordered assemblies consisting of molecules having an anisotropic shape is a self-assembly process with great potential for the fabrication of nanotubes with a customized mesoscopic fine structure inside the pores of hard templates. To this end, particularly disc-like molecules that self-organize into columnar stacks, so-called discotics [63], are promising building blocks. For their anchoring to the surface of a substrate, two limiting cases can be formulated. "Edge-on" orientation means that the molecular planes are oriented normal to the surface of the substrate. The columns formed by the disk-shaped molecules then have a so-called "planar" orientation. "Face-on" anchoring means that the molecular planes are parallel to the substrate surface. Then, the orientation of the columns is called "homeotropic". The way how the discs assemble on a surface depends on the intercolumnar interactions between adjacent discs and the interactions between the disks and the substrate. Particularly polycyclic aromatic hydrocarbons (PAHs) [64] have attracted considerable interest, because their pyrolysis yields nanotubes whose walls consist of graphene layers. For example, Zhi et al. deposited disc-like PAHs of the hexa-*peri*-hexabenzocoronene (HBC) type into AAO hard templates from solutions in dichloromethane [55]. The discs were anchored edge-on, and columns with a planar orientation with respect to the pore walls formed, driven by strong π–π interactions between the HBC discs. Subsequent pyrolysis yielded nanotubes whose walls consisted of highly ordered graphene layers oriented perpendicular to the tube axes. Therefore, the initial orientation of the HBC precursors was conserved during the carbonization step. The properties of the HBC molecules can be engineered by modifying the substitution pattern at the polycyclic aromatic core. Long, branched alkyl side chains as substituents lead to low isotropization temperatures so that melts of correspondingly designed HBC molecules could be infiltrated into AAO hard templates [65]. In this case, the discs formed columns oriented planar with respect to the pore walls in which the plane of the disks was inclined by 45° with respect to the column axis and the long axes of the template pores. HBC molecules bearing acrylate units at the end of six alkyl spacers attached to the polycyclic aromatic core were synthesized by Kastler et al. [66]. The HBC molecules thus modified can easily be cross-linked via the acrylate functions. Deposition of these HBC discs from a solution in dichloromethane into AAO hard templates led to the formation of nanotubes whose walls consisted of long-range ordered stacks of crosslinked HBC molecules aligned with the nanotube axes. Apparently, the discs were anchored edge-on to the pore walls. Cross-linking at a moderate temperature of 170°C fixated the supramolecular columnar architecture without destroying the polycyclic aromatic core of the molecule. The nanotubes thus obtained therefore exhibited high mechanical stability, whereas the supramolecular order, the formation of which had been driven by π–π interactions between the HBC discs, was conserved as revealed by HRTEM and SAED (Fig. 5.10). In the case of hyperbranched tetraphenylcyclopentadienone building blocks deposited from solutions in dichloromethane, crosslinking by a Diels-Alder reaction fixated the supramolecular architecture, and stable tubular nanostructures were obtained. Subsequent carbonization yielded carbonaceous nanotubes exhibiting a highly porous fine structure [67].

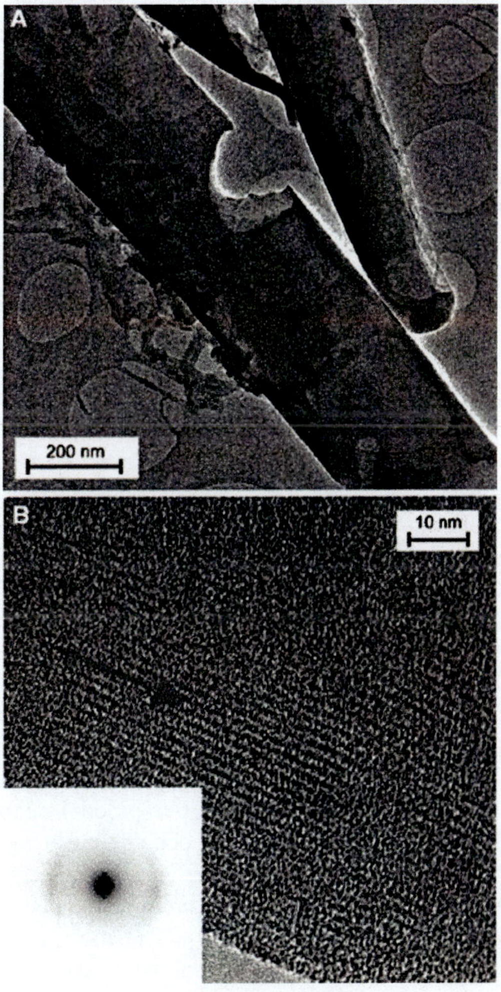

Fig. 5.10 TEM images of carbonaceous nanotubes obtained by assembling HBC molecules inside hard templates. (**a**) Defect evidencing their tubular structure. (**b**) Detail of the wall structure; the *arrow* indicates the directions of the columnar structures and the tube axis; *inset*: electron-diffraction pattern. Reproduced from [66]. Copyright (2007): Wiley-VCH

Self-organization driven by π–π interactions has meanwhile been exploited to assemble nanotubes from even more complex building blocks. To this end, Zhi et al. deposited tetrakis(tert-butyl)-naphthalocyaninato nickel complexes that form columnar structures on the pore walls of AAO hard templates by edge-on π–π stacking from a solution in THF [54]. After thermal annealing, a highly ordered columnar structure was obtained, and stable tetrakis(tert-butyl)-naphthalocyaninato nickel nanotubes could be released from AAO hard templates or converted into graphitic carbon nanotubes containing nitrogen or nickel nanoparticles. Liu et al. prepared nanotubes from sandwich-type (porphyrinato)(phthalocyaninato)europium(III) complexes [68].

The formation of mesophases inside the two-dimensional confinement of nanopores can be complex. Pentacene nanotubes prepared by melt infiltration and slow cooling to room temperature [69] exhibited no long-range order. Also, HBCs that were designed in such a way that they anchor face-on on smooth substrates did not yield nanotubes with uniformly oriented graphene layers when deposited into AAO templates [70]. Thus, it seems that the suppression of face-on anchoring leading to homeotropic orientation of the columns, for example by using molecular building blocks that tend to intercolumnar π–π stracking, results in the formation of highly ordered planar mesophases. However, homeotropic phases inside AAOs are characterized by a high degree of disorder. On the one hand, the curvature of the pore walls prevents a perfect parallel arrangement of the columns along the perimeter of the pores so that growing columns will impinge on their neighbors. On the other hand, the roughness of the pore walls may also introduces disorder.

5.5 Phase Separation in Nanoporous Hard Templates

The decomposition of mixtures in the pores of hard templates is a promising access to nanotubes with tailor-made morphologies. For example, nanotubes with walls exhibiting a microporous fine structure are potential components for storage devices or chromatographic separation processes. Phase boundaries may be crossed if a mixture is subjected to thermal quenching, or if the composition of the mixture changes because of the evaporation of a volatile solvent. Commonly, a spinodal decomposition [71] sets in. Then, periodic composition fluctuations in an initially homogeneous system begin to grow. Simultaneously, ripening of the morphology starts, driven by the tendency to minimize the interfacial area between the coexisting phases, at the initial stage by conformational changes of the polymer chains and subsequently by Ostwald ripening [72]. For a broad composition range, the phase morphology generated by spinodal decomposition is initially a bicontinuous network of the two components that breaks up at later stages of the ripening process. The presence of interfaces modifies the decomposition process in that a surface-induced layered structure forms, a phenomenon known as "surface-directed spinodal decomposition" [73]. The most straightforward way to infiltrate mixtures containing at least one polymeric component into nanoporous hard templates is wetting with homogeneous solutions in a common volatile solvent. However, the presence of a third, evaporating component complicates the understanding of the involved structure formation processes. The analysis of the phase morphology obtained in this way was, up to now, limited to SEM and TEM investigations of nanotubes and nanorods in which the structure evolution was frozen.

As model systems, mixtures of PLA and organometallic complexes containing Pd and Pt dissolved in a common solvent were filled into the pores of hard templates [74]. After the evaporation of the solvent, tubes with a wall thickness of a few tens of nm were formed in which the metal precursors were dispersed. After thermolytic reduction, the evolution of nanoparticles consisting of the elemental metals inside the liquid polymeric matrix was monitored as a model process for spinodal decomposition and morphology ripening. The ripening was stopped by pyrolytic degradation of the PLA. After short ripening times, for example, the walls of Pd nanotubes were rough, highly porous, and had obviously a reticular morphology (Fig. 5.11a, b). At later ripening stages, the nanotube walls had a smoother, layer-like appearance, and the size of the Pd crystals significantly increased (Fig. 5.11c, d). Thus, after short ripening times, the initially homogeneous tube walls are characterized by an interpenetrating morphology with a small spinodal wavelength. Further ripening results in the evolution of a coarser, layered structure that is indicative of surface-induced ordering (Fig. 5.11e). Wetting AAO templates with solutions containing a polymeric wetting carrier and precursors for magnetic metals such as cobald was applied to synthesize magnetic nanotubes [75].

Surface-induced ordering was also observed in nanotubes obtained by deposition of a solution containing PMMA and a discotic liquid crystal of the triphenylene type into AAO templates with a D_p-value of 400 nm. The PMMA segregated to the pore walls, whereas the liquid crystal enriched at the inner surface of the nanotubes. It was assumed that the synergistic interplay of two different physico-chemical phenomena led to a surface-directed phase separation. First, low molecular mass species such as the triphenylene compound enrich at interfaces in the presence of a polymer for entropic reasons [76]. Secondly, taking into account the high compatibility of the liquid crystal used and PS, PMMA should have a higher affinity to the AAO pore walls [77]. Reducing the D_p-value of the AAO hard template to 60 nm resulted in the occurrence of a striking morphological crossover. Solid nanorods were obtained with a disordered segmented morphology. The disappearance of the inner tube surface obviously resulted in a competition of the enthalpic and entropic effects: the PMMA tends to segregate to the pore walls for enthalpic, and the triphenylene compound for entropic reasons. Since only one interface at the pore walls instead of two interfaces in a tubular configuration was available, the formation of a layered structure was prevented. Thus, a confinement-induced transition from a wetting state (one phase of a critical two-phase system in exclusive contact with an interface) to a non-wetting state (both phases in contact with the interface) occurred.

Spinodal decomposition is a straightforward access to mesostructured materials characterized by a certain degree of near order. However, phase separation can even be exploited to fabricate ordered mesostructured materials if amphiphilic species acting as soft templates are involved. Well established syntheses for mesoporous

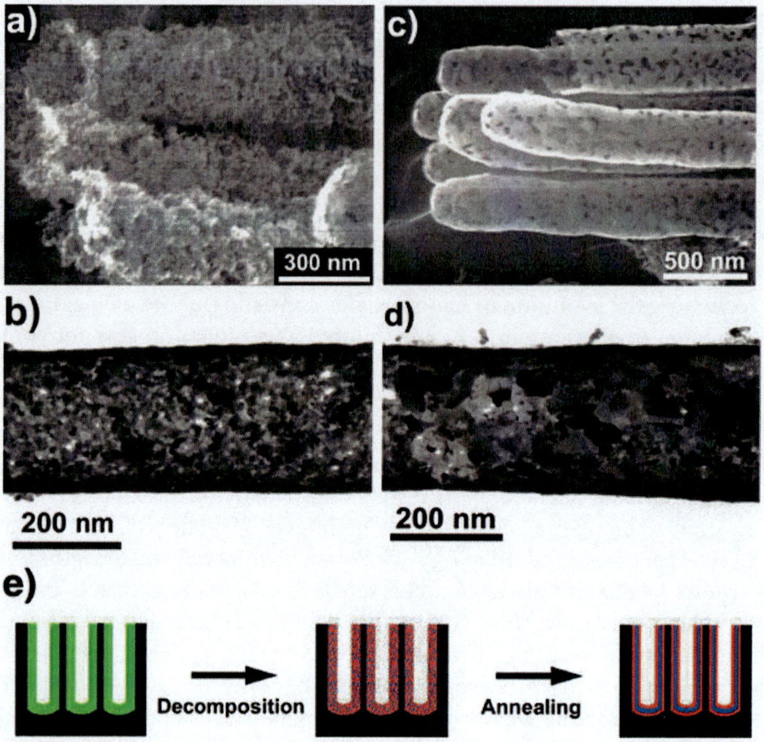

Fig. 5.11 Palladium nanotubes obtained by the ripening of Pd nanoparticles in a PLA matrix after spinodal decomposition of a PLA/palladium acetate/solvent mixture inside AAO and pyrolytic degradation of PLA. (**a**) SEM and (**b**) TEM image after short ripening times; (**c**) SEM and (**d**) TEM image after longer ripening times. (**e**) Schematic diagram of the ripening process. After the decomposition of the initially homogeneous mixture, the walls of the Pd nanotubes were at first rough, highly porous, and had obviously a reticular structure, which is indicative of an interpenetrating morphology with a small spinodal wavelength. At later ripening stages, the nanotube walls had a smoother, layer-like appearance, and the size of the Pd crystals significantly increased, which is indicative of the evolution of a coarser, layered structure and surface-induced ordering. Panels (**a**)–(**d**) are reproduced from. [63] Copyright panels (**a**)–(**d**) (2004): Wiley-VCH

materials with mesopore diameters ranging from a few nanometers up to a few tens of nanometers start with sol solutions containing either low molecular mass surfactants [78] or BCPs containing blocks with different polarity [79, 80] and precursors for scaffolds consisting of inorganic oxides or amorphous carbon [81]. The precursor for the scaffold material commonly segregates into the polar phase defined by the soft template that self-assembles if its concentration is larger than the critical micelle concentration. Subsequently, the morphology thus formed is fixated by a gelation or aging step in which the precursors for the inorganic scaffold materials are crosslinked. Subsequent high-temperature calcination yields inorganic scaffolds containing highly ordered mesopore arrays. Several excellent review articles summarize syntheses for mesoporous materials and their properties [82]. However, they are typically obtained in the form of powders consisting of randomly oriented grains. On solid substrates, the mesopores are arranged parallel to the substrate surfaces, a morphology that is of limited use for applications in the fields of separation, catalysis, and storage. Strategies to overcome this drawback based on surface modifications or the freezing of non-equilibrium structures only yield thin mesoporous layers with mesopores having limited aspect ratios. To address this problem, the self-assembly of BCP soft templates inside the pores of hard templates has emerged as a promising synthetic strategy, taking advantage of the availability of mechanically stable, extended membranes having larger but properly oriented membrane pores.

The first reported procedures for the preparation of mesoporous silica nanofibers by means of BCP soft templates inside hard templates were adaptations of synthetic strategies introduced by Stucky and coworkers [80]. Both tubular and solid entities with a mesoporous fine structure can thus be prepared. In their pioneering work, Yang et al. filled sols containing Pluronic F127 and TEOS into AAO hard templates with a D_p-value of about 250 nm. Inside non-modified AAO the mesoporous silica remained attached to the oxidic, polar pore walls of the hard template so that tubes formed. However, if the hard templates were modified with hydrophobic silane coupling agents, the silica detached from the pore walls of the hard template, and solid rods were obtained [83]. Liang and Susha reported that infiltration of sols into polycarbonate membranes yielded tubular structures if the sols had high ethanol content, whereas solid rods formed when the ethanol content was reduced [84]. Both Liang and Susha, as well as Yao et al. [85] moreover found that aging the sols inside the pores of a hard template in the presence of an external bulk sol reservoir promotes the formation of solid mesoporous silica fibers, whereas the removal of excess sol from the surfaces of the hard templates after the infiltration promotes the formation of tubular structures. Zhu et al. obtained tubular structures in AAO templates with a D_p value of about 200 nm by slow infiltration of mixtures containing prehydrolized TEOS [86]. To this end, a sol solution was cast onto a smooth substrate. After placing AAO hard templates on top of the sol films, the samples were annealed at 100°C. The lengths of the silica nanotubes thus obtained ranged from 500 nm for a heating time of 2 h up to 10 μm for longer heating times. Apparently, the prehydrolized sol could slowly enter the pores but its reduced mobility apparently prevented the filling of the empty space inside the pores formed upon evaporation of the solvent. Thus, it appears that the nature of the pore walls, the composition of the infiltrated sol, and the amount of sol solution that can access the pores are parameters determining whether tubular or solid mesoporous silica structures form inside hard templates.

Gaining control over the orientation of the mesopores inside hard templates has turned out to be a delicate challenge. Subtle changes of parameters such as the composition of the sol or aging conditions lead to striking changes of the mesopore structure, which is in turn affected by the two-dimensional geometric confinement imposed by the pore geometry and the nature of the pore walls. In the case of mesoporous materials, the bulk morphology of which is characterized by hexagonal arrays of aligned mesochannels, the mesopores formed by self-assembly of BCP templates inside the pores of hard templates may align with the long axes of the hard template pores, or they adopt the contour of the pore walls and wind about the long axes of the hard template pores. If the hard templates have pores smaller than about 100 nm, new and unprecedented morphologies are obtained that are substantially different from their bulk counterparts, as discussed below. Self-assembly of different BCP soft templates of the Pluronic type inside the pores of hard templates with D_p-values of a few 100 nm yielded hexagonal arrays of mesopores winding about the long axes of the mesoporous silica nanofibers [83, 84, 87] as well as hexagonal mesochannel arrays [88] and concentric-lamellar structures [89] aligned with the fiber axes. Apparently, the stage at which the assembly of the soft template is frozen determines the morphology of the mesoporous nanofibers that also results from counteracting growth modes governed by surface-induced ordering and the two-dimensional confinement imposed by the geometry of the hard template pores.

Wu et al. systematically studied the self-assembly of $EO_{20}PO_{70}EO_{20}$ inside AAO hard templates having D_p-values from 80 down to 20 nm [90, 91]. Inside such narrow pores, bulk-like morphologies were completely suppressed. For D_p-values between 55 and 73 nm, the mesostructures were composed of a straight core and two more coaxial layers consisting of concentric mesochannels with morphologies as diverse as stacked doughnuts, single helices or double helices (Figs. 5.12, 5.13). For D_p-values ranging from 49 to 54 nm, coaxial double layer helices were found, for D_p-values ranging from 34 to 45 nm a straight inner core was surrounded by one coaxial layer of the helical or stacked doughnut type, for a D_p-value of 31 nm a single helix was observed, and for smaller D_p-values spherical mesopores arranged in one or two rows formed (Figs. 5.12 and 5.13). Analogous to the conceptualization of the structures of carbon nanotubes, Wu et al. suggested a rolling scheme to derive the morphologies obtained in the two-dimensional confinement of the hard template pores from thin-film morphologies. A progression of mesoscopic structures in two-dimensional confinement well in line with the experimental results was obtained by means of self-consistent field calculations with a liquid diblock copolymer/homopolymer mixture as a model system. Moreover, it is remarkable that chiral structures such as helices were obtained from achiral materials, even though a chiral induction leading to enantiomeric excess has not been reported up to now.

Only few efforts have been directed towards the fabrication of mesoporous nanofibers consisting of other inorganic oxides. Wang et al. fabricated nanotubes the walls of which consisted of mesoporous titania exhibiting hexagonal mesopore ordering [92]. The evolution of a tubular structure was attributed to the high affinity of the gel to the alumina pore walls, resulting in volume shrinkage towards the pore walls of the hard template. The first reported procedure for the synthesis of mesoporous amorphous carbon nanofibers involved the preparation of a sacrificial Fe-containing mesoporous silica scaffold inside an AAO hard template with a D_p-value of about 200 nm, exposure to hydrogen at 750°C, incorporation of carbon by supercritical fluid deposition of a xylene/CO_2 mixture, and removal of both the AAO hard template and the sacrificial silica scaffold by etching with hydrofluoric acid [93]. While the mesoscopic fine structure of the mesoporous carbon nanofibers was a perfect replica of the silica scaffold, the AAO hard template is inevitably

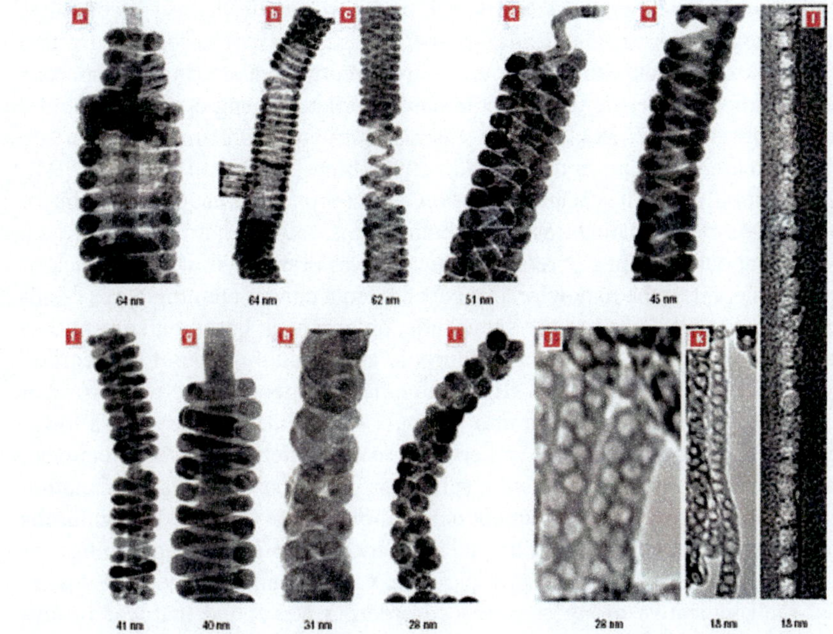

Fig. 5.12 TEM images of mesostructures formed inside AAO with differing confinement dimensions. The confining nanochannel diameter is indicated underneath each image. **a**–**i**, Silver inverted mesostructures prepared by backfilling the confined mesoporous silica; **j**–**k**, free-standing mesoporous silica fibres; **l**, mesoporous silica embedded inside the AAO obtained using a focused ion beam for sample preparation. The structures are (**a**) three-layer stacked doughnuts; (**b**) S-helix; (**c**) core–shell D-helix, in which the core and the shell are both S-helix; (**d**) core–shell triple-helix, in which the shell is a D-helix and the core is a S-helix; (**e**) D-helix; (**f**), (**g**) S-helix with a straight core channel; (**h**) D-helix; (**i**), (**j**) inverted peapod structure with two lines of spherical cages packed along the long axis of the alumina nanochannel; (**k**), (**l**) inverted peapod with one line of cages. Reproduced from [90]. Copyright (2004): Nature Publishing Group

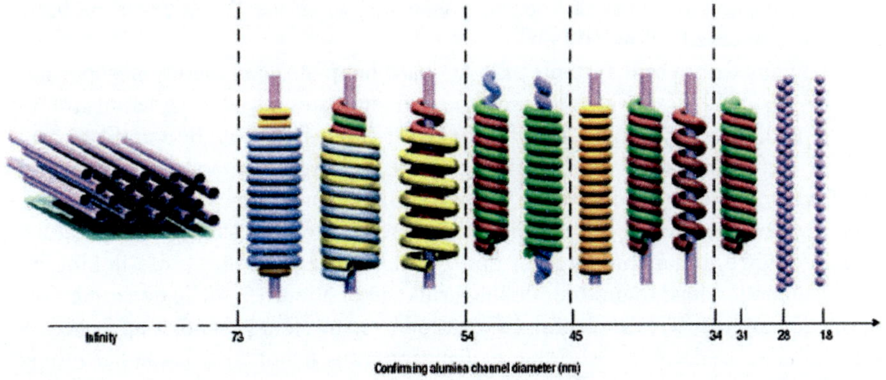

Fig. 5.13 Summary of the experimentally (cf. Fig. 5.12) observed confined mesostructural evolution with varying D_p-value. Reproduced from [90]. Copyright (2004): Nature Publishing Group

destroyed upon removal of the silica scaffold. Consequently, hybrid membranes containing mesoporous carbon inside an AAO matrix are not accessible by this approach. This drawback can be overcome by directly synthesizing mesoporous amorphous carbon inside the pores of the hard template. Zheng et al. infiltrated a mixture of Pluronic F127 ($EO_{106}PO_{70}EO_{106}$) as a structure directing soft template and resol as a carbon precursor dissolved in ethanol into AAO hard templates [94]. After the evaporation of the ethanol, gelation and carbonization at 700°C in nitrogen, mesoporous carbon nanowires were obtained. Inside AAO hard templates with a D_p-value of about 300 nm a core/shell structure was obtained in which a stack of layers perpendicular to the nanowire axis surrounded a core containing pores winding about the nanowire axis. At the same time, an approach based on solvent-free infiltration was reported. A solution of Pluronic F127, phloroglucinol as a carbon source, formaldehyde and traces of HCl in an ethanol/water mixture was stirred at room temperature until a separation into an upper water/ethanol phase and a lower polymer-rich phase occurred. The supernatant solvent-rich phase was removed, and the lower polymer-rich phase was spread on AAO hard templates. Gelation of the infiltrated mixture and subsequent calcination at 500°C yielded mesoporous amorphous nanofibers with a core characterized by a bicontinuous morphology, as desired for applications in the field of separation, catalysis and storage [95]. Again, a layered shell indicative of surface-induced ordering was found that was in turn surrounded by a continuous outermost carbon wall. Whereas the removal of volatile solvents prior to the infiltration may be a measure to minimize volume shrinkage, the low carbonization temperature is important for the fabrication of mesoporous amorphous carbon/AAO hybrid membranes, because the carbonization can be performed while the AAO membrane is still attached to an underlying Al substrate. This configuration is advantageous because the Al substrate stabilizes the AAO layers so that residual material can easily be removed from their surfaces to uncover the pore openings. Selective etching steps can then be applied to remove the Al and to open the pore bottoms. Optionally, an Al ring surrounding the area in which the pore bottoms are open can be conserved to mechanically stabilize the membrane. It is interesting to note that the fiber core vanishes when the D_p-value of the hard template is decreased below 100 nm.

It is obvious that hybrid systems of AAO hard templates containing mesoporous nanofibers obtained by self-assembling BCP soft templates are of considerable interest for a plethora of applications in the fields of catalysis, separation and storage. However, the combination of sol/gel chemistry and high-temperature calcination steps is accompanied by pronounced volume shrinkage of the mesoporous material. Solvent evaporation during the initial gelation step performed at room temperature results in unidirectional shrinkage of up to 20% [96], and further cross-linking as well as calcination lead to unidirectional shrinkage of about 15–40%, depending on the protocol applied [97]. Replacing calcination by extraction is another approach to overcome undesired volume shrinkage of mesoporous nanofibers inside hard templates. Yoo et al. prepared at least partially cubic mesoporous silica inside AAO with a D_p-value of about 200 nm using Brij-56 as a surfactant and removed the latter by extraction with ethanol [98]. Four successive infiltration/drying/surfactant extraction cycles yielded AAO/mesoporous silica hybrid membranes that exhibited excellent helium/nitrogen permselectivity with permselectivity values at the theoretical Knudsen limit. This result indicated the absence of "pinhole defects". Therefore, a defect-free, uniform filling of the AAO pores with mesoporous silica could be evidenced. Another interesting application reported for released mesoporous titania nanotubes, thus insensitive to volume shrinkage, is their use as nanostructured electrode material having outstandingly high specific surface area [92]. Electrons injected into the titania scaffold can rapidly be transferred into electrolyte solutions. The mesoporous titania nanotubes were characterized by large specific capacity and a high charge-discharge rate. Even though a small number of applications for either released mesoporous nanofibers or hybrid membranes containing mesoporous nanofibers synthesized by means of sol-gel chemistry with BCP soft templates have been reported, it appears that these materials still need to be optimized for real-life applications.

5.6 Multilayer Nanotubes by Layer-by-Layer Deposition

In the previous sections, approaches to the generation of mesoscopic fine structures in tubular but also solid one-dimensional nanostructures were discussed that involved deposition of target materials or precursors thereof into porous hard templates by a single infiltration step. Subsequently, the supramolecular organization was guided by the geometric confinement and interfacial interactions with the pore walls. Layer-by-layer deposition [99] is a generic access to nanoscopic multilayer systems with controlled composition that differs from single-step infiltration methods in that a series of successive deposition steps is performed, the number of which determines the properties of the nanotubes thus obtained to a large extent. To attach a new layer to the layers already deposited, specific interactions between the involved species or molecular recognition mechanisms are exploited. Initially, the layer-by-layer technique was applied to consecutively deposit oppositely charged polymeric polyelectrolytes from diluted solutions. Whereas the electrostatic repulsion between equally charged species limits the thickness of a deposited layer, the electrostatic attraction between the alternating oppositely charged layers is the glue holding together the entire assembly. The number of deposition steps determines the number of bilayers formed and therefore the thickness of the entire multilayer. Moreover, it is possible to incorporate inorganic nanoparticles if their surfaces are charged [100]. Therefore, layer-by-layer assembly allows fabricating nanoscopic functional multilayer systems with outstandingly high precision. Whereas at first smooth substrates had been functionalized in this way, the coating of colloidal polymer particles with a multilayer structure consisting of silica nanoparticles and polymeric polyelectrolytes was reported by Caruso et al. in 1998 [101]. It was further shown that hollow capsules with walls consisting of polymeric multilayers can be prepared if polymeric colloidal particles are used as sacrificial templates [102]. Calcination of colloidal polymer particles covered by a multilayer structure in which silica nanoparticles were incorporated led to the formation of hollow silica capsules [103]. Several excellent reviews deal with the fabrication of free-standing and three-dimensional nanostructures by layer-by-layer deposition [104].

It appears to be straightforward to fabricate polymeric tubular structures with complex but well-controlled wall morphologies and adjustable wall thickness by performing layer-by-layer deposition into porous hard templates. To this end, Ai et al. employed a pressure-filter-template technique to deposit poly(allylamine hydrochloride as anionic and poly(styrenesulfonic acid) as cationic component from aqueous solutions also containing NaCl into AAO with a D_p-value of about 200 nm. After starting the deposition sequence with the formation of a poly(allylamine hydrochloride) layer directly on the pore walls, stable but flexible nanotubes consisting of three bilayers with a wall thickness of 50–80 nm were obtained even after etching the hard template with aqueous NaOH solution [105]. The thickness of the nanotube walls was one order of magnitude larger than that of corresponding multilayer structures prepared on smooth substrates in which a bilayer has a thickness of a few nanometers. Using PC membranes with a D_p-value of 400 nm and a T_p-value of 10 μm whose walls were initially coated with poly(ethylenimine), poly(acrylic acid)/poly(allylamine hydrochloride) multilayers were deposited onto the pore walls in the presence of Cu^{2+} and then thermally cross-linked. Moreover, positively charged Au nanoparticles were incorporated in nanotube walls in alternation with four-layer polyelectrolyte structures, and negatively charged semiconductor nanoparticles in alternation with three-layer polyelectrolyte structures. Whereas the wall thickness of the nanotubes thus obtained, which was of the order of several tens of nanometers, could be adjusted by the number of successive deposition cycles, the functionality of the embedded inorganic nanoparticles was preserved [106]. The wall thicknesses of the nanotubes reported in this study were only slightly lager than those in smooth configurations, and the mechanical stability of the nanotubes depended on the number of bilayers their walls consisted of.

Layer-by-layer deposition into porous hard templates has meanwhile been extended to other polyelectrolyte pairs. For example, Ai et al. prepared polypyrrole/poly(allylamine hydrochloride) nanotubes consisting of six or twelve bilayers in PC membranes with a D_p-value of 400 nm that had initially been coated with poly(ethylenimine) [107]. Again, the observed value of the wall thickness of the nanotubes of some tens of nanometers was much larger than that of corresponding multilayer systems deposited on smooth substrates. However, a clear dependence of the wall thickness on the number of deposition cycles was found. Nanotubes consisting of dendrimers were fabricated by Kim and coworkers (Fig. 5.14a) [108]. Dendrimers, synthesized by stepwisely attaching another generation of low-molar mass building blocks to a parent structure, represent a class of functional materials which can be customized with an unrivaled precision. On the one hand, they can be employed as functional nanocontainers. On the other hand, they contain a well-defined number of terminal functional groups residing at their surface [109]. For the preparation of the dendrimer nanotubes, bilayers containing globular shaped, N,N-disubstituted hydrazine phosphorus-containing dendrimers [110] of the fourth generation having 96 terminal functional groups with either cationic

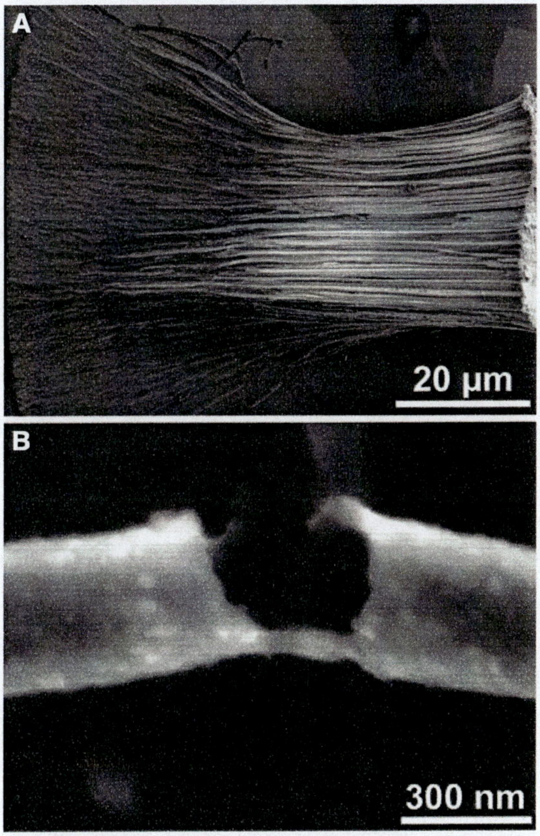

Fig. 5.14 SEM images of dendrimer nanotubes obtained by layer-by-layer deposition. (**a**) Broken nanotube; (**b**) array consisting of nanotubes exhibiting a gradient of their mechanical stability along their long axes. Reproduced from [108]. Copyright (2005): Wiley-VCH

$[G_4(NH^+Et_2Cl^-)_{96}]$ or anionic $[G_4(CH\text{-}COO^-Na^+)_{96}]$ character were deposited on the walls of AAO hard templates with a D_p-value of 400 nm. Since dendrimers can be considered as hard spheres, nanotubes consisting of dendrimeric polyelectrolytes might be useful if swelling or de-swelling needs to be minimized. The mechanical stability of the dendrimer nanotubes increased with the number of deposited bilayers. In the case of dendrimer nanotubes with high aspect ratios of the order of 200 that were prepared in hard templates with closed pore bottoms, their mechanical stability decreased with increasing distance to the pore opening. Whereas the nanotube segments initially located next to the pore openings with a length of about 35 μm were rigid, the nanotube segments farther away from the pore openings were prone to mechanical deformation due to the occurrence of capillary forces, which occur when nanofiber arrays dry after the wet-chemical etching of the hard template (Fig. 5.14b). Lu et al. reported the deposition of the hemoprotein human serum albumin (HSA) into AAO hard templates with a D_p-value of about 200 nm [111, 112]. Inversion of the charges borne by the HAS molecules was achieved by adjusting the pH-value of the solutions used for deposition to values below or above the isoelectric point of HSA. Therefore, in alternating deposition steps HSA could be deposited as a polycation or as a polyanion. Moreover, nanotubes whose walls consisted of HSA/phosholipid multilayers were prepared [111].

Extending the initial approach to exploit electrostatic interactions between polyelectrolytic building blocks for their rational arrangement in nanotube walls by layer-by-layer assembly, a number of modifications of this methodology based on different kinds of interactions has been developed. For example, the formation of hydrogen bonds between hydroxyl groups of poly(acrylic acid) and the nitrogen groups of poly(4-vinylpyridine) was exploited to coat the pore walls of PC membranes with multilayer structures consisting of these polyelectrolytes [113]. An advantage of this approach lies in the fact that solutions in organic solvents can be used to deposit the monolayers. Nanotubes consisting of poly(ethylenimine)/poly(styrene-*alt*-maleic anhydride) multilayers were obtained by connecting the alternating monolayers by amid bonds [114]. A similar approach was applied to incorporate the fluorescent compound 3,4,9,10-perylenetetracarboxylicdianhydride into nanotubes having multilayered walls with poly(ethylenimine) as a second component [115]. Since the fluorescent component retained its fluorescence, single nanotubes could be imaged by fluorescence microscopy, and the formation of the multilayer structure could be monitored by UV absorption spectroscopy. It was found that the absorption linearly increased along with the number of deposition cycles.

Hou et al. reported the fabrication of glutaraldehyde/protein nanotubes [116]. Using phosphorous-containing coupling agents, a first glutaraldehyde layer was grafted onto the pore walls of AAO hard templates with a D_p-value of 200 nm. Subsequently, a protein layer was bond with its free amino sites to the excess aldehyde functions of the glutaraldehyde layer, and in turn another glutaraldehyde layer to free amino sites of the proteins. For example, bioactive nanotubes could be fabricated by the repeated deposition of glutaraldehyde/glucose oxidase bilayers. The activity of the glucose oxidase in the liberated nanotubes increased along with the number of protein layers in their walls. However, inside the AAO hard templates the activity of the nanotubes decreased for more than three bilayers, because the accessibility of the protein molecules through the hollow channel inside the nanotubes became more and more limited as the diameter of the channel decreased with each additional layer. Also, Hou et al. showed that hemoglobin nanotubes that were produced in a similar manner exhibited heme electroactivity. Tian et al. used AAO hard templates activated with a poly(ethylenimine)/poly(sodium-4-styrenesulfonate) bilayer to deposit cytrochrome C/glutaraldehyde bilayers and obtained nanotubes in which the bioactivity and the electronic properties of cytochrome C were retained [117]. Hou et al. fabricated DNA nanotubes employing a hybridization-based layer-by-layer strategy. After the initial grafting of DNA strands on the pore walls with the aid of phosphorous-containing coupling agents, the hard templates were successively immersed into DNA solutions, allowing binding of further DNA strands to those already immobilized by hybridization [118].

Whereas layer-by-layer deposition into porous hard templates has been proven to be a promising access to precisely designed polymeric nanotubes that can functionalize hard templates or that can be released, two problems still need to be addressed. The first one is elucidating the structure formation of the polyelectrolyte layers inside hard templates. The significantly increased thickness of bilayers reported by Ai et al. [105] was also observed by other authors. Lee et al. found that the thickness of poly(allylamine hydrochloride)/poly(sodium-4-styrenesulfonate) multilayers deposited into porous PC membranes exceeded that of corresponding multilayers on smooth substrates obtained after the same number of deposition cycles. For example, 24.5 bilayers had a thickness of 250 nm within a hard template as compared to 155 nm on a smooth silicon wafer [119]. Alem et al. studied the layer-by-layer deposition of a pair of strong polyelectrolytes, namely poly(vinylbenzylammonium chloride) as a polycation and poly(sodium-4-styrenesulfonate) as a polyanion, into track-etched PC membranes with D_p-values ranging from 50 to 850 nm. The end-to-end distances of the polyelectrolyte chains were systematically varied by varying the molecular weight and the ionic strength of the solutions used for deposition [120]. Whereas the bilayers deposited on smooth substrates had a thickness of 1–3 nm, the first bilayer deposited into porous hard templates covering the pore walls had a thickness of 50–120 nm. Further deposition cycles led to only small increases in the thickness of the polyelectrolyte layers, which turned out to be nearly independent of the end-to-end distance of the polyelectrolytes and the ionic strength of the stock solutions used but strongly depended on the D_p-values of the hard templates. For small D_p-values, the thickness of the polyelectrolyte layers was proportional to D_p, whereas progressive deviations from this relationship were found for D_p-values larger than 250 nm. Based on geometric considerations, Alem et al. proposed a mechanism governing the growth of polyelectrolyte layers inside porous hard templates that involves the enrichment of the polyelectrolytes inside the pores. Hence, a dense, swollen polyelectrolyte gel fills pores and collapses upon drying.

The second issue that needs to be further addressed is the development of strategies for the anchoring of the first deposited layer onto the pore walls of the hard templates. This is particularly the case for the widely used AAO membranes, the pore walls of which consist of amorphous alumina containing water, electrolyte anions and positively charged defects (Section 5.2). Moreover, composition and distribution of the contaminations across the pore walls are inhomogeneous (see, for example, [19]). Therefore, isotropic etching steps performed to widen the pores of as-anodized, self-ordered AAO with an initial porosity of 10% [23] or below will change the properties of the pore walls and affect their reactivity. In the case of commercially available disordered AAO membranes with a D_p-value of 200 nm positively charged polyelectrolytes such as poly(ethylenimine) [114, 117] or human serum albumin at a pH value of 3.8 [111] could directly be deposited as the first layer. However, Dai et al. reported a procedure to coat the same type of hard templates that started with the deposition of poly(acrylic acid), hence with a polyanion, at a pH value of 4.0 [121]. The reports dealing with the surface chemistry of AAO are to a large extent inconsistent, and it appears that the surface properties of the pore walls largely depend on the anodization conditions and post-anodization treatments. Strategies to overcome the problems associated with the lack of knowledge of the properties of the hard templates are based on their modification by grafting anchor layers onto the pore walls. For example, Kim et al. used 3-aminopropyldimethylethoxysilane, a silane coupling agent, to generate a layer with a high density of positive charges on the walls of self-ordered porous alumina with a D_p-value of 400 nm [108]. Hou et al. [116, 118, 122] adapted a surface modification strategy based on a double layer of phosphorous-containing coupling agents initially introduced by Mallouk and coworkers [123] that allowed further layer-by-layer deposition mediated by specific chemical interactions.

First applications of polymeric nanotubes fabricated by layer-by-layer deposition have already been reported. Using track-etched PC membranes with D_p-values ranging from 400 to 800 nm functionalized with poly(allylamine hydrochloride)/poly(sodium-4-styrenesulfonate) bilayers, Lee et al. demonstrated reversible pH-induced hysteretic gating [119]. Membrane pores, the walls of which were covered with 18.5 bilayers, could be closed a pH-value of 2.5 by swelling the polyelectrolytes. The pores thus closed retained their closed state up to pH 9. At higher pH values, the swollen polyelectrolyte layer collapsed, and the pores switched to the open state. The switching behavior of the system could be customized by the number of bilayers deposited on the pore walls. The flux of pH-adjusted water through membrane was studied and indicated discontinuous swelling/deswelling behavior but lesser swelling in pores than on smooth substrates. Lee et al. suggested that swelling in the pores of hard templates is suppressed because of the curvature-induced stress generated by the volume expansion in a curved geometry.

Dai et al. reported a strategy for analyzing proteins by selective binding to antibodies in such a way that nonspecific adsorption and protein denaturation could be prevented. To this end, poly(acrylic acid)/protonated poly(allylamine) multilayers that are well known to resist nonspecific adsorption of proteins and to allow for covalent immobilization of arrays of active antibodies were coated on the walls of AAO hard templates with a D_p-value of 200 nm. Activation of the surface carboxyl groups of the poly(acrylic acid) with *N*-(3-dimethylaminopropyl)-*N*'-ethylcarbodiimide and *N*-hydroxysuccinimide enabled the covalent attachment of antibodies. A 500-fold increase in the surface area as compared to thin film configurations decreased the protein-microarray detection limit by two orders of magnitude [121]. Feng et al. adapted the procedure for the fabrication of dendrimer nanotubes reported by Kim et al. [108] to incorporate a graded bandgap structure similar to that previously reported by Franzl et al. for thin-film configurations on smooth substrates [124] into the walls of dendrimer nanotubes. The dendrimer layers acted as a rigid scaffold for the engineering of a multilayer configuration of inorganic semiconductor quantum dots having different diameters. Taking advantage of a fluorescence resonance energy transfer cascade from donor nanoparticles located near the outer surface on the nanotube walls to acceptor particles located near the inner surface of the nanotube walls, the hybridization of DNA strands grafted on the inner tube surface with complementary labeled DNA strands could be detected with significantly increased sensitivity [125].

5.7 Conclusion

Nanostubes consisting of soft matter or obtained by exploiting self-assembly processes in soft matter are being considered as functional building blocks for a broad range of device architectures, and some promising applications have already been reported. Control over their internal mesoscopic fine structure is the most crucial means of tailoring their properties. This is all the more the case as the supramolecular organization on mesoscopic length scales determines the properties of soft matter to a large extent. Nanoporous, shape-defining hard templates provide a two-dimensionally confined space in which self-organization processes such as crystallization, mesophase formation, and phase separation may result in supramolecular fine structures fundamentally different from those obtained in thin film configurations and in the bulk. A particular advantage of hard templates is the possibility to induce and manipulate self-assembly inside the shape-defining pores. Therefore, much more parameters allow tailoring the mesoscopic morphology of the nanofibers in hard-template based preparation processes than in procedures for their production not relying on the rigidity of confining pore walls. The supramolecular organization in the two-dimensional confinement of nanopores can be manipulated by the pore diameter, the nature of the pore walls, the composition of the infiltrated material, environmental conditions, and the thermal history of the sample. Moreover, nanotubes characterized by complex, multilayered walls are accessible by successive deposition steps into the hard templates, thus exploiting specific interactions between the deposited species and the material already deposited. The preparative approaches reviewed in this chapter allow generating complex and functional supramolecular structures in the walls of nanotubes, such as ordered mesoporous structures and microphase morphologies as well as uniform crystalline and liquid crystalline textures. The control over the supramolecular organization of the materials the nanotubes consist of is the prerequisite for the rational tailoring of peculiar mechanical, optical and electronic properties.

References

1. Wehrspohn, R. B., *Ordered porous nanostructures and applications.* (Springer, New York; [Great Britain], 2005).
2. Martin, C. R., *Science* **266**, 1961 (1994); Martin, C. R., *Chemistry of Materials* **8**, 1739 (1996); Huczko, A., *Applied Physics A-Materials Science & Processing* **70**, 365 (2000).
3. Cai, Z. H. and Martin, C. R., *Journal of the American Chemical Society* **111**, 4138 (1989).
4. Liang, W. B. and Martin, C. R., *Journal of the American Chemical Society* **112**, 9666 (1990).
5. Martin, C. R., Vandyke, L. S., Cai, Z. H., and Liang, W. B., *Journal of the American Chemical Society* **112**, 8976 (1990); Martin, C. R., *Advanced Materials* **3**, 457 (1991); Penner, R. M. and Martin, C. R., *Journal of the Electrochemical Society* **133**, 2206 (1986).
6. Parthasarathy, R. V. and Martin, C. R., *Chemistry of Materials* **6**, 1627 (1994).
7. Martin, C. R., *Accounts of Chemical Research* **28**, 61 (1995).
8. Menon, V. P., Lei, J. T., and Martin, C. R., *Chemistry of Materials* **8**, 2382 (1996).
9. Cepak, V. M. and Martin, C. R., *Chemistry of Materials* **11**, 1363 (1999).
10. Kohli, P. et al., *Science* **305**, 984 (2004).

11. Lakshmi, B. B. and Martin, C. R., *Nature* **388**, 758 (1997).
12. Jin, M. H. et al., *Advanced Materials* **17**, 1977 (2005); Xu, J. H., Li, M., Zhao, Y., and Lu, Q. H., *Colloids and Surfaces A-Physicochemical and Engineering Aspects* **302**, 136 (2007).
13. Chen, X., Steinhart, M., Hess, C., and Gösele, U., *Advanced Materials* **18**, 2153 (2006); Grimm, S. et al., *Small* **3**, 993 (2007).
14. Ott, A. W. et al., *Chemistry of Materials* **9**, 707 (1997); Xiong, G. et al., *Journal of Physical Chemistry B* **109**, 14059 (2005).
15. Apel, P., *Radiation Measurements* **34**, 559 (2001).
16. Crawford, G. P. et al., *Journal of Chemical Physics* **96**, 7788 (1992).
17. Fox, H. W., Hare, E. F., and Zisman, W. A., *Journal of Physical Chemistry* **59**, 1097 (1955).
18. Thompson, G. E. and Wood, G. C., *Nature* **290**, 230 (1981).
19. Chen, W., Yuan, J. H., and Xia, X. H., *Analytical Chemistry* **77**, 8102 (2005); Choi, J. et al., *Journal of Applied Physics* **94**, 4757 (2003); Vrublevsky, I., Jagminas, A., Schreckenbach, J., and Goedel, W. A., *Applied Surface Science* **253**, 4680 (2007).
20. Masuda, H. and Fukuda, K., *Science* **268**, 1466 (1995).
21. Masuda, H., Hasegwa, F., and Ono, S., *Journal of the Electrochemical Society* **144**, L127 (1997).
22. Masuda, H., Yada, K., and Osaka, A., *Japanese Journal of Applied Physics Part 2-Letters* **37**, L1340 (1998); Li, A. P. et al., *Journal of Applied Physics* **84**, 6023 (1998).
23. Nielsch, K. et al., *Nano Letters* **2**, 677 (2002).
24. Lee, W., Ji, R., Gösele, U., and Nielsch, K., *Nature Materials* **5**, 741 (2006).
25. De Gennes, P. G., *Reviews of Modern Physics* **57**, 827 (1985); Ausserré, D., Picard, A. M., and Léger, L., *Physical Review Letters* **57**, 2671 (1986); Léger, L. and Joanny, J. F., *Reports on Progress in Physics* **55**, 431 (1992).
26. Derjaguin, B. V. and Churaev, N. V., *Journal of Colloid and Interface Science* **49**, 249 (1974); Mate, C. M. and Novotny, V. J., *Journal of Chemical Physics* **94**, 8420 (1991); Israelachvili, J. N., *Intermolecular and surface forces*, 2nd ed. (Academic Press, New York, 1991).
27. Bernadiner, M. G., *Transport in Porous Media* **30**, 251 (1998); Everett, D. H. and Haynes, J. M., *Journal of Colloid and Interface Science* **38**, 125 (1972); Hammond, P. S., *Journal of Fluid Mechanics* **137**, 363 (1983); Lenormand, R., *Journal of Physics-Condensed Matter* **2**, SA79 (1990).
28. Steinhart, M. et al., *Science* **296**, 1997 (2002).
29. Steinhart, M., Wendorff, J. H., and Wehrspohn, R. B., *Chemphyschem* **4**, 1171 (2003); Steinhart, M., Wehrspohn, R. B., Gösele, U., and Wendorff, J. H., *Angewandte Chemie-International Edition* **43**, 1334 (2004).
30. Steinhart, M. et al., *Physical Review Letters* **97**, 027801 (2006).
31. Steinhart, M. et al., *Advanced Functional Materials* **15**, 1656 (2005).
32. Lau, S., Zheng, R. K., Chan, H. L. W., and Choy, C. L., *Materials Letters* **60**, 2357 (2006); Zheng, R. K. et al., *Chemical Communications*, 1447 (2005).
33. Zhang, M. F. et al., *Nano Letters* **6**, 1075 (2006).
34. She, X. L. et al., *Journal of Materials Research* **21**, 1209 (2006); She, X. L. et al., *Polymer Journal* **38**, 639 (2006).
35. Kim, E., Xia, Y. N., and Whitesides, G. M., *Nature* **376**, 581 (1995); Suh, K. Y., Kim, Y. S., and Lee, H. H., *Advanced Materials* **13**, 1386 (2001).
36. Moon, S. I. and McCarthy, T. J., *Macromolecules* **36**, 4253 (2003).
37. Kriha, O. et al., *Advanced Functional Materials* **17**, 1327 (2007).
38. Xiang, H. Q. et al., *Macromolecules* **37**, 5660 (2004).
39. Sun, Y. M. et al., *Macromolecular Rapid Communications* **26**, 369 (2005).
40. Song, G. J., She, X. L., Fu, Z. F., and Li, J. J., *Journal of Materials Research* **19**, 3324 (2004).
41. Li, J. J. et al., *Polymer Journal* **38**, 554 (2006).
42. Steinhart, M. et al., *Macromolecules* **36**, 3646 (2003).
43. Primak, S. V. et al., *Physical Review E* **65**, 4 (2002).

44. Ai, S. F. et al., *Colloids and Surfaces A-Physicochemical and Engineering Aspects* **275**, 218 (2006).
45. Binder, K., *Journal of Non-Equilibrium Thermodynamics* **23**, 1 (1998); Gelb, L. D., Gubbins, K. E., Radhakrishnan, R., and Sliwinska-Bartkowiak, M., *Reports on Progress in Physics* **62**, 1573 (1999).
46. Liu, A. J., Durian, D. J., Herbolzheimer, E., and Safran, S. A., *Physical Review Letters* **65**, 1897 (1990).
47. De Gennes, P. G., *Macromolecules* **14**, 1637 (1981); Lai, P. Y. and Binder, K., *Journal of Chemical Physics* **97**, 586 (1992); O'Shaughnessy, B. and Vavylonis, D., *Journal of Physics-Condensed Matter* **17**, R63 (2005).
48. Chen, J. T. et al., *Advanced Functional Materials* **16**, 1476 (2006).
49. Rodriguez, A. T. et al., *Journal of the American Chemical Society* **128**, 9276 (2006).
50. Rayleigh, L., *Philosophical Magazine Serié 5* **34**, 177 (1892); Quéré, D., Dimeglio, J. M., and Brochard-Wyart, F., *Science* **249**, 1256 (1990).
51. Chen, J. T., Zhang, M. F., and Russell, T. P., *Nano Letters* **7**, 183 (2007).
52. Shin, K. et al., *Macromolecules* **40** (18), 6617 (2007).
53. Wu, H., Wang, W., Yang, H. X., and Su, Z. H., *Macromolecules* **40**, 4244 (2007).
54. Zhi, L. J. et al., *Journal of the American Chemical Society* **127**, 12792 (2005).
55. Zhi, L. J. et al., *Angewandte Chemie-International Edition* **44**, 2120 (2005).
56. Woo, E., Huh, J., Jeong, Y. G., and Shin, K., *Physical Review Letters* **98**, 4 (2007).
57. Koenig, J. L., *Spectroscopy of polymers*, 2nd ed. (Elsevier, Amsterdam; Oxford, 1999).
58. Martin, C. R., Parthasarathy, R., and Menon, V., *Synthetic Metals* **55**, 1165 (1993).
59. Cai, Z. H. et al., *Chemistry of Materials* **3**, 960 (1991).
60. Keller, A., *Philosophical Magazine* **2**, 1171 (1957); Reiter, G. and Strobl, G. R., *Progress in understanding of polymer crystallization.* (Springer, Berlin; London, 2007); Strobl, G. R., *The physics of polymers: concepts for understanding their structures and behavior*, 3rd ed. (Springer, Berlin, 2007).
61. Keith, H. D. and Padden, F. J., *Journal of Applied Physics* **34**, 2409 (1963); Bassett, D. C. and Vaughan, A. S., *Polymer* **26**, 717 (1985); Granasy, L. et al., *Physical Review E* **72**, 15 (2005).
62. Turnbull, D., *Journal of Applied Physics* **21**, 1022 (1950); Turnbull, D., *Journal of Chemical Physics* **18**, 198 (1950).
63. Laschat, S. et al., *Angewandte Chemie-International Edition* **46**, 4832 (2007).
64. Wu, J. S., Pisula, W., and Müllen, K., *Chemical Reviews* **107**, 718 (2007).
65. Pisula, W. et al., *Journal of the American Chemical Society* **128**, 14424 (2006).
66. Kastler, M. et al., *Small* **3**, 1438 (2007).
67. Zhi, L. J. et al., *Advanced Materials* **17**, 1492 (2005).
68. Liu, Q. Y. et al., *Journal of Physical Chemistry C* **111**, 7298 (2007).
69. Barrett, C. et al., *Chemistry of Materials* **19**, 338 (2007).
70. Palermo, V. et al., *Philosophical Transactions of the Royal Society A-Mathematical Physical and Engineering Sciences* **365**, 1577 (2007).
71. Cahn, J. W., *Acta Metallurgica* **9**, 795 (1961); Hashimoto, T., in *Structure and properties of polymers*, edited by E. L. Thomas (Wiley-VCH, Weinheim, 1993), Vol. 12, p. 251.
72. De Gennes, P. G., *Journal of Chemical Physics* **72**, 4756 (1980).
73. Jones, R. A. L. et al., *Physical Review Letters* **66**, 1326 (1991).
74. Steinhart, M. et al., *Advanced Materials* **15**, 706 (2003); Luo, Y. et al., *Nano Letters* **4**, 143 (2004).
75. Nielsch, K., Castano, F. J., Ross, C. A., and Krishnan, R., *Journal of Applied Physics* **98**, 6 (2005); Nielsch, K. et al., *Advanced Engineering Materials* **7**, 217 (2005).
76. Tanaka, K., Takahara, A., and Kajiyama, T., *Macromolecules* **31**, 863 (1998); Tanaka, K., Kajiyama, T., Takahara, A., and Tasaki, S., *Macromolecules* **35**, 4702 (2002).
77. Walheim, S. et al., *Macromolecules* **30**, 4995 (1997); Buck, E. and Fuhrmann, J., *Macromolecules* **34**, 2172 (2001); Harris, M., Appel, G., and Ade, H., *Macromolecules* **36**, 3307 (2003).

78. Kresge, C. T. et al., *Nature* **359**, 710 (1992); Attard, G. S., Glyde, J. C., and Göltner, C. G., *Nature* **378**, 366 (1995); Lu, Y. F. et al., *Nature* **389**, 364 (1997).
79. Templin, M. et al., *Science* **278**, 1795 (1997).
80. Zhao, D. Y. et al., *Science* **279**, 548 (1998).
81. Joo, S. H. et al., *Nature* **412**, 169 (2001); Ryoo, R., Joo, S. H., Kruk, M., and Jaroniec, M., *Advanced Materials* **13**, 677 (2001); Liang, C. D. et al., *Angewandte Chemie-International Edition* **43**, 5785 (2004); Zhang, F. Q. et al., *Journal of the American Chemical Society* **127**, 13508 (2005); Liang, C. D. and Dai, S., *Journal of the American Chemical Society* **128**, 5316 (2006); Zhang, F. Q. et al., *Journal of the American Chemical Society* **129**, 7746 (2007).
82. Ying, J. Y., Mehnert, C. P., and Wong, M. S., *Angewandte Chemie-International Edition* **38**, 56 (1999); Davis, M. E., *Nature* **417**, 813 (2002); Kickelbick, G., *Small* **1**, 168 (2005); Lu, A. H. and Schüth, F., *Advanced Materials* **18**, 1793 (2006); Kanatzidis, M. G., *Advanced Materials* **19**, 1165 (2007); Wan, Y., Shi, Y. F., and Zhao, D. Y., *Chemical Communications*, 897 (2007); Wan, Y. and Zhao, D. Y., *Chemical Reviews* **107**, 2821 (2007).
83. Yang, Z. L. et al., *Angewandte Chemie-International Edition* **42**, 4201 (2003).
84. Liang, Z. J. and Susha, A. S., *Chemistry-A European Journal* **10**, 4910 (2004).
85. Yao, B., Fleming, D., Morris, M. A., and Lawrence, S. E., *Chemistry of Materials* **16**, 4851 (2004).
86. Zhu, W. P., Han, Y. C., and An, L. J., *Microporous and Mesoporous Materials* **84**, 69 (2005).
87. Chae, W. S. et al., *Chemistry of Materials* **17**, 5651 (2005).
88. Lu, Q. Y., Gao, F., Komarneni, S., and Mallouk, T. E., *Journal of the American Chemical Society* **126**, 8650 (2004).
89. Wang, D. H. et al., *Chemical Communications*, 166 (2005).
90. Wu, Y. Y. et al., *Nature Materials* **3**, 816 (2004).
91. Wu, Y. Y. et al., *Nano Letters* **4**, 2337 (2004).
92. Wang, K. et al., *Advanced Materials* **19**, 3016 (2007).
93. Cott, D. J. et al., *Journal of the American Chemical Society* **128**, 3920 (2006).
94. Zheng, M. et al., *Carbon* **45**, 1111 (2007).
95. Steinhart, M. et al., *Chemistry of Materials* **19**, 2383 (2007).
96. Klotz, M. et al., *Chemistry of Materials* **12**, 1721 (2000).
97. Grosso, D. et al., *Chemistry of Materials* **13**, 1848 (2001).
98. Yoo, S. J., Ford, D. M., and Shantz, D. F., *Langmuir* **22**, 1839 (2006).
99. Decher, G. and Hong, J. D., *Makromolekulare Chemie-Macromolecular Symposia* **46**, 321 (1991); Lvov, Y., Decher, G., and Möhwald, H., *Langmuir* **9**, 481 (1993); Decher, G., *Science* **277**, 1232 (1997).
100. Feldheim, D. L. et al., *Journal of the American Chemical Society* **118**, 7640 (1996); Schmitt, J. et al., *Advanced Materials* **9**, 61 (1997).
101. Caruso, F. et al., *Journal of the American Chemical Society* **120**, 8523 (1998).
102. Donath, E. et al., *Angewandte Chemie-International Edition* **37**, 2202 (1998).
103. Caruso, F., Caruso, R. A., and Möhwald, H., *Science* **282** (5391), 1111 (1998).
104. Jiang, C. Y. and Tsukruk, V. V., *Advanced Materials* **18**, 829 (2006); Caruso, F., *Advanced Materials* **13** (1), 11 (2001); Li, J. B. and Cui, Y., *Journal of Nanoscience and Nanotechnology* **6**, 1552 (2006).
105. Ai, S. F., Lu, G., He, Q., and Li, J. B., *Journal of the American Chemical Society* **125**, 11140 (2003).
106. Liang, Z. J., Susha, A. S., Yu, A. M., and Caruso, F., *Advanced Materials* **15**, 1849 (2003).
107. Ai, S. F. et al., *Macromolecular Rapid Communications* **26**, 1965 (2005).
108. Kim, D. H. et al., *Small* **1**, 99 (2005).
109. Frechet, J. M. J., *Science* **263**, 1710 (1994); Tomalia, D. A., Naylor, A. M., and Goddard, W. A., *Angewandte Chemie-International Edition in English* **29**, 138 (1990); Bosman, A. W., Janssen, H. M., and Meijer, E. W., *Chemical Reviews* **99**, 1665 (1999).
110. Majoral, J. P. and Caminade, A. M., *Chemical Reviews* **99**, 845 (1999).
111. Lu, G., Al, S. F., and Li, J. B., *Langmuir* **21**, 1679 (2005).

112. Lu, G., Komatsu, T., and Tsuchida, E., *Chemical Communications*, 2980 (2007).
113. Tian, Y. et al., *Chemistry-A European Journal* **12**, 4808 (2006).
114. Tian, Y. et al., *Journal of Nanoscience and Nanotechnology* **6**, 2072 (2006).
115. Tian, Y., He, Q., Tao, C., and Li, J. B., *Langmuir* **22**, 360 (2006).
116. Hou, S. F., Wang, J. H., and Martin, C. R., *Nano Letters* **5**, 231 (2005).
117. Tian, Y., He, Q., Cui, Y., and Li, J. B., *Biomacromolecules* **7**, 2539 (2006).
118. Hou, S. F., Wang, J. H., and Martin, C. R., *Journal of the American Chemical Society* **127**, 8586 (2005).
119. Lee, D. et al., *Journal of the American Chemical Society* **128**, 8521 (2006).
120. Alem, H. et al., *Macromolecules* **40**, 3366 (2007).
121. Dai, J. H., Baker, G. L., and Bruening, M. L., *Analytical Chemistry* **78**, 135 (2006).
122. Hou, S. F. et al., *Journal of the American Chemical Society* **126**, 5674 (2004).
123. Lee, H. et al., *Journal of the American Chemical Society* **110**, 618 (1988).
124. Franzl, T. et al., *Nano Letters* **4**, 1599 (2004).
125. Feng, C. L. et al., *Advanced Materials* **19**, 1933 (2007).

Chapter 6
Biosensing with Nanopores and Nanotubes

Lindsay T. Sexton, Lloyd P. Horne, and Charles R. Martin

Abbreviations

A	Avogadro's number
C	Concentration
D	Diffusion coefficient
d_a	Diameter of analyte
d_p	Diameter of pore
d_b	Base opening diameter of a conical nanopore
d_{bf}	Final base diameter of a conical nanopore, after the second etch step
d_{b1}	Base diameter of a conical nanopore after the first etch step
d_t	Tip opening diameter of a conical nanopore
d_{tf}	Final tip diameter of a conical nanopore, after the second etch step
d_{t1}	Tip diameter of a conical nanopore after the first etch step
E	Electric field strength
E_{app}	Applied transmembrane potential
e	Electronic charge
F	Faraday's constant
f_p	Current-pulse frequency
G	Ionic conductance (in Siemens, S)
G_f	Ionic conductance of the nanopore after the second etch step (S)
I_f	Ion current value at which the second etch step is stopped
i	Steady-state nanopore current
Δi	Current pulse magnitude
J	Electrophoretic flux
L	Length of nanopore/thickness of membrane
L_{eff}	Effective length
η	Solution viscosity

C.R. Martin (✉)
Department of Chemistry, Center for Research at the Bio/Nano Interface, University of Florida, Gainesville, FL 32611-7200, USA
e-mail: crmartin@chem.ufl.edu

O. Hayden, K. Nielsch (eds.), *Molecular- and Nano-Tubes*,
DOI 10.1007/978-1-4419-9443-1_6,

R	Gas constant
σ	Specific conductivity (S/cm)
T	Temperature
τ	Current-pulse duration
τ_b	Time required for blockage of ion current
ν	Electrophoretic velocity
ν_B	Bulk etch rate
ν_T	Track etch rate
Δx	Change in base/tip diameter
z	Charge

6.1 Introduction

Recent advances in nanotechnology have led to the development of new methods for fabricating membranes containing single, nanometer-sized pores. One potential application for such single-nanopore membranes is in biosensing. In particular, there has been a great deal of recent interest in using nanopores as the sensing element in resistive-pulse sensors [1–92]. The resistive pulse sensing method [1–4], which is sometimes referred to as stochastic sensing [1–19], entails mounting a membrane containing a single nanopore between two halves of an electrochemical cell filled with an electrolyte solution. A transmembrane potential is applied, and the resulting ion current flowing through the electrolyte-filled nanopore is recorded versus time. As an analyte, with dimensions comparable to the nanopore diameter, is driven through the pore a momentary block in the ion current is observed, yielding a downward current-pulse. The concentration of the analyte can be determined from the frequency of these current-pulse events and the identity of the analyte is encoded in the magnitude and duration of the current pulse [1–4].

Current work in the field of resistive-pulse sensing is aimed at the detection and characterization of molecules, ions and biopolymers [1–92]. Sensing of such molecular-sized analytes is possible if the diameter of the nanopore sensor element is of molecular dimensions. Nanopores in both biological [6–33] and artificial [65–93] membranes have been used to sense such analytes. In particular, a number of prototype resistive-pulse sensing devices have been developed from biological nanopores [6–33]. This biological-nanopore work will be briefly reviewed in a later section. The techniques that are currently being used to prepare artificial nanopores in synthetic membranes will also be reviewed.

The Martin Group is developing resistive-pulse sensors based on conically shaped nanopores embedded in chemically and mechanically stable polymeric membranes. This chapter provides an overview of this work. The first criterion for designing a practical sensing device from an artificial nanopore is developing methods for reproducibly preparing the nanopore. A two-step track-etching process has been developed that allows for single conical nanopores to be fabricated reproducibly. Means for controllably varying the conical nanopore geometry (e.g., cone angle) by varying parameters of the etching process have also been developed.

Conical nanopores exhibit unique properties that make them well suited for the resistive-pulse sensing application. These properties as they apply to sensor design and development will be discussed. Methods for modifying conical nanopores so that they function as biological ion channel mimics will also be discussed. Prototype sensing devices based on conical nanopores will be described. The chapter will conclude by considering the advantages of, and the future outlook for, this type of nanopore sensor.

6.2 Why Nanopores?

The growing interest surrounding nanopores arises impart because of the critical roles biological nanopores play in many physiological processes of living organisms [51, 52]. Biological nanopores and nanochannels are present in the cellular membranes of all living cells. These channels are formed by membrane proteins that span the entire thickness of the cell membrane (~4 nm). They are the primary device used by cells to communicate with other cells [51, 52]. Intercellular communication occurs through transport of ions and neutral molecules through the channels [51, 52]. The transport of ions by way of the protein channels is often a highly selective and controlled process. Controlled transport can be achieved because biological ion channels are often selective for certain ions and open and close through gated mechanisms. Gating of ion channels occurs when they open and close in response to certain stimuli, such as deformations in the cell membrane, the presence of a ligand or some other signaling molecule, or changes in membrane potential [51, 52]. For example, with voltage-gated ion channels the opened and closed states are dependent on membrane potential (Fig. 6.1) [53]. When ion channels are in the opened or

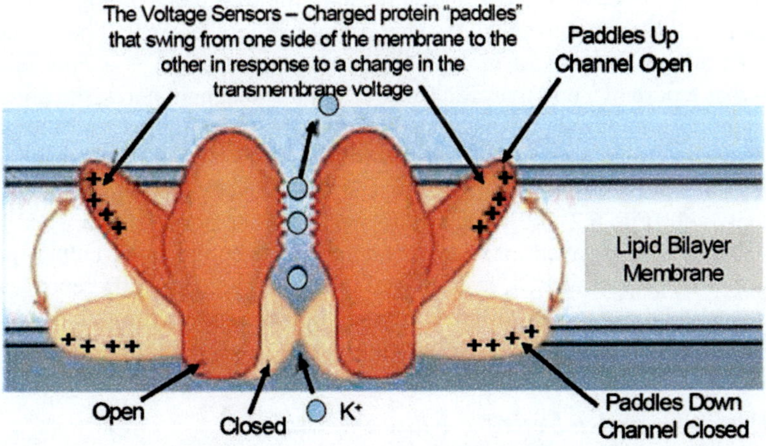

Fig. 6.1 An example of a voltage-gated biological ion channel. The electromechanical gating mechanism of the potassium voltage-gated ion channel. Reprinted by permission from Macmillan Publishers Ltd: *Nature* [53], copyright 2003

"on" state, ions are allowed to pass through the channels and when in the closed or "off" state, ion transport is blocked.

Many of the transport properties of these biological ion channels are not well understood. Investigating these properties can often be a challenge with biological systems due to their fragile nature. Constructing artificial nanopore devices with transport properties similar to those of biological ion channels could give new insight into the physical and chemical principles of biological ion channel operations [5]. Already, very recent advances in single artificial nanopore design are shedding light on the mechanisms by which naturally occurring ion channels function [54–64].

Another motivation for studying transport in nanopore membranes comes from the possible implementation of these membranes into single-molecule chemical and biochemical sensing devices. In previous work, it has been shown that single biological transmembrane protein nanopores embedded in lipid bilayer membranes can function as a single-molecule sensing device using the resistive-pulse sensing method (*vide infra*) [6–33]. Biological nanopores have proven to function as extremely versatile and selective resistive-pulse sensors. It has been recently shown that artificial single nanopore systems can also be used as platforms for single-molecule resistive-pulse sensing devices [2–5, 36–39, 66–92]. Currently, there is a large research effort focused towards fabricating single nanopores in synthetic materials [40–48, 65–88].

6.3 Operating Principles of Nanopore-Based Sensors

6.3.1 General Experimental Setup

In general, nanopore-based resistive-pulse sensors employ either a biological or artificial nanopore which is embedded in a supporting membrane. As will be discussed in detail below, in the case of biological pores, the membrane is a lipid bilayer coated across a pore in a polymer film [6–35]. For artificial nanopores, the pore can be fabricated in a multitude of materials including polymers, glass, and inorganic materials, such as silicon nitride and silicon wafers [36–50, 65–93]. The nanopore-containing membrane is immobilized between two halves of an electrolyte-filled conductivity cell (Fig. 6.2) [43], and electrolyte floods the pore. An electrode, typically Ag/AgCl, is immersed into each chamber of the conductivity cell. A patch clamp amplifier, or potentiostat, is employed to apply a potential difference across the nanopore and measure the current carried by ionic migration through the pore.

6.3.2 Resistive-Pulse Sensing Fundamentals

In the absence of analyte molecules a steady-state ion current is achieved, which is free of any resistive-current pulses (Fig. 6.3a). Analyte molecules are then added to

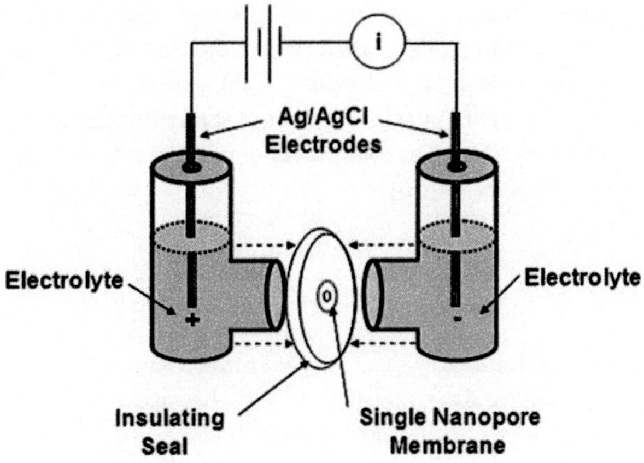

Fig. 6.2 General experimental setup for resistive-pulse sensing

one of the electrolyte solutions, and if the analyte is charged, it is driven by electrophoresis through the nanopore. If the analyte is uncharged, diffusion will drive it into the pore. If we assume a simple Coulter-counter model [1, 4], then, while in the pore, the molecule displaces a volumetric fraction of electrolyte that is equivalent to the molecular volume. This displacement results in a transient increase in nanopore

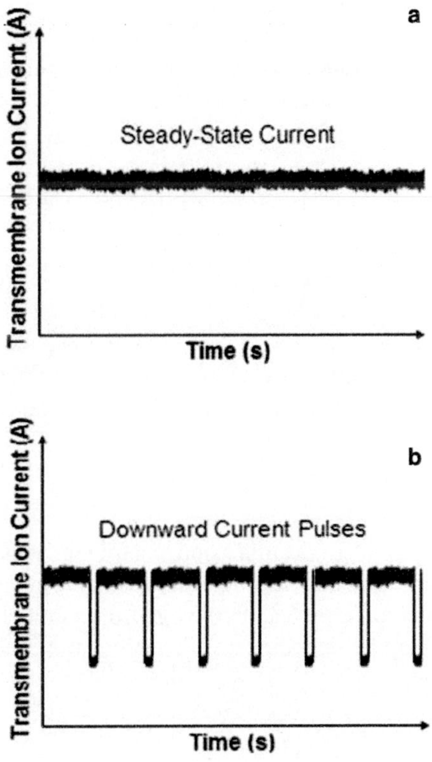

Fig. 6.3 Generalized, schematic drawing of (**a**) steady-state background current in the absence of analyte, and (**b**) downward current pulses due to analyte translocation and detection

resistance, and a corresponding transient decrease in transmembrane ion current, for the duration of the molecule's residence in the pore (Fig. 6.3b). The magnitude of the decrease in ion current is determined by the volume of solution displaced and hence reveals information about the volume of the translocating analyte. Again, assuming the Coulter-counter model, the magnitude of the resulting current pulse (Δi) is given by [4, 36],

$$\frac{\Delta i}{i} = S(d_p, d_a)\frac{{d_a}^3}{L{d_p}^2} \tag{6.1}$$

where, i is the steady-state current of the nanopore, d_a is the diameter of the translocating analyte, d_p is the diameter of the nanopore, L is the length of the pore, and $S(d_p, d_a)$ is a correlation factor that depends on the relative values of d_p and d_a [3, 36].

If the analyte is charged and driven through the nanopore by electrophoresis, electrophoretic transport theory can be used to make predictions about the current-pulse duration and frequency [90]. For example, the electrophoretic velocity, ν, of a charged analyte molecule is given by [94],

$$\nu = \frac{|z|\, eE}{6\pi\eta r} \tag{6.2}$$

where z is the charge, e is the electronic charge, E is the electric field strength, η is the solution viscosity, and r is the radius of the molecule. The translocation time, or duration of the resistive pulse, τ, can be calculated by taking the inverse of both sides of this equation and then multiplying by the length of the pore, L [90].

$$\tau = \frac{6\pi\eta r L}{zeE} \tag{6.3}$$

Equation (6.3) indicates that the current-pulse duration is inversely proportional to the electric field strength, and this has been observed experimentally for a DNA analyte in a track-etched (*vide infra*) nanopore sensor.

The current-pulse frequency, f_p (molecules translocating the pore per second) for a charged analyte can be calculated from the electrophoretic flux, J [90],

$$J = \frac{-zFDCE}{RT} \tag{6.4}$$

where D is the diffusion coefficient and C is the concentration of the ion, F is Faraday's constant, R is the gas constant and T is the temperature. Multiplying both sides of Eq. (6.4) by Avogadro's number (A) and the cross sectional area of the nanopore opening ($\pi {r_t}^2$) converts this equation to [90].

$$f_p = \frac{\mathrm{zFDCE}\pi {r_t}^2\mathrm{A}}{RT} \tag{6.5}$$

Equation (6.5) predicts that the current-pulse frequency is linearly related to analyte concentration, which has also been observed experimentally [13, 14, 90]. For a fixed concentration of analyte, Eq. (6.5) indicates that current-pulse frequency increases linearly with E. Examples of these linear relationships are shown in the applications section.

6.4 Nanopore-Based Sensors

6.4.1 Biological Nanopore Resistive-Pulse Sensors

A large body of work exists in which biological nanopores are utilized as the sensing element in electrical detection systems, namely, resistive-pulse sensing devices. Such resistive-pulse sensing devices have served as the impetus for the development of artificial nanopore sensors. The biological nanopore resistive-pulse sensors are typically comprised of a single transmembrane protein embedded in a planar lipid bilayer support. Membrane proteins that have been utilized as the sensing element in resistive-pulse sensing devices include α-hemolysin [6–30], maltoporin [31, 32], and outer membrane protein F (OmpF) [33], with α-hemolysin being the most common [6–30].

The α-hemolysin protein channel, seen in Fig. 6.4, is commonly used in biological nanopore resistive-pulse sensors for several reasons. The first reason being that the membrane protein has been well studied and characterized [34]. The protein

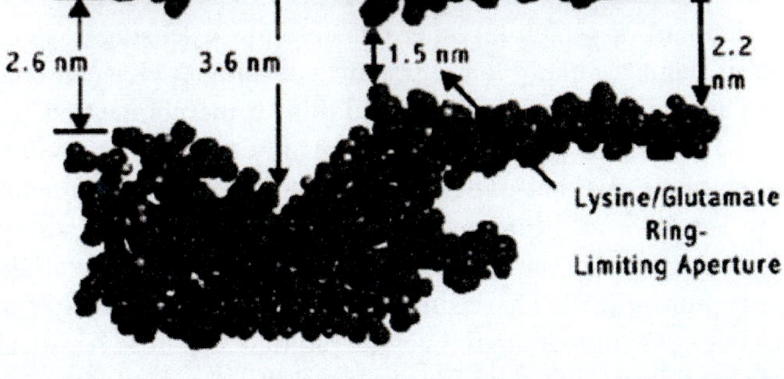

Fig. 6.4 Cross-section of the α-hemolysin protein nanopore showing the varying dimensions throughout the length of the pore. Reprinted with permission from [30]. Copyright 2002 American Chemical Society

channel also self-assembles into lipid bilayer membranes, which results in the formation of reproducible nanopore sensors. Another advantage of α-hemolysin is that it stands up well to modifications. The channel formed through the lipid bilayer is very stable, as is the protein itself, which is capable of withstanding temperatures up to 65°C [6]. These features have allowed for the α-hemolysin pore to be designed and engineered to detect and characterize specific analytes [6–20]. The conductance through the α-hemolysin pore is also large enough so that analytes passing through the channel can be easily detected [7].

A diverse number of analytes have been detected and characterized with biological nanopores. α-Hemolysin nanopores, either in their wild state or engineered form, have been used to detect enantiomers of drug molecules [8], DNA [9, 10, 23, 25–30], nitroaromatic compounds [11], metal ions [12, 13], small organic molecules [14], anions [15], proteins [16, 17], and polymers [22]. Maltoporin channels have been used to investigate the translocation of sugar molecules [31, 32], and the OmpF protein channels have been used to study the mode of antibiotic transport with the molecule ampicillin [33].

The α-hemolysin-based resistive-pulse detection system offers two main advantages. The first advantage is in the reproducibility in nanopore dimensions. Since α-hemolysin self-assembles into the lipid bilayer support, the dimensions and geometry of the channel are the same from sample to sample. This allows for consistent and reproducible sensing from sample to sample. The second advantage offered by the α-hemolysin nanopore is the selectivity imparted through engineering of the protein. Bayley and coworkers have performed numerous modifications to α-hemolysin through genetic engineering and chemical modification [6–20]. These modifications have allowed for selective sensors to be designed. For instance, Braha et al. prepared a sensor selective for divalent metal ions via genetic engineering of the α-hemolysin protein [13]. The engineered nanopore contained histidine residues, which projected into the lumen of the pore. Histidine residues can form complexes with a variety of divalent metal ions. When incorporated into the nanopore, each metal ion yielded a unique signal that was characteristic of the binding between the surface-bound histidine residues and the translocating metal ion.

The work done with the α-hemolysin nanopore has been extremely influential in the development of nanopore-based resistive-pulse sensors. This work currently stands as the benchmark to which all other resistive-pulse sensing devices are evaluated. However, it seems unlikely that a practical sensing device can be developed from this pore. This is because the planar lipid bilayer membrane that supports the nanopore is very fragile and typically stable for only a few hours before rupture. Such bilayer membranes cannot endure a wide range of pHs, temperatures, and solvents and are sensitive to vibrations [3, 7]. Another drawback suffered by the bilayers is that they can only support applied transmembrane potentials of about 200 mV before rupturing [35]. These limitations have sparked a major interest for researchers to develop synthetic analogues of biological pores. An ideal artificial nanopore sensor would have the same sensing capabilities and offer the same advantages as the biological nanopores, but show chemical and mechanical stability over a wide range of conditions.

6.4.2 Artificial Nanopores from Microlithographic Methods

A variety of methods and technologies have been employed to fabricate single, artificial nanopore sensors in synthetic materials. Currently, the most widely used methods for single nanopore fabrication utilize ion or electron milling techniques [65–77]. Ion or electron milling is a process in which a high energy ion or electron beam bombards a surface and causes the material to be sputtered away. The materials most commonly used in this fabrication technique include silicon nitride and silicon oxide [65–77]. Resistive-pulse sensors based on such nanopores have been used to study mainly DNA [66–69, 71–76], but some protein sensing [77] work has also been reported. However, nanopores prepared with this technique often have ill-defined geometries [72–74] and surface properties [68]. Reproducibly fabricating these pores can also be challenging [4]. This particular fabrication method also requires the use of expensive and specialized equipment, and can be labor intensive and time consuming.

Classic lithographic methods, specifically photolithography and electron beam lithography, combined with micromolding techniques is another approach that is currently being used to fabricate single nanopores and channels [78–81]. Saleh and Sohn have utilized this fabrication scheme to produce nanochannels in both a quartz and poly(dimethylsiloxane) (PDMS) substrate. These channels have been used to sense colloidal particles [78] and DNA [79] via resistive-pulse sensing. Size-based selectivity has also been demonstrated with nanopores created using this technique. These pores were used to directly detect the binding of antigens to antibody-coated colloidal particles [80, 81]. Devices prepared with this technique offer the advantages that they are stable, reusable, reproducible and the pore dimensions can be controlled. The technique is also less expensive and faster than ion and electron milling techniques. The main drawback encountered with this technique is the minimum pore size available and the length of the nanopores (3–10 μm). The smallest diameter pores that have been produced are around 150 nm in diameter, which limits the size of analytes that can be detected.

Single carbon nanotubes embedded in an epoxy membrane have also been used to form single nanochannels [4, 82–86]. These devices have been used as the sensing element for the detection of nanoparticles and DNA [4, 82–86]. The nanopores prepared with this method have well-defined physical and chemical characteristics, and there is little variability in the size and shape of pores prepared from the same carbon nanotube. The embedded carbon nanotubes are also free from surface charge, which means transport of analytes through the nanotube sensor is only due to electrophoresis, and not due to electroosmotic flow [83]. While it has not yet been demonstrated, it should be possible to modify the surface of the carbon nanotubes in order to introduce selectivity into these devices. The main drawback to this technique is again in the limiting diameter of the pore. With this technique it is difficult to achieve single nanotube assemblies with diameters smaller than about 100 nm. This again limits the size of analytes that can be detected with this device. This process for single nanotube fabrication also requires a great deal of skill and is labor intensive.

A femtosecond-pulsed laser-based technique has been developed to create single pores in glass, which have been used to detect immune complexes [87]. This method produces stable nanopores in a robust membrane system. Such nanopores have been used to detect analytes in the low nanomolar range. The primary disadvantage of this method is that it yields pores with relatively large diameters, >500 nm.

6.4.3 Artificial Nanopores from the Track-Etch Method

Base etching of silicon wafers [88] and track etching of polymer membranes [40–48] are other methods that have been used to fabricate single nanopores. The nanopore work discussed here has utilized the track-etch method to prepare conically-shaped nanopores in polymer membranes. Compared to other single-nanopore fabrication techniques, the track-etch method is relatively simple and straightforward. Most importantly, the tracked membrane that is used to make the nanopore (see below) can be obtained commercially from Gesellschaft fur Schwerionenforschung (GSI) in Darmstadt, Germany.

6.5 Fabrication of Artificial Nanopores in Polymer Membranes

The Track-Etch Method. The track-etch method has been practiced commercially for decades to prepare multi-pore membranes that are used, for example, for filtration applications or as templates for the deposition of other materials. This method allows for micro- and nano-meter sized pores to be prepared with various dimensions and geometries. Membranes prepared by the track-etch method are created by first bombarding the membranes with a beam of high-energy particles (>1 MeV/nucleon) from a nuclear reactor or cyclotron (Fig. 6.5a). This process creates linear damage tracks that span the entire thickness of the membrane, which is typically 5–10 μm (Fig. 6.5b). The damage tracks are then chemically etched to create the pores (Fig. 6.5c). In the commercial process, the ion-tracked membrane is simply immersed into the etching solution, and the damage tracks are etched

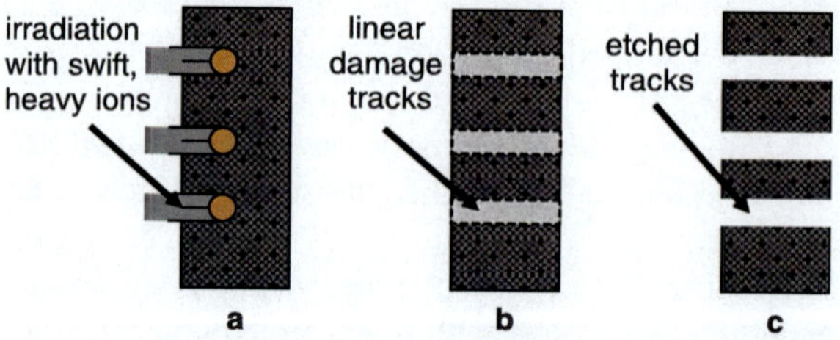

Fig. 6.5 Principle of the ion-track etching technique. (**a**) Swift, heavy ions accelerated through membrane material. (**b**) Heavy ions form damage tracks as they pass through the material. (**c**) Selective, chemical etching results in a membrane containing pores or channels [40]

from both faces of the membrane (isotropic etching). This yields cylindrical pores through the membrane; the pore diameter is determined by the type of material, concentration of etchant, etching time and etchant temperature. The pore density is determined by the exposure time to the particle beam. Multi-pore membranes with pore diameters ranging from 10 nm to 10 μm and pore densities ranging from 10^5 to 10^9 pores/cm^2 are commercially available [95].

For resistive-pulse sensing applications it is necessary to obtain membranes containing only a single damage track. A procedure for preparing such single-track membranes was developed at Gesellschaft fur Schwerionenforschung (GSI) in Darmstadt, Germany. They use a defocused heavy ion beam to irradiate the polymer membrane with a single ion [42]. Single-ion irradiation is achieved by placing a shutter in between the ion beam and the membrane and an ion detector behind the membrane. When the ion detector registers that a single ion has traversed the membrane the shutter is closed, thus precluding any further exposure of the membrane to the beam.

For reasons that will be discussed in detail below, conically-shaped nanopores are particularly advantageous for resistive-pulse sensing. The etching process for preparing such conically-shaped nanopores was first developed by Apel et al. [43]. In this process, the single ion-tracked membrane is first placed between two halves of the same electrochemical cell described above (Fig. 6.2) [43]. An etching solution is added to one side of the cell and a stopping solution is added to the other side. The damage track is preferentially etched from the face of the membrane in contact with the etching solution. When the etchant breaks through to the other side of the membrane the stopping solution neutralizes the etchant. The etching process is stopped by placing the nanopore membrane in water or the stopping solution. The resulting nanopore is conical in shape with the large-diameter (base) opening facing the etch solution and the small-diameter (tip) opening facing the stopping solution (Fig. 6.6).

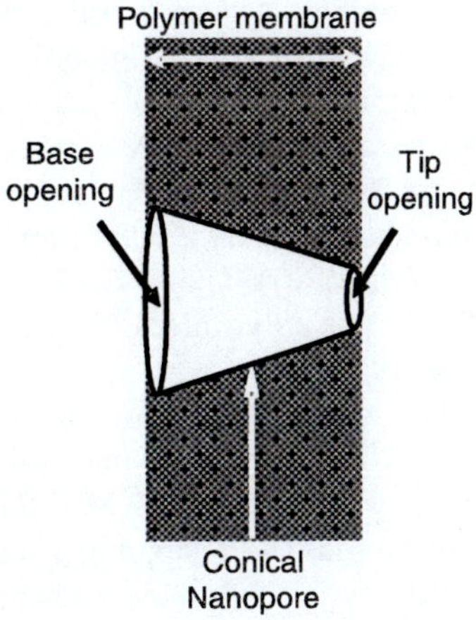

Fig. 6.6 Schematic of a conical nanopore in a polymer membrane showing the base diameter and tip diameter (*drawing not to scale*) [46]

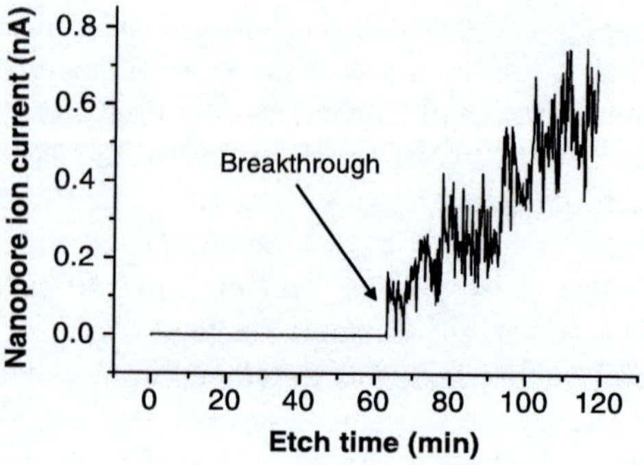

Fig. 6.7 Plot of current versus time during the first step, anisotropic etching of a conical nanopore. The moment of breakthrough is marked by the sudden rise in current

During the etching process platinum wire electrodes are placed on either side of the membrane and a potential is applied. The electrodes are arranged so that the anode is in the half-cell containing the etch solution. The applied transmembrane potential serves several purposes. It provides a means for monitoring the transmembrane current during etching, thereby allowing the determination of when the etchant has broken through the polymer membrane. Initially the current is zero, however, breakthrough is marked by a sudden increase in the ionic current (Fig. 6.7). The steady increase in transmembrane current is associated with an increase in the diameter of the tip opening as the etch proceeds.

After breakthrough, the applied potential also causes an electro-stopping process to occur [43]. To understand this process, consider the case of poly(ethylene terephthalate) (PET). The etching solution used for PET is 9 M NaOH and the stopping solution is 1 M formic acid and 1 M KCl. Placing the anode in the etch solution causes the OH^- etchant to be electrophoretically driven away from the nanopore tip opening – electro-stopping. This process further promotes the asymmetry, or conical shape of the nanopore, thus allowing for nanopores with very small tip diameters to be prepared (<5 nm) [43].

Materials. A variety of membrane materials are well-suited for the track-etch technique. However, polymer membranes have seen the greatest use due to their chemical and mechanical robustness and high susceptibility to selective ion-track etching [5]. A number of polymer materials are suitable for preparing ion track-etched conical nanopores, including poly(carbonate) (PC), poly(ethylene terephthalate) (PET), poly(propylene) (PP), poly(vinylidenefluoride) (PVDF), and poly(imide) (PI). The work discussed in the forthcoming sections has utilized either PET, PC or poly(imide) (Kapton) (Fig. 6.8) to prepare conical nanopores via the track-etch method.

The ideal etching parameters, such as etchant composition, and etching temperature differ for each material. For example, ion tracks in PET membranes are typically etched with a 9 M NaOH etchant and a 1 M formic acid plus 1 M potassium chloride (KCl) stopping solution, at room temperature [43, 45]. Upon nanopore breakthrough, hydroxide etchant is simply neutralized by the formic acid. The

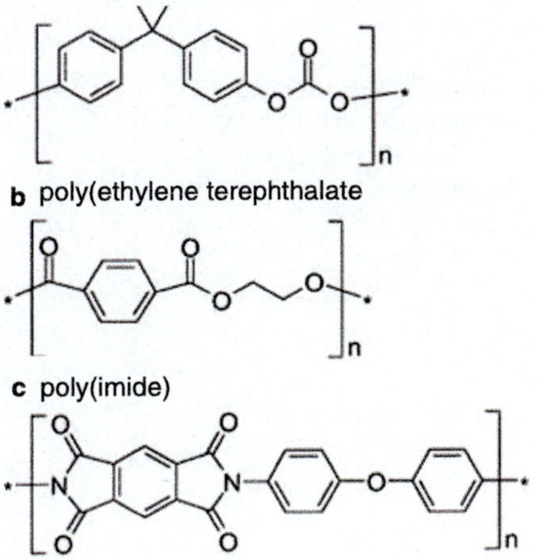

Fig. 6.8 Chemical structures of polymers commonly used in the track-etch method [2]. Reproduced by permission of the PCCP Owner Societies

NaOH hydrolyzes the ester bonds in PET resulting in the formation of carboxylate and hydroxyl groups inside the pore [96].

In contrast, ion tracks in Kapton membranes are etched with a NaOCl etchant with an active chlorine content of 13%, and a 1 M potassium iodide (KI) stopping solution, at a temperature of 50°C [44, 45, 56]. Upon etchant breakthrough at the nanopore tip opening, an oxidation-reduction reaction occurs, whereby iodide ions catalyze the reduction of hypochlorite ions to produce chloride ions. This stop-etch reaction yields iodine, yellow in color, which provides a colorimetric indicator of breakthrough. Etching of Kapton results in the formation of carboxylate groups inside the pore via hydrolysis of imide bonds [56, 97].

Etching rates also differ from one material to the other. The two etching properties that have the greatest influence on the shape of conical nanopores are the bulk etch rate, v_B, and the track etch rate, v_T, of the material. The bulk etch rate for PET has been determined to be ~2.17 nm/min [43, 46]. The track etch rate is ~10 μm/h for 12 μm thick PET [43]. Kapton has a bulk etch rate of 0.42±0.04 μm/h and a track etch rate of 3.12 ± 0.65 μm/h for 12 μm thick Kapton [56]. For a conically shaped pore, the ratio of v_B/v_T determines the cone angle of the pores. Kapton has a much higher v_B/v_T ratio than PET, which results in conical pores with much larger base diameters and cone angles [43–45, 56].

Controlling the Geometry of Conical Nanopores. The geometry of a conical pore (i.e., cone angle) has a significant impact on the transport properties of the pore. It is therefore important to control the geometry of conical nanopores. The two key parameters that have been investigated as a means for varying the cone angle are the etching potential and the etchant composition.

Effects of Etching Potential on Conical Nanopore Geometry. It has been demonstrated that the base diameter of conical nanopores can be systematically varied by

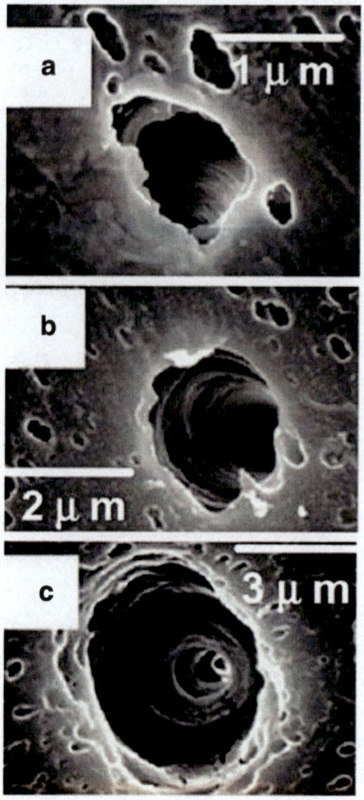

Fig. 6.9 FE-SEM images of the base openings of conical nanopores in PC, anisotropically track-etched with applied transmembrane potentials of (**a**) 0 V, (**b**) 15 V and (**c**) 30 V. Figure reprinted with permission from [47]

adjusting the potential applied during the anisotropic etching process [47]. Field-emission scanning electron microscope (FE-SEM) images of the base openings of conical nanopores etched in PC are shown in Fig. 6.9. All three of the conical nanopores shown were etched for a period of 5 h with the only parameter changed being that of the applied potential. Potentials were applied between 0 and 30 V during etching, resulting in base diameters that could be systematically varied between 1 to 6 μm (Fig. 6.10). The base diameters of the nanopores etched at 0, 15 and 30 V were 1.0, 2.7 and 5.5 μm, respectively with corresponding tip diameters of 10, 17, and 28 nm [47].

Figure 6.11 shows FE-SEM images of gold replicas of the conical nanopores etched at 0, 15 and 30 V. The replicas of the nanopores were prepared by filling the pores with gold using the electroless deposition method (*vide infra*) and subsequently etching away the polymer membranes. The electron micrographs clearly show that the cone angle of the conical pores increases as well with increasing applied transmembrane potential. It is interesting to note that when etching with no applied potential a pore that is nearly cylindrical through most of its length with a conical segment of ~3 μm at one end is obtained [47].

A mechanism has been proposed to account for the effect of the etching potential on track etching. As the applied potential is increased during the etching process, the

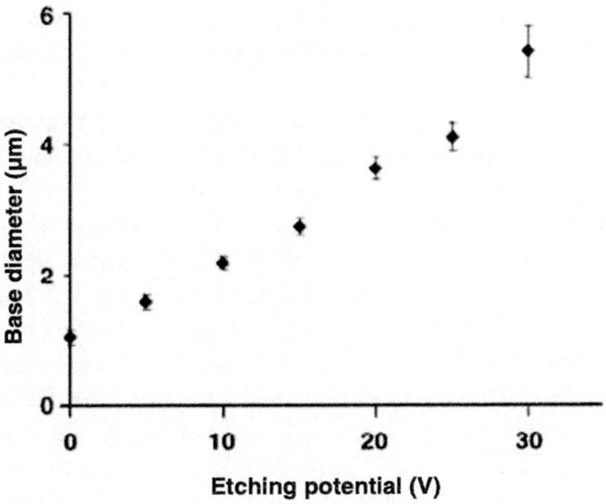

Fig. 6.10 Plot of the base diameter opening of conical nanopores, obtained from FE-SEM images, versus the etching potential applied during anisotropic etching. Figure reprinted with permission from [47]

ion current flowing through the nanopore is increased. The higher ion currents result in resistive heating of the solution inside the nanopore [47]. It is well known that etching rates increase with increasing temperature [43]. Therefore, local resistive heating causes the face of the membrane that is in contact with the etching solution

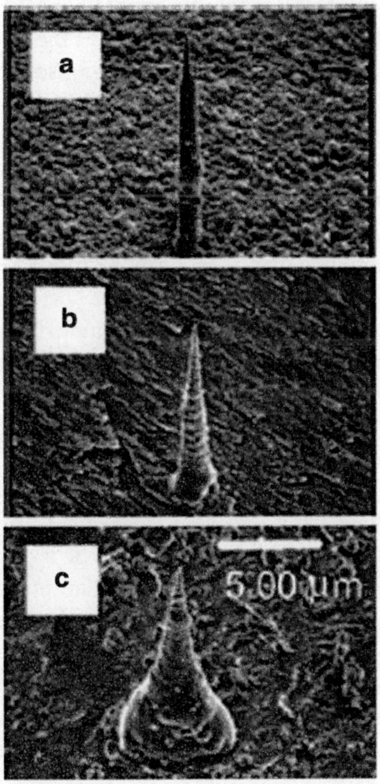

Fig. 6.11 FE-SEM images of gold nanocone replicas obtained by electrolessly plating gold into conical nanopores that where anisotropically track-etched with applied transmembrane potentials of (**a**) 0 V, (**b**) 15 V and (**c**) 30 V and then removing the polymer membrane. Figure reprinted with permission from [47]

to etch at a faster rate. This results in larger base diameters and cone angles as a function of increased applied transmembrane potentials.

Effect of Solvent Concentration in Etching Solution. Another parameter that greatly influences the track-etching process is etchant composition. The addition of solvent to the etching solution results in conical nanopores with increased cone angles. This effect has been demonstrated through the anisotropic etching of multi-pore membranes of PET with varying concentrations of ethanol in the etchant solution [48]. Addition of ethanol to the etchant causes the ratio of ν_B/ν_T to change, thereby influencing the shape of the conical nanopores. Increasing the ethanol concentration increases the ratio ν_B/ν_T, which results in conical nanopores with increased base diameters, and consequently larger cone angles.

This effect can be observed by examining FE-SEM images of gold replicas of conical nanopores etched with various concentrations of ethanol in the etchant (Fig. 6.12). The conical nanopores, of which these replicas were produced, were etched with the etchant solution containing only water, a 3:7 mix of ethanol/water, a 1:1 mix of ethanol/water, a 7:3 mix of ethanol/water, and only ethanol. The temperature, applied transmembrane potential and concentration of base in the etchant were kept constant, and the etching process was stopped when the ionic current flowing through the nanopores reached a predetermined value. As the ethanol concentration in the etching solution was increased the amount of time that it took for the predetermined current value to be reached decreased dramatically (i.e., 41 min with H_2O and 2 min with ethanol) [48]. This is believed to occur because the alcohol increases the solvation of the polymer chains in the membrane and makes them more accessible to the etching solution [48].

It is obvious from the electron micrograph images that as the volume of ethanol in the etching solution is increased, the cone angle and base diameter are also increased. Analysis of the gold nanopore replicas show that from the etching solution containing only water to that of the solution containing only ethanol, the cone angle increases from 0.5° to 4.4° [48]. A comparison of the conical geometries obtained using the various ratios of ethanol/water in the etching solutions are shown in Table 6.1.

The etchant solvent composition and the etching potential have proven to be valuable means for controlling the etching process. The manipulation of these parameters has allowed for methodic control of base diameter, cone angle, and etching time. The outcome is that conical nanopore geometry can be rationally and reproducibly varied through careful consideration of etching conditions.

Reproducible Fabrication of Conically-Shaped Nanopores. A key issue in artificial nanopore sensor design is reproducible fabrication of the nanopore sensor element. This has proven to be a challenge with artificial nanopore fabrication techniques [3, 4, 68, 72–74]. The anisotropic chemical etching of conical nanopores gives good reproducibility and control in the base diameter of conical track-etched nanopores [46–48]. Reproducibility is achieved by stopping the first etch step after a predetermined amount of time. To measure the reproducibility of the base diameter, membranes containing multiple conical nanopores were imaged using a FE-SEM. Multi-pore membranes were used because it is difficult to locate a single nanopore

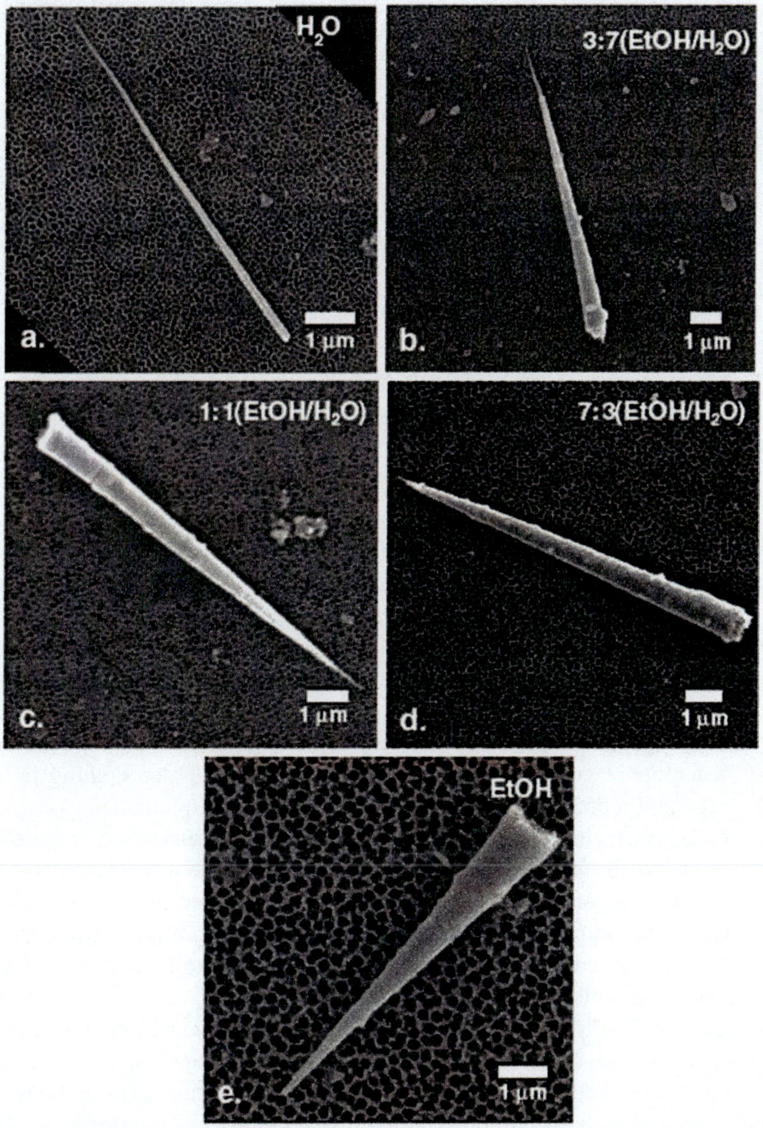

Fig. 6.12 FE-SEM images of liberated Au nanocones showing the effect of the volume fraction of ethanol in the etch solution on cone angle. Etch solutions were prepared with (**a**) water, (**b**) 3:7 ethanol/water, (**c**) 1:1 water/ethanol, (**d**) 7:3 ethanol/water and (**e**) ethanol [48]. Figure reprinted with permission from IOP Publishing, Inc

with the FE-SEM. It has also been demonstrated that the pore diameter obtained for track-etched membranes is independent of the track density [98]. This means that conical nanopores in single track-etched membranes will have the same base diameter as the multi-pore membranes if the same etching conditions are used.

Table 6.1 Effect of ethanol/water ratio on conical nanopore geometry and effect of cone angle on nanopore resistance [48]. Figure reprinted with permission from IOP Publishing, Inc

5 M KOH etch solution	Length[a] (μm)	Base width[a] (nm)	Cone half-angle[a] θ (deg)	Time to etch (min)	Pore resistance[b] (MΩ)
H_2O	7.4 ± 0.9	190 ± 20	0.5 ± 0.1	41	690
3:7 EtOH/H_2O	9.1 ± 0.6	720 ± 40	1.8 ± 0.2	35	200
1:1 EtOH/H_2O	9.2 ± 0.7	820 ± 40	2.3 ± 0.2	22	160
7:3 EtOH/H_2O	10.2 ± 0.6	840 ± 50	2.5 ± 0.2	18	140
EtOH	7.5 ± 0.7	1270 ± 110	4.4 ± 0.5	2	82

[a] Measured from SEM image.
[b] Effect of cone angle on resistance calculated from a modified version of Eq. (6.6), where $r_{base} = r_{tip} + L\tan\sigma$. See [48] for details.

However, it has been found that if the track-etching process is stopped after a predetermined amount of time, the tip diameter will vary greatly from sample to sample. This is believed to be due to the nature of the anisotropic etching process. The mixing of etching and stopping solutions in the nanopore tip makes it difficult to control the etch rate in this region [46]. Recently, however, a second, isotropic etching step was developed to fine tune, and accurately tailor, the nanopore tip opening [46]. Addition of this second isotropic etch step has proven to give good reproducibility in nanopore geometry. In particular, this new two-step etching process allows for excellent reproducibility in the nanopore tip opening diameter [46].

Single-track PET membranes were used to develop the two-step etching method. The first etch step is the anisotropic etch discussed above. The second isotropic etch step involves an analogous setup to anisotropic etching, except that a more diluted etchant is placed on both sides of the membrane. This isotropic chemical etching process allows for the entire length of the conical nanopore to be uniformly etched at a slower, more controllable, rate.

In the case of PET, 1 M NaOH is placed on both sides of the nanopore membrane during the isotropic etch step. A transmembrane potential is again applied with platinum wire electrodes, and the current is monitored as a function of time. Instead of stopping the second etch at a predetermined time, the process is stopped at a predetermined current value (Fig. 6.13). If the second etch is stopped after a certain amount of time, the variability in tip diameter, resulting from the first etch step, is retained. However, it has been demonstrated that the current flowing through the nanopore can be correlated to the tip opening diameter [46]. Therefore, by stopping this isotropic etching process at a pre-determined current value, accurate and highly reproducible tip diameters are obtained.

Figure 6.13 shows the current-time etching plots during the second step etch for three membranes that were subjected to the same first step etch. The etching process was stopped when a final nanopore ion current (I_f) of 40 nA was obtained.

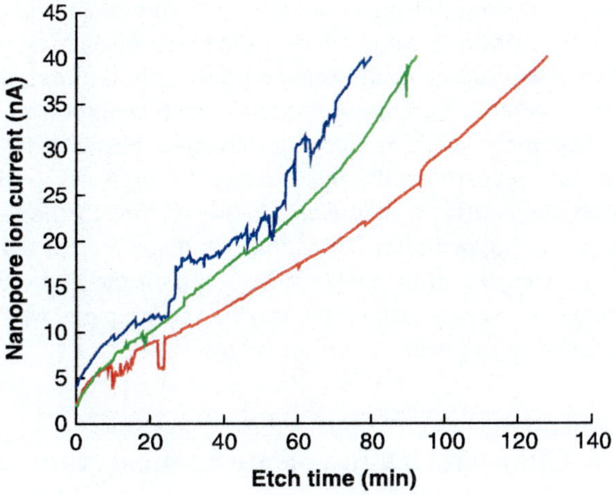

Fig. 6.13 Current-time transients obtained during the second, isotropic etch step for three membranes that were subjected to the same first etch. The second etch was stopped in each case when a final nanopore ion current (I_f) of 40 nA was obtained [46]

The time required for each membrane to reach this final ion current varied from sample to sample, however, the resulting conical nanopores all had very similar tip opening diameters (~64±1 nm) (Fig. 6.14) [46]. Etching during this step is again stopped by replacing the diluted etchant on both sides of the membrane with water or stopping solution. The second etching step is slow enough that the time required for the stopping process will not cause variation in the tip diameter. As will be shown in the following section, there is an exact mathematical relationship between the ion

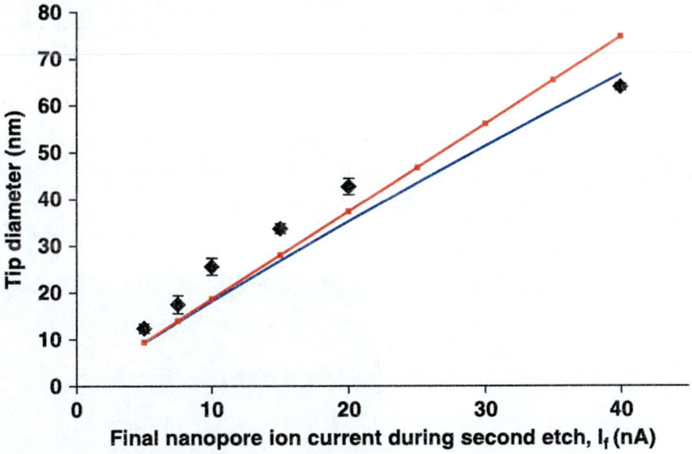

Fig. 6.14 Plot of the nanopore tip diameter opening after the second etch step versus the nanopore ion current at which the second etch was stopped (I_f). The points are the experimentally-measured tip diameters. The error bars are associated with measurements on three different membrane samples prepared identically. The *solid curves* were calculated using the simplified equation (Eq. 6.13, *red curve*) and the exact equation (Eq. 6.12, *blue curve*) [46]

current flowing through the nanopore when the second etch is stopped (I_f) and the diameter of tip opening. Hence, the two-step etching procedure allows for the base diameter of the conical pores to be fixed in the first, anisotropic, etch step and the tip diameter in the second, isotropic, etch step [46].

Nanopores with tip opening diameters between 10 and 60 nm have been reproducibly prepared with this method (Fig. 6.14) [46]. It is possible to prepare nanopores with tip diameters below 10 nm. In the work presented, the first etch step was stopped after 2 h and the tip diameters varied between 1 to 7 nm. Smaller tip diameters can be achieved by stopping the first etch step sooner. Nanopore tip diameters greater than 60 nm can also be prepared by stopping the second etch step at higher nanopore ion current values.

6.6 Characterization of the Conical Nanopores

6.6.1 Electron Microscopy

The three-dimensional shape of conical nanopores is generally characterized from SEM images of gold nanopore replicas (see e.g. Figs. 6.9, 6.11, 6.12, and 6.15). The gold replicas are obtained by electrolessly depositing gold inside the empty nanopores. The electroless deposition method (*vide infra*) also leaves a layer of gold along both faces of the nanopore membrane surface. In order to liberate the gold replicas from the nanopore membrane, the surface layers of gold are first removed from both faces of the membrane using either an ethanol wetted cotton swab or a tape strip. The polymer membrane is then removed by dissolution in hexafluro isopropanol. The nanopore replicas are then filtered through an alumina filter and imaged via SEM. The images in Fig. 6.12 were obtained using this technique.

Using a similar technique, it is also possible to image an array of standing nanopore replicas. This is achieved by removing only the surface layer of gold on the tip opening face of the gold plated conical nanopore membrane. The polymer membrane is then removed using reactive-ion etching in an oxygen plasma. This

Fig. 6.15 SEM image of standing gold nanocone array obtained after oxygen plasma etching [48]. Figure reprinted with permission from IOP Publishing, Inc

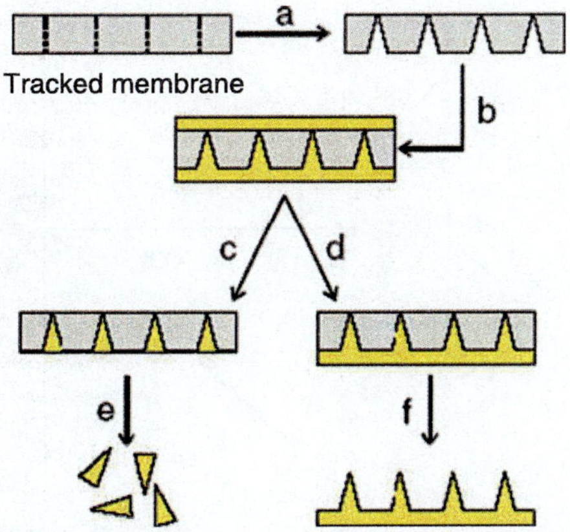

Fig. 6.16 Schematic diagram of the template synthesis method used to prepare gold nanocone replicas of track-etched conical nanopores. (**a**) asymmetric track-etch, (**b**) electroless gold deposition, (**c**) removal of gold surface layers from both faces of the membrane, (**d**) removal of gold surface layer on tip face of the membrane, (**e**) dissolution of polymer membrane and filtration of nanocones, (**f**) oxygen plasma etch of the front side of the polymer membrane [48]. Figure reprinted with permission from IOP Publishing, Inc

process yields an array of nanopore replicas standing on their bases (Fig. 6.15). The gold nanopore replicas seen in Figs. 6.11, 6.12 and 6.15 show that the track-etched nanopores have a nearly ideal conical shape. A schematic diagram of these two processes can be seen in Fig. 6.16.

As previously discussed, the base diameters, d_b, of conical nanopores obtained after the first etch step have also been characterized and measured with FE-SEM. Electron micrographs of the base openings of conical pores (Fig. 6.9) in track-etched multi-pore membranes can be used to determine the bulk etch rate, ν_B, of the material being etched. The bulk etch rate of the material determines the diameter of the base opening in conical nanopores. If the bulk etch rate is known, then the base diameter of single conical nanopores can be calculated by multiplying ν_B by the total etching time of the first step etch. For example, the bulk etch rate of PET was determined to be 2.17 nm/min from FE-SEM images [43, 46]. If the anisotropic etching process is stopped after 2 h then the base diameter of a conical nanopore in PET will be $\sim$520$\pm$50 nm.

6.6.2 Current-Voltage Curves

Current-voltage curves are typically used to determine the tip opening diameter, d_t, of single conical track-etched nanopores. This electrochemical method for tip size determination entails mounting the membrane containing the conical nanopore in the same cell setup used for the etching process (Fig. 6.2). An electrolyte solution, typically 1 M KCl, of known ionic conductivity is introduced into both sides of

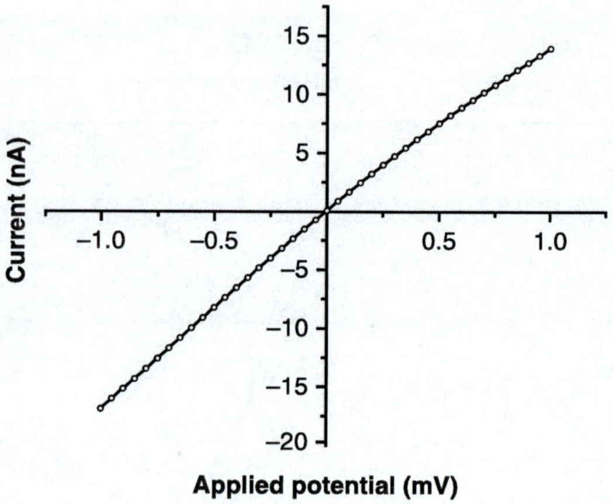

Fig. 6.17 A typical current-voltage curve used to calculate the tip diameter of the conical nanopore

the cell along with Ag/AgCl electrodes. A current-voltage curve for the electrolyte-filled nanopore is then obtained via a linear scan of the transmembrane potential from –1.0 to +1.0 V, while measuring the resulting ion current flowing through the nanopore (Fig. 6.17). The slope of this current-voltage curve is equal to the ionic conductance, G (in Siemens, S) of the electrolyte-filled nanopore. Since conical nanopores with small tip diameters rectify ion current, the linear portion of the current-voltage curve (between –200 and +200 mV) is used to calculate the nanopore tip diameter [90].

The equation for the ionic conductance of a conical pore is [43, 46, 89, 90],

$$G = \frac{\sigma \pi \mathrm{d_b d_t}}{4L} \tag{6.6}$$

where σ is the specific conductivity of the electrolyte solution (S/cm), L is the length of the nanopore (thickness of the membrane), d_b is the experimentally determined base opening diameter, and d_t is the diameter of the tip opening. Since all other parameters except for d_t in Eq. (6.6) are known, d_t can be calculated. Post-anisotropic etching tip opening diameters typically range from 1 to 7 nm [46]. It is important to note that this equation can only be applied to conical nanopores after the first etching step [46].

In order to determine the tip diameter after the second, isotropic etch step, one must take into account the change in both base and tip diameters, Δx, that occur during the second etch step. The change in base and tip diameters results in a nanopore with a final base diameter, d_{bf}, and a final tip diameter, d_{tf}, given as [46],

$$\mathrm{d_{bf}} = \Delta \mathrm{x} + \mathrm{d_{bl}} \tag{6.7}$$

and,

$$\mathrm{d_{tf}} = \Delta \mathrm{x} + \mathrm{d_{tl}} \tag{6.8}$$

where, d_{b1} is the base diameter after the first etch step and d_{t1} is the tip diameter after the first etch step. Substituting the values of the final base and tip diameters into Eq. (6.6) gives [46],

$$G_f = \frac{(\sigma\pi(d_{b1} + \Delta x)(d_{t1} + \Delta x))}{4L} \tag{6.9}$$

where G_f is the linear slope of the current-voltage curve for the nanopore taken after the second etch step. If in Eq. (6.9), Δx is substituted with $(d_{tf} - d_{t1})$, then the equation can be rearranged and solved for d_{tf} using the quadratic formula, the solution to which is [46],

$$d_{tf} = \frac{\left(-(d_{b1} - d_{t1}) + \left[(d_{b1} - d_{t1})^2 + \frac{4G_f}{M}\right]^{1/2}\right)}{2} \tag{6.10}$$

where $M = \sigma\pi/4L$. Equation (6.10) can be used to calculate the final tip diameter obtained after the second etch step, since d_{b1} and d_{t1} are determined after the first etching step, and all other parameters are known.

A mathematical model that relates the current flowing through the nanopore during the second etch to the nanopore tip diameter has also been developed. This model allows one to predict the diameter of the nanopore tip after the second etch step for any current value at which the second etch is stopped, I_f. In this model, the conductance of the nanopore after the second etch, G_f, is defined as the second etch stopping current, I_f, divided by the applied transmembrane potential during the second etch, E_{app}. This value for G_f can be substituted into Eq. (6.9), and rearranged to give [46],

$$I_f = \frac{E_{app}\,[\sigma_{etch}\pi(d_{b1} + \Delta x)(d_{t1} + \Delta x)]}{4L} \tag{6.11}$$

where σ_{etch} is the conductivity of the etchant used in the second etch step. Equation (6.11) is also a quadratic equation in d_{tf} for which the solution is [46],

$$d_{tf} = \frac{-(d_{b1} - d_{t1}) + \sqrt{(d_{b1} - d_{t1})^2 + 4I_f/K}}{2} \tag{6.12}$$

where $K = E_{app}\sigma_{etch}\pi/4L$. As shown in Eq. (6.12), the tip diameter of the conical pore during the second etch is uniquely related to the ionic current flowing through the nanopore.

For final tip diameters under ~50 nm the change in the base diameter $(d_{bf} - d_{b1})$, is negligible compared to the change in the tip diameter $(d_{tf} - d_{t1})$ [46]. This allows for a simplified version of Eq. (6.11) to be used to calculate the final tip diameter [46].

$$d_{tf} = \frac{I_f 4L}{E_{app}\sigma_{etch}\pi d_{b1}} \tag{6.13}$$

In Eq. (6.13), the conductance through the nanopore, G_f, is again substituted with I_f/E_{app} and rearranged to solve for d_{tf}. Figure 6.14 shows plots of d_{tf} versus I_f, calculated using Eqs. (6.12) and (6.13), as the two solid lines. The calculated tip diameters from the simplified and exact equations are identical for tips below ~20 nm. The tip diameters predicted by the mathematical model are also in good agreement with experimental values [46].

6.7 Tailoring the Surface Chemistry of Artificial Nanopores

A key advantage of the α-hemolysin-based sensor is the ability to chemically functionalize the inside walls of the pore, so that selective sensors can be obtained. It is likewise important to be able to control the surface chemistry of artificial nanopore sensors. One way to accomplish this is to deposit a correspondingly conical gold nanotube within the pore [50, 92, 93]. This is advantageous because the gold surface can be easily modified using simple thiol-based chemistry. We have used this approach to attach biochemical molecular-recognition agents to such conical nanotube sensors to make the sensor selective to, for example, proteins that bind to these agents [92]. The electroless deposition method used to prepare the conical gold nanotubes is reviewed here.

6.7.1 Electroless Gold Deposition

The electroless gold deposition method is a form of template synthesis. The template synthesis method [99–110] for the preparation of nanomaterials was pioneered by the Martin Group, and involves depositing a desired material into the nanopores of a solid host. This method is useful for preparing hollow nanotubes and solid nanowires. In the electroless deposition of metals [50], a reducing agent and a catalyst are used to plate metals from solution onto a solid surface.

Electroless deposition of gold is achieved by first sensitizing the membrane with Sn(II). This is done by immersing the nanopore membranes in methanol for five minutes and then into a 50/50 water/methanol solution that is 0.026 M in $SnCl_2$ and 0.07 M in triflouroacetic acid for 45 min. The Sn(II) sensitizer will bind to the pore walls and membrane surface via electrostatic complexation with the negatively-charged surface functional groups of the polymer formed during chemical etching [50].

The membrane is then washed again for 5 min in methanol and placed in an aqueous ammoniac solution that is 0.029 M in $AgNO_3$ for 7.5 min. During this step a surface redox reaction occurs and the surface bound Sn(II) is oxidized to Sn(IV) and Ag(I) is reduced to elemental Ag. This deposits silver nanoparticles on the pore walls [50]. The membrane is again washed in methanol for 5 min before being placed in a gold-plating bath that is 7.9×10^{-3} M in $Na_3Au(SO_3)_2$, 0.127 M Na_2SO_3, 0.025 M in $NaHCO_3$ and 0.625 M in formaldehyde at 4°C. The pH of this solution is adjusted to 10 by dropwise addition of 1 M H_2SO_4. While in this gold-plating bath another surface redox reaction occurs. Since the standard reduction potential of gold is more positive than that of silver, gold galvanically displaces the silver to yield gold nanoparticles on the pore surface. These nanoscopic gold particles catalyze the subsequent reduction of Au(I) to Au(0) using formaldehyde as the reducing agent [50].

$$2\,\mathrm{Au(I)} + \mathrm{HCHO} + 3\mathrm{OH}^- \longrightarrow \mathrm{HCOO}^- + 2\mathrm{H_2O} + 2\,\mathrm{Au(0)} \qquad (6.14)$$

This process occurs spontaneously and produces elemental gold via redox chemistry without using electrodes, hence the name "electroless". Electroless deposition yields the conical gold nanotubes lining the pore walls as well as gold surface layers that cover both faces of the polymer membrane. These surface layers are typically to thin to block the openings of the conical gold nanotube at the membrane surfaces. The surface layers can be removed via a tape-peel method or by swabbing the surfaces with an ethanol-wetted swab. A schematic of the plating process is shown in Fig. 6.18.

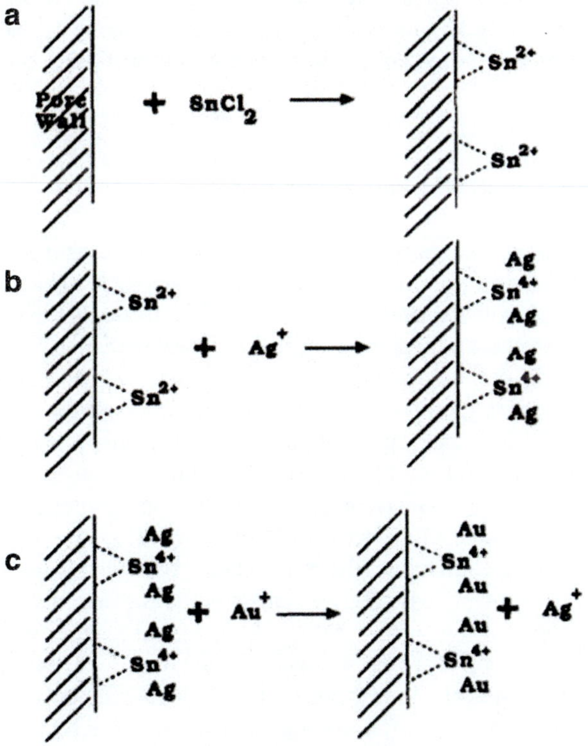

Fig. 6.18 Schematic diagram of the electroless gold plating procedure: (**a**) sensitize surface with Sn(II), (**b**) surface redox reaction with Ag(I), and (**c**) Ag particles displaced by Au(I). Reprinted with permission from [50]. Copyright 1995 American Chemical Society

The thickness of the walls of the gold nanotube can be varied by varying the gold plating time, and this provides another means for controlling the diameter of the tip opening of the nanotubes. Gold nanotubes with tip diameters of molecular dimensions (1 nm) can be obtained. Plating for longer periods of time will result in the nanopores being completely filled with gold to yield solid gold nanocones. As previously mentioned the resulting nanocones can be liberated from the membrane and imaged via SEM.

6.7.2 Generating Biocompatible Nanopore Surfaces

If these sensors are to be used for biological samples, a means for preventing non-specific protein adsorption to the nanotube walls must be developed. This has been accomplished by chemisorbing a poly(ethylene glycol) (PEG) thiol to the gold nanotubes [49, 92, 111, 112]. Typically, a commercially-available, 5 kD thiolated-PEG is used for this purpose [49, 92]. PEG surfaces are ideal for preventing non-specific adsorption because they are hydrophilic and non-ionic. The process of chemisorption entails immersion of the membrane containing the gold nanopore in a PEG-thiol solution at 4°C for a minimum of 15 h. The PEG-functionalized gold nanopore is then removed and rinsed with high purity water. The tip opening diameter after functionalization can be measured electrochemically, as previously described. After chemical functionalization, the conical nanopore can be utilized for resistive-pulse sensing of biomolecules, in particular proteins [92].

6.8 Electric Field Focusing

An important feature of the conical nanopore sensor is that the voltage drop caused by the ion current flowing through the nanopore is focused at the nanopore tip [54]. Indeed, calculations done by Lee et al. indicate that the field strength in the solution just inside the nanopore tip can be greater than 10^6 V/m, when the total voltage drop across the nanopore membrane is only 1 V (Fig. 6.19) [54]. A consequence of this focusing effect is that the ion current is extremely sensitive to analyte species present in or near the nanopore tip. That is, there is an analyte "sensing zone" just inside the tip [54, 89, 90]. This focusing effect makes conically-shaped nanopores better suited for resistive-pulse sensing than cylindrical nanopores.

Because of this electric-field focusing effect it is not the total length of the pore (membrane thickness) that is relevant to the sensing application but rather the "effective length." This effective length is the distance from the tip opening where the majority of the electric field is dropped, put another way, the length of the analyte-sensing zone. If we use the criterion that the effective length is that length over which 80% of the voltage is dropped, then conical nanopores with bases of 5 μm and tips of 20 nm have an effective length, as determined by the finite-element method [54],

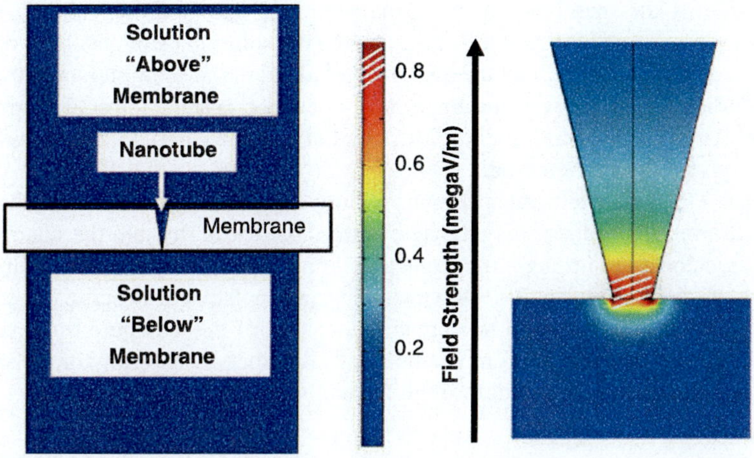

Fig. 6.19 Distribution of the electric field across a conical nanopore with a base diameter opening of 2.5 μm, a tip opening diameter of 60 nm and a thickness of 6 μm with 1 V applied across the nanopore in 1 M KCl. Because figure may be reprinted without color white hash marks are added to the section of the nanopore where the majority of the electric field is focused. Reprinted with permission from [54]. Copyright 2004 American Chemical Society

of 50 nm. Furthermore, such simulations show that the effective length can be controlled by varying the cone angle of the conical nanopore [54]. This is important because it allows for the length of the sensing zone to be tailored to the size of the analyte species top be detected.

6.9 Sensors Based on Conical Nanopores and Tubes

Conical nanopores in Kapton, PC and PET have been used as prototype sensing devices for the detection of small molecules [89], DNA [90] and proteins [93]. In the first two applications, a single conical nanopore functioned as the sensing element using the traditional resistive-pulse paradigm. In the third application, a conical gold nanotube sensor was used, and it functioned in an on/off fashion, much like a ligand-gated ion channel.

6.9.1 Resistive-Pulse Detection of Small Molecules

We have recently reported the first example of resistive-pulse sensing of a small-molecule analyte with an artificial nanopore [89]. Single conical nanopores anisotropically track-etched in Kapton membranes were used as the sensing element in this work, and the analyte used was a porphyrin, 4,4',4'',4'''-(porphine-5,10,15,20-tetrayl)-tetrakis(benzenesulfonic acid) (TPPS). The cell shown in Fig. 6.2 was used. An electrolyte solution and a Ag/AgCl electrode were placed on each side of the membrane and a transmembrane potential was applied. The electrodes were arranged so that the anode was in the solution facing the base opening of the conical nanopore, and the analyte was added to the solution facing the tip opening. When no analyte was present, a steady-state ion current flowed through the nanopore. As the applied transmembrane potential was increased, the magnitude of the steady-state current increased.

Because TPPS is a tetravalent anion, adding it to the solution facing the tip opening allowed us to drive this analyte electrophoretically through the nanopore. Analyte translocation through the detection zone at the tip resulted in downward current-pulses (Fig. 6.20). Downward current-pulses are observed because the diameter of TPPS (~2 nm) is comparable to that of the nanopore tip opening (~4.5 nm). Therefore, the TPPS molecules partially block the pathway of the small electrolyte ions that carry current through the nanopore [89].

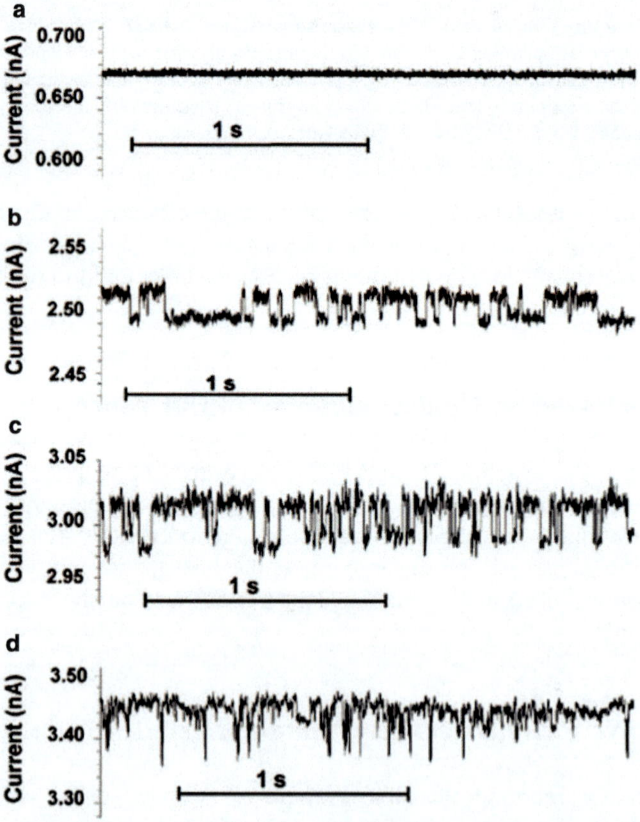

Fig. 6.20 Current-time traces for a Kapton conical nanopore with a tip opening diameter of ~4 nm in the presence of 60 nM TPPS, taken in an electrolyte solution of 1 M KCl, pH=8. The applied transmembrane potential was (**a**) 100 mV, (**b**) 400 mV, (**c**) 500 mV and (**d**) 600 mV. Reprinted with permission from [89]. Copyright 2005 American Chemical Society

The current-time traces for a solution 60 nM in TPPS, at various applied transmembrane potentials, are shown in Fig. 6.20. Current-pulse translocation events for the TPPS analyte were not observed at applied transmembrane potentials below ~300 mV. This threshold voltage phenomenon is observed for two reasons, the first being that the nanopore walls are negatively charged in the electrolyte solution used, which causes the anionic TPPS analyte to be electrostatically repelled from the nanopore [89]. The second reason is that the analyte pays an entropic penalty for entering such a narrow pore [89]. Only at larger values of applied positive transmembrane potentials can these electrostatic and entropic barriers be overcome by electrophoretic force acting on the charged TPPS molecule. We have also observed this threshold voltage phenomenon with DNA [90] and protein analytes [92]. Others have observed this phenomenon with the translocation of DNA through the α-hemolysin nanopore [25].

The current-pulse duration is determined by the time it takes for an analyte molecule to translocate the sensing zone [89, 90]. Current-pulse duration data for TPPS were analyzed with histograms showing the number of times a current pulse of a particular duration was observed versus the pulse duration (Fig. 6.21). The histograms show that as the applied transmembrane potential is increased from 400 to 600 mV, the average current-pulse duration decreases, and the distribution of pulse

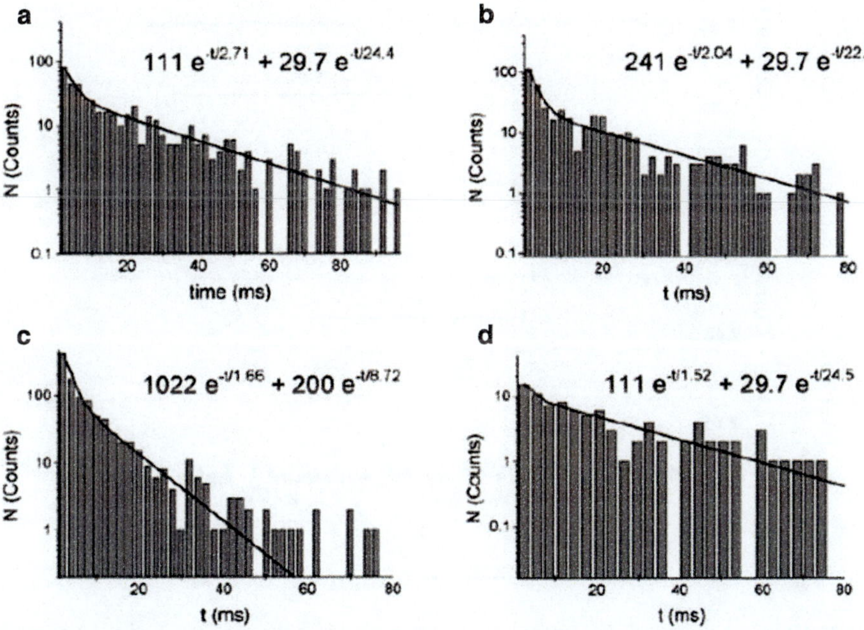

Fig. 6.21 Histograms of the current-pulse duration data for the Kapton conical nanopore in the presence of 60 nM TPPS at an applied transmembrane potential of (**a**) 400 mV, (**b**) 500 mV and (**c**) 600 mV and in the presence of 20 nM TPPS at an applied transmembrane potential of (**d**) 400 mV. Reprinted with permission from [89]. Copyright 2005 American Chemical Society

durations decreases. The decrease in pulse duration with increasing applied potential can be explained by examining the equation for current-pulse duration (Eq. 6.3), which predicts that duration decreases with increasing field strength. This ultimately results from an increase in electrophoretic velocity with increasing field strength (Eq. 6.2).

The frequency of the TPPS current-pulses was also investigated. Current-pulse frequency was found to be dependent on both the applied transmembrane potential and the analyte concentration. The pulse frequency increases with increasing applied transmembrane potential because the electrophoretic flux, J, of an ion is related to the electric field strength by Eq. (6.4). The concentration dependence can also be explained by examining Eqs. (6.4) and (6.5), as the electrophoretic flux is also dependent on the concentration of analyte. As the concentration of TPPS was decreased a decrease in event frequency was also observed (Fig. 6.22).

The electrophoretic theory and experimental results presented here illustrate an important feature in resistive-pulse sensing. The concentration dependence of current-pulse events means that at increasingly lower concentrations the time

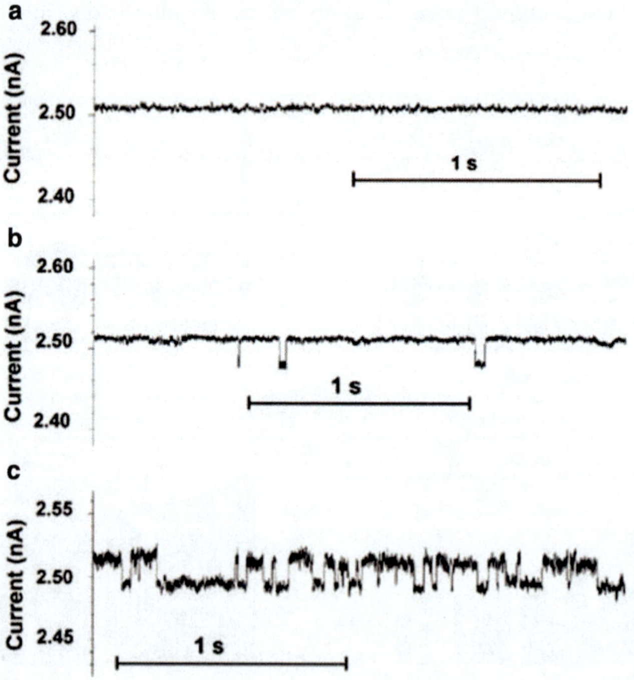

Fig. 6.22 Current-time traces for a Kapton conical nanopore with a tip opening diameter of ~4 nm, taken in an electrolyte solution of 1 M KCl, pH=8 at an applied transmembrane potential of 400 mV with TPPS concentrations of (**a**) 0 nM, (**b**) 20 nM and (**c**) 60 nM. The current-pulse frequency increases with increasing TPPS concentration. Reprinted with permission from [89]. Copyright 2005 American Chemical Society

between the events becomes increasingly longer. This concentration dependence means that the limit of detection in the resistive-pulse sensing method must be defined by the amount of time the analyst is willing to wait to observe current-pulse events [90]. However, as shown in Eq. (6.5), the event frequency is also dependent on the electric field strength [90]. Therefore, by increasing the applied transmembrane potential the event frequency should increase, and thereby improve the sensitivity of the method.

This study successfully demonstrated that the principles of resistive-pulse sensing can be applied to artificial conical nanopores in polymer membranes. This work also demonstrated many of the important operating principles of resistive-pulse sensing, for instance the voltage and concentration dependence of current-pulse events, and the dependence of current-pulse duration on the applied potential.

6.9.2 Resistive-Pulse Sensing of DNA Analytes

Similar resistive-pulse sensing studies have been performed using both single-stranded and double-stranded DNA as the analyte species [90]. For the DNA sensing work, single conical nanopores in polycarbonate membranes were used as the sensing element. These pores had typical base diameters of 1.5 μm and tip diameters of 40 nm. The two DNA analytes that were sensed were a single-stranded phage DNA (7,250 bases) and a double-stranded plasmid DNA (6,600 base pairs). The DNA chains are anionic, so they were again driven through the conical nanopores via electrophoresis. In a similar fashion to the sensing done with TPPS, the DNA analytes were added to the solution facing the tip opening of the nanopore. Therefore at positive applied transmembrane potentials (anode in solution facing base opening) the DNA was driven through the nanopore from the tip side to the base side. Translocation through the sensing zone again resulted in downward current-pulse events.

In the absence of DNA, no current-pulse events were observed, and at an applied transmembrane potential of 900 mV, the steady-state current was 4.8 nA. After addition of the single-stranded phage DNA (ssDNA) to the tip side of the nanopore, a continuous string of current-pulse events was observed (Fig. 6.23). As per the TPPS work, we observed a linear increase in the current-pulse frequency with increasing DNA concentration (Fig. 6.24). A threshold voltage was also observed with the ssDNA, as no current-pulse events were observed at applied potentials below 500 mV. However, at sufficiently high applied transmembrane potentials a linear dependence of the current-pulse frequency on the applied potential was observed (Fig. 6.25). As per the TPPS studies, the current-pulse duration for ssDNA decreased with increasing applied transmembrane potential.

The ssDNA resistive-pulse data were also analyzed through scatter plots of the current-pulse magnitude (Δi) versus the current-pulse duration (τ) (Fig. 6.26). The scatter plot showed that the distribution in Δi values was much less than the distribution in τ values. The reason for the larger spread in τ values is due to the

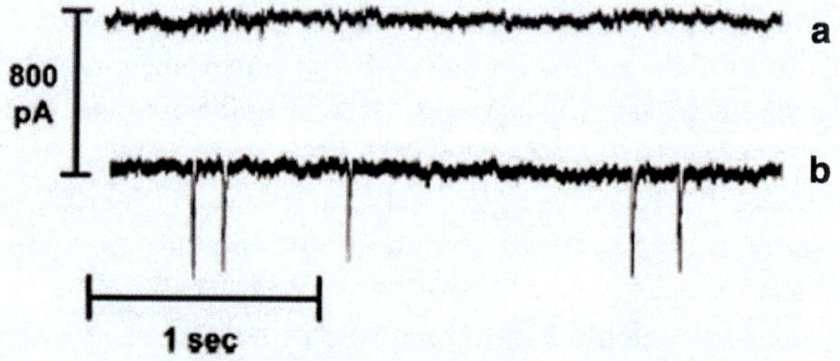

Fig. 6.23 Current-time traces for a single conical nanopore in polycarbonate at an applied transmembrane potential of 900 mV. (a) Steady-state current with no analyte present. (b) In the presence of 10 nM ssDNA, added to the tip side of the nanopore membrane. Reprinted with permission from [90]. Copyright 2006 American Chemical Society

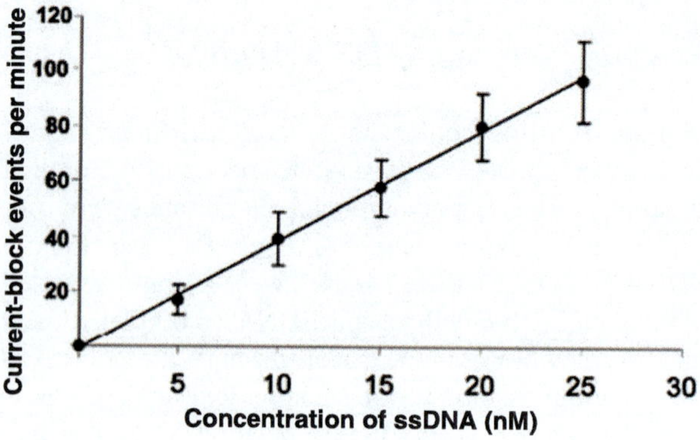

Fig. 6.24 Plot of current-pulse frequency versus the concentration of ssDNA. The applied transmembrane potential was kept constant at 900 mV, while the concentration was varied. The error bars represent averages of three 30 s time windows of current-pulse events. Reprinted with permission from [90]. Copyright 2006 American Chemical Society

Fig. 6.25 Plot of current-pulse frequency of ssDNA versus the applied transmembrane potential. The concentration of ssDNA was constant at 20 nM while the applied potential was varied. Reprinted with permission from [90]. Copyright 2006 American Chemical Society

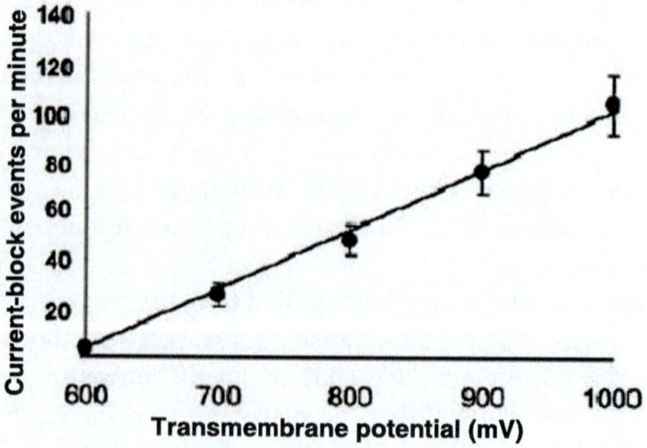

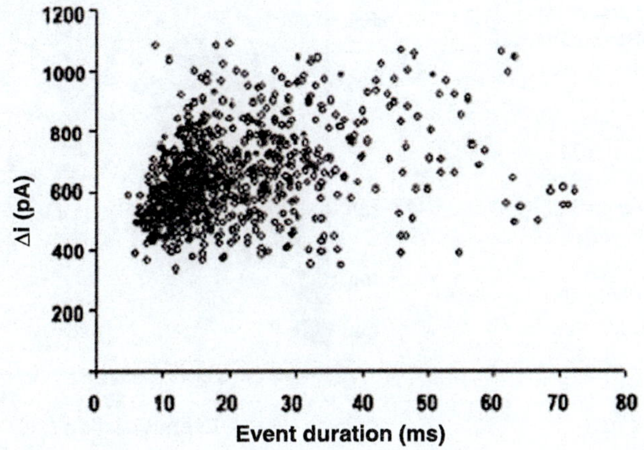

Fig. 6.26 Scatter plot of the current-pulse magnitude, Δi, versus the current-pulse duration for the current-pulse events associated with the translocation of a solution 10 nM in ssDNA at an applied transmembrane potential of 900 mV. The range in Δi values is about a factor of three, while the range in current-pulse durations is greater than a factor of 8. Reprinted with permission from [90]. Copyright 2006 American Chemical Society

conformational changes that must occur in order for the ssDNA to translocate the nanopore. The diameter of this particular ssDNA was calculated to be ~128 nm [90], which is over three times larger than the diameter of the nanopore tip for this sensor. Thus in order to translocate the nanopore, the ssDNA must adopt an extended conformation and reptate through the tip [90]. The distribution in the τ values is a reflection of the time required for each individual DNA strand to adopt the necessary conformation for translocation [90]. The distribution in Δi values is not as large because the current-pulse magnitude is determined after the necessary conformation for translocation is adopted.

Addition of the double-stranded plasmid DNA (ds-pDNA) to the solution facing the nanopore tip also resulted in the appearance of current-pulse events; however, the duration and magnitude of these events were much smaller than those for the ssDNA. This is because the ds-pDNA does not actually translocate the nanopore but instead "bumps" into the tip and bounces off. Light scattering data gave a radius of gyration for the ds-pDNA of 104 nm, or a diameter of 208 nm, which is 4 times larger than the tip opening diameter. dsDNA is also less flexible than ssDNA and therefore cannot easily adopt the conformations needed to translocate the nanopore [90].

Since the current-pulse events associated with the ds-pDNA "bumping" were very different in duration and magnitude than the events associated with ssDNA translocation, it was possible to distinguish between them in a solution containing a mixture of the two analytes. Figure 6.27 shows the scatter plot obtained from the current-pulse data of a solution that was 10 nM in both DNAs. Two distinct regions are seen, with the bottom left-hand cluster of events corresponding to the ds-pDNA "bumping" and the top right-hand cluster corresponding to the ssDNA translocation

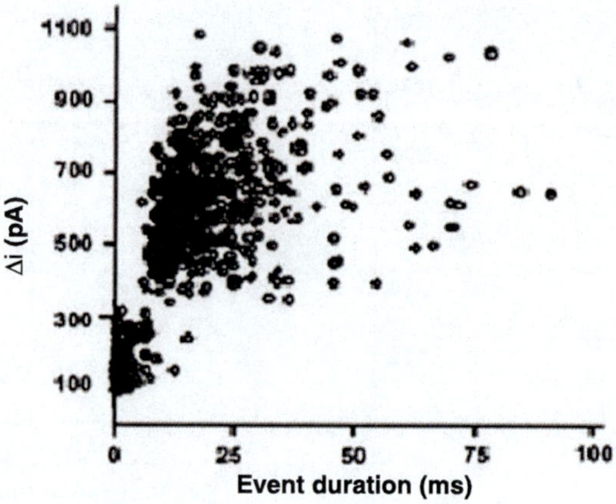

Fig. 6.27 Scatter plot of Δi versus current-pulse duration for the current-pulse events associated with the translocation of a mixture of 10 nM ssDNA and 10 nM ds-pDNA at an applied transmembrane potential of 900 mV. Reprinted with permission from [90]. Copyright 2006 American Chemical Society

events. These results demonstrate that size-based selectivity can be observed with artificial conical nanopore resistive-pulse sensors.

Conical nanopores in Kapton membranes prepared by the track-etch method have also been used by others to study the translocation of DNA molecules via the resistive-pulse sensing method [91]. This work was successful in showing that conical nanopore sensors are capable of differentiating between different lengths of DNA, and that the current-pulse durations scale linearly with DNA length. The observed current-pulse durations and the applied driving potentials were similar to those found in α-hemolysin based resistive-pulse DNA detection.

The data discussed above demonstrate the sensing capabilities of synthetic conical nanopores prepared by the track-etch method. However, in order for artificial nanopores to compete with the biological-nanopore sensors, it must be shown that greater selectivity than the size-based selectivity observed with DNA can be obtained. As discussed in the next section, this can be accomplished by attaching a molecular-recognition agent that selectively binds the analyte to the nanopore or nanotube sensor element.

6.9.3 Attaching Molecular-Recognition Agents to Make Highly Selective Protein Sensors

Single conical Au nanotubes in PET membranes were used to design a new class of protein biosensors. The nanotubes were modified with various biochemical molecular recognition agents (MRAs) that selectively bind specific protein analytes [93]. The nanotube tip was fined-tuned to match the size of the desired protein analyte. With this sensing paradigm, current pulses are not observed. Rather, when the analyte protein is bound by the MRA in the nanotube tip, it occludes the tip. As a result selective binding of the analyte by the MRA shut off the ion current. Because proteins that are not bound by the MRA had no effect on the ion current, highly selective sensors were obtained.

As described above, conical nanopores in PET were electrolessly plated with gold to yield conical gold nanotubes. The base diameter of the nanotubes was ~600 nm and the tip diameters were varied between 5 and 9 nm, depending on the protein analyte to be detected. Attachment of the MRAs to the nanotube surface was carried out using gold-thiol chemistry. Three MRA/analyte systems were used to evaluate this type of sensor – biotin/streptavidin, protein-G/immunoglobin G (IgG), and anti-ricin/ricin.

In the first system biotin was immobilized onto the gold nanotube surface through chemisorption of a thiolated biotin derivative. The MRAs in the other two systems were attached to the surface using biotin-streptavidin chemistry [93]. After functionalization all three nanotube sensors had tip diameters of ~5 nm. The functionalized nanotube membranes were mounted in the cell (Fig. 6.2) and a transmembrane potential was applied. The ion current flowing through the nanotube was monitored by obtaining current-voltage curves.

Current-voltage curves were first taken before exposure to any protein. Control experiments were then performed by taking current-voltage curves in the presence of a protein that did not bind to the attached MRA. For example, when biotin was used as the MRA, control experiments were performed with the proteins lysozyme and bovine serum albumin (BSA). These non-binding proteins had no effect on the current-voltage curve (Fig. 6.28). However when the biotin modified nanotubes were exposed to a solution 100 pM in streptavidin, the ion current was completely shut off (Fig. 6.28); hence the MRA-modified sensor only responded to the protein that was bound by the MRA.

The time required to achieve blockage of the ion current, τ_B, was found to be inversely related to the streptavidin concentration. This time dependence (Fig. 6.29) is observed because as the concentration of analyte is decreased the number of molecules that encounter the membrane surface per second likewise decreases.

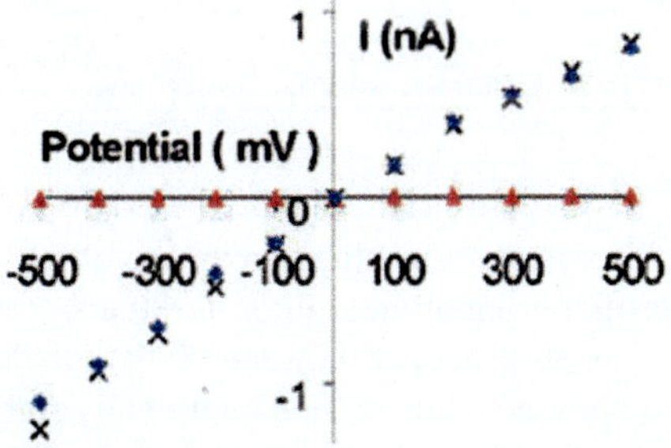

Fig. 6.28 Current-voltage curves for the conical Au nanotube functionalized with biotin. In the presence of no protein (x) and in the presence of a non-binding protein, 100 nM lysozyme (*diamond*), the current-voltage curves are the same. However, in the presence of 180 pM streptavidin (*triangle*) the current is completely shut off. Reprinted with permission from [93]. Copyright 2005 American Chemical Society

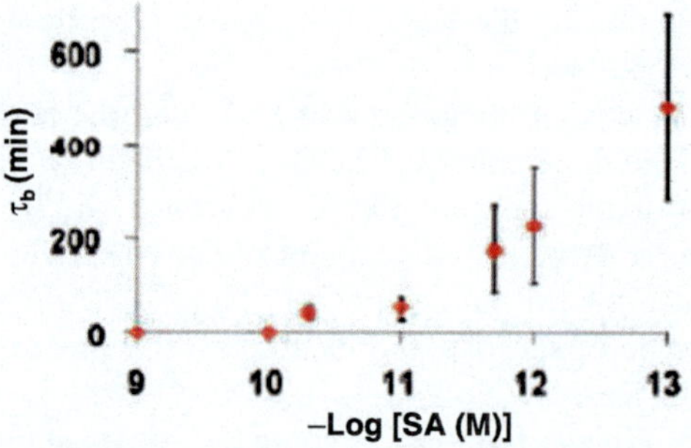

Fig. 6.29 Plot of the time required for blockage of the ion current to occur, τ_b, versus the –log of the concentration of analyte. The error bars represent the averages of three measurements done with three different nanotube sensors of the same dimensions. Reprinted with permission from [93]. Copyright 2005 American Chemical Society

At the lowest concentration the τ_B values are quite long. However, in these experiments the protein analyte reached the membrane surface only by diffusion. This is because a transmembrane potential was not applied while the analyte was equilibrating with the nanotube membrane. By applying a transmembrane potential, the analyte can be driven to the membrane surface by electrophoresis (as we have shown above), and this should lower the response time, τ_B. Decreasing τ_B by this route should also lower the error in the measurement of τ_B.

Similar results for the protein-G/IgG and anti-ricin/ricin sensor were obtained. The protein-G modified nanotube showed complete blockage of the ion current in the presence of 10 nM of horse IgG and the anti-ricin modified nanotube was blocked in the presence of 100 nM ricin. This "on/off" sensing paradigm has been shown to be a highly sensitive and selective. Furthermore, it should be possible to modify the Au nanotube surface with a wide range of MRAs to selectively detect a wide variety of analytes. Hence, this is a very general sensing paradigm.

6.10 Advantages of Track-Etched Conical Nanopores/Nanotube Sensors

The track-etch method offers many advantages over the other techniques currently being used to fabricate artificial nanopores. First, the track-etch method is a well-known technique that has been practiced commercially for decades. Second, the polymer membranes used are mechanically and chemically stable, and a wide variety of polymeric materials can be tracked and etched. Furthermore, in order for artificial nanopores to compete with biological nanopores as sensing devices, one must be able to reproducibly prepare the artificial nanopore. Reproducibly preparing single artificial nanopores has proven to be challenging with some of the fabrication techniques in use today [3, 68, 72–75]. However, as was reviewed here, the track-etched conical nanopore sensors can be prepared with good reproducibility. It has also been demonstrated that selectivity can be added to conical nanopore sensors through attachment of MRAs to the nanopore walls [93].

The track-etched synthetic nanopores discussed here also offer several advantages over the biological nanopore-based sensors. The main advantage is that the fragile lipid bilayer is eliminated and replaced with a chemically and mechanically robust membrane. This significantly increases the lifetime of sensors developed from artificial conical nanopores compared to the biological nanopores. Another advantage that comes with synthetic nanopore sensors is that lower limits of detection can be achieved. The sensitivity of biological nanopore sensors has proven to be somewhat modest. The lowest analyte concentrations that have been detected with biological nanopores were achieved with α-hemolysin-based sensors and were in the low nanomolar range [9, 13, 17, 23, 27]. In fact, much of the sensing done with biological nanopores has been with analyte concentrations in the micromolar range [8, 10, 11, 14–16, 19, 26, 29]. This limited sensitivity is in part due to the breakdown of the lipid bilayer membranes at potentials above ~200 mV [35]. The track-etched polymer membranes, on the other hand, are capable of withstanding applied potentials up to ~20 V [49]. As discussed previously, by increasing the applied potential the sensitivity of the device can be increased.

Another disadvantage encountered when using biological nanopores for resistive-pulse detection is the restriction in the nanopore geometry. The smallest constriction within the α-hemolysin pore has a diameter of ~1.5 nm. As a result, analyte species larger than 1.5 nm cannot be detected by the simple translocation-based current-pulse method. In contrast, as we have demonstrated here, with the conical nanopore/nanotube sensors, the size of the tip opening can be matched to the size of the analyte to be detected, and we have made sensors with tip diameters varying from 1 to 60 nm.

6.11 Summary

This field of artificial nanopore sensors is currently in its infancy. However, with the track-etch method we have the ability to reproducibly prepare the nanopore sensor element, and this is an essential prerequisite to making practical sensing devices. In addition, plating gold nanotubes within the pores provides a simple means for attaching analyte-selective molecular-recognition agents to the pore walls. As we have shown here, this provides a route for making analyte-selective sensors. Prototype conical nanopore and nanotube sensor have been used to detect small molecules, DNA and protein analytes. No other artificial nanopore technology has yet to show such diversity with regard to analyte type. In addition, we have demonstrated that artificial conical nanopores can function in the more traditional resistive-pulse mode and in an on/off fashion, similar to that of ligand-gated ion channels. These various results, reviewed here, suggest that track-etched conical nanopores show great promise for development into practical biosensing devices.

Acknowledgements We kindly thank the Martin Group members whose work contributed to this chapter. Parts of this work were funded by The National Science Foundation and the Air Force Office of Scientific Research.

References

1. Bayley, H.; Martin, C.R., Resistive-pulse sensing – from microbes to molecules. *Chem. Rev.* **100**, 2575–2594 (2000).
2. Choi, Y.; Baker, L.A.; Hillebrenner, H.; Martin, C.R., Biosensing with conically-shaped nanopores and nanotubes. *Phys. Chem. Chem. Phys.* **8**, 4976–4988 (2006).
3. Schmidt, J., Stochastic sensors. *J. Mater. Chem.* **15**, 831–840 (2005).
4. Henriquez, R.R.; Ito, T.; Sun, L.; Crooks, R.M., The resurgence of coulter counting for analyzing nanoscale objects. *Analyst* **129**, 478–482 (2004).
5. Harrell, C.C., *Designing Abiotic Single Nanotube Membranes for Bioanalytical Biomedical Applications.* (PhD Dissertation, University of Florida, Gainesville, 2004).
6. Bayley, H.; Braha, O.; Gu, L., Stochastic sensing with protein pores. *Adv. Mater.* **12**, 139–142 (2000).
7. Bayley, H.; Cremer, P.S., Stochastic sensors inspired by biology. *Nature* **413**, 226–230 (2001).
8. Kang, X.; Cheley, S.; Guan, X.; Bayley, H., Stochastic detection of enantiomers. *J. Am. Chem. Soc.* **128**, 10684–10685 (2006).
9. Howorka, S.; Cheley, S.; Bayley, H., Sequence-specific detection of individual DNA strands using engineered nanopores. *Nat. Biotech.* **19**, 636–639 (2001).
10. Astier, Y.; Braha, O.; Bayley, H., Toward single molecule DNA sequencing: Direct identification of ribonucleoside and deoxyribonucleoside 5′-monophosphates by using an engineered protein nanopore equipped with a molecular adapter. *J. Am. Chem. Soc.* **128**, 1705–1710 (2006).
11. Guan, X.; Gu, L.; Cheley, S.; Braha, O.; Bayley, H., Stochastic sensing of TNT with a genetically engineered pore. *Chem. Biol. Chem.* **6**, 1875–1881 (2005).
12. Kasianowicz, J.J.; Burden, D.L.; Han, L.C.; Cheley, S.; Bayley, H., Genetically engineered metal binding sites on the outside of a channel's transmembrane β-barrel. *Biophys. J.* **76**, 837–845 (1999).
13. Braha, O.; Gu, L.; Zhou, L.; Lu, X.; Cheley, S.; Bayley, H., Simultaneous stochastic sensing of divalent metal ions. *Nat. Biotech.* **8**, 1005–1007 (2000).
14. Gu, L.; Braha, O.; Conlan, S.; Cheley, S.; Bayley, H., Stochastic sensing of organic analytes by a pore-forming protein containing a molecular adapter. *Nature* **398**, 686–690 (1999).
15. Cheley, S.; Gu, L.; Bayley, H., Stochastic sensing of nanomolar inositol 1,4,5-trisphosphate with an engineered pore. *Chem. Biol.* **9**, 829–838 (2002).
16. Howorka, S.; Nam, J.; Bayley, H.; Kahne, D., Stochastic detection of monovalent and bivalent protein-ligand interactions. *Angew. Chem. Int. Ed.* **43**, 842–846 (2004).
17. Movileanu, L.; Howorka, S.; Braha, O.; Bayley, H., Detecting protein analytes that modulate transmembrane movement of a polymer chain within a single protein pore. *Nat. Biotech.* **18**, 1091–1095 (2000).
18. Shin, S.; Bayley, H., Stepwise growth of a single polymer chain. *J. Am. Chem. Soc.* **127**, 10462–10463 (2005).
19. Gu, L.; Cheley, S.; Bayley, H., Prolonged residence time of a noncovalent molecular adapter, β-cyclodextrin, within the lumen of a mutant α-hemolysin pore. *J. Gen. Physiol.* **118**, 481–493 (2001).
20. Bayley, H.; Jayasinghe, L., Functional engineered channels and pores. *Mol. Membr. Biol.* **21**, 209–220 (2004).

21. Kasianowicz, J.J.; Bezrukov, S.M., Dynamics and free energy of polymers partitioning into a nanoscale pore. *Biophys. J.* **69**, 94–105 (1995).
22. Bezrukov, S.M.; Vodyanoy, I.; Brutyan, R.A.; Kaianowicz, J.J., Dynamics and free energy of polymers partitioning into a nanoscale pore. *Macromolecules* **29**, 8517–8522 (1996).
23. Kasianowicz, J.J.; Henrickson, S.E.; Weetall, H.H.; Robertson, B., Simultaneous multianalyte detection wth a nanometer-scale pore. *Anal. Chem.* **73**, 2268–2272 (2001).
24. Halverson, K.M.; Panchal, R.G.; Nguyen, T.L.; Gussio, R.; Little, S.F.; Misakian, M.; Bavari, S.; Kasianowicz, J.J., Anthrax biosensor, protective antigen ion channel asymmetric blockade. *J. Biol. Chem.* **280**, 34056–34062 (2005).
25. Henrickson, S.E.; Misakian, M.; Robertson, B.; Kasianowicz, J.J., Driven DNA transport into an asymmetric nanometer-scale pore. *Phys. Rev. Lett.* **85**, 3057–3060 (2000).
26. Kasianowicz, J.J.; Brandin, E.; Branton, D.; Deamer, D.W., Characterization of individual polynucleotide molecules using a membrane channel. *Proc. Natl. Acad. Sci. USA* **93**, 13770–13773 (1996).
27. Meller, A.; Nivon, L.; Brandin, E.; Golovchenko, J.; Branton, D., Rapid nanopore discrimination between single polynucleotide molecules. *Proc. Natl. Acad. Sci. USA* **97**, 1079–1084 (2000).
28. Meller, A.; Nivon, L.; Branton, D., Voltage-driven DNA translocations through a nanopore. *Phys. Rev. Lett.* **86**, 3435–3438 (2001).
29. Meller, A.; Branton, D., Single molecule measurement of DNA transport through a nanopore. *Electrophoresis* **23**, 2583–2591 (2002).
30. Deamer, D.W. ; Branton, D., Characterization of nucleic acids by nanopore analysis. *Acc. Chem. Res.* **35**, 817–825 (2002).
31. Kullman, L.; Winterhalter, M.; Bezrukov, S.M., Transport of maltodextrins through maltoporin: A single-channel study. *Biophys. J.* **82**, 803–812 (2002).
32. Bezrukov, S.M.; Kullman, L.; Winterhalter, M., Probing sugar translocation through maltoporin at the single channel level. *FEBS Lett.* **476**, 224–228 (2000).
33. Danelon, C.; Lindermann, M.; Borin, C.; Fournier, D.; Winterhalter, M., Channel-forming membrane proteins as molecular sensors. *IEEE Trans. Nanobiosci.* **3**, 46–48 (2004).
34. Gouaux, J.E.; Graha, O.; Hobaugh, M.R.; Song, L.; Cheley, S.; Shustak, C.; Bayley, H., Subunit stoichiometry of staphylococcal α-hemolysin in crystals and on membranes: A heptameric transmembrane pore. *Proc. Natl. Acad. Sci. USA* **91**, 12828–12831 (1994).
35. Mayer, M.; Kriebel, J.K.; Tosteson, M.T.; Whitesides, G.M., Microfabricated Teflon membranes for low-noise recordings of ion channels in planar lipid bilayers. *Biophys. J.* **85**, 2684–2695 (2003).
36. DeBlois, R.W.; Bean, C.P., Counting and sizing of submicron particles by the resistive pulse technique. *Rev. Sci. Instrum.* **41**, 909–916 (1970).
37. DeBlois, R.W.; Uzgiris, E.E.; Cluxton, D.H.; Mazzone, H.M., Comparative measurements of size and polydispersity of several insect viruses. *Anal. Biochem.* **90**, 273–288 (1978).
38. DeBlois, R.W.; Wesley, R.K.A., Sizes and concentrations of several type C oncornaviruses and bacteriophage T2 by the resistive-pulse technique. *J. Virol.* **23**, 227–233 (1977).
39. DeBlois, R.W.; Bean, C.P.; Wesley, R.K.A., Electrokinetic measurements with submicron particles and pores by the resistive pulse technique. *J. Colloid Interface Sci.* **61**, 323–335 (1977).
40. Spohr, R., *Ions Tracks and Microtechnology Principles and Application.* (Friedr. Vieweg & Sohn: Verlegsgschaft mbH, Braunschweig, 1990).
41. Apel, P., Swift ion effects in polymers: Industrial applications. *Nucl. Instr. Meth. Phys. Res. B* **208**, 11 (2003).
42. Spohr, R., *German Patent DE 2 951 376 C2*, (1983). b) Spohr, R., *US Patent 4 369 370*, (1983).
43. Apel, P.Y.; Korchev, Y.E.; Siwy, Z.; Spohr, R.; Yoshida, M., Diode-like single-ion track membrane prepared by electro-stopping. *Nucl. Instr. Meth. Phys. Res. B* **184**, (3), 337–346 (2001).

44. Trautmann, C.; Bouffard, S.; Spohr, R., Etching threshold for ion tracks in polyimide. *Nucl. Instrum. Methods Phys. Res. B* **116**, 429–433 (1996).
45. Siwy, Z.; Apel, P.; Baur, D.; Dobrev, D.D.; Korchev, Y.E.; Neumann, R.; Spohr, R.; Trautmann, C.; Voss, K.-O., Preparation of synthetic nanopores with transport properties analogous to biological channels. *Surf. Sci.* **532–535**, 1061–1066 (2003).
46. Wharton, J.E.; Jin, P.; Sexton, L.T.; Horne, L.P.; Sherrill, S.A.; Martin, C.R., A method for reproducibly preparing synthetic nanopores for resistive-pulse biosensors. *Small* accepted (2007).
47. Harrell, C.C.; Siwy, Z.S.; Martin, C.R., Conical nanopore membranes: Controlling the nanopore shape. *Small* **2**, 194–198 (2006).
48. Scopece, P.; Baker, L.A.; Ugo, P.; Martin, C.R., Conical nanopore membranes: Solvent shaping of nanopores. *Nanotechnology* **17**, 3951–3956 (2006).
49. Yu, S.; Lee, S.B.; Martin, C.R., Electrophoretic protein transport in gold nanotube membranes. *Anal. Chem.* **75**, 1239–1244 (2003).
50. Menon, V.P.; Martin, C.R., Fabrication and evaluation of nanoelectrode ensembles. *Anal. Chem.* **67**, 1920–1928 (1995).
51. Hille, B. *Ion Channels of Excitable Membranes*, 3rd ed. (Sinauer, Sunderland, MA, 2001).
52. Voet, D.; Voet, J.D. *Biochemistry* (Wiley, New York, 2004).
53. Sigworth, F.J., Structural biology: Life's transistors. *Nature* **423**, (6935), 21–22 (2003).
54. Lee, S.; Zhang, Y.; White, H.S.; Harrell, C.C.; Martin, C.R., Electrophoretic capture and detection of nanoparticles at the opening of a membrane pore using scanning electrochemical microscopy. *Anal. Chem.* **76**, 6108–6115 (2004).
55. Siwy, Z.; Gu, Y.; Spohr, H.A.; Baur, D.; Wolf-Reber, A.; Spohr, R.; Apel, P.; Korchev, Y.E., Rectification and voltage gating of ion currents in a nanofabricated pore. *Europhys. Lett.* **60**, (3), 349–355 (2002).
56. Siwy, Z.; Dobrev, D.; Neumann, R.; Trautmann, C.; Voss, K., Electro-responsive asymmetric nanopores in polyimide with stable ion-current signal. *Appl. Phys. A: Mater. Sci. Process.* **76**, (5), 781–785 (2003).
57. Siwy, Z.; Apel, P.; Dobrev, D.; Neumann, R.; Spohr, R.; Trautmann, C.; Voss, K., Ion transport through asymmetric nanopores prepared by ion track etching. *Nucl. Instrum. Methods Phys. Res. B* **208**, 143–148 (2003).
58. Siwy, Z.; Mercik, S.; Weron, K.; Spohr, R.; Wolf, A.; Grzywna, Z., Comparison of single channel potassium current in biological and synthetic systems – dependence on voltage. *Acta Physica Polonica, B* **31**, (5), 1125–1141 (2000).
59. Siwy, Z.; Fulinski, A., A nanodevice for rectification and pumping ions. *Am. J. Phys.* **72**, 567–574 (2004).
60. Baker, L.; Jin, P.; Martin, C.R., Biomaterials and biotechnologies base on nanotube membranes. *Crit. Rev. Solid State* **30**, 183–205 (2005).
61. Siwy, Z.; Heins, E.; Harrell, C.C.; Kohli, P.; Martin, C.R., Conical-nanotube ion-current rectifiers: The role of surface charge. *J. Am. Chem. Soc.* **126**, 10850–10851 (2004).
62. Harrell, C.C.; Kohli, P.; Siwy, Z.; Martin, C.R., DNA-nanotube artificial ion channels. *J. Am. Chem. Soc.* **126**, 15646–15647 (2004).
63. Heins, E.A.; Baker, L.A.; Siwy, Z.S.; Mota, M.O.; Martin, C.R., Effect of crown ether on ion currents through synthetic membranes containing a single conically shaped nanopore. *J. Phys. Chem.* **109**, 18400–18407 (2005).
64. Siwy, Z.S.; Powell, M.R.; Petrov, A.; Kalman, E.; Trautmann, C.; Eisenberg, R.S., Calcium-induced voltage gating in single conical nanopores. *Nano Lett.* **6**, 1729–1734 (2006).
65. Li, J.; Stein, D.; McMullan, C.; Branton, D.; Aziz, M.J.; Golovchenko, J.A., Ion-beam sculpting at nanometre length scale. *Nature* **412**, 166–169 (2001).
66. Li, J.; Gershow, M.; Stein, D.; Brandin, E.; Golovchenko, J.A., DNA molecules and configurations in a solid-state nanopore microscope. *Nat. Mater.* **2**, 611–615 (2003).
67. Fologea, D.; Gershow, M.; Ledden, B.; McNabb, D.S.; Golovchenko, J.A.; Li, J., Detecting single stranded DNA with a solid state nanopore. *Nano Lett.* **5**, 1905–1909 (2005).

68. Chen, P.; Mitsui, T.; Farmer, D.B.; Golovchenko, J.; Gordon, R.G.; Branton, D., Atomic layer deposition to fine-tune the surface properties and diameters of fabricated nanopores. *Nano Lett.* **4,** 1333–1337 (2004).
69. Chen, P.; Gu, J.; Brandin, E.; Kin, Y.R.; Wang, Q.; Branton, D., Probing single DNA molecule transport using fabricated nanopores. *Nano Lett.* **4,** 2293–2298 (2004).
70. Storm, A.J.; Chen, J.H.; Ling, X.S.; Zandbergen, H.W.; Dekker, C., Fabrication of solid-state nanopores with single-nanometre precision. *Nat. Mater.* **2**, 537–540 (2003).
71. Storm, A.J.; Chen, J.H.; Zandbergen, H.W.; Dekker, C., Translocation of double-stranded DNA through a silicon oxide nanopore. *Phys. Rev.* **71**, 051903 (2005).
72. Ho, C.; Qiao, R.; Heng, J.B.; Chatterjee, A.; Timp, R.J.; Aluru, N.R.; Timp, G., Electrolytic transport through a synthetic nanometer-diameter pore. *Proc. Natl. Acad. Sci. USA* **102,** 10445–10450 (2005).
73. Heng, J.B.; Ho, C.; Kim, T.; Timp, R.; Aksimentiew, A.; Grinkova, Y.V.; Sligar, S.; Schulten, K.; Timp, G., Sizing DNA using a nanometer-diameter pore. *Biophys. J.* **87**, 2905–2911 (2004).
74. Heng, J.B.; Aksimentiev, A.; Ho, C.; Marks, P.; Grinkova, Y.V.; Sligar, S.; Schulten, K.; Timp, G., The electromechanics of DNA in a synthetic nanopore. *Biophys. J.* **90**, 1098–1106 (2006).
75. Chang, H.; Kosari, F.; Andreadakis, G.; Alam, M.A.; Vasmatzis, G.; Bashir, R., DNA-mediated fluctuations in ionic current through silicon oxide nanopore channels. *Nano Lett.* **4**, 1551–1556 (2004).
76. Iqbal, S.M.; Akin, D.; Bashir, R., Solid-state nanopore channels with DNA selectivity. *Nat. Nanotech.* **2**, 243–248 (2007).
77. Han, A.; Schurmann, G.; Mondin, G.; Bitterli, R.A.; Hegelbach, N.G.; de Rooij, N.F.; Staufer, U., Sensing protein molecules using nanofabricated pores. *Appl. Phys. Lett.* **88**, 093901 (2006).
78. Saleh, O.A.; Sohn; L.L., Quantitative sensing of nanoscale colloids using a microchip Coulter counter. *Rev. Sci. Instrum.* **72**, 4449–4451 (2001).
79. Saleh, O.A.; Sohn, L.L., An artificial nanopore for molecular sensing. *Nano Lett.* **3**, 37–38 (2003).
80. Saleh, O.A.; Sohn, L.L., Direct detection of antibody-antigen binding using an on-chip artificial pore. *Proc. Natl. Acad. Sci. USA* **100**, 820–824 (2003).
81. Carbonaro, A.; Sohn, L.L., A resistive-pulse sensor chip for multianalyte immunoassays. *Lab on a Chip* **5**, 1155–1160 (2005).
82. Sun, L.; Crooks, R.M, Fabrication and characterization of single pores for modeling mass transport across porous membranes. *Langmuir* **15**, 738–741 (1999).
83. Sun, L.; Crooks, R.M., Single carbon nanotube membranes: A well-defined model for studying mass transport through nanoporous materials. *J. Am. Chem. Soc.* **122**, 12340–12345 (2000).
84. Ito, T.; Sun, L.; Crooks, R.M., Simultaneous determination of size and surface charge of individual nanoparticles using a carbon nanotube-based Coulter counter. *Anal. Chem.* **75**, 2399–2406 (2003).
85. Ito, T.; Sun, L.; Henriquez, R.R.; Crooks, R.M., A carbon nanotube-based Coulter nanoparticle counter. *Acc. Chem. Res.* **37**, 937–945 (2004).
86. Ito, T.; Sun, L.; Crooks, R.M., Observation of DNA transport through a single carbon nanotube channel using fluorescence microscopy. *Chem. Commun.* **13,** 1482–1483 (2003).
87. Uram, J.D.; Ke, K.; Hunt, A.J.; Mayer, M., Label-free affinity assays by rapid detection of immune complexes in submicrometer pores. *Angew. Chem. Int. Ed.* **45**, 2281–2285 (2006).
88. Park, S.R.; Peng, H.; Ling, X.S., Fabrication of nanopores in silicon chips using feedback chemical etching. *Small* **3**, 116–119 (2007).
89. Heins, E.A.; Siwy, Z.S.; Baker, L.A.; Martin, C.R., Detecting single porphyrin molecules in a conically-shaped synthetic nanopore. *Nano Lett.* **5**, 1824–1829 (2005).

90. Harrell, C.C.; Choi, Y.; Horne, L.P.; Baker, L.A.; Siwy, Z.S.; Martin, C.R., Resistive-pulse DNA detection with a conical nanopore sensor. *Langmuir* **22**, 10837–10843 (2006).
91. Mara, A.; Siwy, Z.; Trautmann, C.; Wan, J.; Kamme, F., An asymmetric polymer nanopore for single molecule detection. *Nano Lett.* **4**, 497–501 (2004).
92. Sexton, L.T.; Horne, L.P.; Sherrill, S.; Bishop, G.W.; Baker, L.A.; Martin, C.R., Resistive pulse investigations of proteins and protein/antibody complexes using a conical nanotube sensor. *J. Am. Chem. Soc.* submitted (2007).
93. Siwy, Z.; Trofin, L.; Kohli, P.; Baker, L.A.; Trautmann, C.; Martin, C.R., Protein biosensors based on biofunctionalized conical gold nanotubes. *J. Am. Chem. Soc.* **127**, 5000–5001 (2005).
94. Bard, A.J.; Faulkner, L.R. *Electrochemical Methods*, 2nd ed. (John Wiley and Sons, New York, 2001).
95. Poretics, www.sterlitech.com
96. Wolf, A.; Reber, N.; Apel, P.Y.; Fischer, B.E.; Spohr, R., Electrolyte transport in charged single ion track capillaries. *Nucl. Instr. Methods B* **105**, 291–293 (1995).
97. March, J. *Advanced Organic Chemistry: Reactions, Mechanisms, and Structure* (McGraw-Hill Book Company, New York, p. 383, 1992).
98. Harrell, C.; Lee, S.; Martin, C.R., Synthetic single-nanopore and nanotube membranes. *Anal. Chem.* **75**, 6861–6867 (2003).
99. Foss, C.A., Jr.; Hornyak, G.L.; Stockert, J.A.; Martin, C.R., Template-synthesized nanoscopic gold particles: Optical spectra and effects of particle size and shape. *J. Phys. Chem.* **98**, 2963–2971 (1994).
100. Brumlik, C.J.; Menon, V.P.; Martin, C.R., Template synthesis of metal microtubule ensembles utilizing chemical, electrochemical, and vacuum deposition techniques. *J. Mater. Res.* **9**, 1174–1183 (1994).
101. Martin, C.R., Template synthesis of electronically conductive polymer nanostructures. *Acc. Chem. Res.* **28**, 61–68 (1995).
102. Parthasarathy, R.V.; Phani, K.L.N.; Martin, C.R., Template synthesis of graphitic nanotubules. *Adv. Mater.* **7**, 896–897 (1995).
103. Nishizawa, M.; Menon, V.P.; Martin, C.R., Metal nanotubule membranes with electrochemically switchable ion-transport selectivity. *Science (Washington, D.C.)* **268**, 700–702 (1995).
104. Lakshmi, B.B.; Dorhout, P.K.; Martin, C.R., Sol-gel template synthesis of semiconductor nanostructures. *Chem. Mater.* **9**, 857–862 (1997).
105. Hulteen, J.C.; Martin, C.R., Template synthesis of carbon nanotubule and nanofiber arrays. *Nanoparticles Nanostructured Films* **9**, 235–262 (1998).
106. Ang, L.-M.; Hor, T.S.A.; Xu, G.-Q.; Tung, C.-H.; Zhao, S.; Wang, J.L.S., Electroless plating of metals onto carbon nanotubes activated by single-step activation method. *Chem. Mater.* **11**, 2115–2118 (1999).
107. Cepak, V.M.; Martin, C.R., Preparation of polymeric micro- and nanostructures using a template-based deposition method. *Chem. Mater.* **11**, (5), 1363–1367 (1999).
108. Hou, S.; Wang, J.; Martin, C.R., Template-synthesized DNA nanotubes. *J. Am. Chem. Soc.* **127**, 8586–8587 (2005).
109. Hou, S.; Wang, J.; Martin, C.R., Template-synthesized protein nanotubes. *Nano Lett.* **5**, 231–234 (2005).
110. Martin, C.R.; Nishizawa, M.; Jirage, K.B.; Kang, M., Investigations of the transport properties of gold nanotubule membranes. *J. Phys. Chem. B* **105**, 1925–1934 (2001).
111. Lu, H.B.; Campbell, C.T.; Castner, D.G., Attachment of functionalized poly(ethylene glycol) films to gold surfaces. *Langmuir* **16**, 1711–1718 (2000).
112. Finklea, H.O.; Avery, S.; Lynch, M.; Furtsch, T., Blocking oriented monolayers of alkyl mercaptans on gold electrodes. *Langmuir* **3**, 409–413 (1987).

113. Yang, Z.; Galloway, J.A.; Yu, H., Protein interactions with poly(ethylene glycol) self-assembled monolayers on glass substrates: Diffusion and adsorption. *Langmuir* **15**, 8405–8411 (1999).
114. Bezrukov, S.M.; Krasilnikov, O.V.; Yuldasheva, L.N.; Berezhkovskii, A.M.; Rogrigues, C.J., Field-dependent effect of crown ether (18-crown-6) on ionic conductance of α-hemolysin channels. *Biophys. J.* **87**, 3162–3171 (2004).

Chapter 7
Tunable Elastomeric Nanopores

G.R. Willmott, M.F. Broom, M.L. Jansen, R.M. Young, and W.M. Arnold

Abstract A general introduction to tunable elastomeric nanopore technology is presented. Tunable nanopores have been used to detect particles ranging from dsDNA to 10 μm diameter spheres using the resistive pulse method. These nanopores are formed in thermoplastic polyurethane by mechanically puncturing a membrane with a sharpened tungsten probe. Optical microscopy, AFM and SEM demonstrate that the pores are near-circular cones. The fabrication process, actuation mechanism and key material properties such as viscoelasticity and failure mechanisms are described. Ionic current-voltage experiments, in which membranes are stretched to ∼30% strain and relaxed, are used to demonstrate reversible actuation of the estimated smaller pore radius over an order of magnitude. Actuation characteristics are compared with simple analyses and modelling of the stretching process. These nanopores are efficient and versatile when compared with similar technologies. Virus sensing is the most immediately appealing application; this and other applications are discussed, along with directions for future work.

Abbreviations and Symbols

α HL	Alpha-haemolysin
AFM	Atomic force microscope
ASTM	American Society for Testing and Materials
dsDNA	double-stranded DNA
ssDNA	single-stranded DNA
PDMS	Polydimethylsiloxane
PNP	Poisson-Nernst-Planck
PS	Polystyrene
SEM	Scanning electron microscope
TPU	Thermoplastic polyurethane
a	Pore radius, smaller radius of a conical pore
a_0	Pore radius at fabrication
a_1	Pore radius following fabrication

G.R. Willmott (✉)
Industrial Research Limited, Lower Hutt, New Zealand; The MacDiarmid Institute for Advanced Materials and Nanotechnology, Lower Hutt, New Zealand
e-mail: g.willmott@irl.cri.nz

O. Hayden, K. Nielsch (eds.), *Molecular- and Nano-Tubes*,
DOI 10.1007/978-1-4419-9443-1_7,

a_2	Pore radius in relaxed membrane
a'	Radius of spherical particle
A	A constant
b	Larger radius of a conical pore
c	Elliptical semi-major axis
C	Membrane capacitance
C_0	Zero-stretch membrane capacitance
d	Elliptical semi-minor axis
D	Diffusion constant
e	Electronic charge
$\mathbf{e}$	Strain tensor
e_i	i-component of $\mathbf{e}$
E	Young's modulus
$\mathbf{E}$	Electric vector field
E_i	i-component of $\mathbf{E}$
f	Frequency in an AC circuit
f_c	Charge factor affecting resistive pulse size
f_s	Shape factor affecting resistive pulse size
F	Faraday constant
G	Conductance
I	Current through a nanopore
I_0	Current through a nanopore that does not contain a particle
I_{ep}	Electrophoretic current through a nanopore
I'	AC peak current in phase with applied voltage
I''	AC peak current leading applied voltage by 90°
$\mathbf{J}$	Current density
$\mathbf{J}_{\mathbf{diff}}$	Current density due to diffusion
$\mathbf{J}_{\mathbf{eo}}$	Current density due to electroosmosis
$\mathbf{J}_{\mathbf{ep}}$	Current density due to electrophoresis
k_B	Boltzmann's constant
K_s	Surface conductance
l	Length of nanopore in z-direction
m	A constant
n	An empirical constant
N_A	Avagadro's number
r	Cylindrical polar co-ordinate
R	The gas constant, resistance of a nanopore
R_0	Nanopore resistance when unstretched
P	Pressure
Q	Quality factor of an AC circuit
Q_{vol}	Fluid volume flow rate
t	Thickness of a membrane
T	Temperature
u_r	Radial displacement
v_0	Orifice volume

v_p	Particle volume
V_0	Potential between fluid cell electrodes
V_{pk}	Peak voltage in an AC circuit
x	Cartesian co-ordinate
X	Distance between holes at distal ends of opposing cruciform arms
X_0	Value of X in an unstretched membrane
y	Cartesian co-ordinate
z	Cartesian co-ordinate, cylindrical polar co-ordinate
α	Extensional strain applied to cruciform arms
ε_0	Permittivity of a vacuum
ε_r	Relative permittivity
$\mathbf{\Phi}_{\mathbf{pd}}$	Particle flux due to pressure-driven flow
η	Fluid viscosity
ν	Poisson's ratio
θ	Cylindrical polar co-ordinate
ρ	Electrical resistivity
$\boldsymbol{\sigma}$	Stress tensor
σ_i	i-component of $\boldsymbol{\sigma}$
$\sigma_{1,2}$	Azimuthal stresses at the edge of an elliptical void
σ_∞	Isotropic radial stress applied to a membrane far from a pore
ζ	Effective electronic charge (e) per particle

7.1 Introduction

In this chapter we describe the recent development of tunable elastomeric nanopores. The technology is based on simple underlying concepts. A nanopore is made by puncturing a hole in an elastomeric membrane, so that when that membrane is "tuned" by biaxially stretching and relaxing on a macroscopic (millimetre) scale, the nanopore is reversibly opened and closed. Pores are typically used in an aqueous ionic environment where the size of the pore, and passage of particles and fluid through the membrane, can be studied by applying a potential across the membrane and measuring the current through the pore. Despite the simplicity of fabrication and operation, or perhaps because of it, tunable nanopores have very high potential for applications in nanoscale sensing, fluidics and biotechnology.

Tunable nanopore technology has been developed by Izon Science Ltd (Christchurch, New Zealand; formerly Australo Ltd), who have developed specialist expertise in producing individual nanopores in thermoplastic polyurethane (TPU) membranes. Izon uses specially etched tungsten needles to control membrane puncture and create single apertures within an elastomeric "cruciform", so-called because of its cross-like shape. This configuration enables precise, dynamic control of aperture size via biaxial stretching and relaxation of the distal ends of the cruciform. The reversible actuating capability of an elastomeric nanopore is demonstrated in Fig. 7.1.

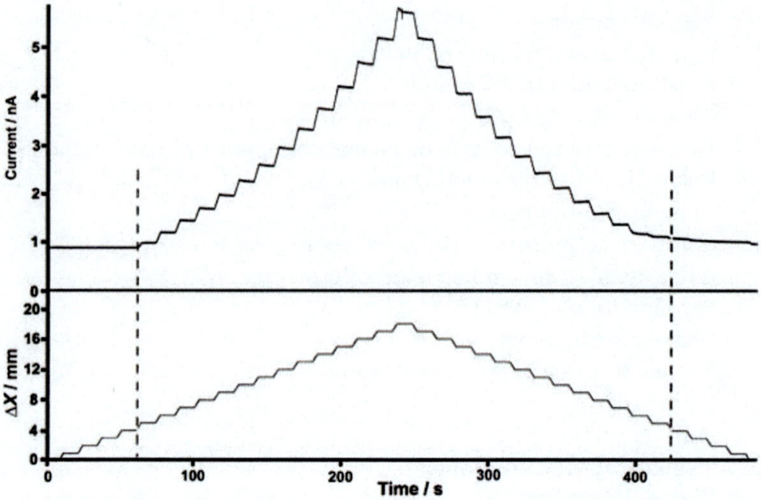

Fig. 7.1 Plots against the same time base showing the current measured through a tunable nanopore (*top*) and macroscopic extension of a cruciform, ΔX (*bottom*). The nanopore was surrounded and wet by 0.1 M KCl solution and applied transmembrane potential was 100 mV. The cruciform was mechanically tensioned in 1 mm steps from a relaxed state ($\Delta X = 0$ mm) to $\Delta X =$ 18 mm, and back to a relaxed state. This particular aperture has a residual (open) current of 0.8 nA when unstretched. Deformation up to $\Delta X = 4$ mm (denoted by *vertical dashed lines*) does not alter the aperture current; above $\Delta X = 4$ mm, stepwise incremental changes in aperture current match the cruciform extension

Nanopores enable nanoscale particle detection based on the "resistive pulse" method (also known as the Coulter Principle™). While a particle is within an opening, the electrophoretic current passing through that opening is blocked. The passage of the particle, known as a "translocation" event, is observed as a resistive pulse, or a brief reduction in the observed current. Tunable nanopores represent a significant extension of nanopore technology. They have demonstrated capability of sensing and gating individual DNA oligonucleotides [1], while reversible actuation is possible over an order of magnitude of effective linear pore size [2, 3]. Recent work has focussed on sensing dispersed nanoparticles [4–8], including measurements of particle concentration [4, 5], size [6] and resistive pulse shape [7]. Particles have also been distinguished from each other based on their size or biofunctionalization [9].

7.1.1 Comparison with Similar Technologies

Controlled, reversible alteration of nanopore size in situ is a unique concept within the literature [10]. Related work can be considered either in terms of polymer actuation in micro- and nanofluidic devices, or studies of individual static nanopores. The former is an area of rich potential which has not been widely studied, whereas the latter is an established topic of intense interest in nanotechnology.

Polymers are becoming more widely used in the development and construction of microfluidic devices because they are cheaper and easier to process and seal than more traditional microfabrication materials such as glass or silicon, whilst the analytical devices that result can attain comparable resolution [11]. Actuation of non-glassy polymers, such as elastomers, has not been widely studied in this context. There have been reports describing stretchable microfluidic channels and valves in elastomeric polydimethylsiloxane (PDMS) for size-sorting of microparticles [12, 13] and actuation of membranes by altering the size of carbon nanofibres [14]. There is extensive contemporary research into nano- and microelectronic machinery, so it is likely that actuation of polymers in small-scale fluidics will be of increasing interest.

The science of static, solid-state nanopores provides a well-studied point of comparison for tunable nanopore work. Static pores are usually fabricated using advanced silicon-based techniques or by track-etching of hard polymers [15, 16]. A large body of nanopore research has been carried out using static biological pores. Typically, alpha-haemolysin (α-HL) pores [17] have been studied in phospholipid bilayer membranes [15, 18–26].

By comparison with static solid-state or biological pores, there are a number of advantages of tunable nanopores. The fabrication technique is extremely time and cost efficient when compared with focused ion beams, electron beams or track-etching. A single fabrication unit can produce several tunable nanopores per hour using commercially available TPUs. Light, portable actuation and analysis systems have been developed, which offer advantages for applications in the field as well as saving space in the laboratory.

Once fabricated, tunable nanopores are physically and chemically robust. There are no exceptional conditions required for long-term storage and effective specimen reuse at near-neutral pH, although care should be taken to avoid contamination, extremes of temperature and some polar solvents and chlorinated hydrocarbons [27]. Wear and tear can be expected for actuation over an extended time period. Crucially, contaminant particles caught within an aperture can be dislodged by simply opening the aperture, especially in conjunction with a sharp increase or reversal in applied pressure or potential. Elastomeric pores are stable at high transmembrane potentials, which can adversely affect biological nanopores.

In terms of utility, there are a number of simple, novel functionalities facilitated by a nanopore which can resize in situ. The effective diameter of any individual pore can be reversibly controlled over an order of magnitude. The exact range of operation depends on the individual pore or series of pores. Gating capability has been observed for particles ranging from 10 μm diameter spheres to DNA molecules, using different nanopore specimens. In some cases, pores close entirely when the membrane is relaxed, so that fluid and ionic species may also be gated. The ability to alter pore size greatly aids analysis of different nanoparticle populations which are distinguishable on the basis of size.

Tunable nanopores are so-called due to their capability for fine in situ adjustments to the experimental system. Nanopore science necessarily involves delicate experimental work, which may require careful consideration of a specific chemical and physical environment. The tunable pore approach allows the user to first set up an experiment within loose constraints, then obtain the precise configuration or properties required by actuating the pore. Real-time control electronics allow the user to identify whether, for example, the required electrophoretic current or translocation characteristics are observed. Tunable nanopores will also facilitate the use of further instrumentation in conjunction with individual nanopores.

There are several other advantages to tunable nanopores. Particular parameters of a test solution can be resolved with increased precision and accuracy by repeating an experiment at several different aperture sizes. Gating could be used to mix solutions within the pore, and therefore localise and study an interaction or reaction. A closing pore might be used to trap an elongated particle, such as a polypeptide or nanotube, within a pore. The main disadvantage of tunable nanopores is that manufactured pores are not entirely uniform in geometry. This issue can be addressed using tuning, calibration and microscopy.

A number of specific applications have been proposed for tunable nanopores. Probably the most immediately promising involves the detection, counting and analysis of virus particles on length scales of tens to hundreds of nanometres. General protocols have been developed for analysing the size distributions and concentrations of dispersed particles. Particle interactions, aggregation and surface group reactions can also be initiated in situ, and monitored as they progress. The simplicity of the concept suggests that this technology could become an important, fundamental tool in nanotechnology. Tunable pores are likely to facilitate improvements over static nanopores in any particular application area. The application that has created the most widespread general interest in nanopores is high speed DNA sequencing. Further nanopore applications in biosensing, biotechnology, industry and nanometrology seem inevitable.

7.1.2 Structure of Chapter

The actuation apparatus, cruciforms and pores that have been developed by Izon are introduced in Section 7.2. TPU, the key material, is discussed in some detail, with special attention paid to the relevant mechanical, structure and failure properties. Section 7.2 also introduces the nanopore fabrication methodology and presents imaging of the tunable nanopores. Materials other than TPU elastomers and different actuation platforms are not explicitly considered here. However, characterisation of isotropic strain in the membrane surrounding the nanopore ("local" strain), which will be relevant to tunable nanopores regardless of apparatus, is described at the end of Section 7.2.

Section 7.3 presents analytical descriptions of actuation for idealised pore shapes. These results are compared with actual changes in pore characteristics using ionic current measurements and imaging techniques. Although there is a clear correlation between cruciform deformation and changes in ionic current, determination of absolute aperture sizes is difficult. The effects of viscoelasticity are demonstrated using a model experimental system and AC electronic measurements, and pore shape variations are discussed.

Translocations are the subject of Section 7.4. Concepts relating to resistive pulse signals, applied electric fields and particle flux are introduced. A number of translocation experiments are then presented, demonstrating the influence of variables such as pore size and particle size, type and concentration. The final parts of this chapter introduce and discuss applications (Section 7.5) and future research directions (Section 7.6).

Development of tunable nanopore technology is continuous and ongoing. Experiments presented should be considered "snapshots" which demonstrate key features of the technology, even though some specific experimental conditions may be difficult to reproduce. For example, some of the precise TPU materials and elastomeric cruciform geometries used in experiments are obsolete in the manufacturing process. Most tunable nanopore experiments to date have used pores of the order of tens to hundreds of nanometres in diameter. However, many of the concepts and trends presented apply to both larger and smaller pores. A notable exception is the neglect of electroosmotic effects in parts of Sections 7.3 and 7.4. For molecular scale pores, more sophisticated theoretical treatments should be employed.

7.2 Apparatus, Materials and Nanopores

7.2.1 Cruciforms

The platforms used to support and actuate nanopores are cross-shaped, injection-moulded polyether-based TPU planar units known as "cruciforms". Cruciform geometry is shown in Fig. 7.2. Pores typically taper from a smaller opening on the upper ("trans") surface to a larger opening on the lower ("cis") surface of each cruciform, as viewed in 7.2b. The TPU presently used in Izon's manufacturing process is Elastollan® 1160D (BASF), but cruciforms have previously been fabricated in another TPU (Elastollan® 1195A, BASF) [1, 2, 8]. In another version of the technology [2], the central area of the injection-moulded template was absent and a urethane cement was used to affix a patch of similar, but not injection-moulded, TPU (e.g. Stevens polyurethane ST-1522FS-85, thickness 200 μm). A single nanopore was then formed in each patch.

Basic material properties of the materials used are shown in Table 7.1. Poisson's ratio ν is poorly characterised for TPUs, with the literature suggesting values such as $\nu > 0.49$ [28] and $\nu \cong 0.45$ [29]. A value of $\nu = 0.49$ was used in modelling and analysis described here.

7.2.2 Actuation Devices and Fluid Cells

Resizing of pores is achieved by macroscopic actuation of the cruciform using pegs placed in the holes at the ends of the four arms. In experiments to date, actuation

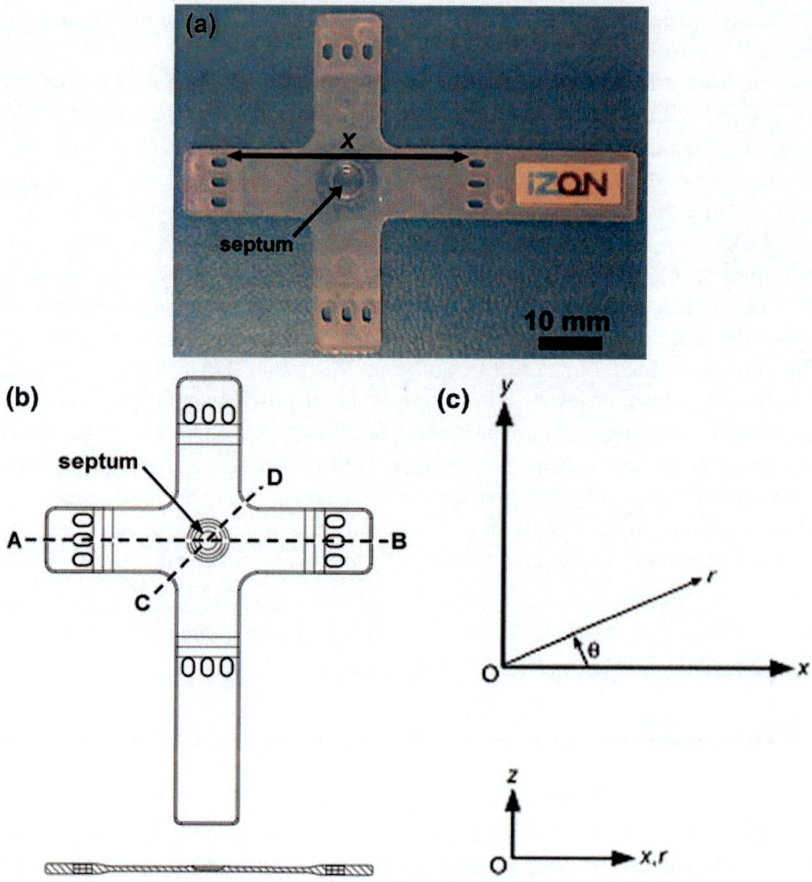

Fig. 7.2 (**a**) Photograph of a cruciform. (**b**) Schematic plan (*upper*) and sectional (*lower*) views of a cruciform, as manufactured at the time of writing. Cruciforms are mostly 0.8 mm thick. Material within ~5 mm of the end of the cruciform arms, near the holes used for actuation, is ~1.5 mm thick. The central septum has diameter 2.8 mm and is 250 μm thick. (**c**) Cartesian (*x*, *y*, *z*) and cylindrical polar (*r*, θ, *z*) co-ordinates, defined in the same plane as shown in the plan and sectional (*lower*) views. For both co-ordinate systems, the origin is at the nanopore opening on the lower surface of the cruciform

has been symmetric: displacement of the pegs was equal for the four arms. Internal hole surfaces that are furthest from the centre of the cruciform are separated by a distance X across opposite arms. The value of X at zero applied stress is X_0. For a pristine, resting cruciform, $X = X_0 = 41.5$ mm. Extensional strain α is defined as

$$\alpha = \frac{X}{X_0} - 1. \tag{7.1}$$

Actuation units can be hand-powered, or automated using a motorised drive. Hand-powered units (such as the Izon *qNano*) can be small and portable, and produce

Table 7.1 Material properties of elastomeric TPUs used in tunable nanopore manufacture, as listed by the manufacturers [30, 31] and converted to SI units. All three polymers are polyether-based TPUs

Property	ASTM[a] test method	Unit	Elastollan 1160D	Elastollan 1195A	Stevens ST-1522FS-85
Specific gravity	D792	g cm^{-3}	1.17	1.14	1.12
Hardness	D2240	Shore A	~100[b]	95	85
Tensile stress					
@ 100% elongation	D412	MPa	22	12	5.5[c]
@ 300% elongation	D412	MPa	33	21	11[c]
At break:					
Tensile stress	D412	MPa	40	36	45[c]
Elongation	D412	–	415%	490%	500%[c]
Set	D412	–	60%	65%	25%[c]
E-Modulus	D638	MPa	200[d]	51.7	–

[a]American Society for Testing and Materials.
[b]Shore D hardness of 60.
[c]Obtained using ASTM method D638.
[d]Obtained using ISO method 527.

less electronic noise. Nanopores are typically used in a liquid environment, requiring a fluid cell (e.g. Fig. 7.3) which prevents evaporation and contamination, while facilitating actuation. High resolution ionic current measurement is required for nanopore experiments. For example, the typical current change caused by a DNA translocation is ~100 pA. Currents can be measured using precision instrumentation such as a patch clamp amplifier in conjunction with a digitiser unit. Commercial

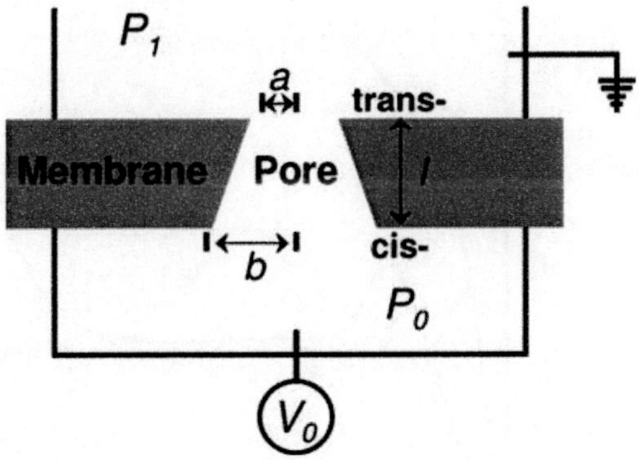

Fig. 7.3 Schematic cross section of a typical fluid cell used with tunable nanopores, in which solution can be added to the upper fluid cell. Pores can inverted (cis-surface upwards), and surface tension prevents leakage along the membrane surfaces. Labelled parameters (applied pressure ΔP $(= P_1 - P_0)$, applied voltage V_0, pore length l and conical radii a and b) apply for the conical geometry assumed and discussed in Section 7.3.3

Izon equipment includes Ag/AgCl electrodes built into shielded fluid cells, coupled with electronics so that the typical RMS noise during operation is less than 10 pA. An important recent addition to commercial Izon apparatus is the variable pressure module (VPM), which enables a precisely measurable pressure difference to be applied across the membrane [4, 5].

7.2.3 Mechanical Properties of Thermoplastic Polyurethanes

Material and mechanical properties are a key consideration for tunable nanopores. The most relevant aspects of this complex topic will be introduced here. Texts providing detailed descriptions of these properties range from high-level summaries and mechanical toolboxes (e.g. [32]) to modern texts dealing with advanced mechanical models of polymer microstructure [33, 34].

Elastomers approximate the simple behaviour of an ideal rubber, which is elastic, isotropic and incompressible. If deformation is independent of strain rate, this idealized deformation is "hyperelastic", and the stress-strain curve may be non-linear. Schematic stress-strain curves for elastomers and similar materials are shown in Fig. 7.4. Rubbers and elastomers (including TPUs) typically have a continuous, near-linear stress-strain relation which does not indicate yield. When considering actuation of an elastomeric cruciform at strains up to ~30%, much less than strain at failure (Table 7.1), yield and failure are not relevant. The curve shape can be described using a linear elastic (Hookean) relation with Young's modulus E,

$$\sigma = E\mathbf{e}. \tag{7.2}$$

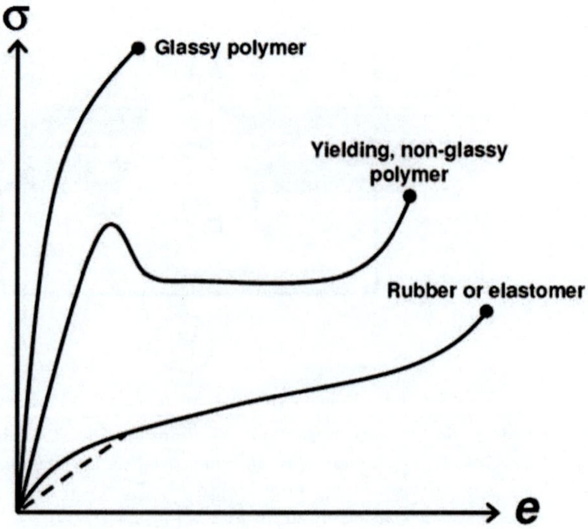

Fig. 7.4 Schematic stress–strain curves for tensile extension of pristine material from zero strain at constant strain rate, following Painter and Coleman [32]. The local maximum in the *curve* for a non-glassy polymer (e.g. polythene) indicates yield, at which point irreversible damage weakens the material. Terminal failure occurs at the end of each curve. TPUs fail at ~500% strain (Table 7.1); the *dotted line* represents the Hookean approximation over ~100% elongation

Use of Eq. (7.2) to model the stress-strain behaviour in Fig. 7.4 is clearly a simplistic approximation. In most materials, Hooke's law is only applicable for small strains. More complicated hyperelastic models, for example the Mooney-Rivlin model, can give a closer fit to the curve.

First attempts at analysis and modelling presented here and in [3] assume linear elasticity (Eq. 7.2). In practice, elastomers are non-ideal rubbers which exhibit some viscoelasticity, volume change and permanent deformation or "set" (Table 7.1), especially following loading to large strains. For tunable nanopores, deviations from linear elastic behaviour are most important for near-pore material, which is loaded to high strains during fabrication, and critically affects cruciform actuation at low strains.

A viscoelastic material retains elements of both viscous and elastic behaviour. The viscoelasticity of Elastollan® 1195A has been studied in a series of experiments. Apparatus included a light horizontal shaft clamped to the test sample at one end. At the other end was a load cell (S beam, 25 kg, Model 60001, EMC, Auckland) in parallel with a linear displacement transducer (RS Type DC15, LVDT, Radio Spares, Auckland). Samples were held in a constant temperature environment (± 2°C) during each experiment, because TPU rigidity decreases with increasing temperature [32]. Loads were applied using weights suspended from a wire cable.

Changes in cruciform length have been recorded following the application or release of stress. In Fig. 7.5a, a cruciform is stretched to a constant displacement for 2 h, then released to zero stress. Immediately following load release, the displacement falls sharply, but continues to fall after 16 h. The relaxation time constant is about 30 s. Figure 7.5b shows the residual displacement 30 min after load release over a wide range of extension times. This log–log plot reveals two linear

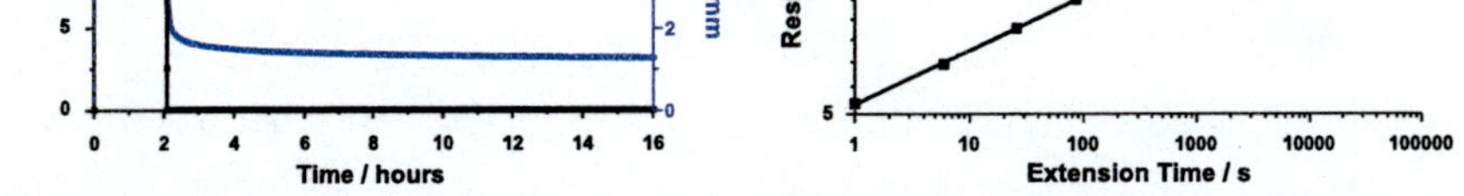

Fig. 7.5 Results from experiments at 20°C in which clamps holding pristine TPU cruciforms, initially at ~32 mm spacing, were stretched by 20 mm in one direction, and held at constant displacement for some time before being released to zero stress. (**a**) Time variation of displacement and load for a particular experiment. (**b**) A log–log plot of residual extension 30 min after load release, following various periods of extension. *Straight lines* are fitted to the data for less than 120 s extension (*solid line*, gradient = 0.091) and greater than 300 s extension (*dotted line*, gradient = 0.152)

portions with separate power law dependencies. The transition occurs at ~100 s extension.

The effects of cyclical extension and relaxation of elastomers are of particular relevance. Permanent deformation is related to the Mullins effect [35–37], an empirical description of stress softening in rubber-like materials. The Mullins effect is seen when an elastomer undergoes stretching (including biaxial stretching [38]) cycles between zero applied stress and a certain maximum strain. After a number of cycles (usually between 3 and 10, depending on the material), the stress-strain curve becomes repeatable, as long as the material is not further extended beyond the maximum strain. This repeatable behaviour can be considered approximately elastic. The Mullins effect "reflects configurational changes within the fine structure of the material which permits subsequent deformation to take place more readily" [39]. Real materials exhibit a non-ideal Mullins effect, with creep and a relatively small degree of inelastic stretching observed in continued cycles. However, if cruciforms are pre-cycled prior to use, a working range of strains is established in which pore actuation is reasonably well characterised and controllable [2].

Measurements to assess the Mullins effect for TPU are shown in Fig. 7.6. The data show that the load required to achieve the target displacement fell by about 5% after the first extension cycle. Following 10 cycles, the load converged to a final value some 10% less than the initial load required for the same displacement. Prior to convergence, the total time at high extension was about 100 s, approximately the same as the crossover extension time in the creep measurement. It would appear that the initial linear portion of the curve in Fig. 7.5b represents irreversible deformation

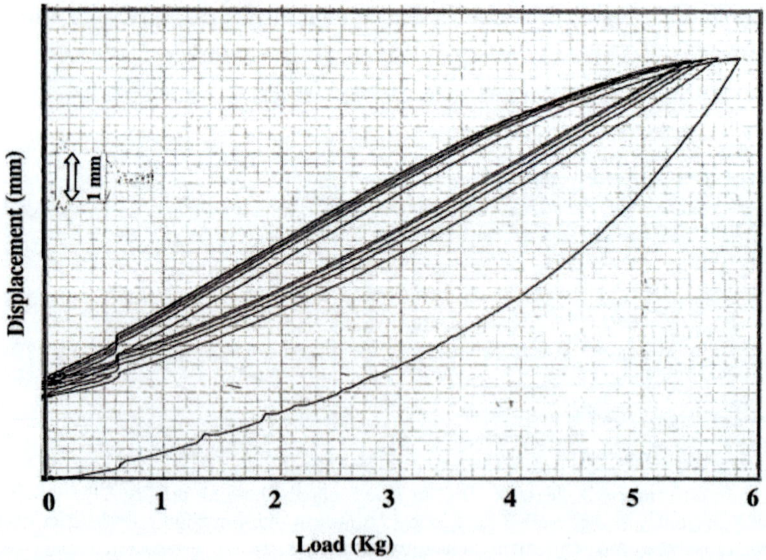

Fig. 7.6 Cyclical stress-strain curves for uniaxial extension across opposite arms of a cruciform. Measurements were obtained using an Instron 1,122 Tensile Testing Machine, fitted with a 10 kg load cell. Crosshead speed was 50 mm min^{-1}, initial jaw separation was 30 mm, and the final tensile set was ~2 mm

equivalent to repeated cycling, while the latter portion represents a different long time-scale behaviour.

To complete this section, we consider the choice of material for elastomeric nanopores. Apart from being readily available, TPUs provide two significant advantages. Firstly, they can be used over a wide range of strains, allowing actuation of estimated pore radius over an order of magnitude in response to ~30% macroscopic strain, as demonstrated in Section 7.3. Secondly, reproducibility can be enhanced by cycling the pristine material prior to use. Nevertheless, actuation of TPUs is difficult to precisely quantify due to viscoelasticity, and creep could be important for some applications. Control of nanopore size, shape and morphology are important for precise and specialized applications, and are also likely to be important in future selection of materials.

7.2.4 Failure and Structure of TPUs

The failure of nanopore materials is a critical consideration, because failure mechanisms determine pore shape and the properties of pore walls. Failure modes are most strongly related to material brittleness (or ductility). Elastomers are very ductile, requiring a large degree of energy to be absorbed prior to failure. On macroscopic length scales, it is sufficient to know that failure occurs via "ductile tearing" rather than brittle fracture. In the present case, we are concerned with micro- to nanoscale failure morphologies in the region adjacent to failure. Failure morphologies for TPUs can vary widely, depending on elastomer composition [40].

Although rubbers and elastomers do not typically yield (Fig. 7.4), the plateau in the stress-strain curve can be attributed to rearrangement of polymer chains and lamellae in the direction of stress [32]. The material can be stretched until localised instabilities form and "necking" (extension to greater strain than surrounding material) occurs [41]. Regions in which there is necking become progressively weaker, leading to ductile tearing. As with any solid, failure preferentially develops in the most energetically favourable direction, which is determined by a combination of high tensile stress and low tensile strength [42], and depends heavily on populations of flaws or impurities.

The microstructure of the material is important when nanopores have dimensions comparable to the size of discrete structural elements. TPU is composed of chains of alternating hard crystalline and soft amorphous regions, with "virtual" crosslinks [27]. The polymers in Table 7.1 are polyether-based and have Shore A hardness greater than 70, which suggests that lengths of soft and hard chain regions are approximately 15–90 and 150 nm respectively [27]. X-ray diffraction imaging of a polyurethane elastomer at 500% elongation [43] shows chains oriented in the direction of applied strain. In this case, lengths of soft and hard chain regions were approximately 1–2 μm and 250 nm respectively. When a stressed, ductile material contains such stress-concentrating inhomogeneities, ruptures or cavities form adjacent to these features, eventually forcing ductile failure of the material between cavities [42]. The resulting fracture surface displays a dimpled or ribbed pattern.

The following further failure morphologies could be expected for TPUs:

- Little permanent deformation (e.g. "crazing" and "shear banding" observed with other polymers), because the deformation process is mostly reversible.
- Folding near the failure surface, where material has been hyperextended prior to the onset of instabilities. Folding is more pronounced in a necking region.
- Cracks, which can propagate relatively short distanced in ductile materials, especially at low temperature or high strain rate [41].

7.2.5 Fabrication Method

Conceptually, nanopore production is simple: apertures are produced by penetrating a membrane with a sharp tip. This process is controlled by an electronic feedback circuit that monitors the degree of penetration. The feedback circuit requires that the penetrating tip conducts electricity, so electrochemically etched tungsten probes are used. These probes (Fig. 7.7a) are produced by standard electrochemical

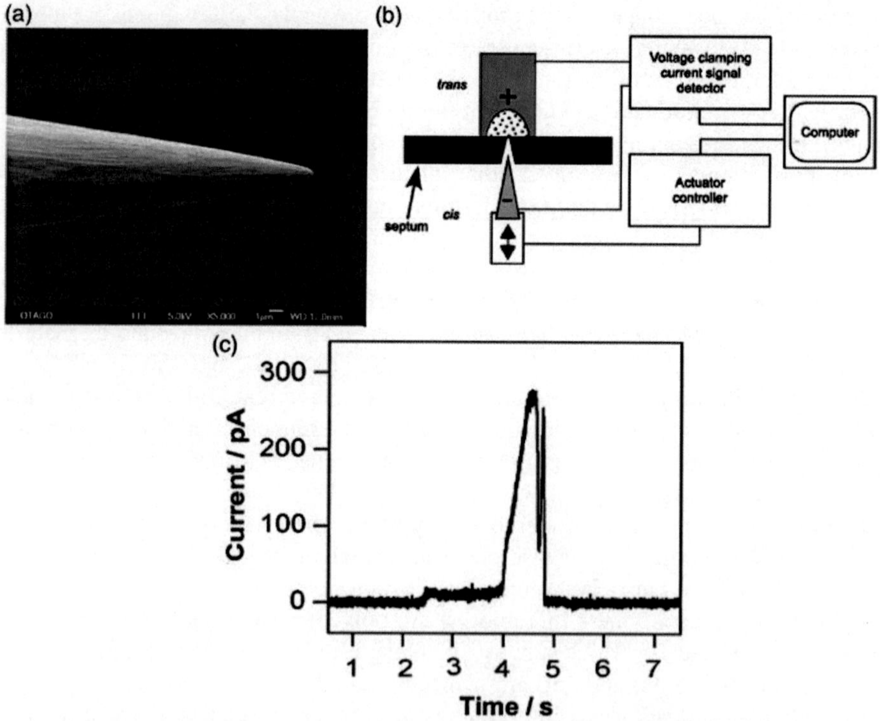

Fig. 7.7 Nanopore fabrication. (**a**) SEM image of a typical tungsten probe used in aperture manufacture. Terminal radius can be at least as low as 50–100 nm. (**b**) A schematic diagram of the electronic feedback system used. (**c**) A record of current against time during penetration. (**b**) and (**c**) are reproduced from [1]

etching of polycrystalline tungsten rod in sodium hydroxide. It is possible to control parameters of the etching process to reproducibly target specific probe geometries. For a regularly-shaped pore, it is critical that these probes are symmetrical and smooth.

Electronic feedback (Fig. 7.7b) is provided by setting up a voltage clamp between two electrodes. The first electrode is the tungsten probe, which is attached to a computer-controlled actuator, and penetrates the membrane perpendicular to the cis face. The second is a stainless steel, electrolyte-filled receptacle on the trans side of the membrane. A user-defined bias voltage is applied across the two electrodes (typically 200 mV) and penetration is performed at a user-defined rate (typically 1.5 μm s^{-1}). A current trace recorded during the fabrication of the aperture (Fig. 7.7c; see also [1]), shows the onset of penetration and the increasing current due to the progressive immersion of the exposed probe in electrolyte. Penetration continues until a set-point current is reached, whereupon probe actuation reverses until the probe is completely withdrawn.

By monitoring the circuit resistance in this way, it is possible to precisely monitor the penetration process and produce apertures of particular sizes – where size is measured in terms of characteristic resistance. However, the level of control is such that nanopores have unique features on the micron to nanometre scale, regardless of the parameters used.

7.2.6 Nanopore Outcomes

Optical microscopy can be used to observe the larger, cis- openings of individual nanopores to resolutions of the order of microns (Fig. 7.8). Optical images of pores in a transparent polymer can also be obtained at an oblique angle to the membrane surfaces, revealing the profile of the pore within the membrane (Fig. 7.8b). This profile appears roughly symmetric about a central axis, with a slightly concave (trumpet-shaped) internal surface.

SEM has been used to investigate the morphology of cis (Fig. 7.9) and trans (Fig. 7.10) nanopore openings. SEM is essentially a destructive technique: the coated pore is contaminated and therefore unsuitable for ionic current measurements.

Figure 7.9a demonstrates that an unstretched pore can be irregularly shaped at the membrane surface. When the membrane is stretched, the pore appears more circular, but aspects of the unstretched shape are retained. Asymmetric failure of the membrane surface about the tungsten tip can occur during pore formation. When unstretched, there is evidence for a zone of ribbed, ductile folding radiating from the pore. This material has been hyperextended during pore formation. Unstretched cis openings typically have diameters between 5 and 20 μm.

Images of another typical pore (Figs. 7.9b and 7.9c) demonstrate that geometrical irregularity is most pronounced at the membrane surface. Pores are regularly shaped, with a near-circular cross section, as they taper into the elastomer below the cis

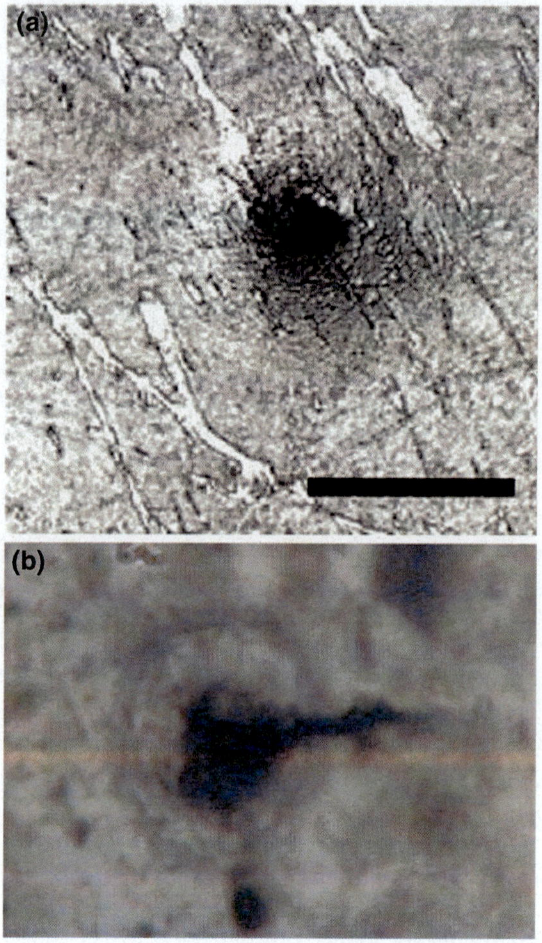

Fig. 7.8 Optical micrographs of nanopores. (**a**) The cis surface of a membrane, reproduced from [1]. The *scale bar* represents 50 μm. (**b**) A dark pore profile viewed with the normal to the membrane surface tilted by ~50° relative to the line-of-sight. Cis-openings of each nanopore are approximately 20 μm in diameter

surface. Any non-circularity arises from azimuthal anisotropy in the stress field, caused by either the applied strain or probe geometry during puncture.

Cis pore images clearly relate to the failure mechanisms discussed above. In Fig. 7.9c, ribbed structure is observed on internal pore walls, beneath the membrane surface. Cavities or instabilities have nucleated with typical spacing of ~1 μm, consistent with the spacing of hard and soft regions in an elongated TPU. The folding mechanism explains the observed morphology of unstretched pores (e.g. Fig. 7.9a), as material must be conserved after the newly-punctured elastomer is relaxed. The appearance of elongated branches for the unstretched pore in Fig. 7.9a is perhaps surprising. Tensile stress near the edge of the pore is intensified during stretching, and may (along with any local stress-concentrating factors) exceed the tensile strength of TPU.

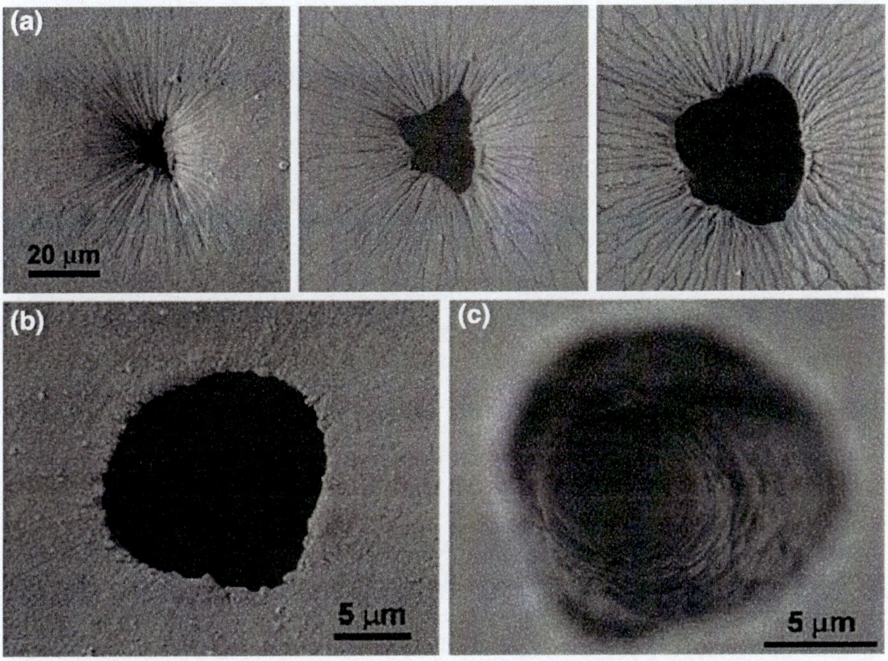

Fig. 7.9 SEM images of the cis aspects of two pores. In (**a**), three images are shown at the same scale. From *left* to *right*, images correspond to stretches of $\alpha = 0$, 0.10 and 0.22 respectively. (**b**) and (**c**) show another pore, with the focus shifted from the surface in (**b**) to the pore walls below the surface in (**c**). In order to image the insulating polyurethane, membranes were sputtered with a ~5 nm thick layer of gold/palladium. Imaging at the lip and interior of the pore are sometimes affected by charging due to uneven application of this coating

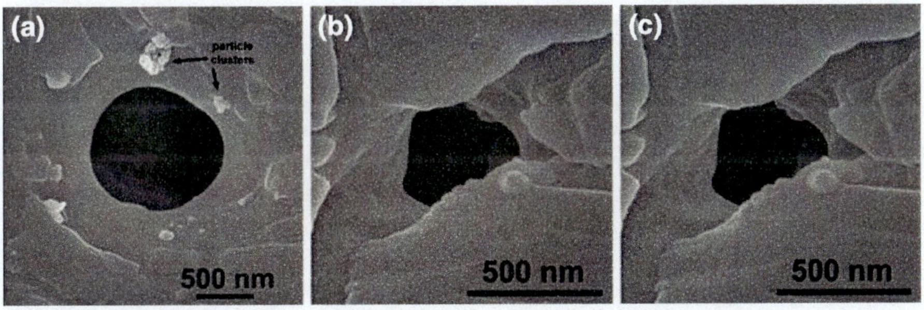

Fig. 7.10 Micrographs of trans pores with surfaces coated as in Fig. 7.9. Prior to imaging, pores were used for translocation experiments using polystyrene spheres. (**a**) A pore at $\alpha = 0.22$, reproduced from [4]. (**b**) A pore at $\alpha = 0.10$, in which the conductive coating partially obscured the opening. (**c**) Another pore at $\alpha = 0.10$, with a near-elliptical shape and polystyrene spheres adhered to the membrane surface

SEM studies of trans surface pores (Fig. 7.10) are difficult: their size and conductive coating make them hard to locate and image. One clear image (Fig. 7.10a) shows a near-circular stretched pore. The opening observed in Fig. 7.10b appears smaller, and is slightly irregularly-shaped. The entrance is also partly obscured by a protruding plate from the conductive gold/palladium coating. A third opening (Fig. 7.10c) has a smooth, elongated shape. Polystyrene spheres, which were used in experiments prior to imaging, are observed as clusters in Fig. 7.10a and as individual particles in Fig. 7.10c. The observed particles were apparently unmoved by stretching, voltage changes or pressure while in use, so it is likely that they have adhered or bound to the pore surface. Particles immobilised in or near the pore entrance could significantly affect current measurements.

Atomic Force Microscopy (AFM) can be used to obtain near-surface nanopore profiles. Resolution is generally lower than for SEM imaging. An AFM cantilever is physically restricted by the pore walls, so the feedback system generates additional noise and, at some depth, the probe loses contact with the surface. The TPU surface is relatively soft, so complicated tip-surface interactions can produce further artefacts in the scan.

Cis and trans pores imaged using AFM (Fig. 7.11) show some evidence of directional failure or tearing (e.g. Fig. 7.11a), consistent with those imaged using SEM. Depth profiles across trans openings have been measured as a function of α (Fig. 7.12). The pore becomes progressively smaller below the trans surface. The gradient of the pore walls increases with increasing α, so it is apparent that the hole size increases when the membrane is stretched.

Trans side features observed using SEM and AFM appear to be of the order of a micron in size, yet it is demonstrated in the next section that currents measured through the pores suggest a much smaller hole: effectively with a radius of 100 nm or less [2]. This disparity can be explained by the following lines of reasoning, which are consistent with all of the available evidence.

1. SEM and AFM images only reveal surface features. AFM in particular indicates that the narrowest part of the pore opening (the "constriction") lies below the membrane surface.
2. Trans pore openings are difficult to locate and image. It is possible that the particular pores imaged have relatively large surface features.
3. Calculations of effective radius from ionic current measurements (Section 7.3) do not take into account the concavity seen in optical pore profiles (Fig. 7.8b).

7.2.7 "Local" Strain from Cruciform Actuation

A variety of tunable nanopore methods, geometries and materials have been and are being trialled. However, analysis of pore actuation can be simplified to exclude specific geometry by considering an azimuthally symmetric, radial strain e_r in the surrounding isotropic membrane. We label this strain the "local" strain to denote that it is defined without reference to geometric or mechanical properties external to

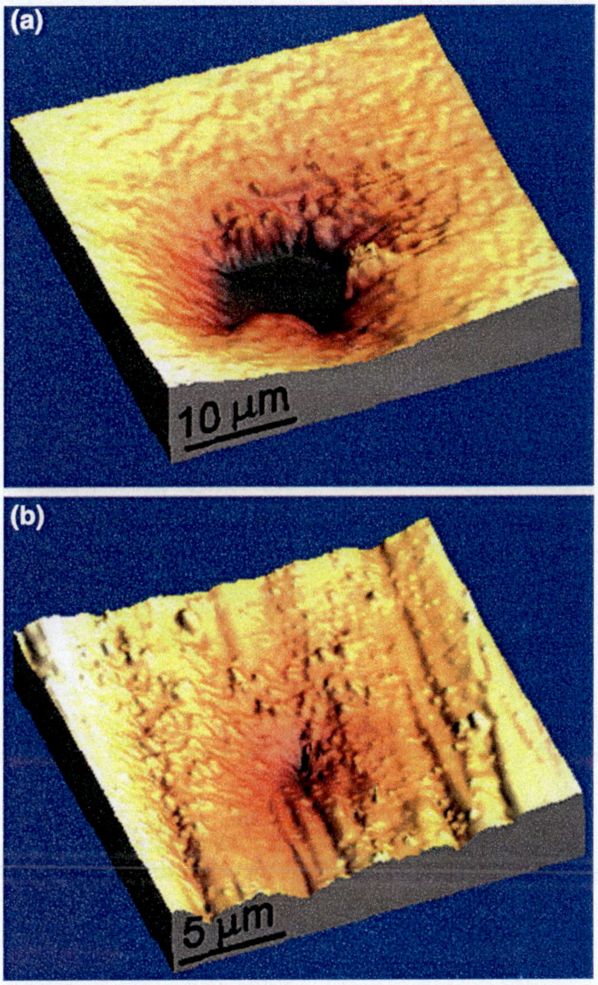

Fig. 7.11 Tapping mode AFM scans from showing 3D projections of (**a**) a cis pore at $\alpha = 0$, and (**b**) a trans pore at $\alpha = 0$, reproduced from [2]. The z scaling is the same as the horizontal scaling in these images

the membrane. It is applied far enough from the pore that strain is invariant relative to azimuthal or radial position, and through the membrane thickness.

In the case of a particular cruciform geometry [3] similar to Fig. 7.2, local strain has been studied both by modelling and by experiment. Figure 7.13 shows modelled strain profiles in the plane of a cruciform at $\alpha \cong 0.24$. Strain within the septum is isotropic and homogeneous to better than 2%. The largest strains are between the cruciform arms, at the edge of the elastomer. Along the cruciform arms, the largest strain occurs in extension of the arms, while the arm width decreases. The distribution of stresses closely follows that of the strains.

The strain field near a nanopore was measured from optical micrographs [2]. Data confirm that it is reasonable to assume that e_r is isotropic over the approximate range 50–250 μm from the pore (measured at zero stretch). Local strain

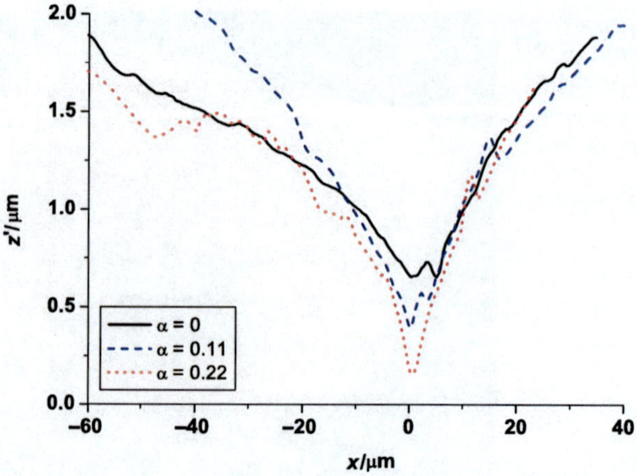

Fig. 7.12 AFM profiles along a line parallel to scanning direction, passing through the centre of a trans opening, reproduced from [2]. Data has been smoothed for clarity. The scans have been shifted in the *x*-direction to align pore positions. Alignment in the *z'* direction is arbitrary, although an attempt has been made to align the position of surfaces far from the pore. Note that $z' = -z + 2.1\ \mu m$

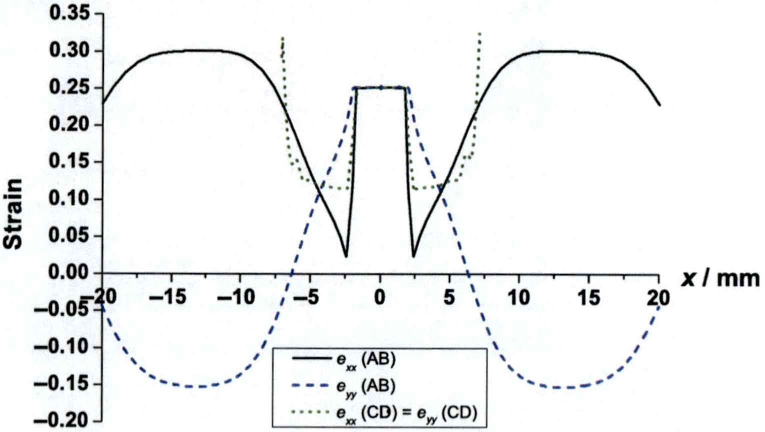

Fig. 7.13 Modelled strain distribution in the plane of a cruciform at $\alpha \cong 0.24$, reproduced from [3]. The lines AB and CD are defined in Fig. 7.2b. By symmetry, $e_{yy} = e_{xx}$ along the line CD. Finite element modelling used the Ansys v.11.1 Multiphysics package with plane stress or axially symmetric linear elastic elements ($\nu = 0.49$ and $E = 51.7$ MPa). The free meshing option gave good resolution over a range which included both the centimetre (macroscopic) scale and the sub-micron scale (around the pore). Modelling did not include detailed reproduction of pegs and cruciform grips. The nanopore is too small to significantly affect the result on this scale

increases linearly with α, and therefore with X, from below $\alpha = 0.1$ up to maximum extension (Fig. 7.14). The gradient over this part of the data was 0.99, close to the value obtained by modelling (1.06). The non-linear experimental response at low strains, which indicates increased responsiveness of the local strain to extension, is not reproduced using the linear elastic model.

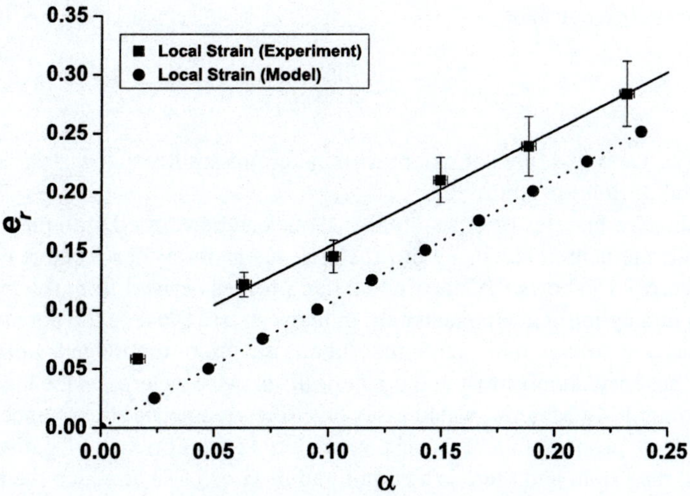

Fig. 7.14 Experimental and modelled plots of the local strain as a function of α, each with a linear fit to the data. The fit to the experimental local strain data ($R^2 > 0.98$) ignores the data point at the origin and the first stretched data point. Data reproduced from [2] and [3]

7.3 Actuation

In the first part of this section, the actuation of linear elastic holes with circular cylindrical and circular conical geometry is analysed. The previous section has indicated that this approach ignores some practical issues. However, the simple approach allows important concepts of pore actuation to be introduced. The ionic current measured passing through the pore is then used to calculate an effective pore size, using a series of analytic assumptions. Sections 7.3.4–7.3.7 discuss further important considerations for nanopore actuation.

7.3.1 Ideal Cylindrical Pore

A suitable starting point for analysis of elastomeric nanopore actuation is a cylindrical hole in a linear elastic (Hookean) sheet. For a Hookean material in equilibrium, the azimuthally-symmetric stress field around the pore can be described using simple potentials (Appendix, [3]). The radial and hoop strains are then given by

$$\begin{aligned} e_r &= \frac{\sigma_\infty}{E}\left(-(1+v)\frac{a^2}{r^2}+1-v\right) \text{ and} \\ e_\theta &= \frac{\sigma_\infty}{E}\left((1+v)\frac{a^2}{r^2}+1-v\right) \end{aligned} \tag{7.3}$$

and the z-strain is a constant,

$$e_z = -\frac{2\nu}{E}\sigma_\infty. \tag{7.4}$$

Equations (7.3) and (7.4) are consistent with incompressibility to first order in strain for a material of Poisson's ratio 0.5.

The qualitative features of this simple analysis can be verified with the aid of an ANSYS finite element model using linear elastic elements, as described in Fig. 7.13 and [3]. Figure 7.15 shows the stress and strain profiles derived from this model in the vicinity of a cylindrical Hookean pore in plane stress. Close to the pore wall, the large stresses and strains in the azimuthal (hoop) direction reproduce a well-known result, that the tensile hoop stress at the pore walls is twice as large as the local stress in the elastomer [44]. Material weaknesses or further geometric stress concentrators in this region are prone to failure. Near-pore material is compressed (negative strain) in the radial direction, and there is a certain radius (Eq. 7.23) at which the material does not experience radial strain for any extension.

This simple approach can be extended to describe pore fabrication. The cruciform is first extended, then a hole of radius a_0 is introduced into the membrane. The inner boundary is then released so that the pore assumes a new radius, a_1. The cruciform is then allowed to relax so that there is no far-field stress and the new radius is $a_1 > a_0 > a_2$, and suggests that, under the Hookean assumption, the pore radius can be reduced to zero. Ionic current measurements have indicated that unstretched elastomeric nanopores can have dramatically reduced radii [2] or close completely [1].

7.3.2 Ideal Circular Conical Pore

Real elastomeric nanopores taper from a relatively large opening on the cis side of the membrane to a sub-micron constriction near the trans surface. We now consider

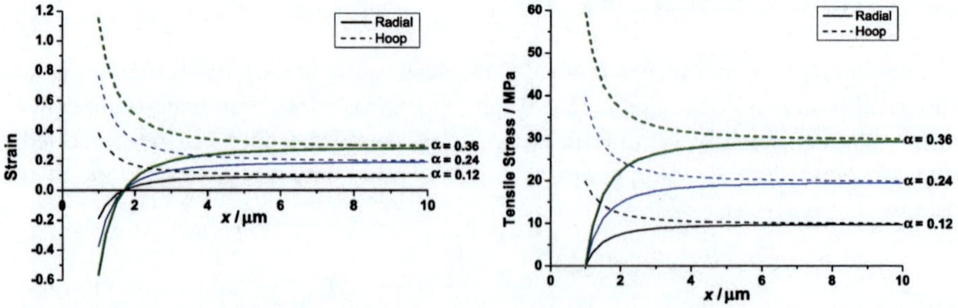

Fig. 7.15 Radial and hoop (azimuthal) strains (**a**) and stresses (**b**) plotted as a function of radial position for local strains characterised by α. The results have been calculated for a linear elastic cylindrical pore of stretched radius 1 µm using ANSYS modelling (see Fig. 7.13 for details)

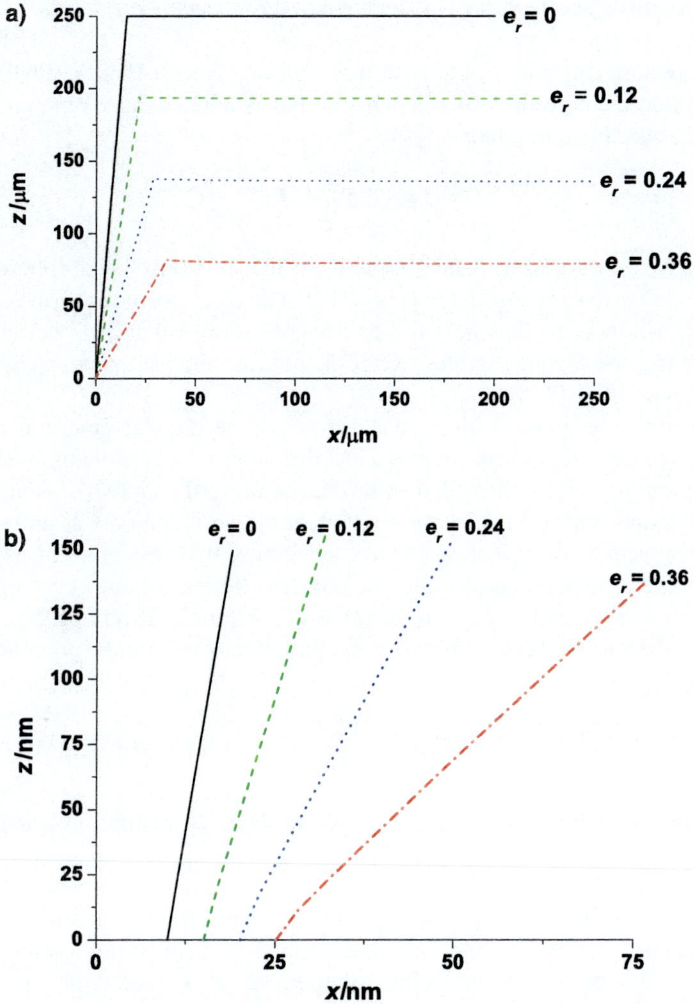

Fig. 7.16 (**a**) Pore profiles in the *x-z* plane for various local strains. (**b**) A magnified view of (**a**) near the smaller (trans) opening of the nanopore, reproduced from [3]. Unstretched cone radii are 10 nm and 15 μm, and membrane thickness is 250 μm. See Fig. 7.13 for details of modelling

a right regular conical nanopore in a linear elastic material. The model is restricted to the *r-z* plane, for a membrane with isotropically applied local strain. Calculated pore profiles are shown in Fig. 7.16. The trend is intuitive, although the stretched membrane is slightly thicker near the pore. A magnified view of the trans side of the membrane (Fig. 7.16b) indicates that, even for a linear elastic material, there is curvature of the pore wall when the membrane is stretched.

7.3.3 Nanopore Actuation – Electronic Measurements

Previous nanopore studies [45, 46] have used Poisson-Nernst-Planck (PNP) theory to establish that the current density **J** when a bias is applied across the membrane has three contributing components,

$$\mathbf{J} = \mathbf{J}_{\mathbf{diff}} + \mathbf{J}_{\mathbf{ep}} + \mathbf{J}_{\mathbf{eo}}, \tag{7.5}$$

where the terms on the right hand side refer to the diffusion, electrophoretic and electroosmotic currents respectively (Fig. 7.17). The electroosmotic term is usually neglected in nanopore studies [47–49], as it can be shown that the electroosmotic contribution to current is negligible, even if the constriction is of similar scale to the electrical double layer [4, 8, 50, 51].

In some studies of pores with similar dimensions to the elastomeric nanopores being discussed here, it has been assumed that diffusion current is negligible [16, 52, 53]. In this case the current through the pore can be derived simply by assuming that the electrolyte has uniform resistivity ρ and treating the membrane as an insulator. The pore than acts as a conical conductor connecting two reservoirs of negligible resistance – the immersed surface area of both electrodes is large relative to pore size. If the cone has smaller radius a, larger radius b and the length of the cone is l (Fig. 7.3), the electrophoretic current due to applied transmembrane potential V_0 is

$$I_{ep} = \frac{V_0 \pi ab}{\rho l}. \tag{7.6}$$

More detailed simulations [47, 48, 52, 53] have applied fewer simplifications to PNP theory.

Measurements of current-voltage (*I-V*) characteristics for elastomeric nanopores typically yield results as shown in Fig. 7.18 [2]. The small anisotropy between current response at positive and negative applied bias in current-voltage measurements could be consistent with Ramirez et al.'s prediction of current rectification effects [53]. Current anisotropy would also be observed if pore morphology is altered by voltage-dependent hydrodynamic or electrostatic forces.

I-V measurements can be used to estimate the smaller pore radius using Eq. (7.6), with the larger cone radius b is estimated from microscopy of various specimens. Effective resting pore radii in typical elastomeric specimens range from tens to hundreds of nanometres, and can be actuated over an order of magnitude between $\alpha = 0$ and $\alpha = 0.3$ [2]. Other pores have resting radii of molecular dimensions [1].

Figure 7.19 compares experimental values of pore resistance [2] with various models [3]. Data from typical experiments include a low stretch response ($\alpha < 0.1$) for which current is low and does not strongly depend on stretch. At greater stretch, a "working range" is found. The empirical relation,

$$R = R_0 \alpha^{-n}, \tag{7.7}$$

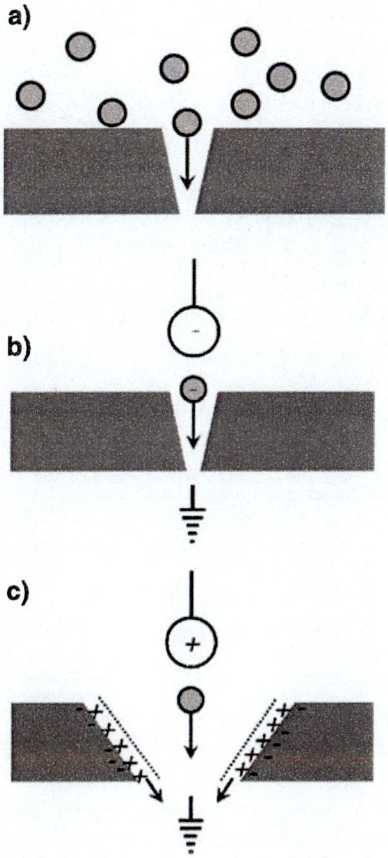

Fig. 7.17 Schematic diagrams demonstrating three effects which drive particle translocation through nanopores. Each effect is present for any type of particle, including ions, nanospheres and polymers. (**a**) Thermal motion drives particle diffusion when there is a concentration gradient across the membrane. (**b**) When a potential is applied across the membrane, an electrophoretic force acts on any charge-carrying particle. In (**c**), the pore is drawn larger for clarity. An elastomeric polymer typically has a slight negative surface charge. A double layer (*dotted lines*) containing mostly positive ions is formed adjacent to the surface. When potential is applied, the double layer ions pull the bulk fluid with them. This creates an electrosmotic flow which also affects any other particles

describes experimental data within the working range well, using values of R_0 and n which are particular to each pore. A physical interpretation of Eq. (7.7) is that the pore is closed when relaxed ($a = 0$, $\alpha = 0$). When the membrane is stretched, the membrane undergoes affine expansion (scaling linearly with e_r in two dimensions) beyond some radial distance from the edge of the pore. This distance remains constant, regardless of a and α, and volume is not conserved.

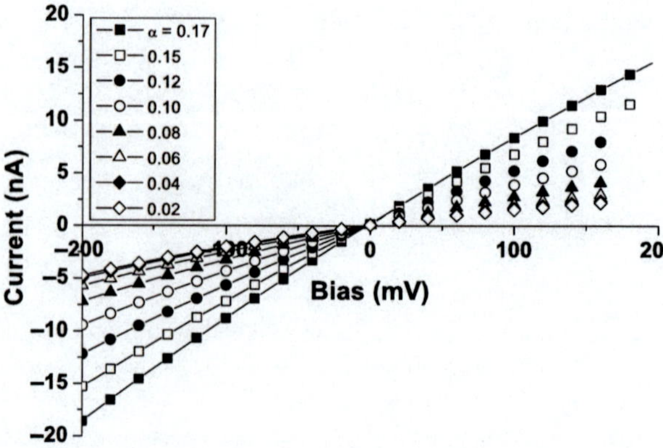

Fig. 7.18 Typical current-voltage data [2] over a full relaxation cycle, with lines drawn to assist interpretation. Data was collected with an ambient bias of ± 200 mV, moving through successive 20 mV voltage steps. Viscoelastic effects were minimised by using stress-softened cruciforms and carrying out measurements with consistent timing relative to stretching events

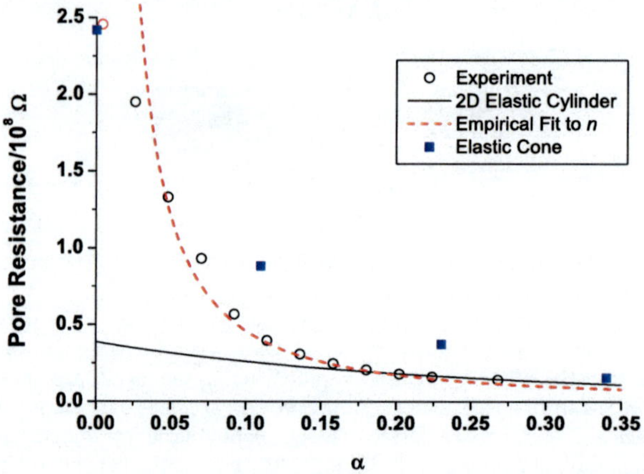

Fig. 7.19 Approaches for determining nanopore resistance as a function of α, using data from [2, 3]. Theoretical models are as described in the text, and the experiment is trial 1 using specimen B2 in [2]. Resistances are calculated using Eqs. (7.6) and (7.7) and Fig. 7.16 as appropriate, with resistivity equal to that of 0.1 M KCl, as used in the experiment

The models which employ an elastic 2D cylinder and an elastic cone do not describe the data as successfully as Eq. (7.7). Plots for these cases are scaled to fit the experimental larger cone radius b and membrane thickness, and require only one parameter (smaller pore radius) to be fitted to the data. Of these models, the conical model is closer to the shape of the experimental data. Pore actuation at the onset of stretching ($\alpha < 0.075$) is not well described by modelling. Viscoelastic and morphological effects associated with the extreme fabrication event determine the response in this range.

7.3.4 Viscoelasticity in Real Pores

Viscoelastic properties of TPU (Section 7.2) are an important consideration for pore actuation. An example of viscoelastic pore behaviour is shown in Fig. 7.20. When a pristine nanopore is stretched through a number of cycles, the current at maximum extension and zero stress gradually changes from cycle to cycle. This example demonstrates why it is beneficial to stress-soften a nanopore prior to repeated measurements.

To further study viscoelastic nanopore actuation, experiments have been carried out using a model system consisting of relatively large holes (130 μm diameter) in 200 μm–thick TPU membranes. While there are some drawbacks to this model system (especially the large ratio of pore size to membrane thickness), it provides the opportunity to study gross changes in hole size. Stress softening of the membrane prior to fabrication resulted in about 30% greater hole diameter during the first minutes of extension. This result can be understood in terms of irreversible deformation, which leaves a residual "set" at the time of fabrication. Holes made in cruciforms while in the stretched condition were larger than unstretched pristine holes by up to 20%.

Figure 7.21a shows that, following relaxation of a cruciform after a few minutes' extension, the model pore rapidly decreases in size (about 10% over 100 s) and then continues to decrease by a further 2% over several hours. Pristine model cruciforms were observed during cyclic opening and closing (Fig. 7.21b). The timing of this

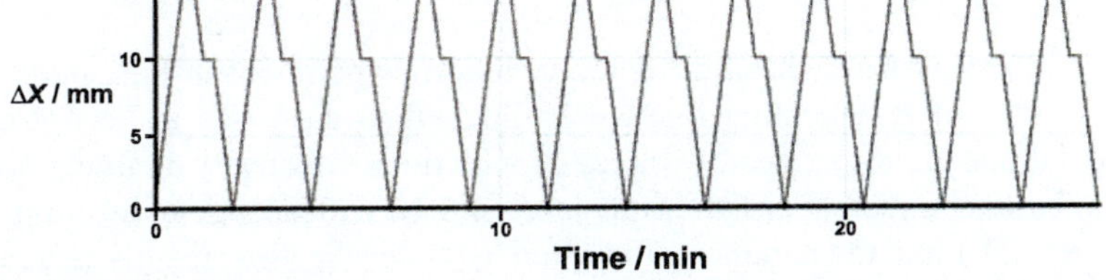

Fig. 7.20 Current measured through a pristine specimen (*top*) and cyclical cruciform extension (*bottom*) plotted against time, with 500 mV applied potential. The 0.1 M KCl solution wets the nanopore near maximum extension during the fourth cycle

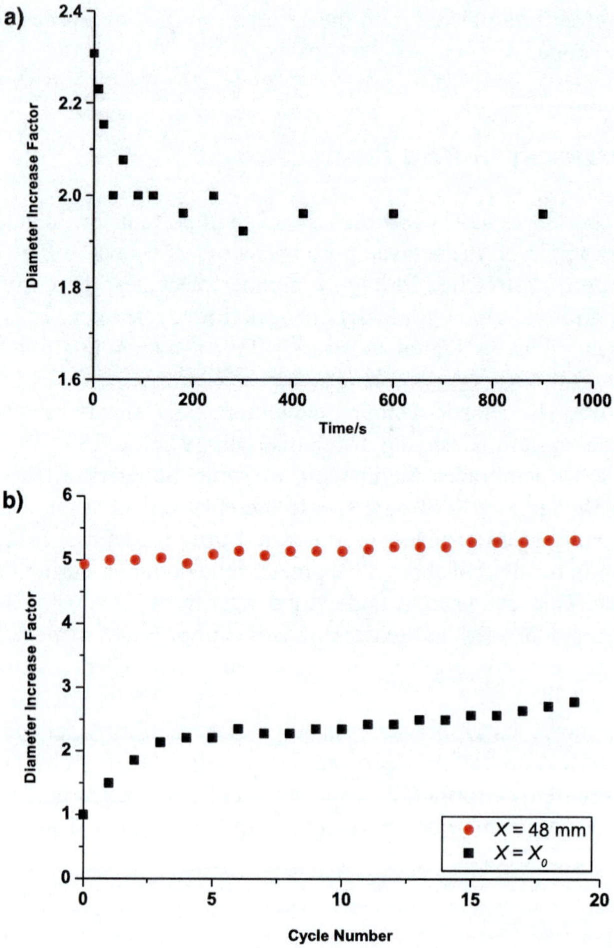

Fig. 7.21 (**a**) Relaxation of a hole in a membrane after extension to $\alpha = 0.3$ for ~200 s. Measurements of hole diameter were made using a projection microscope at 100 times magnification. (**b**) Hole size as the cruciform was repeatedly extended to $X = 58$ mm, held at that extension for 10 s, then closed to X_0. Rise and return times are both 60 s. The traces show the stretched hole diameter (*upper*) and the relaxed hole diameter (*lower*)

sequence reproduces the cycles during the nanopore experiment in Fig. 7.20, and both Figs. 7.20 and 7.21 demonstrate the Mullins effect discussed in Section 2.3. The relaxed model hole size steadily increases with time, effectively doubling in the first few minutes and thereafter increasing more slowly, with no sign of levelling off after 20 cycles. Unlike the nanopore, the extended hole size also steadily increases over the same period.

These experiments show that the complexity of viscoelastic elastomeric behaviour should not be underestimated. Nanopore size at any given time will depend primarily on initial stress softening, the degree of membrane extension and the short-term creep response, variables which can be controlled in ionic current experiments. Additionally, there is a small amount of creep or relaxation over several hours following a change in extension. Therefore the precise history, including time held in any extended position and time spent relaxing, cannot be ignored entirely. Temperature affects both the rate at which equilibrium is attained and the dimensions reached for any given extension.

7.3.5 Capacitance and AC Measurements

It is useful to use an alternating current (AC) to further measure the viscoelastic properties of elastomeric membranes and nanopores. Such measurements encompass both conductivity and capacitance. They are sensitive to the stretching and thickness change of the membrane, as well as the opening of larger pores: the dynamics of all of these processes can be acquired.

When a sinusoidal voltage of cyclic frequency f and peak value V_{pk} is applied across a membrane made of an insulator such as TPU, the AC that flows will also be sinusoidal. The in-phase AC is proportional to the conductance $G(f)$,

$$I' = V_{pk}G(f). \tag{7.8}$$

There will also be an AC component that is 90° in advance of the voltage, and this component is proportional to the membrane capacitance $C(f)$ and to the frequency:

$$I'' = V_{pk}2\pi fC(f), \tag{7.9}$$

For a good insulator, $I'' \gg I'$, and the ratio $I''/I' = 2\pi f\, C(f)/G(f)$ is often referred to as the quality factor $Q(f)$ of the assembly. Unlike the current seen in DC measurements, $G(f)$ reflects a variety of lossy (heat-generating) processes: not only ionic flow through a pore but also electric field-induced electronic, atomic and molecular reorientations within the polymer [54]. Capacitance causes the applied voltage to store energy within the polymer, and is proportional to the area of the membrane in contact with electrodes (A_{mem}):

$$C(f) = \frac{A_{mem}\varepsilon_r\varepsilon_0}{t}, \tag{7.10}$$

where t is the membrane thickness, $\varepsilon_r\varepsilon_0$ is the product of the relative and vacuum permittivities, and ε_r has a value of about 7.0 for TPU [55]. The corresponding value for water is ~80 at room temperature, but a single aqueous nanopore is so small that it is not expected to make a significant contribution to the capacitance. Hence an observed increase in $C(f)$ may be due to an increase in A_{mem} or a decrease in t, or both. For an isotropic radial strain field in an incompressible membrane, it is expected that $C(f)$ increases as $(1+e_r)^4$. Therefore, using the assumption that $\alpha \cong e_r$, the cruciform membrane capacitance response to an extension α might be expected to fit some law:

$$\frac{C_0(f) + \Delta C(f)}{C_0(f)} = (1+\alpha)^m, \tag{7.11}$$

where m is a constant and $C_0(f)$ is the capacitance with no stretch applied.

Measurements of capacitance and conductance across a cruciform membrane with patch-type geometry (see Section 7.2.1) are shown in Fig. 7.22. As X (equal to $X_0 + \Delta X$) is increased in regular steps (Fig. 7.22a), the capacitance and conductance both increase at greater than a linear rate with respect to ΔX (Fig. 7.22b). This cruciform had been repeatedly cycled between a relaxed state and $\Delta X = 16$ mm, yet exhibited relaxations in capacitance and conductance measurements following 1.5 mm step-changes in X (Fig. 7.22a). Although measurements were consistent from cycle to cycle, the relaxations at each extension continued over numerous cycles, with no further gain in stability.

This cruciform contained a large pore: the upper conductance limit of 1.8 μS seen in Fig. 7.22b corresponds to a right cylindrical pore of 4 μm radius and 200 μm length (Eq. 7.6). The conductance exhibited a similar percentage relaxation to that of the capacitance following each step-change in X. It is interesting that the capacitance relaxations are more pronounced after increase of X, whereas the conductance changes are more pronounced after decrease of X.

Figure 7.23 shows capacitance and conductance data for a second membrane. In this case, the membrane was fully moulded from Elastollan® 1195. The membrane had no pore apparent in DC measurements; the conductance is almost 100 times lower than that of the cruciform containing a pore. It can therefore be assumed that the conductance seen in Fig. 7.22 is dominated by that of the pore, whereas the conductance in Fig. 7.23 is due to AC losses in the polymer. This cruciform had been repeatedly cycled to $\Delta X = 16$ mm, and shows stable values of capacitance and conductance up to that extension, with no apparent relaxation following stretching. The difference in behaviour when compared with Fig. 7.22 may be related to the different material properties of the patch (Table 7.1) or the mechanical properties of the bonding epoxy.

Once $\Delta X = 16$ mm is exceeded, conductance and especially capacitance values clearly decrease during the 120 s pauses after positive, but not negative, incremental changes in X. This trend can be interpreted as being due to creep in the arms of the cruciform causing a decrease in the stress applied to the central membrane. Q is close to 20 and decreases with increasing X, indicating that conductance increases more rapidly than does capacitance.

Two aspects of the Mullins effect (Section 7.2.3) are evident here:

1. Extension beyond the historical limit in Fig. 7.23b causes hysteresis between extension and recovery traces for both capacitance and conductance measurements. The first cruciform exhibited a degree of hysteresis over the full

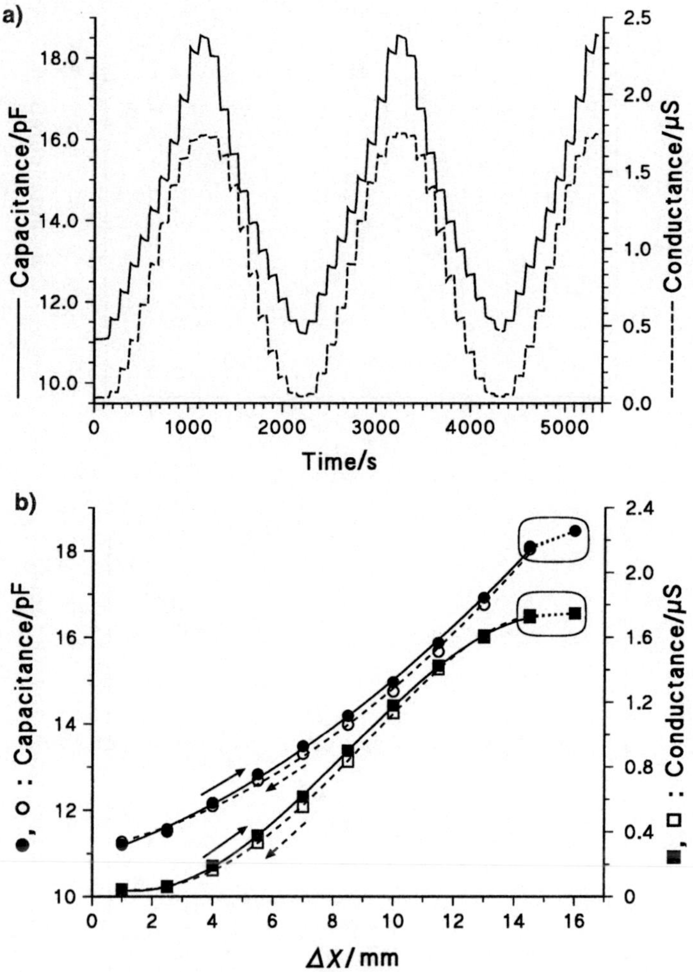

Fig. 7.22 (**a**) Extension and relaxation of a cruciform to $\Delta X = 16.0$ mm, with measurement of capacitance and conductance at 10 kHz using an impedance analyzer (HP 4194A), which applied a peak voltage of 0.5 V. ΔX was increased in 1.5 mm steps from 1.0 to 16.0 mm, with a pause of 120 s at each extension. The cruciform was placed with its lower face in contact with highly conductive (6.9 S m^{-1}) electrolyte. A hemispherical droplet of electrolyte formed the upper electrical contact. Electrodes (silver wire and stainless gauze) were inserted into the two liquid compartments. (**b**) Capacitance and conductance values derived from (**a**). *Continuous curves* connect data points from steps of increasing *X*; *dashed curves* connect data points from relaxation steps. The capacitance curves are 2nd order, and the conductance curves 4th order, polynomial best fits (method of splines). *Dotted lines* within *oval boxes* for the most extended step were not part of the regressions used to fit these curves

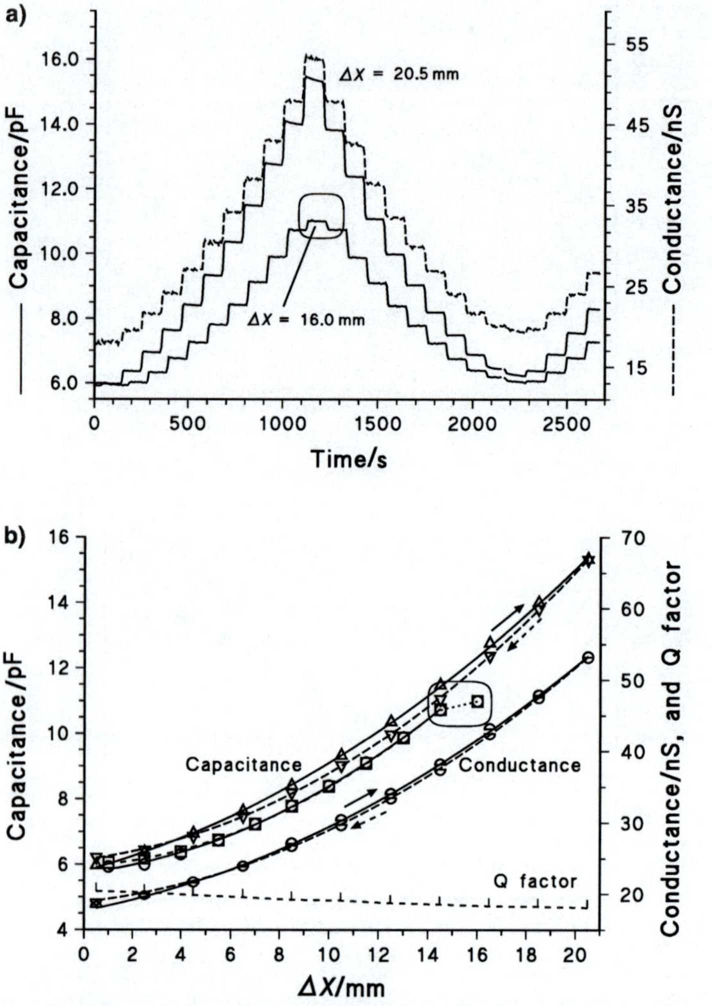

Fig. 7.23 (**a**) Extension and relaxation of a moulded cruciform from $\Delta X = 1.0$ to 16.0 mm in ten steps of 1.5 mm and then from $\Delta X = 0.5$ to 20.5 mm in ten steps of 2 mm. Experimental details were otherwise the same as those in Fig. 7.22. (**b**) Capacitance, conductance and Q values derived from (**a**). *Continuous curves* connect data points from steps of increasing ΔX; *dashed curves* connect data points from relaxation steps. All curves are 2nd order polynomial best fits (method of splines). The *dotted line* in the *oval box* was not part of the regressions used to fit the other curves

range of extensions (Fig. 7.22b), consistent with the trend of creep immediately following extension, as discussed above.

2. For both cruciforms, the polynomial fits are only good up to values less than the historical cyclical extension limit. The last 1.5 mm of extension (outlined within the oval box) does not conform with the trend from lower ΔX values.

Values of the exponent m derived from fits to Eq. (7.11) are between 1.9 and 2.2 [56]. These values are close to or below the limiting value of 2 expected for a non-thinning membrane, suggesting that the local response to macroscopic stretching is reasonably unresponsive.

7.3.6 Pores with Variable Azimuthal Geometry

SEM images suggest that pores can be significantly non-circular, and may be near-elliptical (Fig. 7.10c). Initial insight into the importance of azimuthal anisotropy can be gained by considering stress concentration factors for a linear elastic elliptical hole under biaxial stress. For the ellipse defined in Fig. 7.24, the classical solution [44] gives

$$\frac{\sigma_1}{\sigma_\infty} = \frac{2c}{d} \text{ and}$$
$$\frac{\sigma_2}{\sigma_\infty} = \frac{2d}{c}. \tag{7.12}$$

The greatest stresses and strains occur at the sharper end of the ellipse. Conversely, when the plate is relaxed, strain relaxation will be greater around the sharp edge of the ellipse and the eccentricity will increase.

Anisotropic folding is also evident in SEM images of unstretched pores. Even if fabrication is largely azimuthally symmetric, localised anisotropy will be present due to the growth of necking instabilities. Overextended material near the hole must be conserved as the elastomer is relaxed, and will compress and fold according to these local features.

7.3.7 Conclusions on Actuation

There are three key differences between idealised pores and those characterised experimentally. Firstly, and most importantly, extreme deformation occurs during penetration in the material adjacent to the pore. Near-pore material stretched towards failure undergoes significant inelastic deformation and becomes less responsive to

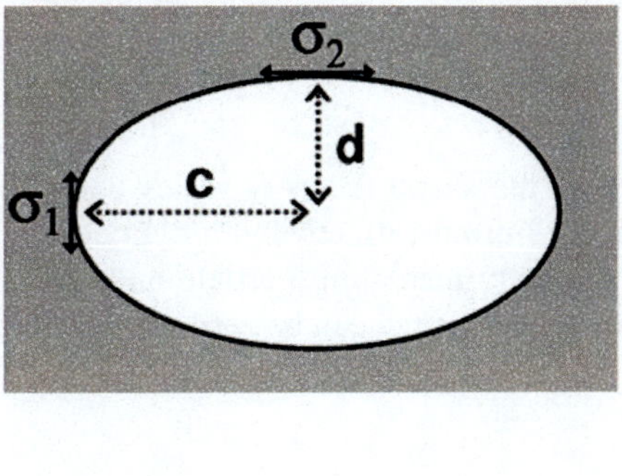

Fig. 7.24 Schematic diagram of a two-dimensional elliptical pore, with semi-major and semi-minor axes c and d respectively. Tensile stresses σ_1 and σ_2 are defined at the edge of the ellipse

further tensile actuation. The fabrication process also involves z-direction compression or bending. These effects combine to produce complicated, history-dependent low strain experimental data. As a result, reconciliation of experimental ionic current data with model systems to date (Fig. 7.19) is poor, especially at low strains.

The second key difference is the occurrence of pores of varying three-dimensional shape, increasing the difficulty of matching experimental results with simple models. Ongoing improvements to fabrication techniques are producing more regular pores. Theory and simulations can be developed to incorporate non-circular cross sections (e.g. ellipses) and variable cone-like shapes.

Finally, the macroscopic mechanical response of TPU is nonlinear and viscoelastic. This behaviour can be experimentally characterised using microscopy and electronic techniques, and future models should be able to describe bulk viscoelasticity relatively accurately.

The assumptions underlying analysis of current measurement experiments should be revisited for nanopores manufactured on smaller scales. Electronic effects such as rectification will become more significant in current measurement experiments. The electric field strength is highly intensified at small apertures. The membrane itself may have significant dielectric properties and support surface charge.

7.4 Translocations

7.4.1 Resistive Pulse Signals

When a particle passes through a nanopore, there is a brief change in the measured current while the particle remains within the pore. This resistive pulse is easily understood qualitatively, in terms of the pore's electrical resistance to charge carriers. An insulating particle within the pore decreases the volume available for carrier transport, thereby increasing pore resistance and (for constant applied bias) decreasing the current.

Accurate prediction of the pulse size is more difficult. Jelstch and Zimmermann [57] analysed resistive pulses in the context of cells transiting the central orifice in a Coulter counter. They considered an ellipsoidal particle with bulk dielectric characteristics in a homogeneous electric field. The translocation current change ΔI is related to the baseline current I_0 and the volumes of the particle v_p and the orifice v_0 using

$$\Delta I = f_s f_c \frac{v_p}{v_0} I_0, \tag{7.13}$$

where the shape factor f_s varies between 1.5 for a sphere and 1 for an infinitely long ellipsoid (effectively, a cylinder). The charge factor f_c varies between 1 for a perfectly insulating particle and $-(f_s-1)^{-1}$ for a conducting particle. Fractional resistance change can be zero, change sign and even diverge (for $f_s = 1$) depending on the shape and electronic properties of the translocating particle. A similar simple method, developed for use with Coulter technology [58], considers the translocating particle to be a perfect insulator, so that the resistivity change is proportional to the volume of the particle (or particles) within the pore. These methods are often sufficient for measurement of particle size in an application context (Section 7.5), and Eq. (7.13) can be used in a first approach to the interpretation of nanopore translocation data.

More complex effects can be important for nanoscale resistive pulses. Particles of interest, including biological molecules, typically incorporate molecular-scale charge distributions in addition to bulk dielectric properties. Examples include the negatively charged surface of carboxylated polystyrene (PS) spheres and the charged "backbone" of a DNA molecule. In solution, the arrangement of mobile ions around a particle will depend on molecular charge and shape characteristics; any inherent charge on a particle is partially "screened" by mobile ions of opposite charge. As an example of the complexities involved, consider the screening of the DNA sugar-phosphate backbone by counter-ions. Different sources [59, 60] have suggested that the fractional effective charge following screening is between 0.1 and 0.5, depending on the precise electronic conditions of the solution, the nanopore walls and the applied voltage. There is evidence that DNA translocations can produce a positive current pulse in certain ionic solutions [59, 60], suggesting that the factor f_c in Eq. (7.13) is solution-dependent for DNA.

It is usually most practical to consider the "effective" charge, which is derived from the mobility of a particle in an electric field, and incorporates a constant screening effect. Screening can be understood using the concept of double layer conductance. The enhanced concentration of mobile ions in the double layer surrounding a charged particle gives rise to an electrical conductance which is practically independent of the ionic strength of the solution. Values of this "surface" conductance K_s close to 1 nS have been found for highly charged polymer particles [61]. For a particle of radius a', this conductance increases the effective conductivity by $2K_s/a'$ [62]. Hence a highly-charged particle of radius 100 nm can be expected to have an effective conductivity of perhaps 0.02 S m^{-1}, which is far less than that of 100 mM KCl solution but above that of 1 mM KCl. A similar mechanism should give a moderate conductivity for DNA in aqueous solution, and indeed dsDNA from lambda phage has been found by an AC dielectrophoretic method to have an effective conductivity of 0.03 S m^{-1} [63].

Detailed understanding of translocation dynamics provides a further challenge. A simple picture of transport, based on particle flux in PNP theory (c.f. Eq. (7.5)), will be sufficient for many experiments [4, 5, 8]. More generally, measurements of dwell time within the pore, translocation frequency and thresholds for initiating translocation are all affected by numerous variables, including pore area, length and material properties, applied voltage, ionic concentration and pH. DNA translocation thresholds have been studied for molecular-scale nanopores, and the relationship between applied bias, translocation frequency and dwell time is only loosely understood [22, 49, 64–66].

7.4.2 Electric Field and Particle Flux

Analytic expressions for the electric field in and around the nanopore can be derived subject to the same simple assumptions used for Eq. (7.6). The electric field in a cylindrical pore of length l is independent of r and z, and is entirely parallel to the z-axis,

$$E_z = \frac{V_0}{l} \quad (0 < z < l). \tag{7.14}$$

If pore geometry is a right circular cone having smaller and larger openings of radii a and b respectively, the electric field is again independent of r, but varies with z. As long as $(b - a) << l$, the field can be considered parallel to the z-axis, so that

$$E_z(z) = \frac{V_0 abl}{(al + (b - a)z)^2} \quad (0 < z < l). \tag{7.15}$$

For small particles, the electrophoretic current density $\mathbf{J_{ep}}$ typically dominates electroosmosis and diffusion [4, 5, 8] (see Section 7.3.3 and Fig. 7.17). For charged particles of concentration C [46],

$$\mathbf{J_{ep}} = \frac{\zeta e}{k_B T} DC\mathbf{E}, \tag{7.16}$$

where k_B is the Boltzmann constant, T is temperature, $\mathbf{E}$ is the electric field and D is the diffusion constant. Equivalent notations use the gas constant $R = N_A k_B$ and the Faraday constant $F = N_A e$, where N_A is Avagadro's number. The particle flux density is simply $\mathbf{J_{ep}}$ divided by the effective charge per particle, ζe. Eq. (7.16) predicts that electrophoretic particle flux through a nanopore of consistent geometry is linearly proportional to particle concentration.

To verify that electrophoresis dominates diffusion, the ratio of energy associated with electrophoresis (ζeV) to thermal energy ($k_B T$) can be calculated. This ratio is approximately 3 if $\zeta = 1$ and $V = 100$ mV. For the carboxylate-terminated nanospheres often used in translocation experiments, ζ is at least of the order of a few tens (see below), so we expect a ratio of at least 50. It has been shown that diffusion is not significant when using DNA plasmids in translocation events [46].

7.4.3 Translocations Observed Using Elastomeric Nanopores

The remainder of this section presents examples of translocation experiments driven by electrophoresis. Most data have been obtained using carboxylated polystyrene (PS) spheres (Polysciences), with nominal diameters ranging from tens to hundreds of nanometres. The carboxylate surface groups result in a negative effective charge, so spheres move in the opposite direction to the applied electric field. Current traces can be positive or negative depending on the direction of applied bias required to observe translocations. The following critical operational conditions should be met:

1. The pore should have suitable geometry for resolving the particles of interest.
2. A high purity aqueous phase solution containing a buffered electrolyte and surfactant that wets the hydrophobic surfaces of the pore should be used. The solution used in experiments below was 0.1 M KCl containing 10 mM tris base, 3 mM EDTA to reduce particle aggregation and 0.01 vol% Triton X100 (critical micelle concentration 0.033%) at pH 8.0, filtered through 0.22 μm filters and stored in frozen aliquots.
3. Non-aggregating, charged particles of interest should be added to this solution at a suitable concentration.
4. There should be an appropriate applied voltage.
5. The electronic sampling rate should be high enough to resolve translocation events. Data below were captured at 10 or 20 kHz.

If these criteria are met, brief translocation events are detected soon after an analyte is injected into one half of the fluid cell. Blockages of extended duration can be observed if conditions are not ideal. Figure 7.25 shows sample injection leading to initial perturbation of the baseline current, followed by rapid detection of particles.

Figure 7.26 demonstrates an aperture resolving 84 nm particles and 150 nm particles at the same operating conditions. Translocations are significantly clearer and larger for the latter particles. The same nanopore could resolve larger particles (800 nm diameter) when the cruciform was further stretched, covering a larger range of particle sizes than a nanopore of fixed dimensions. The lower size limit for conventional Coulter counters, which utilise apertures of 15–20 μm or greater diameter, is typically 500–1,000 nm.

Electron microscopy and nanopore analysis of a polydispersed nanoparticle sample are compared in Fig. 7.27. Whereas electron microscopy can be time consuming, nanopore data is captured in a few minutes and analysed in real time. Raw data can

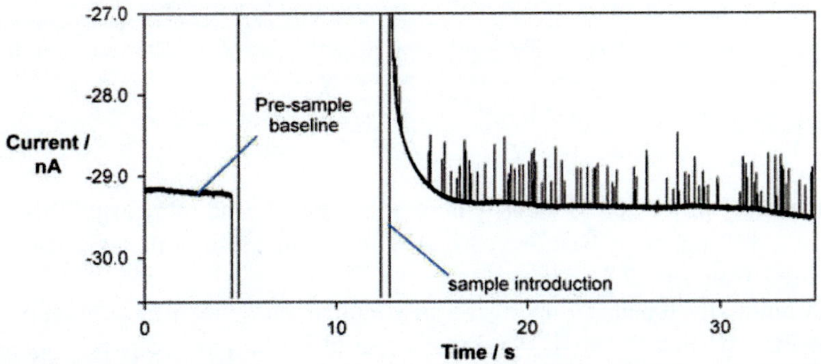

Fig. 7.25 Translocation events. Ionic current plotted against time as a sample containing 400 nm PS nanospheres at a concentration of 1.36×10^{10} mL^{-1} is added to one half of the fluid cell. The cruciform was extended to $\Delta X = 15$ mm ($\alpha = 0.36$) and applied bias was 100 mV

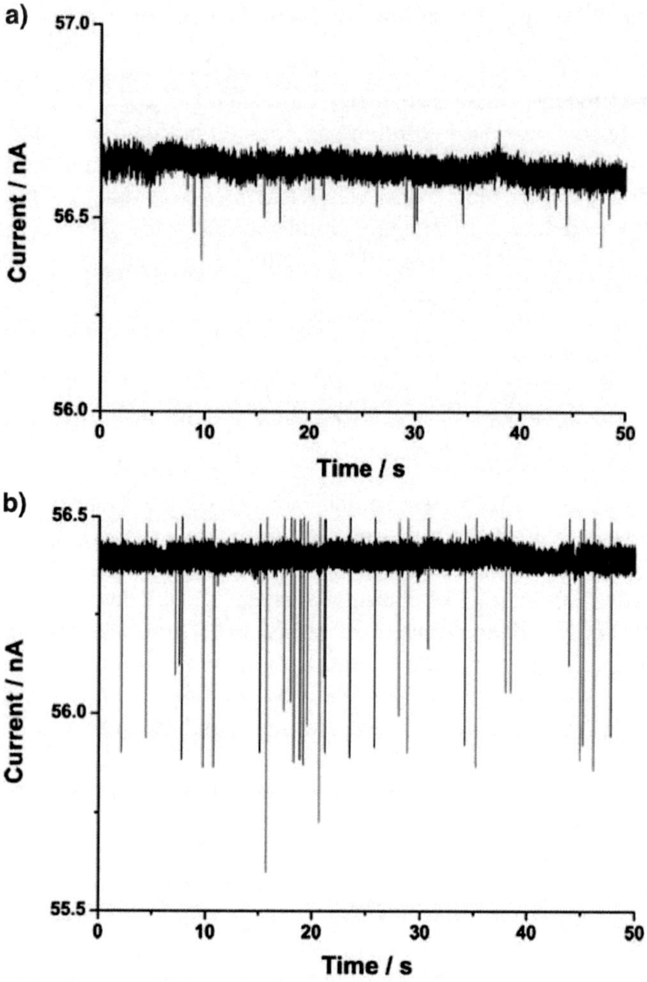

Fig. 7.26 Translocation events observed using polydispersed PS particles of average diameter (**a**) 84 nm at 1.52×10^{10} mL^{-1} and (**b**) 150 nm at 1.36×10^{9} mL^{-1}. Horizontal and vertical scaling is the same in both plots. In both cases, the applied bias was 300 mV

be automatically processed to identify current blockades, and to measure their size relative to a shifting, noisy baseline current. The absolute size of the pore itself can be estimated using Eq. (7.6).

When optimally tuned, an aperture can simultaneously resolve populations of different-sized particles. This is particularly useful if two particle populations are reasonably close in size. Figure 7.28 shows blockade event traces for polydispersed particle sets characterised by average diameters 84 and 150 nm. The data show two clear peaks in the histogram for the solution containing a mixture of the two particle

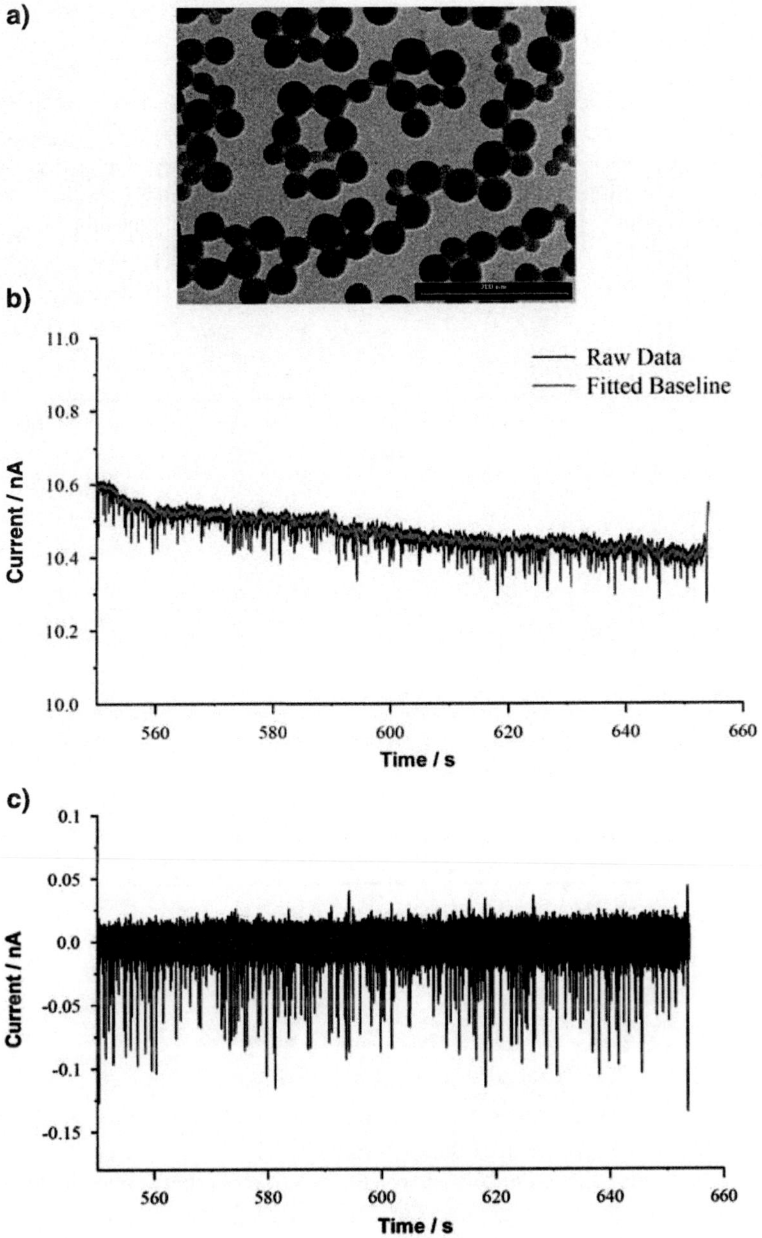

Fig. 7.27 Particle sizing using particle blockade analysis. (**a**) An SEM image of the polydispersed PS nanoparticles. (**b**) Resistive pulse data using these particles at a concentration of 1.64 × 10^{11} mL^{-1} and applied voltage 300 mV. (**c**) Data normalised relative to the baseline current. (**d**) Identification of individual events. (**e**) A histogram of resistive pulse amplitudes

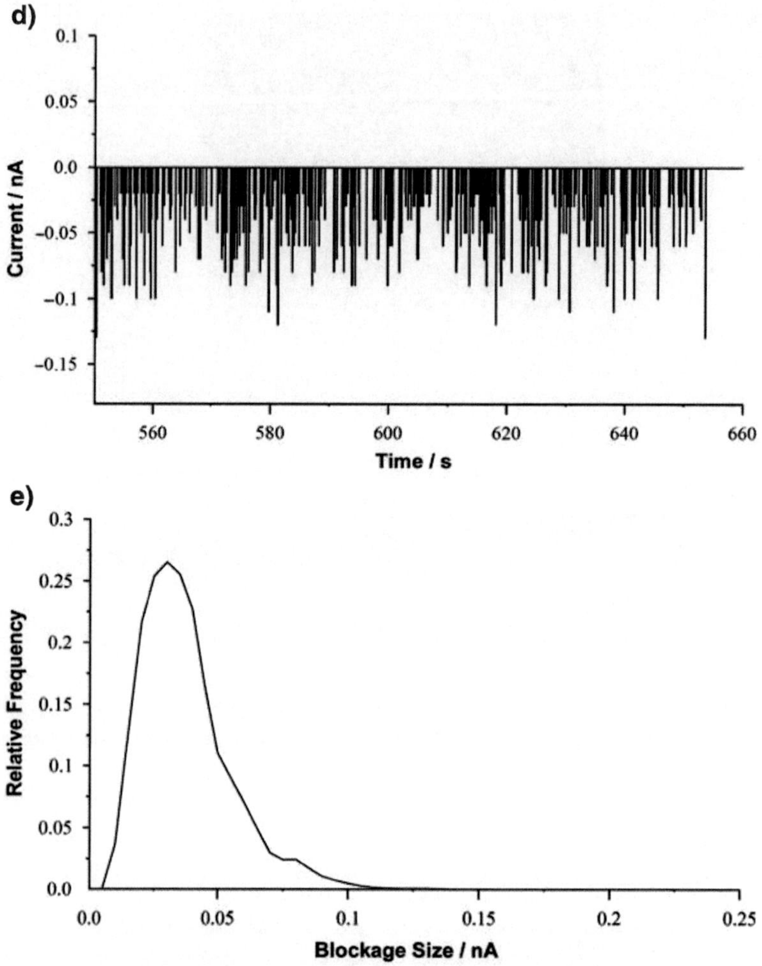

Fig. 7.27 (continued)

populations. It is interesting that blockade magnitude peaks are closer together for the mixed solution than for the individual particle sets.

Further experiments have confirmed that the relationship between translocation frequency and particle concentration (Fig. 7.29) is predominantly linear, as predicted by Eq. (7.16), at two aperture sizes for a single nanopore. Apart from surface charge and electroosmotic effects described previously, deviations from linearity might be further influenced by interactions and clustering of particles. The translocation rate was also significantly reduced with the smaller aperture. The effectively useful concentration range for nanoparticles in this experiment was from 2×10^9 to 3×10^{10} mL^{-1}.

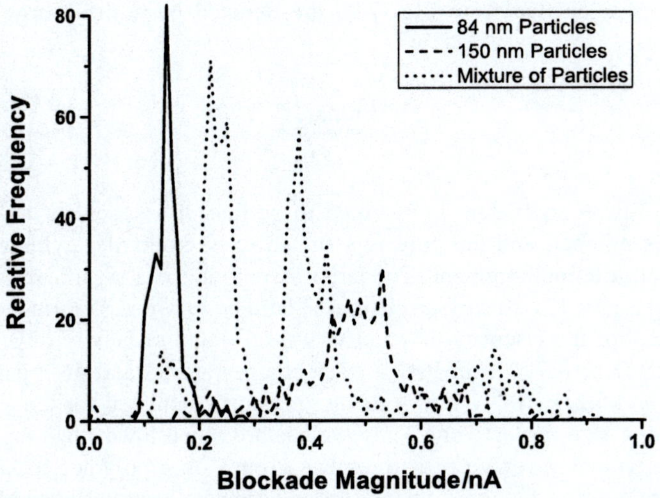

Fig. 7.28 Histogram showing resistive pulse amplitude distributions for translocation of PS nanoparticles through a single elastomeric aperture held at constant cruciform extension with 300 mV applied bias. The plots show the population distributions for particles of average size 84 nm at 1.52×10^{10} mL^{-1}, 150 nm at 1.36×10^{9} mL^{-1} and a mixture of these two solutions

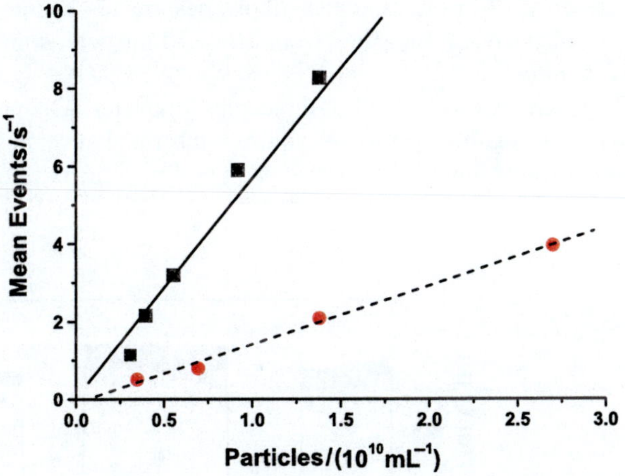

Fig. 7.29 Results of experiments in which various dilutions of 84 nm PS nanoparticles were analysed using a single elastomeric nanopore at two different aperture sizes. Translocation events were recorded over 10 min to derive the mean number of events per second for each solution. For 300 mV applied bias, baseline current was initially maintained at 70 ± 2 nA (*black squares*). The *solid straight line* is a linear fit to this data, with gradient $(5.6 \pm 0.3) \times 10^{-10}$ mL s^{-1} and $R^2 = 0.991$. The cruciform was then relaxed and the experiment was repeated with a smaller aperture, for which the baseline current was 34 ± 2 nA (*circles*), yielding a linear fit (*dashed line*) for which the gradient is $(1.50 \pm 0.08) \times 10^{-10}$ mL s^{-1} and $R^2 = 0.994$

If gradients calculated from Fig. 7.29 are denoted by *f*, the charge on each particle is

$$\zeta e = \frac{6\pi \eta a' f}{\rho I}, \tag{7.17}$$

where the diffusion coefficient is assumed to be that of a sphere of radius *a'* at low Reynolds number, and the pore is assumed to be a circular cylinder for the purposes of current measurement. The latter assumption is a significant source of error in using Eq. (7.17), so surface charges should be regarded as estimates. Using suitable values for the viscosity of water (0.9 mPa s) and resistivity of 0.1 M KCl solution (0.85 Ω m [67]), estimated surface charges are 83 and 46 *e* per particle (9.4×10^{-4} and 5.2×10^{-4} *e* nm^{-2}) using gradients calculated for the larger and smaller aperture size respectively. These values are much lower than expected, if based on the typical coverage of carboxylate groups on a polymer surface ($O(1)$ *e* nm^{-2}). This confirms that ionic charges around a sphere heavily screen the overall particle charge. Moreover, there is evidence that screening is affected by the size of the pore constriction.

Much of the recent interest in nanopore science has centred around analysis of biological particles. Subcellular biological particles are typically smaller than commercial PS nanospheres. Even extremely large molecular weight protein complexes such as ferritin are only 12 nm in diameter, ribosomes are 17–25 nm in size, and virus particles typically have dimensions from 10 to 70 nm with some exhibiting lengths of up to 200 nm.

Sowerby et al. [1] have resolved dsDNA molecules (diameter ~2.2 nm, Fig. 7.30) using a molecular-size tunable nanopore, and demonstrated reversible gating of DNA. Translocation events, initially observed at a baseline ionic current of 2.8 nA,

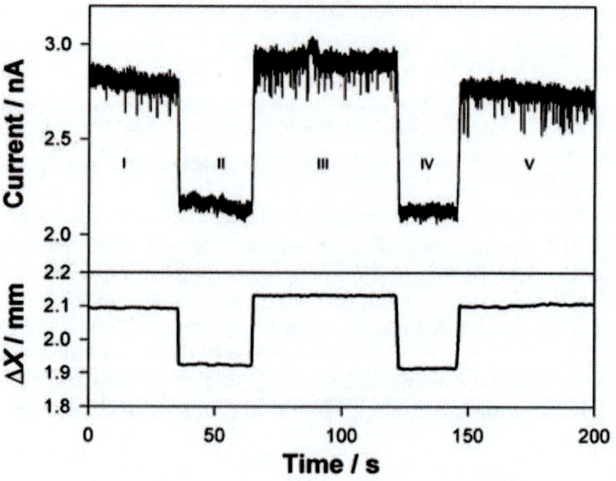

Fig. 7.30 Translocations and gating of dsDNA molecules using an elastomeric nanopore [1]. Ionic current through a nanopore is plotted as a function of time for a transmembrane potential of 200 mV (*top*). The change in the cruciform extension is plotted over the same time base (*bottom*)

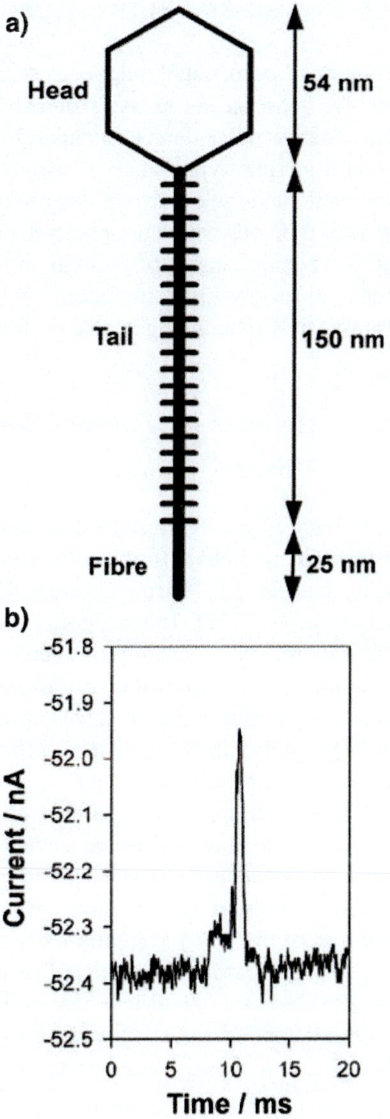

Fig. 7.31 Translocation of lambda phage particles using an elastomeric nanopore. (**a**) Sketch showing the size and shape of a lambda phage particle. (**b**) A typical resistive pulse corresponding to a translocation event

discontinued when the aperture was reduced in size to a current of 2.2 nA, and resumed when the aperture was re-opened to 2.8 nA.

Tunable nanopores have also been used to observe lambda phage translocation events (Fig. 7.31). Individual translocation signals exhibited two plateaus in the event, suggesting that the nanopore is able to resolve head and tail signatures. This result suggests that the length of the pore constriction, which is most sensitive to current blockages, is of similar size to the phage head (~50 nm) or smaller.

7.5 Present and Future Applications

The most immediately appealing applications of tunable nanopore technology involve sensing and analysis of nanometre-sized objects. In the previous section, the resistive pulse detection capability of tunable nanopores was demonstrated for several particle types, each of which are smaller than can easily be resolved using conventional single-particle detection techniques. Here, we briefly introduce some general methods which use resistive pulse data to derive key information for applications. Specific classes of particle and applications presently under investigation are noted. A review and commentary on the possible contribution of tunable nanopores to fast DNA sequencing is also included.

7.5.1 Methods for Particle Concentration, Size and Interaction Analysis

The frequency, magnitude and duration of translocation events constitute three key measurement metrics for dispersed particles. Algorithms are being developed which analyse digitized electronic current records in order to automatically generate these data (see Fig. 7.27). These algorithms have been built-in to successive Izon Science software releases. The current release (v1.2, May 2010) allows the user to set amplitude and duration thresholds for event detection, whether the pulse is conductive or resistive. Measurement protocols have been developed which use the three data metrics to provide useful high-level information for applications. Here, some of these protocols are briefly described.

The concentration of a sample of biological or synthetic particles can be measured over a range spanning several orders of magnitude ($\sim 10^5$–10^{12} particles mL^{-1}). Fast, accurate concentration measurements are enabled by controlled application of pressure-driven flow in combination with an applied potential. With addition of the VPM system, this combination now comes as standard to commercial Izon apparatus. The particle flux in pressure-driven flow ($\boldsymbol{\Phi}_{\mathbf{pd}}$) can be estimated by considering a cylindrical nanopore of length l and radius a with pressure difference ΔP applied across its ends. Using the Navier-Stokes equation for steady-state flow at low Reynolds' number and no slip length, we find ($l \gg a$)

$$\left|\boldsymbol{\Phi}_{\mathbf{pd}}\right| = \frac{C\Delta P}{l}\frac{a^2}{8\eta}. \tag{7.18}$$

If the pressure term dominates the electrokinetic terms (see Eq. (7.16)), then ΔP should be a linear function of blockade frequency. By comparing the gradient of a pressure-frequency plot for a particle set of known concentration with a sample of interest, the concentration of the latter can be derived. This recently demonstrated method [4, 5] is independent of particle size, shape and charge, and is therefore particularly suited to particle sets about which little is known.

Similar measurement methods can be derived. For example, the event frequency should be proportional to particle concentration C, regardless of whether particles are driven by pressure, electrokinesis, or both – as demonstrated in Fig. 7.29. If it is possible to make up solutions containing a known concentration of the particle of interest, then an unknown concentration can be measured. The solutions at known concentration can be used to define the linear frequency-concentration relation at constant pressure, voltage and membrane stretch. When the unknown particle set is introduced, the measured frequency measurement allows particle concentration to be derived. If electrophoresis is the dominant transport mechanism, Eq. (7.16) suggests that a calibration particle set with similar size to surface charge ratio to the unknown particle set is sufficient. Calculation of particle charge from Fig. 7.29 and Eq. (7.17) is another example of a measurement protocol derived from electrophoresis alone (Eq. (7.16)).

A further protocol allows particle sizes, and size distributions, to be measured [6]. A key associated application area is measurement of aggregation and polydispersity in synthetic particle populations. Simple but effective analyses [57, 58] suggest that resistance pulse amplitude should be proportional to particle volume. Using reference particles of known size (preferably monodispersed), the magnitude histogram for an unknown population sample, obtained under the same operating conditions, can be scaled to obtain a histogram of particle volumes, or equivalent diameters. Figs. 7.27 and 7.28 show size-dependent translocation measurements, although the latter demonstrates that further research is required concerning the effects on blockade size of geometric and electronic properties of both pore and particle, as well as the solution chemistry.

Time-dependent monitoring of particle interactions is a key application area. Such interactions can be between two particles, a particle and a biomolecule, or could involve functionalisation or aggregation. Antibody-antigen reactions and binding of ligands to receptor complexes are obviously of high interest. Components can be mixed in the fluid cell, so that data are continuously recorded, or mixed elsewhere and periodically introduced, for more discrete measurements over extended time periods. The products of an interaction can have different size, shape and electrophoretic properties in comparison with the initial components, and a direct "before" and "after" comparison is possible. A change in a certain metric (frequency, size or duration) is of interest, rather than a very precise measurement of that metric. In general terms, a decrease in event frequency (or increase in duration) under electrophoresis can indicate decreased particle mobility and hence effective charge – this might be detected even if the change in physical particle size is insignificant.

7.5.2 Detecting Viruses, Other Biological Particles and Synthetic Nanoparticles

Virus detection is the application that is perhaps most suited to this technology at the present time. Most viruses have linear dimensions in the range of tens to hundreds of nanometres, very comparable with the range that is easily detected by tunable nanopores. Suitable pores, which are larger than molecular scale, are relatively easy to reproducibly fabricate. It will therefore be possible to distinguish virus populations on the basis of resistive pulse amplitude (Fig. 7.28), or distinctive resistive pulse signatures (Fig. 7.31).

The viral content in drinking water, seawater and groundwater can be counted, and individual virus particles observed. The technology enables fundamental research into the biology of ubiquitous viruses such as influenza and norovirus, and unknown viral pathogens. At present, the only direct method for counting virus particles is electron microscopy, which is costly, time-consuming and inactivates the virus. Using a nanopore method, virus particles might remain viable after counting. There are often inherent problems and assumptions associated with indirect methods for counting virus particles, such as virus plaque assay.

Apart from viruses, projects have commenced in several application areas. Tunable nanopores are emerging as a key measurement tool for synthetic nanoparticle size distributions and effective charges. Apart from aiding synthesis, the technology is of interest to the size standards and metrology communities. One possible mode of tunable nanopore operation involves "gating" particles by actuating the pore to allow translocations as appropriate. A controlled number of particles and biomolecules could be dispensed on to one side of the membrane, perhaps with the pore tuned to separate one particle type from others. There is also the prospect of using actuation to clamp single molecules with high aspect ratios (e.g. nanotubes, DNA and proteins) as they pass through the pore, to make them available for further interrogation.

More generally, tunable nanopores provide advantages for sensing of many nanoparticle types, whether biological or synthetic, naturally occurring or introduced. There is growing concern about the potential effects of nanoparticles in food and water supplies, as well as populated or occupational environments [68]. Border security is an obvious application area. Nanoanalysis will become important for quantitative monitoring and regulation of particle-based industrial nanotechnologies.

Tunable nanopore technology provides a high degree of hands-on practicality. An actuation unit can be relatively small, robust and portable, and is therefore suited to operation in remote locations, while adding convenience for laboratory-based work. The tunable nature of the technology allows optimisation of experimental resolution while in the field. A further practical advantage is that apertures can easily be replaced if they cease to function properly through contamination or overuse.

7.5.3 Sequencing and Single Polymer Interactions

It is worth reviewing research efforts towards fast sequencing of DNA, the application that has generated more attention for nanopores than any other, in the context of tunable nanopores. The first sequencing of a human genome in 2001 followed more than a decade of research by numerous collaborators [69, 70]. A cheap, rapid method of sequencing would be invaluable in the burgeoning field of genomics. A typical efficiency target [71, 72] is sequencing of a single genome within 24 h at a cost of less than $US 1,000, requiring a sequencing rate of 3.5×10^4 bp s^{-1} (base pairs per second). Perhaps the most promising of any proposed sequencing technique uses a waveguide system to identify tagged nucleotides as DNA is replicated [73]. Nanopore translocation rates (between 10^4 and 10^7 bp s^{-1} [15, 64]) are easily fast enough to meet efficiency targets, prompting intense interest in the prospect of a nanopore-based sequencing technique [74, 75].

Single-stranded DNA translocations through an α-HL nanopore (diameter 2.6 nm) were reported in 1996 [18], and polymer translocations in α-HL have since been well studied [15, 19–26]. DNA translocations through solid-state pores can be observed using resistive pulse sensing [15, 65, 76–78] and methods such as "sandwich" membranes [51, 79]. Solid-state pores can translocate strands up to 100 kbp long and dsDNA if the pore is wide enough. Further study is required to understand the arrangement of charges around a DNA molecule and on the pore surface, the effects of lateral pore confinement, and physical limits for structural resolution [15]. It is extremely difficult to distinguish resistive pulses of individual bases using the small number of ions that pass through a pore while a single base pair is within the pore [15]. A viable sequencing technique requires improved base-recognition methodology, slower translocation, or both. The translocation rate can be slowed by an order of magnitude using any one of greater salt concentration, lower applied bias, higher fluid viscosity or low temperature. Further techniques employ electric fields and bias reversal [80–82], DNA "hairpins" and polymer chain "unzipping" [25, 83], binding of nanobeads [24] and manipulation using optical tweezers [84, 85] or magnetic beads. At slow translocation rates, parallelized analysis of genomic fragments, followed by computational resequencing, is required to sequence an entire genome within 24 h.

Specific advantages of tunable nanopores for sequencing are as follows:

1. Economy of manufacture. Many pores may be required for parallel processing.
2. Robustness. Clogging is less of an issue for tunable pores than for any static pore, while biological pores are relatively sensitive to environmental factors.

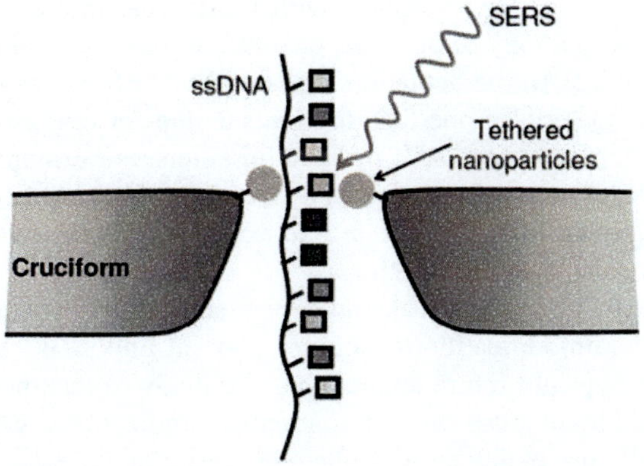

Fig. 7.32 Schematic cartoon of a possible route to fast sequencing [86]. Individual bases, represented by *square boxes*, could be spectroscopically identified using Surface Enhanced Raman Spectroscopy (SERS). The function of the tunable nanopore would be to confine the DNA and tune the plasmonic resonance. SERS is capable of resolving single dye molecules on a substrate of metallic nanospheres

3. Possible use of nanopore actuation to interact with polymers.
4. Versatility, particularly in-situ signal optimization and ease of integration.

As an illustrative example of point 4, membrane stretching could be used to tune Surface Enhanced Raman Spectroscopy (SERS) for DNA sequencing at a nanopore [86, 87] (Fig. 7.32). SERS can resolve single dye molecules [88], but obtaining the correct positioning of metallic colloids and dye molecules is a stochastic process. Using translocation, individual DNA strands could be localised relative to surface colloids, and membrane actuation could be used to tune colloidal separation for an optimised spectral response.

7.6 Future Research Directions

It is an intensely interesting and exciting time for tunable nanopore technology. An array of possible applications reflects the simplicity and universality of the concept. Future applications will require more information regarding optimal materials, pore sizes, pore performance characteristics, actuation techniques and methods of interfacing with other instrumentation. There are some general research directions which will significantly develop the tunable nanopore concept.

Electronic properties are largely unstudied for these nanopores. Charges will influence ionic current and resistive pulse measurements on all length scales, but especially when fluid is confined within a pore of radius similar to the electrical double layer. The obvious characterisation measurement for membrane materials is that of surface (or zeta) potential, which determines the physics of electroosmosis, and will depend on surface treatments (physical and chemical) and the composition of the electrolyte (e.g. pH, concentration).

The mechanical properties of TPU and other elastomers will be of continuing interest for actuation and manufacturing. Mechanical and failure properties are important over a range of length scales, with most acute interest falling on the failure-dependent morphology of the pore, changes in pore shape during actuation, and irreversible changes to the actuation characteristics of the region adjacent to the pore. Particular material properties that are suitable for one aspect of tunable nanopore operation are not necessarily optimal for others. For example, viscoelastic properties that enable a nanopore to change size also affect mechanical response times. Characterisation of internal pore profile remains difficult, although confocal microscopy shows some promise [4, 9].

There are several other key areas for pore characterization and applications development. Surface modification techniques play an important role in contemporary nanotechnology, and future applications are likely to incorporate particular chemistries on membrane surfaces. For particular applications, capacitance and noise reduction will pose engineering challenges. Use of a nanopore as a confined, localized volume for a chemical reaction has not been studied. Such a reaction could be probed using the electrical characteristics of the pore or instrumentation aligned with the pore. Finally, it is worth speculating on a role for tunable nanopores in the study of biological ion channels. Membrane deformation may be one of the most important variables in the regulation of biological particles passing through cell membranes [16]. Much could be deduced by mimicking ion channel functionality using flexible pores.

Acknowledgements Tunable nanopore studies have been supported by The MacDiarmid Institute for Advanced Materials and Nanotechnology and New Zealand's Foundation for Research, Science and Technology. The authors would like to thank Prof. Jeff Tallon, Julien Boizot and Kay Card of Industrial Research Limited and Hans van der Voorn, Paul Atkins, Martin Jones and Linda Groenewegen of Izon Science Ltd for helpful discussions and contributions.

Appendix: Analysis for Ideal Cylindrical Pore

A suitable starting point for analysis of elastomeric nanopore actuation is a cylindrical hole in a linear elastic (Hookean) sheet. For a Hookean material in equilibrium, the stress field around the pore can be described using simple potentials. Assuming rotational symmetry about the z-axis,

$$\begin{aligned} \sigma_r &= \tfrac{A}{r^2} + \sigma_\infty \text{ and} \\ \sigma_\theta &= -\tfrac{A}{r^2} + \sigma_\infty, \end{aligned} \tag{7.19}$$

where A is a constant and σ_∞ can be identified with the far field pressure. To determine A we require $\sigma_r = 0$ at $r = a$ (the pore radius). Then $A = -\sigma_\infty a^2$, the radial and hoop strains are given by

$$\begin{aligned} e_r &= \frac{1}{E}(\sigma_r - \nu\sigma_\theta) = \frac{\sigma_\theta}{E}\left(-(1+\nu)\frac{a^2}{r^2} + 1 - \nu\right) \text{ and} \\ e_\theta &= \frac{1}{E}(\sigma_\theta - \nu\sigma_\theta) = \frac{\sigma_\theta}{E}\left((1+\nu)\frac{a^2}{r^2} + 1 - \nu\right) \end{aligned} \tag{7.20}$$

and the z-strain is a constant,

$$e_z = -\frac{2\nu}{E}\sigma_\infty. \tag{7.21}$$

Equations (20 and 21) are consistent with incompressibility to first order in strain for a material of Poisson's ratio 0.5. Displacement is purely radial, and is given by

$$u_r = \frac{\sigma_\infty}{E}\left((1+\nu)\frac{a^2}{r} + (1-\nu)r\right). \tag{7.22}$$

At the radius

$$\frac{r}{a} = \sqrt{\frac{1+\nu}{1-\nu}}, \tag{7.23}$$

the radial strain is zero and displacement is minimized.

The simple analytical approach (Eqs. 19, 20, 21, 22, and 23) can be extended to describe pore fabrication, using the further assumption that there is no deformation during puncturing. For pore fabrication, the cruciform is first extended so that a far-field stress of magnitude σ_∞ is established. Then a hole of radius a_0 is introduced into the membrane, initially without allowing the inner radius to move. The inner boundary is then released so that the pore assumes a new radius, which can be found by applying the condition $r = a = a_0$ in Eq. (7.22). The new radius of the pore a_1, which does not depend on ν, is

$$a_1 = \left(1 + \frac{2\sigma_\infty}{E}\right) a_0. \tag{7.24}$$

The cruciform is then allowed to relax so that there is no far-field stress. To find the new radius a_2, an additional stress ($-\sigma_\infty$) is imposed in the far-field, reducing the total stress there to zero. Equation (22) now implies

$$a_2 = \left(1 - \frac{2\sigma_\infty}{E}\right) a_1, \tag{7.25}$$

so that

$$a_2 = \left(1 - \frac{4{\sigma_\infty}^2}{E^2}\right) a_0. \tag{7.26}$$

For positive (extensional) far field strain, this analysis reveals that $a_1 > a_0 > a_2$. Furthermore, $a_2 < 0$ if $\sigma_\infty > (E/2)$, suggesting that the pore radius can be reduced to zero under the Hookean assumption underlying this analysis.

References

1. Sowerby, S. J., Broom, M. F. & Petersen, G. B. Dynamically resizable nanometre-scale apertures for molecular sensing. *Sens. Actu. B* **123**, 325–330 (2007).
2. Willmott, G. R. & Moore, P. W. Reversible mechanical actuation of elastomeric nanopores. *Nanotechnology* **19**, 475504 (2008).
3. Willmott, G. R. & Young, R. M. Analysis and finite element modelling of resizable nanopores. *AIP Conf. Proc.* **1151**, 153–156 (2009).
4. Willmott, G. R., et al., Use of tunable nanopore blockade rates to investigate colloidal dispersions. *J. Phys.-Condens. Matt.* **22**, 454116 (2010).
5. Willmott, G. R., Yu, S. S. C. & Vogel, R. Pressure dependence of particle transport through resizable nanopores (Proceedings of ICONN 2010, accepted).
6. Vogel, R., et al., Quantitative sizing of nano/ microparticles with a tunable elastic pore sensor. *Anal. Chem.* (submitted). DOI: 10.1021/ac200195n.
7. Willmott, G. R. & Parry, B. E. T. Resistive pulse asymmetry for nanospheres passing through tunable nanopores. *J. Appl. Phys.* (in Press). DOI: 10.1063/1.3580283.
8. Willmott, G. R. & Bauerfeind, L. H. Detection of polystyrene sphere translocations using resizable elastomeric nanopores (Industrial Research Limited report no. 2385, arXiv:1002.0611v1, 2009).

9. Roberts, G. S., et al., Tunable nano/micropores for particle detection and discrimination: scanning ion occlusion spectroscopy. *Small* **6**, 2653–2658 (2010).
10. Sowerby, S. J., et al., Detecting, measuring and controlling particles and electromagnetic radiation. Patent GB2421303 (2006); Sowerby, S. J. et al. Fabrication of a deformable aperture in deformable elastomeric material. Patent GB2429524 (2007); Sowerby, S. J. et al. Detecting, measuring and controlling particles and electromagnetic radiation. Patent NZ537147 (2007); Sowerby, S. J. et al. Detecting, measuring and controlling particles and electromagnetic radiation. Patent application PCT/EP2005/053366 (2006); Sowerby, S. J. et al. Particle sensitive/radiation sensitive devices. Patent application PCT/GB2007/050340 (2007).
11. Becker, H. & Locascio, L. E., Polymer microfluidic devices. *Talanta* **56**, 267287 (2002).
12. Beech, J. P. & Tegenfeldt, J. O., Tuneable separation in elastomeric microfluidics devices. *Lab. Chip.* **8**, 657–659 (2008).
13. Huh, D., et al., Tuneable elastomeric nanochannels for nanofluidic manipulation. *Nat. Mater.* **6**, 424–428 (2007).
14. Fletcher, B. L., et al., Actuatable membranes based on polypyrrole-coated vertically aligned carbon nanofibers. *ACS Nano.* **2**, 247–254 (2008).
15. Dekker, C., Solid-state nanopores. *Nat. Nanotechnol.* **2**, 209–215 (2007).
16. Sexton, L. T., Horne, L. P. & Martin, C. R., Developing synthetic conical nanopores for biosensing applications. *Mol. BioSyst.* **3**, 667685 (2007).
17. Song, L., et al., Structure of staphylococcal alpha-hemolysin, a heptameric transmembrane pore. *Science* **274**, 1859–1866 (1996).
18. Kasianowicz, J. J., et al., Characterization of individual polynucleotide molecules using a membrane channel. *Proc. Natl. Acad. Sci. USA* **93**, 13770–13773 (1996).
19. Benner, S., et al., Sequence-specific detection of individual DNA polymerase complexes in real time using a nanopore. *Nat. Nanotechnol.* **2**, 718–724 (2007).
20. Hornblower, B., et al., Single-molecule analysis of DNA-protein complexes using nanopores. *Nat. Methods* **4**, 315–317 (2007).
21. Gu, L.-Q., Cheley, S. & Bayley, H. Electroosmotic enhancement of the binding of a neutral molecule to a transmembrane pore. *Proc. Natl. Acad. Sci. USA* **100**, 15498–15503 (2003).
22. Meller, A. & Branton, D., Single molecule measurements of DNA transport through a nanopore. *Electrophoresis* **23**, 2583–2591 (2002).
23. Mohammad, M. M., et al., Controlling a single protein in a nanopore through electrostatic traps. *J. Am. Chem. Soc.* **130**, 4081–4088 (2008).
24. Cockroft, S. L., et al., A single-molecule nanopore device detects DNA polymerase activity with single-nucleotide resolution. *J. Am. Chem. Soc.* **130**, 818–820 (2008).
25. Sauer-Budge, A. F., et al., Unzipping kinetics of double-stranded DNA in a nanopore. *Phys. Rev. Lett.* **90**, 238101 (2003).
26. Krasilnikov, O. V., Rodrigues, C. G. & Bezrukov, S. M., Single polymer molecules in a protein nanopore in the limit of a strong polymer-pore attraction. *Phys. Rev. Lett.* **97**, 018301 (2006).
27. Meckel, W., et al., in *Thermoplastic elastomers* (eds Holden, G., Kricheldorf, H. R. & Quirk R. P.) 15–44 (Carl Hanser Verlag, Munich, 2004)
28. Pritz, T., The Poisson's loss factor of solid viscoelastic materials. *J. Sound Vib.* **306**, 790–802 (2007).
29. Mailhot, B., et al., Mechanical and friction properties of thermoplastic polyurethanes determined by scanning force microscopy. *J Appl. Phys.* **89**, 5712–5719 (2001).
30. BASF Corporation. Technical information for Elastollan® 1195D and Elastollan® 1195A (2010).
31. Stevens. Polyurethane ST-1522FS-85 data sheet (2007).
32. Painter, P. C. & Coleman, M. M., *Fundamentals of polymer science* (Technomic, Lancaster, 1997).
33. Doi, M. & Edwards, S. F., *The theory of polymer dynamics* (Clarendon Press, Oxford, 1988).

34. de Gennes, P.-G., *Scaling concepts in polymer physics* (Cornell University Press, Ithaca, 1979).
35. Mullins, L., Softening of rubber by deformation. *Rubber Chem. Technol.* **42**, 339–362 (1969).
36. Ogden, R. W. & Roxburgh, D. G., A pseudo-elastic model for the Mullins effect in filled rubber. *Proc. R. Soc. Lond. A* **455**, 2861–2877 (1999).
37. Dorfmann, A. & Ogden, R. W., A constitutive model for the Mullins effect with permanent set in particle-reinforced rubber. *Int. J. Solids Struct* **41**, 1855–1878 (2004).
38. Beatty, M. F. & Krishnaswamy, S., The Mullins effect in equibiaxial deformation. *Z. Angew. Math. Phys.* **51**, 984–1015 (2000).
39. Mullins, L. in *Elastomers: criteria for engineering design* (eds Hepburn, C. & Reynolds, R. J. W.) 1–18 (Applied Science Publishers, London, 1979).
40. Kuriakose, B. & De, S. K., Elastomer tests. *J. Mat. Sci.* **20**, 1864–1872 (1985).
41. Lawn, B. R., *Fracture of brittle solids* (Cambridge University Press, Cambridge, 1993).
42. Lawn, B. R. & Wilshaw, T. R., *Fracture of brittle solids* (Cambridge University Press, Cambridge, 1974).
43. Hepburn, C., *Polyurethane elastomers* (Applied Science Publishers, Barking, 1982).
44. Young, W. C. & Budynas, R. G., *Roark's formulas for stress and strain* (McGraw Hill, New York, 2002).
45. Bath, B. D., White, H. S. & Scott, E. R. Electrically facilitated molecular transport. *Anal. Chem.* **72**, 433–442 (2000).
46. Gasparac, R., Mitchell, D. T. & Martin, D. R., Electrokinetic DNA transport in a nanopore membrane. *Electrochimica Acta* **49**, 847850 (2004).
47. Constantin, D. & Siwy, Z. S., Poisson-Nernst-Planck model of ion current rectification through a nanofluidic diode. *Phys. Rev. E* **76**, 041202 (2007).
48. Karnik, R., et al., Rectification of ionic current in a nanofluidic diode. *Nano Lett.* **7**, 547–551 (2007).
49. Harrell, C. C., et al., Resistive-pulse DNA detection with a conical nanopore sensor. *Langmuir* **22**, 10837–10843 (2006).
50. Rice, C. L. & Whitehead, R., Electrokinetic flow in a narrow cylindrical capillary. *J. Phys. Chem.* **69**, 4017–4025 (1965).
51. Gracheva, M. E., et al., Simulation of the electric response of DNA translocation through a semiconductor nanopore-capacitor. *Nanotechnology* **17**, 622–633 (2006).
52. Cervera, J., et al., Asymmetric selectivity of synthetic conical nanopores probed by reversal potential measurements. *J. Phys. Chem. C* **111**, 12265–12273 (2007).
53. Ramírez, P., et al., Pore structure and function of synthetic nanopores with fixed charges: Tip shape and rectification properties. *Nanotechnology* **19**, 315707 (2008).
54. Pethig, R., *Dielectric and electronic properties of biological materials* (Wiley, Chichester, 1979).
55. Zhou, Y., Wong, Y. W. & Shin, F. G., Mechanism of bending electrostriction in thermoplastic polyurethane. *J. Appl. Phys.* **96**, 294–299 (2004).
56. Arnold, W. M. Unpublished results. (2009).
57. Jeltsch, E. & Zimmermann, U., Particles in a homogeneous electrical field: A model for the electric breakdown of living cells in a Coulter counter. *Bioelect. Bioenerg.* **6**, 349–384 (1979).
58. DeBlois, R. W. & Bean, C. P., Counting and sizing of submicron particles by the resistive pulse technique. *Rev. Sci. Instrum.* **41**, 909–916 (1970).
59. Smeets, R. M. M., et al., Salt dependence of ion transport and DNA translocation through solid-state nanopores. *Nano Lett.* **6**, 89–95 (2006).
60. Chang, H., et al., DNA-mediated fluctuations in ionic current through silicon oxide nanopore channels. *Nano Lett.* **4**, 1551–1556 (2004).
61. Arnold, W. M., Schwan, H. P. & Zimmermann, U. Surface conductance and other properties of latex particles measured by electrorotation. *J. Phys. Chem.* **91**, 5093–5098 (1987).
62. Pauly, H. & Schwan, H. P. Über die impedanz einer suspension von kugelförmigen teilchen mit einer schale. *Z. Naturforsch* **146**, 125–131 (1959).

63. Hoeb, M., et al., Light-induced dielectrophoretic manipulation of DNA. *Biophys. J.* **93**, 1032–1038 (2007).
64. Meller, A., Nivon, L. & Branton, D., Voltage-driven DNA translocations through a nanopore. *Phys. Rev. Lett.* **86**, 3435–3438 (2001).
65. Heng, J. B., et al., The electromechanics of DNA in a synthetic nanopore. *Biophys. J.* **90**, 1098–1106 (2006).
66. Berezhkovskii, A. M., Pustovoit, M. A. & Bezrukov, S. M. Channel-facilitated membrane transport: Average lifetimes in the channel. *J. Chem. Phys.* **119**, 3943–3951 (2003).
67. Grossman, P. D. in *Capillary electrophoresis* (eds Grossman, P. D. & Colburn, J. C.) 3–44 (Academic Press, San Diego 1992).
68. OECD. Publications relating to nanotechnology and nanosafety at www.oecd.org/sti/nano&www.oecd.org/env/nanosafety (OECD, Paris, accessed 15 October 2008).
69. The International Human Genome Sequencing Consortium. Initial sequencing and analysis of the human genome. *Nature* **409**, 860–921 (2001).
70. Venter, J. C., et al., The sequence of the human genome. *Science* **291**, 1304–1351 (2001).
71. Collins, F. S., et al., A vision for the future of genomics research. *Nature* **422**, 835–847 (2003).
72. Service, R., The race for the $1000 genome. *Science* **311**, 1544–1546 (2006).
73. Foquet, M., et al., Improved fabrication of zero-mode waveguides for single-molecule detection. *J. Appl. Phys.* **103**, 034301 (2008).
74. Deamer, D. W. & Akeson, M., Nanopores and nucleic acids: Prospects for ultrarapid sequencing. *Trends Biotechnol.* **18**, 147–151 (2000).
75. Branton, D., et al., The potential and challenges of nanopore sequencing. *Nat. Biotechnol.* **26**, 1146–1153 (2008).
76. Heng, J. B., et al., Sizing DNA using a nanometer-diameter pore. *Biophys. J.* **87**, 2905–2911 (2004).
77. Oukhaled, G., et al., Unfolding of proteins and long transient conformations detected by single nanopore recording. *Phys. Rev. Lett.* **98**, 158101 (2007).
78. Storm, A. J., Storm, C. & Chen, J., Fast DNA translocation through a solid-state nanopore. *Nano Lett.* **5**, 1193–1197 (2005).
79. Lagerqvist, J., Zwolaky, M. & Di Ventra, M., Influence of the environment and probes on rapid DNA sequencing via transverse electronic transport. *Biophys. J.* **93**, 2384–2390 (2007).
80. Polonsky, S., Rossnagel, S., & Stolovitzky, G., Nanopore in metal-dielectric sandwich for DNA position control. *Appl. Phys. Lett.* **91**, 153103 (2007).
81. Gershow, M. & Golovchenko, J. A., Recapturing and trapping single molecules with a solid-state nanopore. *Nat. Nanotechnol.* **2**, 775–779 (2007).
82. Stein, D., Molecular ping-pong. *Nat. Nanotechnol.* **2**, 741–742 (2007).
83. Mathe, J., et al., Nanopore unzipping of individual DNA hairpin molecules. *Biophys. J.* **87**, 3205–3212 (2004).
84. Keyser, U. F., et al., Optical tweezers for force measurements on DNA in nanopores. *Rev. Sci. Instrum.* **77**, 105105 (2006).
85. Keyser, U. F., et al., Direct force measurements on DNA in a solid-state nanopore. *Nat. Phys.* **2**, 473–477 (2006).
86. Tallon, J. L. Nanotechnologies for single-molecule or single-ligand spectroscopy. Patent USSN 61/129,293 (2008).
87. Hossain, M. K. et al., Gold nanoaggregates on elastomer: a substrate for tunable SERS. *Proceedings of ICONN 2010 Conference* (IEE explore, Sydney, Australia, 2010).
88. Le Ru, E. C., Meyer, M. & Etchegoin, P. G., Proof of single-molecule sensitivity in surface enhanced Raman scattering (SERS) by means of a two-analyte technique. *J. Phys. Chem. B* **110**, 1944–1948 (2006).

Chapter 8
Synthesis of Carbon Nanotubes

Nicole Grobert, Siegmar Roth, John Robertson, and Cheol Jin Lee

Abstract A general overview is given over the most common methods to synthesize single-walled and multi-walled carbon nanotubes. In particular carbon arc synthesis, laser ablation synthesis, chemical vapor deposition, and carbon monoxide disproportionation are discussed. A special section is devoted to the role of the catalyst and some ideas on the growth mechanism are presented.

8.1 Introduction

Carbon nanotubes (and their relatives graphene, graphitic monolayers) are often considered as the most exciting materials at the beginning of the new millennium. This excitement comes about from the combination of the possibility to study basic physical phenomena and the hope for opening new frontiers in industrial applications. In Fig. 8.1 we show the computer model of a single-walled carbon nanotube. It is a seamless tube of a graphitic monolayer. The diameter is about 1 nm and the length can be up to several millimetres. So the tube has an aspect ratio (ratio of length to diameter) of well above 1,000. The tube can be envisaged as being made by rolling graphene sheets. Bending the sheet costs elastic energy, but more energy is won by closing the dangling bonds at the edge to form the seamless tube. Tubes with diameters well below 1 nm are unstable because the bending energy is too large, very wide single-walled tubes collapse into double-layer ribbons to saturate the van der Vals forces between the walls. There are single-walled tubes and multi-walled tubes. Multi-walled tubes are concentric arrangements of single-walled tubes.

Carbon nanotubes are a special type of quantum wires. But different from other quantum wires, not only are they hollow, they also have a perfect surface, with no dangling bonds, no surface reconstruction, and no surface scattering of charge carriers. As a consequence charge transport in carbon nanotubes is ballistic even in fairly long tubes. The confinement of the electrons to the surface of a long and thin cylinder leads to specific features in the electronic density of states and to the existence of semiconducting and metallic tubes, respectively, depending on how the tube is rolled (diameter, helicity).

S. Roth (✉)
School of Electrical Engineering, Korea University, Seoul, Korea; Sineurop Nanotech GmbH, Stuttgart, Germany
e-mail: siegmar@korea.ac.kr; sgmrroth@aol.com

O. Hayden, K. Nielsch (eds.), *Molecular- and Nano-Tubes*,
DOI 10.1007/978-1-4419-9443-1_8,

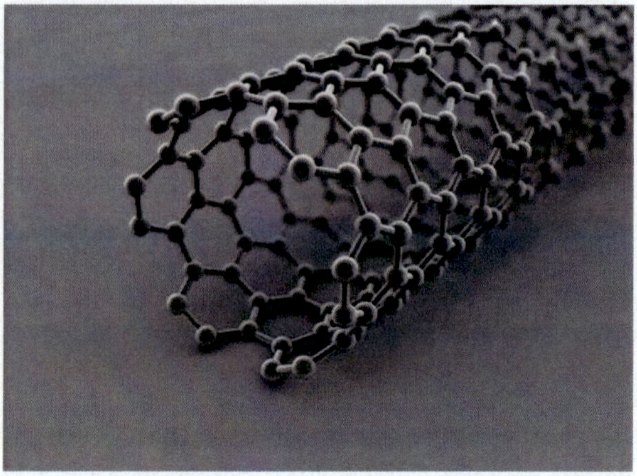

Fig. 8.1 Computer model of a single-walled carbon nanotube

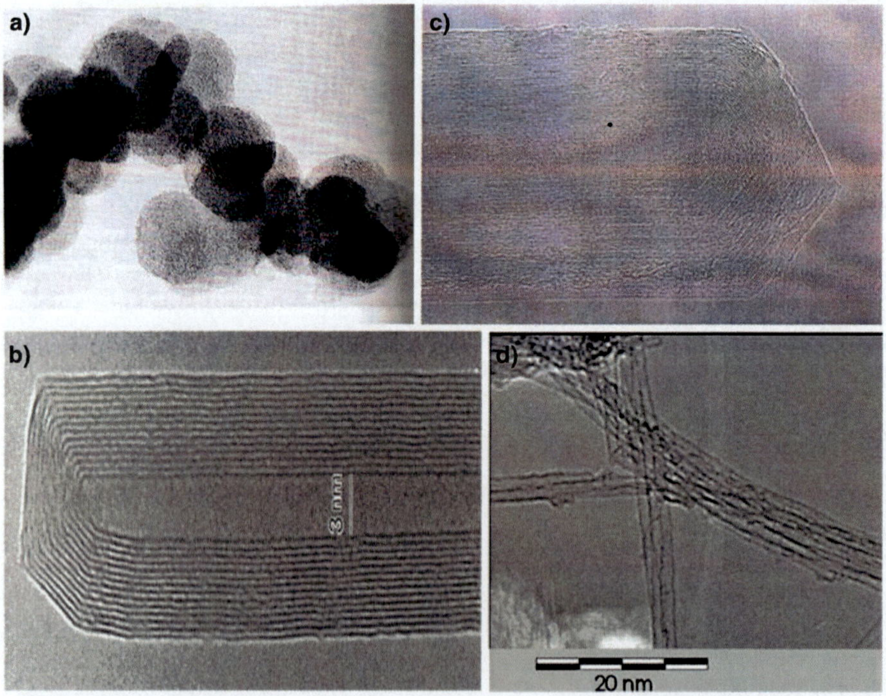

Fig. 8.2 High resolution TEM images (**a**) of ordinary soot (Courtesy R.-G. Gilg) (**b**, **c**) of multi-walled carbon nanotubes (Courtesy Philipp Kohler-Redlich, Courtesy Seamus Curran), and (**d**) of thin bundles of single-walled tubes (Courtesy Shanghai Yangtze Nanomaterials Co., Ltd.)

From the chemical point of view, carbon nanotubes are a sort of soot, albeit a very special sort of soot. Soot consists of carbon nanoparticles. In ordinary soot these particles are globular or flaky. Even in ordinary soot now and then tubular particles are found. The art of synthesizing carbon nanotubes consists of conducting the soot growing process in such a way that predominantly tubes are formed. Figure 8.2 shows high resolution transmission electron microscope images of ordinary soot particles, a multi-walled nanotube with about 14 concentric tubes, a multi-walled tube with 70 walls, and some bundles of single-walled carbon nanotubes.

Conventional soot is produced by the incomplete combustion of carbon-containing materials (wood, coal, gasoline, oil, hydrocarbons etc.). Nanotubes are produced by evaporating graphite in inert atmosphere, by pyrolizing (cracking) hydrocarbons, or by disproportionation of carbon monoxide. The heat for evaporation is generated by arc discharge, laser beams, or solar furnaces. For pyrolysis resistive heating of tubular furnaces is sufficient. In most cases catalysts are used. These are usually metal droplets, in most cases droplets of transition metals like iron, nickel, cobalt, and of their alloys with rare earth metals.

8.2 Carbon Arc Synthesis

Probably the simplest method to grow carbon nanotubes is by the carbon arc technique. This technique was originally used by W. Kraetschmer to synthesize fullerenes [1]. Figure 8.3 shows a student working on a Kraetschmer generator (Courtesy Serhat Sahakalkan). He is just inserting a graphite electrode into the reaction chamber. This chamber is shown schematically in Fig. 8.4 [2]. The chamber is filled with an inert gas, usually helium at about 500 mbar, and an electric arc is ignited between two graphite electrodes. For the electric arc, a simple power supply as from welding machines on construction sites can be used. About 100 A and 50 V

Fig. 8.3 Synthesis of carbon nanotubes using the carbon arc method [1]

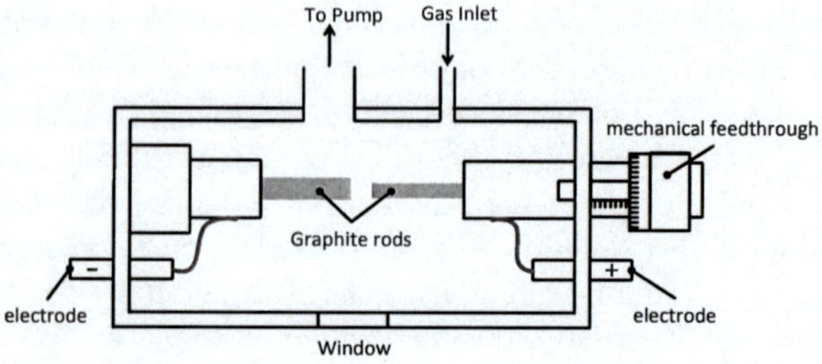

Fig. 8.4 Schematic drawing of Krätschmer generator for nanotube synthesis by the carbon arc method [2]. (Note that here the cathode is a graphite rod, whereas in Figs. 8.3 and 8.5 a large graphite disc has been used)

are needed. The heat produced by the arc evaporates the graphite anode, the carbon vapour cools by collision with helium atoms and condenses into nanoparticles. The wall of the reaction chamber has to be water cooled to remove excess heat. A variety of particles is produced at the same time: fullerenes, nanotubes, ordinary soot (amorphous carbon), and graphitic chiplets. If our goal is to synthesize nanotubes, the other particles are just "by-products".

If the Kraetschmer generator is run without catalyst, the nanotubes obtained are mostly multi-walled tubes. To obtain single-walled tubes, metal particles (nickel, iron, cobalt, rare earth metals [3]) have to be injected into the arc [4]. A convenient method is to drill a hole into the anode and to fill this hole with a mixture of metal powder and graphite powder. Figure 8.5 (Courtesy Bjoern Hornbostel) shows the "harvest" of a typical run in the Kraetschmer generator. The fluffy spider web stuff is an indication of the presence of many single-walled tubes. Usually a run yields some 100 mg of "raw material" where about 30% of the carbon material are in the form of single-walled carbon nanotubes. In addition there are some wt% of metallic catalyst remains in the raw material.

Fig. 8.5 Harvesting a nanotube-rich "spider web" product synthesized by the carbon arc method (Courtesy Serhat Sahakalkan)

8.3 Laser Ablation Synthesis

An alternative method to create carbon vapour is by laser ablation of graphite (Fig. 8.6) (Courtesy Bjoern Hornbostel): A laser beam hits a graphite target into which catalyst particles have been incorporated. The vapour is transported to cooler parts of a reaction tube by an inert carrier gas and a soot, rich in single-walled carbon nanotubes, is deposited [5]. The laser ablation set-up used by the Stuttgart team is shown in Fig. 8.7 (Courtesy Bjoern Hornbostel). Figure 8.8 (Courtesy Bjoern Hornbostel) shows the water cooled collector, where most of the tubes are deposited. Laser ablation tubes are considered as "highest quality" tubes. Usually more than 50% of the carbon material is in tube form, the tubes are longer than in the carbon arc process, they have less defects, and networks of tubes have a higher electrical conductivity. A typical run yields about 100 mg of tubes, and about 10 g can be produced per week.

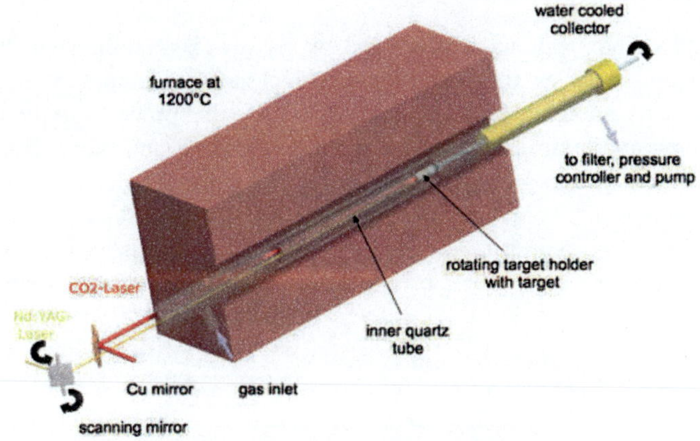

Fig. 8.6 Schematic drawing of equipment for nanotube synthesis by the laser ablation method (Courtesy Bjoern Hornbostel)

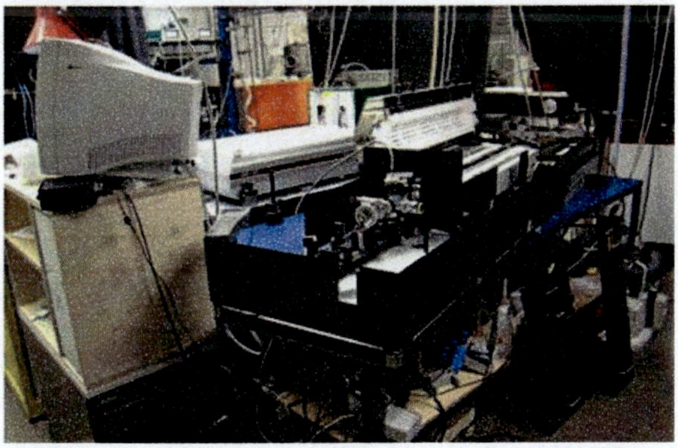

Fig. 8.7 Laser ablation equipment for nanotube synthesis as used by the Stuttgart team (Courtesy Bjoern Hornbostel)

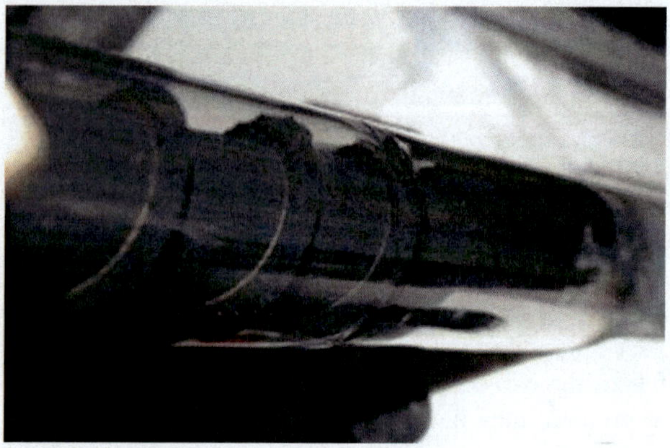

Fig. 8.8 Water cooled collector in Stuttgart laser ablation nanotube synthesis system (Courtesy Bjoern Hornbostel)

8.4 Chemical Vapour Deposition – CVD (Pyrolysis, Cracking)

In this method carbon vapour is produced by pyrolysis (cracking) of hydrocarbons. The set-up is just as simple as the Kraetschmer generator, except that the furnace is perhaps somewhat more expensive than the power supply from the welding machine and hydrocarbon gases are more difficult and more dangerous to handle in a laboratory than graphite electrodes (at least for physicists).

Figure 8.9 shows a schematic drawing of the CVD set-up (Courtesy School of Electrical Engineering, Korea University, Seoul). It just consists of a quartz tube which can be heated up to 1,000°C or a little bit more. A substrate carrying a thin

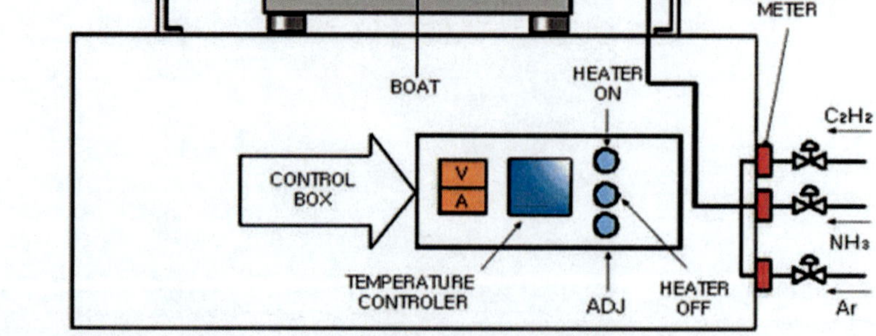

Fig. 8.9 Schematic drawing of reactor for nanotube synthesis by the chemical vapour deposition (CVD) method (Courtesy School of Electrical Engineering, Korea University, Seoul)

metal film (catalyst) is placed in the tube, the tube is heated, and a hydrocarbon gas is blown across. Usually the hydrocarbon gas is diluted by an inert transport gas, such as argon. Flowmeters or mass flow controllers are used at the entrance side and a pressure gauge and a controlled valve at the exit. Methane, ethane, or acetylene are used as hydrocarbon gases ("carbon feedstock"). Sometimes argon is just bubbled through benzene or hexane, and sometimes experimentalists content themselves with bubblers for gas flow control. Very often additional gases, like hydrogen, NH_3, or water vapour are added at special steps of the reaction process: hydrogen for cleaning the substrate or reducing the catalyst, if this had been added in salt form like $FeCl_3$ or $Fe(NO_3)_3$; NH_3 for "etching away" by-products of amorphous carbon; small amounts of water vapour for keeping the catalyst longer alive. The photograph of Fig. 8.10 shows the CVD set-up used by the Stuttgart team (Courtesy Sineurop Nanotech GmbH) and that of Fig. 8.11 shows one of the CVD reactors of the "Nanotube Factory" in Seoul (Courtesy School of Electrical Engineering, Korea

Fig. 8.10 Reactor used by Stuttgart team for carbon nanotube growth by chemical vapour deposition (Courtesy Sineurop Nanotech GmbH)

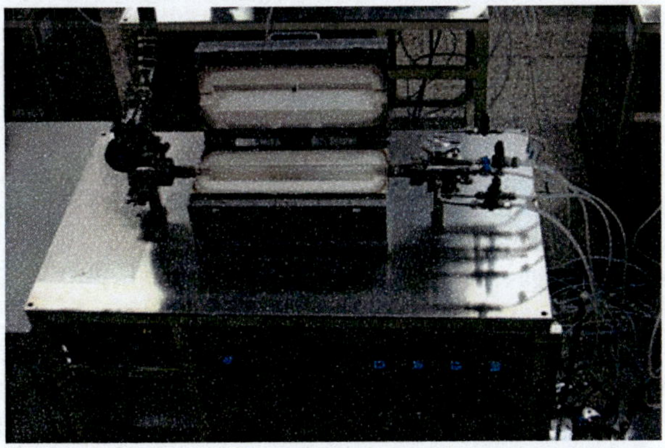

Fig. 8.11 One of the CVD reactors of the "Nanotube Factory" in Seoul (Courtesy School of Electrical Engineering, Korea University, Seoul)

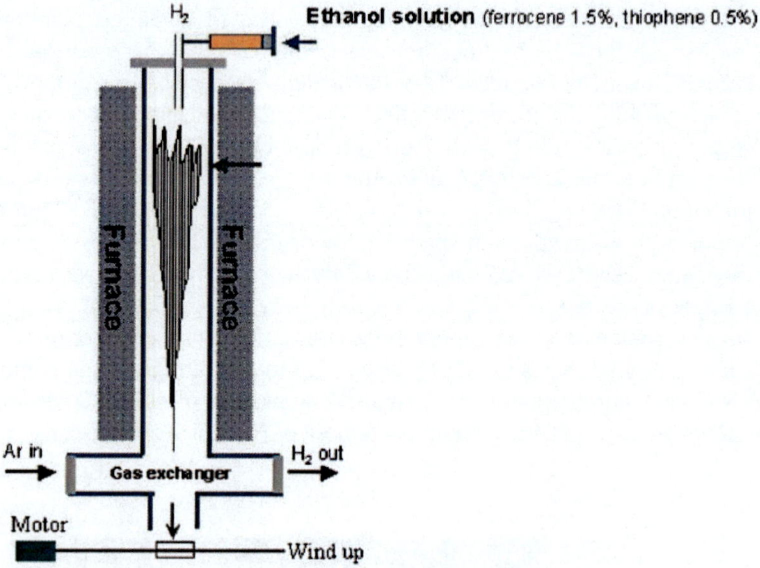

Fig. 8.12 Cambridge CVD reactor for nanotube synthesis and in-situ winding of nanotube yarn [6]

University, Seoul). The Stuttgart set-up is mainly used to grow nanotubes on silicon chips. At temperatures below 900°C predominantly multi-walled tubes are grown, if the temperature is above 1,100°C the tubes are single-walled. To grow single-walled nanotubes at lower temperatures, the carbon activity must be lowered, for example by operating at lower pressures, and the thickness of the catalyst film should be at most 1 nm.

The CVD method is very versatile and can be modified in many ways. Figure 8.12 shows an example where the furnace is kept vertically, where the nanotubes form some spider web material inside the tube, and where this web is grabbed and in-situ spun into a yarn [6]. In this case, the feedstock and catalyst are injected as liquids, such as toluene and ferrocene respectively, and sometimes a promoter of thiophene is also added.

The CVD method can be easily upscaled for large-scale production. Actually, several companies are building plants with production capacities of tons and hundreds of tons of multi-walled carbon nanotubes per year. An important feature in large-scale production is the fluidized bed technique. Here the catalyst is not deposited on a silicon chip but attached to microparticles or nanoparticles of a metallic oxide, say to magnesia (MgO) or of zeolites. In a vertical furnace these particles float on the gas beam entering from the bottom of the tube, thus allowing for intense contact of catalyst and carbon feedstock.

Another variant of CVD is compatible with semiconductor technology. Figure 8.13 shows an example of a "forest" or "lawn" of multi-walled carbon nanotubes on a silicon chip (Courtesy Martti Kaempgen), and Fig. 8.14 shows a network

Fig. 8.13 Nanotube "forest" or "lawn" on a silicon chip as example of CVD grown carbon nanotubes (SEM image) (Courtesy Martti Kaempgen)

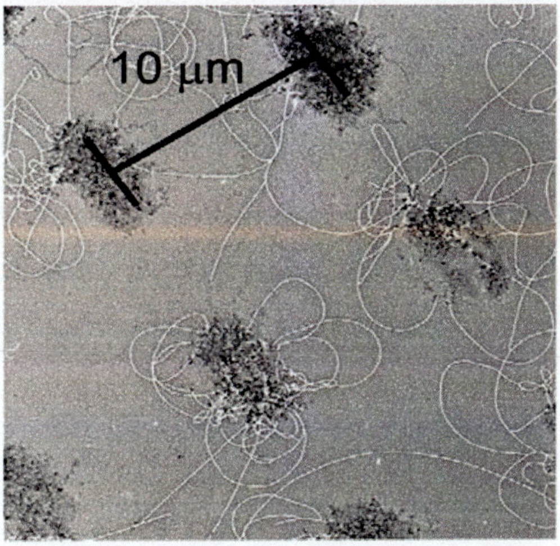

Fig. 8.14 Example of CVD grown nanotubes: Nanotube network in a thin silicon nitride membrane (TEM image) [7]

of single-walled nanotubes an a thin membrane of silicon carbide [7]. This is a particular case where it is desirable to grow high quality multiwalled nanotubes or single walled nanotubes at lower temperatures, by operating at a lower pressure, and careful control of the catalyst [8].

A special example of the compatibility of nanotube growth and semiconductor technology is Fig. 8.15, where we show a 20 nm multi-walled carbon nanotube growing out of a nanohole etched into a silicon chip (Courtesy Infineon) – as a first step towards a VIA, a vertical access interconnect, a vertical conducting lead on a chip, outperforming copper leads because of a much higher current density and because of lower tendency for electromigration. A useful factor of CVD is that

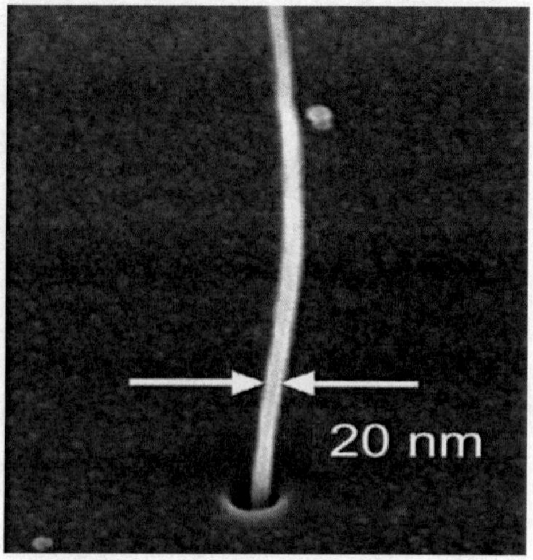

Fig. 8.15 Example of CVD grown nanotubes: A muliwalled carbon nanotube growing out of a nanosized hole etched into a silicon chip (SEM image) (Courtesy Infineon)

the nanotubes only grow where there is catalyst. This allows the patterned grow of nanotube forests, by lithographically patterning the catalyst.

A particular form of CVD which grows rather high forests or carpets is called "supergrowth" [9]. The nanotubes form carpets because their high density means that the nanotubes must grow vertically to avoid bumping into each other. A particular combination of catalyst and supporting oxide is used. Fe on Al_2O_3, gives a very efficient, high density nucleation, which results in the vertically oriented forests. Supergrowth uses a feedstock of ethylene diluted by hydrogen plus a low concentration of water to ensure prolonged growth, so that the forests end up a few mm tall. The water is believed to act as a mild etchant which keeps the catalyst active for longer. Thus supergrowth is the combination of water addition and catalyst, not just water alone. Figure 8.16 shows a side view image of SWNT mats grown by supergrowth [10].

An important aspect of supergrowth is that say 5 mm forests grow from 1 nm layer of catalyst, a yield of $\sim 10^6$ times. Thus the purity of the nanotubes in terms of residual catalyst content is extremely high, compared to other CVD, arc or laser methods. The forests can contain amorphous carbon, which is deposited on the nanotube sidewalls.

A further variant of CVD is to use plasma enhancement, or plasma enhanced chemical vapour deposition (PECVD). The function of the plasma is to partly dissociate the carbon feedstock. This allows growth to occur on less active catalysts, at lower pressures, or at lower temperatures than would typically be possible. However, PECVD will also lead to the uncatalysed deposition of amorphous carbon. It is necessary to stop this parallel reaction by providing a process to etch the amorphous carbon, for example by diluting the hydrocarbon feedstock with hydrogen or ammonia.

Many labs have PECVD systems for other processing. Not all of them are suitable for nanotube growth. They are often RF PECVD which is not so useful, as it

Fig. 8.16 Forest of SWNTs grown on Al-Fe-Al trilayer catalyst by remote microwave assisted PECVD, similar to supergrowth method [10]

often leads to an aggressive atmosphere which etches any nanotubes, or gives ion bombardment which causes disordering. A simple form of PECVD is DC, but operated at a low power. Another is microwave PECVD, where the ion energy is low, so disordering effects are less. DC or RF PECVD does not dissociate molecular hydrogen, so this is why ammonia is traditionally used as the etchant gas in that case. However, it does lead to the accidental incorporation of nitrogen, but this is not usually critical in these applications.

A further form of gentle PECVD is hot-wire assisted deposition. The hot wire (of W or Re at 1,800°C) will dissociate not only the hydrocarbon but also H_2. An important factor in PECVD is that the electric field of the plasma sheath can be used to vertically align the tubes, even when not densely packed [11]. This allows us to produce ordered arrays of vertically aligned tubes for applications such as field emission or sensors (Fig. 8.17). The growth location is defined lithographically, growth only occurs where there is catalyst.

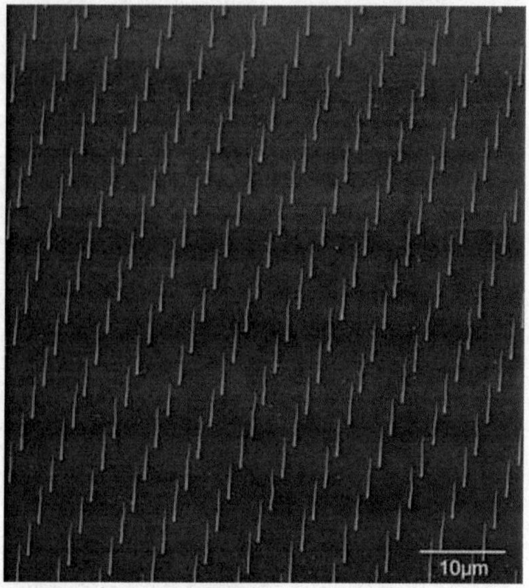

Fig. 8.17 Ordered array of vertically aligned multiwalled nanotubes grown for field emission applications [11]

8.5 Disproportionation of Carbon Monoxide

This method is a special version of the CVD technique. It is usually abbreviated by the acronym HiPCO, which stands for high pressure carbon monoxide. The feedstock is carbon monoxide, which at a certain pressure and temperature regime disproportionates into elementary carbon and carbon dioxide:

$$2\,CO < - > C + CO_2$$

Under the right conditions and with the right catalyst the carbon gas condenses into single-walled carbon nanotubes. The catalyst could be iron, and it could be mixed to the carbon monoxide gas in form of gaseous iron hexacarbonyle, $Fe(CO)_6$. HiPCO carbon nanotubes are already produced on an industrial scale. HiPCO tubes are usually of higher quality (higher electrical conductivity in films and networks) than other mass-produced CVD tubes, but they also are more expensive.

8.6 The Role of the Catalyst

A transition metal catalyst is needed to produce most forms of carbon nanostructures, and is essential to produce SWNTs. In the high temperature processes such as the laser and arc methods, the carbon atoms and metal atoms both condense to form clusters. The ambient gas pressure is set to control a cooling rate so that the catalyst cluster size is in the necessary range when the temperature reaches the growth range of 1,500–1,200°C. The catalyst is liquid in this range. Puretzky et al. [12] have used time-lapse laser diagnostics to measure the evolution of carbon – catalyst

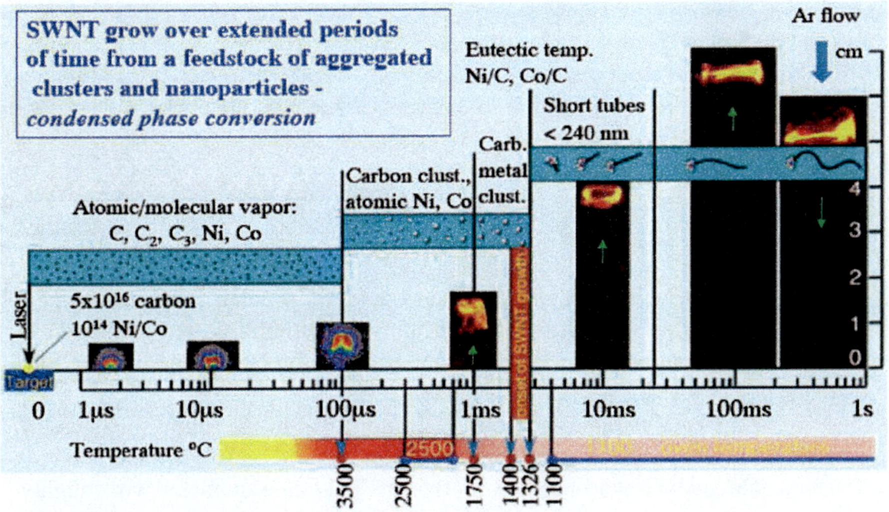

Fig. 8.18 Model formulated by Puretzky et al. [12] for nanotube growth by laser ablation synthesis

plume, and they formulated the model shown in Fig. 8.18. It is likely that a similar mechanism occurs in the arc process but the possibility of measurements is less.

The mechanism is roughly as follows. A limited number of transition metal atoms condense into a cluster or droplet which is still in its liquid form. Carbon atoms have condensed elsewhere into a carbon fullerene cluster. The carbon atoms and clusters land on the metal droplet and dissolve in it. This lowers the metal's melting point to the eutectic point. As the temperature falls, the carbon solubility limit falls, and the carbon droplet becomes super-saturated. The carbon precipitates out as a series of small or single walled nanotubes, looking like a see-urchin. This is seen in post-deposition TEM analyses. Thus many nanotubes grow from a single catalyst droplet [13]. The carbon diffuses across to the growing tube. It is likely that most growth occurs when the catalyst is liquid, as diffusion rates fall sharply when it solidifies.

Nanotube nucleation is believed to occur by a cap forming on top of the catalyst, anchored by C-metal bonds to the catalyst [13]. The cap then lifts off as more carbon atoms are added to the bottom of the cap. This is root growth, which is the dominant growth mechanism for laser and arc single-walled carbon nanotubes (Fig. 8.19).

In laser or arc methods, metal combinations can be more efficient than single metals, because alloying gives a lower melting eutectic. Rare earth metals have also been proposed to lower the surface tension of the traditional transition metal catalysts, Fe, Co and Ni.

Nucleation and growth is slightly different in CVD. Generally, the growth temperature is lower than in laser. The catalyst droplet has not cooled from a much higher temperature, so that there is a sudden super-saturation. Generally, catalysts droplets form and one nanotube grown from each droplet. There are a number of different regimes. The catalyst in most forms of CVD is actually on a "support" which is often an oxide such as SiO_2 or Al_2O_3. The most active catalyst metals,

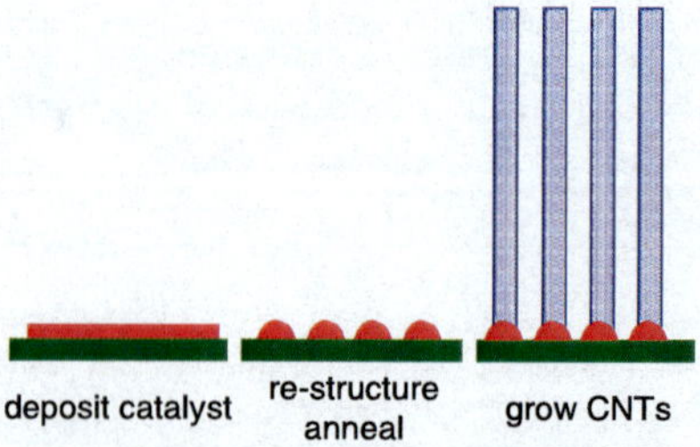

Fig. 8.19 Root growth as dominant mechanism for laser ablation and arc single-walled carbon nanotubes

Fe, Co and Ni, all de-wet these oxides when they are annealed at higher temperatures. Of these metals, Fe will oxidise in ambient conditions, but the oxide is reduced under the reducing growth conditions, and this breaks up the catalyst into nano-particles, because of the volume change.

The size of the catalyst nano-particles is proportional to the initial thickness of the catalyst film. Growth occurs and for the smaller particles, a single nanotube will grow from each catalyst nano-particle, so that the nanotube diameter will roughly equal that of the catalyst particle. So, very small catalyst particles make SWNT, while larger particles make MWNTs. Above a certain limit of about 80 nm, larger catalyst particles will eventually create more than one nanotube. The catalyst is only active when it is in this nanoparticle form, as this gives the large surface area. Thus, low temperature is most successful if the continuous catalyst film can be transformed into nano-particles even at that low temperature [8].

The catalyst acts by dehydrogenating the hydrocarbon feedstock, with the carbon atoms entering the catalyst nano-particle. The nucleation of a nanotube occurs when catalyst is super-saturated with carbon. A single carbon cap forms on one side of the particle [14]. It grows by adding carbon atoms to its base. This will become root growth. Growth continues by a clear surface continuing to dissociate the hydrocarbon, and carbon atoms diffusing across the particle to the growth site. Note that carbon is an interstitial impurity in these metals. The easiest diffusion path is one layer below the surface. Due to size effects, bulk diffusion is harder. SWNTs grow by surface or subsurface diffusion of the carbon atoms [15].

If the catalyst particle is rather larger, bulk diffusion becomes more important because the amount of surface is less. The growth mode changes over to tip growth. The catalyst becomes entrained in the tip of the growing nanotube. Some clear surface still exists to dissociate the feedstock, and the carbon dissolves in the catalyst. The nanotube grows by the catalyst excreting the nanotube behind it as a series of graphitic walls. The strength of the C–C bonds means that the catalyst particle often undergoes distortions as growth continues [16].

There is debate about whether the catalyst is solid or liquid during CVD growth. For bulk CVD of SWNTs, it is likely that the catalyst is liquid. However, direct in-situ experimental measurements show that for lower temperature growth, the catalyst does not need to be liquid; the diffusion rates are high enough to account for growth rates [16].

Finally, in-situ environmental photoemission studies at realistic growth pressures and gases have shown that the catalyst is in the metallic state. The metal

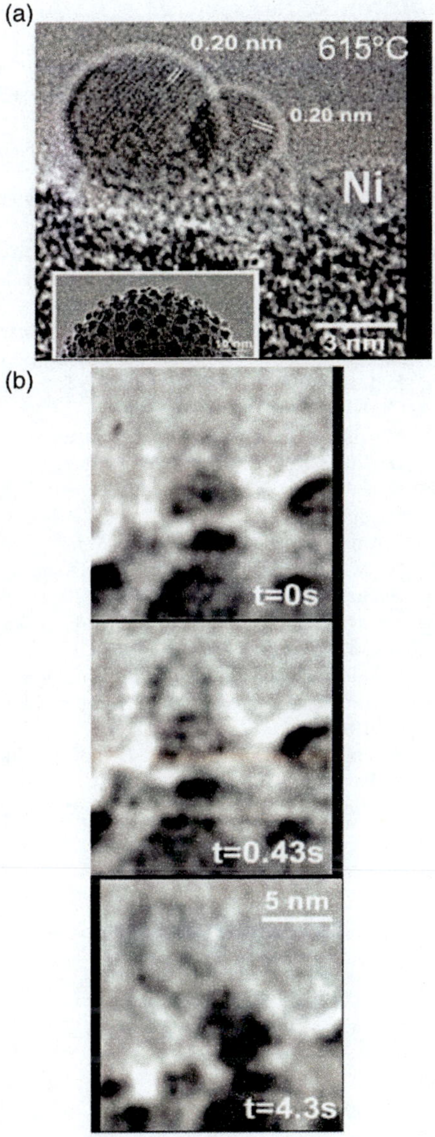

Fig. 8.20 (**a**) In-situ image of catalyst transformed into solid nanoparticle, awaiting to begin growth [16], (**b**) In-situ image of a growing SWNT [16]

oxide is not active. Also, post-growth TEM often finds Ni or Fe carbides. These are not the active catalysts, these metastable phases form during the cool-down (Fig. 8.20a, b) [16].

Acknowledgement This work was supported by World Class University (WCU, R32-2008-000-10082-0) Project of the Korean Ministry of Education, Science and Technology and by the EU 6th Framework Research Project CANAPE

References

1. W. Kraetschmer et al., *Nature* **347**, 354 (1990).
2. R. Saito, G. Dresselhaus, and M. S. Dresselhaus, *Physical Properties of Carbon Nanotubes*, Imperial College Press, London (1998).
3. Y. Saito et al., *J. Phys. Chem.* **99**, 16076 (1995).
4. C. Journet et al., *Nature* **388**, 756 (1997).
5. T. Guo et al., *J. Phys. Chem.* **99**, 10694 (1995); T. Guo et al., *Chem. Phys. Lett.* **243**, 49 (1995); A. Thess et al., *Science* **273**, 483 (1996).
6. Y. L. Li, I. A. Kinloch, and A. H. Windle, *Science* **304**, 276 (2004).
7. G. Gu, G. Philipp, X. Wu, M. Burghard, A. M. Bittner, and S. Roth, *Adv. Funct. Mater.* **11** (4), 295 (2001).
8. M. Cantoro, S. Hofmann, S. Pisana, V. Scardaci, A. Parvez, C. Ducati, A. C. Ferrari, A. M. Blackburn, K. Y. Wang, and J. Robertson, *Nano Lett.* **6**, 1107 (2006).
9. K. Hata, D. N. Futaba, K. Mizuno, T. Namai, M. Yumura, and S. Iijima, *Science* **306**, 1362 (2004).
10. M. Chhowalla, K. B. K. Teo, C. Ducati, N. L. Rupesinghe, G. A. J. Amaratunga, A. C. Ferrari, D. Roy, J. Robertson, and W. I. Milne, *J. Appl. Phys.* **90**, 5308 (2001).
11. K. B. K. Teo, S.-B. Lee, M. Chhowallah, V. Smet, Vu Thien Binh, O. Groening, M. Castignolles, A. Loiseau, G. Piro, P. Leganeux, D. Pribat, D. G. Hasko, H. Ahmed, G. A. J. Amaratunga, and W. I. Milne, *Nanotechnology* **14**, 204 (2003).
12. A. A. Puretzky, H. Schittenhelm, Xudong Fan, M. J. Lance, L. F. Allard Jr., and D. B. Geohegan, *Phys Rev B* **65** 245525 (2002).
13. J. Gavillet, A. Loiseau, C. Journet, F. Willaime, F. Ducastelle, and J.-C. Charlier, *Phys. Rev. Lett.* **87**, 275504 (2001).
14. Y. Shibuta, and S. Murayama, *Chem. Phys. Letts.* **382** 381 (2003).
15. S. Hofmann, G. Csanyi, A. C. Ferrari, M. C. Payne, and J. Robertson, *Phys. Rev. Lett.* **95**, 036101 (2005).
16. S. Hofmann, R. Sharma, C. Ducati, G. Du, C. Mattevi, C. Cepek, M. Cantro, S. Pisana, A. Parvez, F. Cervantes-Sodi, A. C. Ferrari, R. Dunin-Borkowski, S. Lizzit, L. Petaccia, A. Goldini, and J. Robertson, *Nano. Lett.* **7**, 602 (2007).

Chapter 9
Nanotube and Graphene Polymer Composites for Photonics and Optoelectronics

T. Hasan, V. Scardaci, P.H. Tan, F. Bonaccorso, A.G. Rozhin, Z. Sun, and A.C. Ferrari

Abstract Polymer composites are an attractive near-term means to exploit the unique properties of single wall carbon nanotubes and graphene. This is particularly true for composites aimed at photonic and optoelectronic applications, where a number of devices have already been demonstrated. These include transparent conductors, saturable absorbers, electroluminescent and photovoltaic devices. Here, we present an overview of such composites, from solution processing of the raw materials, their sorting, characterization, to their incorporation into polymers, device fabrication and testing.

9.1 Introduction

Incorporation of carbon nanotubes (CNTs) into polymer matrices was first reported in Ref. [5]. Since then, polymer composites of nanostructured carbon materials have developed into a vast research area, mostly focusing on their mechanical applications [72, 82, 88, 237, 281, 382, 486]. A notable difference between CNT/graphene-polymer composites for mechanical applications and optics/photonics is the method by which CNTs/graphene flakes are incorporated into the host matrix. Strong interaction between these carbon nanomaterials and the host polymer is the key for mechanical strength. This is typically attained by functionalization [20, 23, 75, 105–107, 172, 486] and/or in-situ polymerization [117, 237, 281, 327, 349, 449, 475]. These incorporation methods and the mechanical characterizations of the resultant materials will not be covered here. An overview of current progress can be found, e.g., in Refs. [81, 82, 156, 307].

The fabrication and characterization processes of CNT and graphene based composites for photonics and optoelectronics differ from those aimed at mechanical applications. For optical grade composites, fine dispersion, without covalent functionalization, and control of CNT-bundle size are of key importance [96, 125, 220, 358–360, 363, 370, 371, 376, 380, 396, 478]. Different strategies have been developed to produce/grow graphene and transfer it onto flexible substrates [18, 40]. Liquid phase exfoliation (LPE) of graphite [161, 166, 271] is an economic,

A.C. Ferrari (✉)
Department of Engineering, University of Cambridge, Cambridge, CB3 0FA, UK
e-mail: acf26@eng.cam.ac.uk

O. Hayden, K. Nielsch (eds.), *Molecular- and Nano-Tubes*,
DOI 10.1007/978-1-4419-9443-1_9,

up-scalable and promising approach for photonic and optoelectronic applications. A significant effort is ongoing to achieve graphene mono-layer enriched dispersions [160, 161, 166, 271, 284, 411]. For both CNTs and graphene, once the initial dispersion is produced, the fabrication, characterization and integration into devices of the resulting composites follows a similar protocol. We will therefore focus mostly on CNTs, adding additional information on graphene, where necessary.

This chapter is organized as follows. Section 9.3 reviews CNT-dispersion via solution processing, which is essential to overcome the insolubility or near insolubility of unfunctionalized CNTs into various solvents compatible with host polymer matrices. The dispersion of Single Wall Carbon Nanotubes (SWNTs) in aqueous, non-aqueous solvents and liquid crystals are separately discussed in Sections 9.3.1, 9.3.2 and 9.3.3, respectively. Dispersion of graphene in aqueous and non-aqueous media follows a similar principle, though the choice of surfactants is limited due to the two dimensional nature of graphene. This is discussed in Section 9.4. Recent progress in density gradient based sorting of SWNTs, which can be used to fine-tune the optoelectronic device performances is covered in Section 9.5. Section 9.6 discusses inkjet printing of SWNTs. Optical characterization of the SWNT dispersions, with particular focus on the estimation of loading and investigation of the bundle size, is discussed in Sections 9.7.1. and 9.7.2. Section 9.8 covers the preparation of SWNT/graphene-polymer composites for optical applications (Section 9.8.1) and the alignment of SWNTs (Section 9.8.1.1). The desirable characteristics of host polymer matrices with a list of available commercial polymers are presented in Section 9.8.2. The key optical characterizations for such composites, such as Z-scan, photoluminescence, Raman and pump-probe spectroscopies, are summarized in Section 9.9. Particular emphasis is given to Raman spectroscopy, which is one of the most powerful, yet non-destructive characterization techniques available for carbon nanomaterials. Section 9.10 briefly discusses some optical/photonic applications of SWNT-polymer composites. Section 9.11 reviews the application of SWNT-polymer composites as mode-locker in ultra fast lasers. Finally, Section 9.12 discusses graphene based saturable absorbers for ultrafast pulse generation.

9.2 Nanotubes and Graphene for Photonics

SWNTs exhibit strong optical absorption, covering a broad spectral range from UV to near IR [134, 151, 185, 193, 201, 207, 265, 319, 462]. To a first approximation, their band gap varies inversely with the diameter. This can, in principle, be fine-tuned by modifying the growth parameters [191, 200, 249]. Isolated semiconducting SWNTs (s-SWNTs) and small SWNT bundles exhibit photoluminescence (PL) [17, 63, 175, 187, 255–258, 302–304, 319, 324, 342, 421, 433, 463]. PL is quenched for increasing bundle size and the presence of metallic SWNTs (m-SWNTs) [64, 138, 158, 296, 316, 319, 421, 433]. The PL properties of SWNTs have been extensively investigated over the past few years [17, 63, 175, 187, 255–258, 302–304, 319, 324, 330, 342, 353, 421, 433, 439, 463], and the excitonic nature of electronic transitions in SWNTs has been theoretically predicted [10, 66, 197, 335, 402, 492] and experimentally proven [293, 456]. Sub-picosecond carrier relaxation time was also observed in SWNTs [76, 152, 168–170, 178, 230, 246, 283, 329, 351, 427]. In addition, they show significant third-order optical nonlinearities, as theoretically predicted [78, 184, 286, 287] and experimentally confirmed [44, 93, 195, 268, 278, 427] by several groups.

This fast nonlinear optical response is of great technological importance. SWNTs can be used to fabricate ultrafast optoelectronic devices, such as ultrafast sources, optical switches. These are crucial for various applications, for example, high bit rate optical fiber transmission, signal regeneration, dispersion compensator, etc. However, the heterogeneity, impurity and bundling of as-grown SWNTs make it difficult to precisely control the device parameters. Also, it is still difficult to use SWNTs directly grown on substrates [477] to fabricate efficient devices, due to scattering losses [39]. In this context, a more effective solution for the fabrication of SWNT based photonic and optoelectronic devices is to incorporate the processed SWNTs into polymer matrices [96, 125, 220, 358–360, 363, 370, 371, 376, 380, 396, 478]. Wet chemistry processes developed over the past few years can now be readily used. Indeed, it is now possible to untangle [15, 19, 100, 138, 158, 183, 245, 308, 319, 320, 405, 495] and sort SWNTs [12, 13, 41, 86]. A combination of wet chemistry with compatible non-aqueous solvents and polymers of appropriate properties is therefore a viable route for the fabrication of optoelectronic devices. Furthermore, this holds great promise for the mass production of inexpensive photonic devices and their simplified integration into various lightwave systems.

A single layer graphene absorbs 2.3% of incident light [311]. This remains constant from the visible to the near infrared (NIR), due to the linear dispersion of Dirac electrons in graphene [40, 160, 311, 411]. As we discuss later in Section 9.12, such broad absorption band, coupled with ultrafast relaxation dynamics, make graphene one of the most promising candidates for next generation photonic and optoelectronic applications. A viable route for mass fabrication of such graphene based devices also follows a wet chemistry based strategy, similar to SWNTs [160, 411].

9.3 Nanotube Dispersion in Liquid Media

As produced SWNTs usually form entangled networks of bundles or ropes [4, 77, 87, 192, 229, 261, 273, 364, 430, 431, 448] due to strong van der Waals interactions [140, 171, 229, 270, 364, 430]. In such entangled networks, SWNTs do not possess the optimum mechanical, thermal and optoelectronic properties [31, 69, 72, 82, 88, 239, 262, 382]. It is thus important to produce isolated/individual SWNTs from the bundles. Strong ultrasonic treatments in presence of water or non-aqueous solvents and 'dispersants' (e.g. surfactants, polymers etc., aiding the dispersion process) are commonly used to exfoliate highly aggregated SWNT networks into small bundles [100, 138, 157, 158, 183, 279, 299, 308, 319–331, 420, 495]. In the following sections, we present an overview of the methods to achieve dispersions with high concentration of individual SWNTs and small bundles without covalent functionalization. The SWNT dispersions thus obtained are mainly characterized by absorption and photoluminescence spectroscopy, covered in the subsequent sections.

9.3.1 Nanotube Dispersion in Water

The non-polar nature of unfunctionalized (i.e. pristine) SWNTs makes it hard to directly disperse them in a highly polar solvent like water without any functionalization [107, 267, 317, 426]. Therefore, much effort has been devoted to find suitable molecules to interface the non-polar sidewalls of nanotubes with water. To date, stable dispersions of pristine SWNTs in water have been achieved with the aid of ionic and non-ionic surfactants [183, 308, 319, 348, 423, 466], polymers [22, 308, 320], DNA [13, 64, 495], polypeptides [100, 474, 498] and cellulose derivatives [299, 420]. Covalent functionalization of SWNTs disrupts the extended π-network, hence changes their optoelectronic properties [20, 23, 24, 57, 254].

Micelles are aggregated surfactant molecules. The critical micelle concentration (CMC) is the concentration of surfactants in a liquid above which micelles are spontaneously formed [54]. At the CMC, the surface-area between two liquids (e.g., air-water interface) becomes loaded with surfactant molecules. Addition of any more surfactant molecules leads to the formation of micelles [54]. In aqueous solutions, a typical surfactant (e.g. SDBS) micelle arranges its hydrophillic heads in contact with water and the hydrophobic tails in the micelle center [54]. Surfactants, in concentration above the CMC, form micelles around individual SWNTs and small bundles, interfacing their non-polar tail with the tube sidewalls and their polar or ionic end with water, making SWNTs compatible with the aqueous medium [183, 296, 308, 319]. This creates a density difference between individualized and bundled SWNTs. For example, an individual SWNT encapsulated in an SDS micelle is less dense than a 7-tube bundle encapsulated by the same micelles [319]. After centrifugation, heavier bundles precipitate due to higher sedimentation coefficient [46], while the supernatant becomes enriched with individually suspended SWNTs. Several types of surfactants, both ionic and non-ionic, have been reported to stably suspend SWNTs in water [183, 308]. These include SDS, SDBS, SC, SDC, TDC, DTAB, CTAB on the ionic side, and the Triton-X and Brij series on the non-ionic side [183, 308], though the border between surfactant and polymer for the latter is arbitrary. Bile salt surfactants have recently been demonstrated to be very effective in individualizing SWNTs in aqueous dispersions [41]. Bile salts (e.g. SDC, TDC, SC) have a rigid molecular structure, consisting of a cholesterol group with dissimilar sides with a steroid skeleton with a carboxylic acid side-chain and one to three hydroxyl (–OH) groups on the steroid backbone [295, 391]. They are amphiphilic, having both hydrophobic and hydrophilic sides [391], enabling them to individualize SWNTs much more effectively than linear chain surfactants [41]. Figure 9.1 shows some surfactants and polymers commonly used to disperse SWNTs in different solvents.

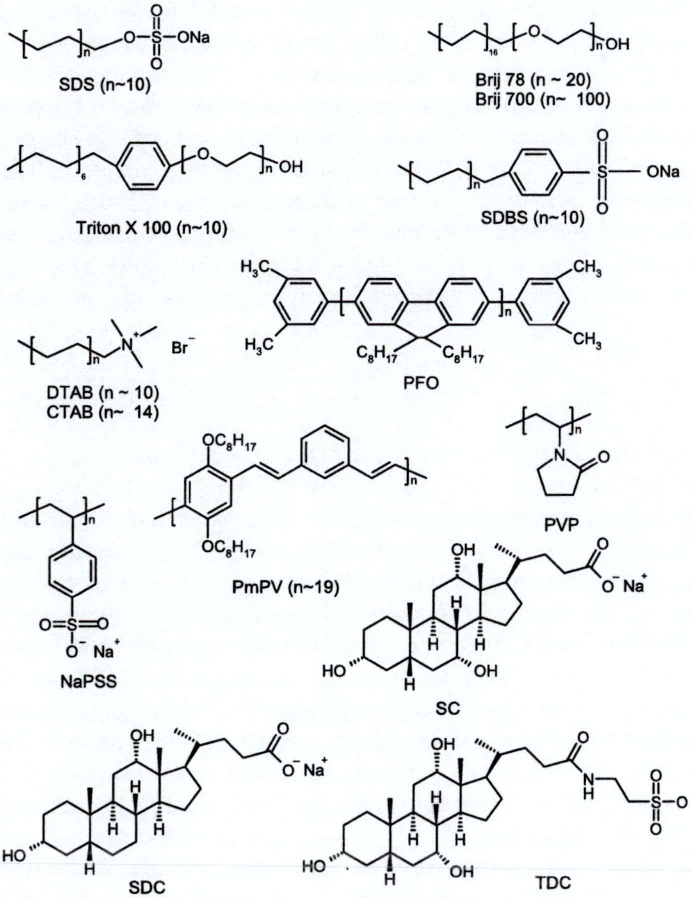

Fig. 9.1 Chemical structures of common surfactants and polymers used to disperse SWNTs in aqueous and organic solvents. SDS: sodium dodecyl sulfate; SDBS: sodium dodecylbenzene-sulphonate; DTAB: dodecyltrimethylammonium bromide; CTAB: cetyltrimethylammonium bromide; PFO: poly(9,9-dioctylfluorenyl-2,7-diyl; PmPV: poly(p-phenlyenevinyleneco-2,5-dioctoxy-m-phenylenevinylene); PVP: polyvinylpyrrolidone; NaPSS: sodium polystyrene sulphonate; SC: sodium cholate; SDC: sodium deoxycholate; TDC: sodium taurodeoxycholate

Water-soluble polymers are reported to wrap around SWNTs [22, 320], thus facilitating their de-bundling and dispersion. In particular, polyvinylpyrrolidone (PVP) and its copolymers with vinyl acetate, acrylic acid, dimethylaminoethyl methacrylate, polystyrene sulphonate, polyvinylsulphate can stably disperse SWNTs in water [320]. The wrapping by water-soluble polymers is thermodynamically favored by the removal of the hydrophobic interface between the nanotube sidewall and the aqueous medium [22, 320]. Cellulose derivatives, e.g. sodium carboxymethylcellulose (Na-CMC) [299, 420] and hydroxyethylcellulose [299] can also disperse a high amount of SWNTs without forming visible aggregations.

DNA is also an excellent dispersant of SWNTs in water [64, 441, 495]. Reference [495] proposed that single-stranded DNA forms a helical wrapping around the tube sidewalls by π-stacking. They also showed that the binding free-energy of single stranded DNA to SWNTs is higher than that between two tubes, facilitating dispersion [495]. Other biomolecules, such as polypeptides, efficiently disperse SWNTs in water [100, 474, 498]. Reference [100] designed a peptide, called nano-1, which folds into an α-helix, whose hydrophobic side interacts with the tube sidewall, whereas the hydrophilic side interacts with water molecules. Increasing the aromatic residues within the peptides improves dispersion [498]. This can be further improved by cross-linking the peptides on the external side of the α-helixes [474].

9.3.2 Nanotube Dispersion in Non-aqueous Solvents

Although the highest concentrations (>1.5 gL^{-1}) of pristine, individualized SWNTs or small bundles have so far been achieved in water, the presence of dispersant molecules is not the best option in view of their integration in devices when preservation of the pristine electronic structure is necessary. Also, the aqueous medium is not suitable for SWNT integration into water-insoluble polymer composites. Therefore, much effort has been devoted to the dispersion of SWNTs in pure non-aqueous solvents, such as N,N-dimethylformamide (DMF), N,N-dimethylacetamide (DMA) and N-methyl-2-pyrrolidone (NMP) [15, 19, 32, 33, 138, 158, 222, 236, 245]. A range of surfactants and polymers has also been investigated as dispersing agents in such solvents to improve the loading of unfunctionalized SWNTs [157–160, 198, 279, 403–405].

Ref. [15] proposed that important criteria for solvents to get good dispersion of SWNTs include high electron-pair donicity, β (hydrogen bond acceptance ability [242, 354]), low hydrogen bond donation parameter, α and high solvatochromic parameter [15], π^*. The latter describes the polarity and polarizability of solvents [285, 354]. Therefore, the Lewis basicity (i.e. electron pair acceptance ability [242, 285, 354]) without hydrogen donors is key to good dispersion of SWNTs [15]. However, this does not cover all the requirements. For example, Ref. [15] showed that, even though DMSO meets all the above criteria, it is only a mediocre solvent for SWNTs. In addition, highly polar alkyl amide solvents with "optimal geometries" are reported to be vital for good SWNT dispersion ability [245]. Amongst the amide solvents, NMP has been described as one of the most effective for dispersing pristine SWNTs [15, 19, 138, 158]. Pure NMP is able to disperse SWNTs with the highest fraction (~70%) of individual tubes at a very low concentration (~0.004 gL^{-1}), with growing average bundle size as the SWNT concentration increases [138]. Individual tubes remain stable in NMP for at least 3 weeks [158].

Dispersion and stabilization of nanoparticles, e.g. nanotubes in pure solvents, can be explained by considering the relative solvent-solvent, solvent-particle and particle-particle interaction strengths [166]. Stable nanoparticle dispersions require the Gibbs free energy of mixing, ΔG_{mix}, to be zero or negative [155]:

$$\Delta G_{mix} = \Delta H_{mix} - T\Delta S_{mix} \tag{9.1}$$

where, T is the absolute temperature, ΔH_{mix} is the enthalpy of mixing and ΔS_{mix} is the entropy change in the mixing process [166]. For large solute particles like graphene and nanotubes, ΔS_{mix} is small [32, 166]. Therefore, for dispersion and stabilization of SWNTs in solvents, ΔH_{mix} needs to be very small. This can be achieved by choosing a solvent whose surface energy is very close to that of SWNTs. This supports the experimental evidence of NMP being the best solvent in dispersing pristine SWNTs [32, 138, 158].

We also investigated non-ionic surfactants such as Triton-X 100, Pluronic F98, Igepal DM-970 in NMP as dispersing agents [158, 159]. Though they are able to disperse a higher amount of SWNTs than pure NMP, they do not help in individualizing nor stabilizing them [158, 159]. Interestingly, if polyvinylpyrrolidone (PVP) is added to SWNTs dispersed in pure NMP, spontaneous de-bundling occurs [158], even after re-aggregation of SWNTs [159]. This process depends on diameter and chirality [158]. In addition, reduction of PVP concentration initiates re-aggregation of dispersed SWNTs, thereby proving PVP to be essential to stabilize the dispersions [157]. Figure 9.2 illustrates the stabilizing effect of PVP concentration on SWNT dispersions in NMP [157].

Amphiphilic block copolymers may also be used to efficiently individualize SWNTs in DMF [198]. Indeed, they act as surfactants, having a hydrophobic block and a hydrophilic one. Polystyrene-block-polyacrilic acid (PS-b-PAA) forms micelles around individual SWNTs by gradual addition of water. The external hydrophilic blocks are finally cross-linked giving stable micelles, see Fig. 9.3 [198].

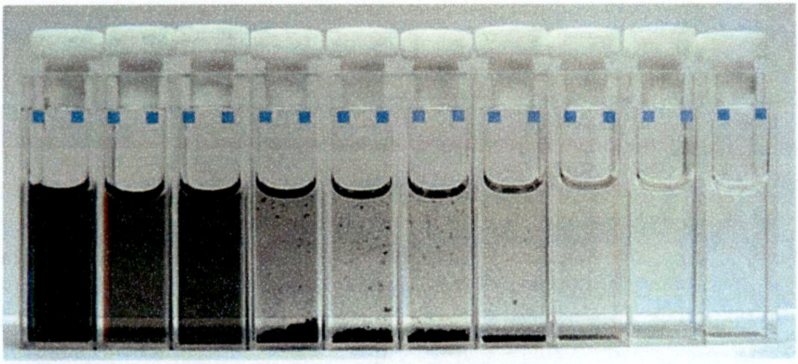

Fig. 9.2 Photograph of SWNT dispersion sets prepared with gradually lower PVP concentration (from *left* to *right*; 8.75, 6.56, 4.38, 2.63, 1.75, 1.31, 0.88, 0.44, 0.22, 0 gL^{-1}) with SWNT concentration (0.02 to ~0.0002 gL^{-1}). Previously dispersed SWNTs (~0.013 gL^{-1}) re-aggregate below ~3 gL^{-1} of PVP concentration [157]

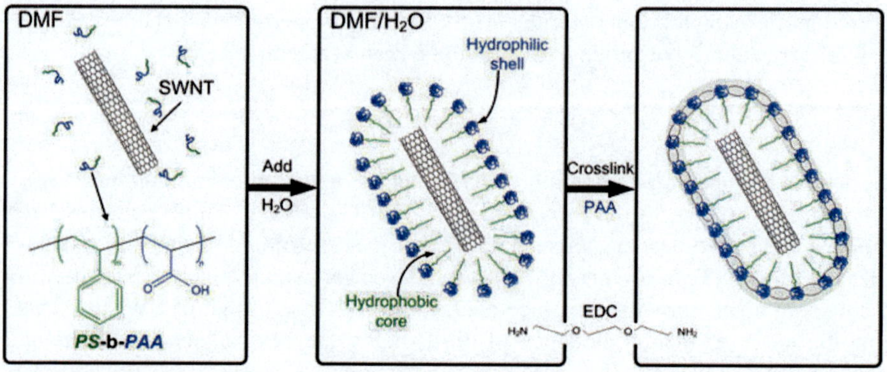

Fig. 9.3 Scheme of the individualization process of a SWNT by the PS-b-PAA block copolymer by creating micelle-like structures by cross-linking. Adapted from [198]

Conjugated polymers can also disperse high amounts of SWNTs in non-aqueous solvents [403–406]. Poly-[(m-phenylenevinylene)-co-(2,5-dioctyloxy-p-phenylenevinylene)] (PmPV) and its derivatives preferentially wrap around bundles rather than individual tubes, when chloroform is used as solvent [405, 406]. PmPV derivatives containing ionic side-groups can also disperse SWNT bundles in protic solvents like ethanol [403]. The dispersion mechanism is proposed to be π-π stacking and van der Waals interaction between PmPV and tube sidewalls [403]. On the other hand, if a hyper-branched variant of PmPV is used, SWNTs can be individually dispersed in chloroform [404]. In this case, the branched structure of the polymer forms cavities that are suitable to host individual SWNTs [404]. Individual SWNTs can also be separated in THF if poly(p-phenylene-1,2-vinylene) (PPV) is co-polymerized with units of p-phenylene-1,1-vinylidene [443]. The resulting copolymer (coPPV) has structural defects (the 1,1-vinylidene units) that allow the polymer backbone to fit the curvature of the nanotube sidewalls better than the homopolymer, thus wrapping around the SWNT by π-stacking [443].

9.3.3 Nanotube Dispersion in Liquid Crystals

The unique structure of Liquid Crystal (LC) molecules enables them to be aligned by surface treatment or by an applied field. The highly anisotropic interaction of SWNTs with light [6, 266, 387] dictates the need of their alignment parallel to the light polarization, so to maximize their absorption with minimal loading [179, 309, 357]. Dispersion of SWNTs in LC is therefore an attractive proposition for SWNT alignment, due to interaction with LCs, for photonic/optical applications [102, 243, 244, 276, 369, 438]. In fact, alignment of multiwall nanotube (MWNT) bundles was reported using E7 LCs and applying an electric field, even higher than the orientational ordering of E7 itself [101, 102]. In addition, even with very small CNT concentrations (e.g. <0.01∼0.05 wt%) [251], enhancement of the nonlinear

[218, 251, 369, 437], electrooptical [74, 176, 252, 253] and dielectric properties [73, 369] of pristine LCs was observed [73, 176, 218, 252, 253, 369]. The resulting high optical intensity of $\sim$100 μWcm^{-2} [218], lower driving voltage [252, 253], suppression of field-screening [73] and faster reorientation of LCs, could improve the stability and response time of LC displays [176, 369]. SWNTs can be dispersed in LCs [369] directly, or using the LC molecules as dispersants [243, 244, 437]. Lyotropic LCs are amphiphillic compounds dissolved in solvents whose LC state or mesophase formation depends on their concentration and on the solvent itself [114]. Lyotropic LCs, in their solvents, have a hydrophobic and hydrophillic part in their structure, much like surfactants, and can be effectively used as dispersants for SWNTs. Reference [243] reported that lyotropic LC molecules arrange in micellar structures around CNTs, similar to the encapsulation of SWNTs by ionic surfactant micelles in aqueous suspensions [17]. Compared to CNT dispersions with thermotropic LCs (whose LC state or mesophase formation depends on temperature [114, 244]) [102, 276, 437, 438], lyotropic LC molecules in solvents allow higher loading (up to 0.15 wt%) [464] and better stability of the resultant dispersion [243, 244, 464]. Even higher loading of SWNTs results in phase separation and consequent decrease in viscosity [464]. It is important to note that for display applications, low loading of SWNTs in LC is desirable to minimize conductive contributions from m-SWNTs [369]. Therefore, lyotropic LCs might be used for CNT alignment only [244, 464]. Nevertheless, in both types of LCs, irrespective of CNT loading, CNTs can be aligned using the LC molecules by applying electric or magnetic fields [101, 102, 244, 464]. Thus, SWNT-LC dispersions are very attractive for large scale CNT alignment [102, 244].

9.4 Graphene Dispersion in Aqueous and Non-Aqueous Solvents

Exfoliation of graphene from graphite is a promising route for up-scalable, graphene based applications. Like SWNTs, graphene/graphite is hydrophobic and requires surfactants to form stable aqueous dispersions. Bile salts are excellent in exfoliating graphite flakes to produce aqueous dispersions of graphene [83, 161, 271, 284, 411]. This is due to the flat molecular structure of bile salts, which readily adsorb on graphitic surfaces [391, 409]. SDC, in particular, forms a large contact area ($\sim$1.8–3 nm^2) per surfactant molecule [368] with its β side compared to linear chain surfactants (e.g. sodium dodecylbenzene sulfonate (SDBS) [94, 271]), which adsorbs on graphitic surfaces through its alkyl chains [183]. The hydrophobicity of bile salts dictates how strongly their molecules are adsorbed on the graphitic surfaces. This is measured by the Hydrophobic Index (HI), defined as [301]:

$$HI = HI_\alpha + HI_\beta \tag{9.2}$$

where,

$$HI_{\alpha(\beta)} = \frac{Hydrophobic\ surface\ area_{\alpha(\beta)}}{Hydrophilic\ surface\ area_{\alpha(\beta)}} \times anomer\% \tag{9.3}$$

The *HI* of deoxycholic acid is 7.27, higher than that of its tri-hydroxy counterpart, cholic acid (6.91) [301]. This is also reflected in the higher effective contact area of SDC molecules with per gram of graphite (6.96 nm^2g^{-1}) compared to that of SC (5.72 nm^2g^{-1}) [368]. This also implies a denser and more regular graphene surface coverage with SC [161]. Di-hydroxy bile salts, e.g. SDC or sodium taurodeoxycholate (TDC), are thus more effective than the trihydroxy ones, e.g. SC or sodium taurocholate (TC), and significantly better than linear chain surfactants, such as SDBS. Thermodynamic parameters of bile salts adsorbing on graphite at 25–30°C show that ΔG_{mix} for SDC adsorbed on graphite (~ -28 kJ mol^{-1} [409]) is more negative than for SC ($\Delta G_{mix} \sim -26$ kJ mol^{-1} [409]). Since Eq. (1) dictates the thermodynamic stability of dispersions, graphene flakes dispersed by SDC in water are expected to be more stable compared to those dispersed by SC-water [409].

For a given graphene volume fraction and flake thickness in non-aqueous solvents, it was shown that [166]:

$$\Delta H_{mix} \propto (\delta_1 - \delta_2)^2 \tag{9.4}$$

where, δ_i is the square root of surface energy of graphene and solvent. This requires the surface energies of graphene and solvent to be very close for a stabilized dispersion, similar to what is empirically observed for pristine SWNTs, as discussed in Section 9.3.2. The surface tensions (γ) of NMP and *ortho*-dichlorobenzene (*o*-DCB) are 35.71 and 44.56 mJ m^{-2} [484]. When converted to solvent surface energy (E_{Sur}) with a generalized surface entropy (S_{Sur}) of 0.1 mJ K^{-1} m^{-2} [137, 275] using the relation $\gamma = (E_{Sur} - TS_{Sur})$ [275], we get ~65–75 mJ m^{-2} [161]. This is within the range of estimated surface energies (~70 mJ m^{-2}) of nanotubes and graphite

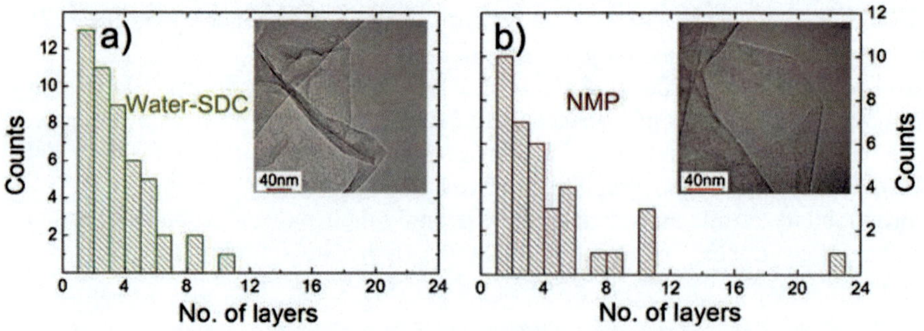

Fig. 9.4 TEM statistics for (**a**) water-SDC graphene dispersions showing ~50% SLG and BLG (**b**) anhydrous NMP dispersions showing ~50% SLG and BLG. Inset:folded SLG. Adapted from Ref. [161]

[32, 141, 166, 374, 436, 488] and explains efficient exfoliation of graphite in certain solvents, e.g. NMP and *o*-DCB. The concentrations of flakes in aqueous and non-aqueous dispersions can be >100 mg L^{-1} [161], with >50% SLG and BLG [161, 166, 271, 284]. Figure 9.4 shows TEM statistics and images of typical exfoliated flakes obtained by LPE.

9.5 Sorting Nanotubes by Chirality and Electronic Type Using Density Gradient Differentiation

The separation of SWNTs by different chiralities from as-grown, heterogeneous mixture of SWNTs would be the ideal way to exploit their full potential in any applications, since one could mix the different chiralities according to the needs. This is particularly important for optoelectronic applications, when SWNTs with defined electronic properties, diameter or chirality are most preferable. To date, different approaches for post-growth selection have been proposed, such as chromatography [496], electrophoresis [104, 235, 259], and conventional [461] or density gradient ultracentrifugation(DGU) [12, 13, 41, 86, 482]. Amongst these techniques, DGU has emerged as the most promising and versatile strategy. Indeed, separation of nanotubes by length [113], number of walls [147], diameter [13], metallic vs semiconducting (m/s) character [12] and chirality [41] has been reported. Here we will use "sorting" to indicate a generic process of postgrowth nanotube selection; "separation" to indicate a process resulting in a sample with diameter in a certain range or to indicate separation of m/s nanotubes; "enrichment" to indicate a process resulting in an increase of the percentage of nanotubes with certain chirality with respect to the pristine material.

Analytical ultracentrifugation is a well established and versatile technique for a wide range of applications [418]. It is, for example, useful to determine molecular weight [85], thermodynamic [361] and hydrodynamic [162] properties of molecules. Preparative ultracentrifugation was historically used in separation and manipulation of biological molecules such as DNA, RNA and different proteins [46, 84, 129, 144, 163, 228, 313]. When a uniform medium is used, this is referred to as differential ultracentrifugation [418]. If density gradient is introduced, this is knows as DGU [30, 337]. Differential ultracentrifugation separates particles based on their sedimentation rate [418], which determines how quickly they sediment out of the fluid in which they are dispersed, in response to a centrifugal force acting on them. In DGU, particles are ultracentrifuged in a preformed density gradient medium (DGM) [30, 337, 467]. During the process, they move along an ultracentrifuge cell, dragged by the centrifugal force, until they reach the corresponding isopycnic point, *i.e.*, the point where their buoyant density equals that of the surrounding DGM [467]. The buoyant density is defined as the density of the medium at the corresponding isopycnic point (measured in g/cm^3), which depends on the dispersion, on the type of surfactant and may also be different in diverse gradient media [180]. Such process depends only on the particles buoyant density, and is also known as isopycnic separation. If the ultracentrifugation is stopped before the particles achieve their isopycnic points, a zonal separation (ZS) is attained [46]. The latter also depends on the particle sedimentation rates [46].

Isopycnic separation thus allows sorting and separation of SWNTs based on their density differences [12, 13, 86]. The DGM choice is largely driven by the material one needs to separate. Salts (such as cesium chloride, lithium chloride, sodium chloride, etc.), sucrose, and Optiprep, i.e. 60% w/v iodixanol (a non-ionic iso-osmotic derivative of tri-iodobenzoic acid [126]) solution in water [12, 41, 86] ($\rho \sim 1.32$ g cm^{-3}), are usually exploited in isopycnic separations. Due to the low viscosity of the DGM, density gradients produced by salts are less stable compared to those produced with Sucrose and Optiprep [41]. Moreover, salts induce strong aggregation on the hydrophobic solutes [55, 70] that sometimes affect the separation process itself [41]. Also, the percentage of sucrose used as DGM can have a significant impact on the separation [41]. Sucrose has high viscosity, increasing at high concentrations, and is mainly used in DGU for ZS rather than for isopycnic separation [41]. On the other hand, Optiprep is better suited for isopycnic separation due to its higher viscosity than salts, and better density tunability than sucrose [41]. Moreover, it has an almost constant viscosity as function of the gradient density [489], low osmolarity [109], is dialyzable and its gradients are self-forming [126]. There are different approaches (linear, non-linear or step gradient [12, 80, 133, 390, 447]) to prepare the density gradient. These are related to how the density of the liquid medium is varied across the length of a centrifuge cell. Step gradients are created by a series of steps increasing in density in order to separate the particles of interest from their undesirable neighbors [390, 447]. Non-linear gradients are formed so that the particles sediment at the same rate over the entire length of the cell [80]. Linear gradients are used to separate materials with very small differences in their buoyant density [46, 133] e.g. in SWNT sorting, as the difference in densities of SWNT is very small. Linear gradients are created directly in centrifuge cells either by using a linear gradient maker or making discrete layers of gradually increasing densities. In the latter case, the discrete layers diffuse into each other and form a linear density gradient [12, 86].

For sorting, a dispersion containing the SWNTs, after sonication and pre-ultracentrifugation, is inserted in the density gradient at the top or at a point where the density of the prepared gradient closely matches that of the solution. This can be determined by measuring the density of the SWNT dispersions and comparing it to that of different layers before their diffusion. The dispersion is then ultracentrifuged until equilibrium is reached. Because of the different densities of the gradient medium along the centrifuge cell, SWNTs are redistributed at their respective isopycnic points [12, 167]. A schematic of the process is presented in Fig. 9.5a. The appearance of different color bands is an indication of SWNTs sorting [12, 86]. The colors depend on the peak optical absorption. Thus, e.g., the purple color (non spectral red-blue combination) of the top band in Fig. 9.5b is due to (6,5) tubes that absorb at ~570 nm (eh_{22}) (yellow).

After ultracentrifugation, the sorted SWNTs are removed from the ultracentrifuge cells, layer by layer, using a fractionation technique. This is a widely used

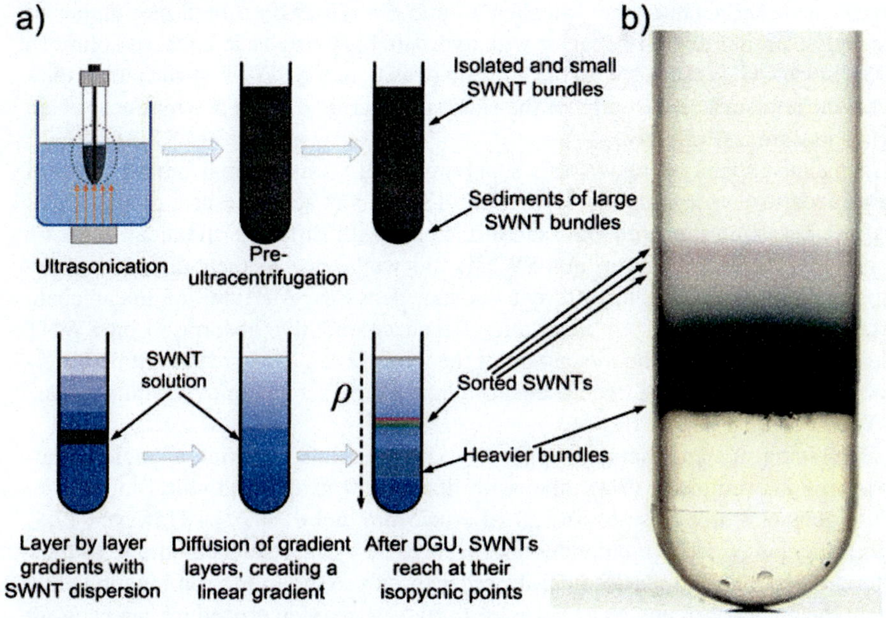

Fig. 9.5 Sorting of SWNTs using isopycnic separation: (**a**) Schematic of the process. (**b**) Sorted SWNTs at their isopycnic points. Adapted from Ref. [12]

process in life sciences by which certain quantities of a mixture are separately extracted to a large number of aliquots, whose composition varies according to the density gradient of the original mixture [407]. Fractionation methods are classified into three main categories: piston [47], down [48] and upward [47] displacement. For SWNT separation, upward displacement fractionation can be used to extract small aliquots [41, 86]. A dense solution, Fluorinert FC-40, is inserted with a needle at the bottom, pushing the gradient up into an inverted collection needle [7]. Because the density of the tubes changes with their diameter, considering a uniform surfactant coverage, this immediately links buoyant density to tube diameter, thus enabling an effective diameter sorting by isopycnic separation [12, 13, 41, 86].

Natural bile salts [356], such as SC, SDC or TDC are the most suitable surfactants for isopycnic separation, due to their steroid skeleton polar tail. These, in aqueous environments, expose their hydrophilic sides to the water and the hydrophobic sides to the SWNTs [465]. This is a key requirement in the adsorption of flat molecules onto the hydrophobic graphitic surface whose structure is composed of carbon atoms [301]. This also allows the exfoliation [161, 271, 284, 411] and separation of graphitic flakes by number of layers [148]. In contrast, linear chain surfactants, such as SDBS, have a flexible cylindrical body and inefficiently form micelles around SWNTs [295]. We demonstrated that the poor performance of linear chain surfactants in diameter separation, with respect to bile salts, is related, other than to inefficient de-bundling, to their surface coverage of SWNTs [41].

Linear chain surfactants form micelles around SWNTs with a random number of molecules, similar to their behavior with hydrophobic particles in aqueous solutions [295]. Since DGU is sensitive to the buoyant density of the SWNT-surfactant assembly, a uniform surface coverage of the sidewalls is critical. This is why linear chain surfactants are less effective.

Isopycnic separation allows m/s separation [12] with minimal modification of the protocol used for diameter separation [13]. This is achieved in a co-surfactant mixture, based on the principle that surfactants with different chemical structure adsorb in a different way on m/s-SWNTs sidewalls, due to their different polarizability [289], resulting in different buoyant densities. Mixtures of linear chain surfactants and bile salts are ideal due to their competitive absorption on SWNT sidewalls depending on the m/s nature of the tubes [272]. This creates subtle differences in the density of the micelle encapsulated SWNTs, enough to separate m- and s-SWNTs [12, 167].

Combining m/s and diameter separation, it is possible to enrich a single chirality [41]. For example, a (7,4) tube, with diameter 0.75 nm and chiral angle $\theta = 21°$ is geometrically close to a (6,5) (d = 0.75 nm and $\theta = 27°$). However, (7,4) is metallic and (6,5) is semiconducting. Hence, they can be separated due to their different electronic properties [12]. In order to reduce the (n,m) combinations and obtain the highest percentage of a single chirality, a two step procedure can be used: m/s separation exploiting a co-surfactant mixture (TDC-SDS) followed by diameter separation (SC). This allows, e.g., to selectively enrich (6,5) with respect to (7,4) and (6,6). Figures 9.6a, b demonstrate the enhancement of (6,5) [41]. Note that for an absolute population measurement, the PL cross section of individual species of SWNTs must be taken into account [323, 330, 353].

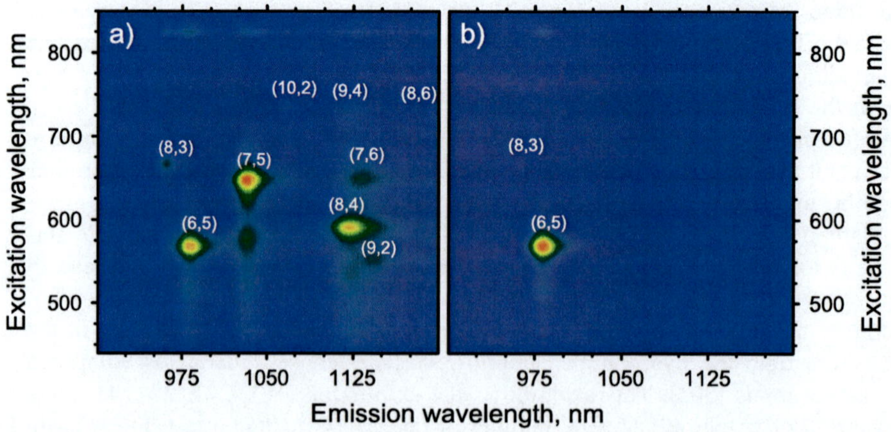

Fig. 9.6 Photoluminescence excitation map of (**a**) initial CoMoCAT dispersion showing the exciton-exciton resonances from different SWNTs; (**b**) After isopycnic separation. The enrichment of (6,5) is shown by the strong (eh_{22}, eh_{11}) resonance at ca. (982, 569) nm

9.6 Inkjet Printing of Nanotube and Graphene Dispersions

Deposition of fluidic droplets to form patterns directly on substrates using inkjet printing offers a mask-less, inexpensive and scalable low-temperature process for large area fabrication [34, 132, 204, 336, 388, 389, 408]. The core technology is very similar to that of consumer-level inkjet printers. Solution processable conjugated polymers are ideal to fully exploit such an inexpensive and flexible printing technique to realize all-polymer devices on a variety of substrates [34, 204, 389, 408]. In fact, a variety of devices for different applications, ranging from all-plastic electronics [204, 389, 408] to organic light emitting displays [386] has so far been demonstrated using this technology. The resolution of inkjet printing can be enhanced by pre-patterning the substrates, so that the functionalized patterns can act as barriers for the deposited droplets [203, 458]. Device channel lengths as small as 500 nm can be obtained [458]. Even higher resolution of 100–400 nm was recently demonstrated by a self-aligned inkjet printing method [318]. Using inkjet printing in conjunction with micro-embossing, self-aligned, vertical channel all-polymer thin film transistors were also reported [408].

Several inkjet printed devices on various un-patterned and pre-patterned substrates have thus far been demonstrated, using conjugated polymers, nanomaterials (e.g. pigments [231], microemulsions [196], magnetic nanoparticle-based inks [65], diamond [128] or metallic nanoparticles [315, 494]), CNTs [28, 174, 392], and graphene dispersed in carrier solvents. To minimize sedimentation, the particle sizes must have dimensions $< 1\ \mu$m [56]. This limits the maximum nanoparticle volume fraction due to increased viscosity [56]. Direct inkjet printing of soluble organic precursors for making metal contacts in organic field effect transistors was recently reported [131, 471]. However, inkjet printing of silver-copper nanoparticle-based solutions yielded lower contact resistance compared to the organic precursor-based approach [131]. The advantages of inkjet printing of nanoparticles suspended in carrier solvents include greater ease of selective deposition and high concentration of materials for partially soluble compounds [56].

Printing nanotubes, graphene and nanotube-polymer composites directly on a range of substrates is thus an attractive technological proposition, due to its flexibility and selective deposition on a small target area [28, 40, 174, 233, 388, 392, 419]. This is even more attractive as it could result in improved mobility, environmental stability and lifetime compared to organic electronic devices [14, 28, 38]. Figure 9.7a, b show a printing head and a high speed close-up of a fluidic droplet being expelled from the 50 μm nozzle. Figure 9.7c is an optical micrograph of an array of inkjet printed SWNT-TFT devices [28] without any boundary patterns to restrict the spreading of the deposited droplets. Nevertheless, this demonstrates the effectiveness of selective deposition. As for nanoparticle-based solutions, it is vital to obtain an effective dispersion of SWNTs for inkjet printing. This is because aggregation of SWNTs, frequently encountered in organic solvents, easily clogs the print nozzles. Dispersions of pristine SWNTs in pure NMP are not stable. The aggregation is triggered by the tendency of NMP to absorb moisture. That is why using even a week-old SWNT dispersion for inkjet printing results in lumps

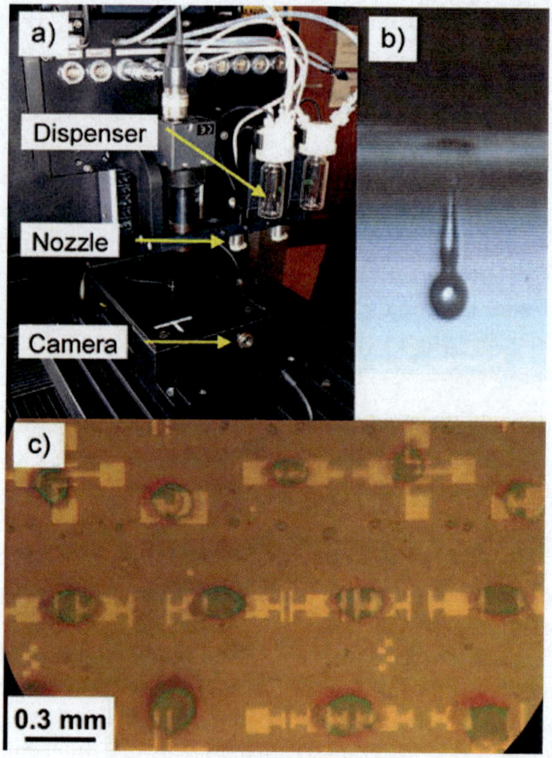

Fig. 9.7 (**a**) Inkjet printing head (**b**) High speed photographic close-up of the nozzle showing a droplet being expelled (**c**) Optical micrograph TFT-SWNT devices printed using inkjet [28]

on the substrate [28]. Printing percolated networks of SWNTs as transparent conductors replacements is also being investigated [98] (see Section 9.10.1). Though mostly unfunctionalized SWNTs have been used in literature, using different side-wall functionalizations, it was shown that the general characteristics of SWNT based inkjet printed transistors may be modified [142]. SWNTs stabilized by a water-soluble conducting polymer, namely poly(2-methoxyaniline-5-sulfonic acid) (PMAS), were also ink-jet printed on plastic substrates to yield transparent, conducting films [392]. Also, exclusively inkjet-printed SWNT thin film transistors with low-voltage operation were recently demonstrated using high-capacitance ionic-liquid dielectrics [326]. We also prepared a graphene-ink suitable for inkjet printing, achieving graphene TFTs with up to 95 $cm^2V^{-1}s^{-1}$ mobility and 80% optical transmittance, paving the way to high mobility all-inkjet printed graphene-based optoelectronics.

9.7 Optical Characterizations of Nanotube and Graphene in Dispersions

Significant efforts have been devoted to understanding the electronic and optical properties of nanotubes [17, 201, 293, 319, 335, 456]. Their quasi-one dimensionality gives rise to sharp spikes (van Hove singularities) in their electronic density of states. The nature of the electronic transitions responsible for the optical properties in SWNTs has been the subject of intense debate. Many authors have discussed their experimental observations in terms of band-band transitions [201] involving free electron-hole pairs, Fig. 9.8a. The ensemble of transition energies E_{ii} between i-th van Hove peaks on opposite sides of the Fermi level vs tube diameter form the so-called *Kataura plot* [201]. This is widely used to understand absorption spectroscopy and resonant Raman spectroscopy of SWNTs. However, the electron-hole interaction in nanotubes is very strong [335]. The exciton binding energies of SWNTs are very large, ranging from tens of meV to 1 eV, depending on diameter, chirality, and dielectric screening [293, 335, 456]. Therefore, an incoming photon creates an exciton formed by bound electron(e)-hole(h) pairs in the i-th sub-band, Fig. 9.8b.

SWNT dispersions are usually characterized by UV-Vis absorption and photoluminescence excitation (PLE) spectroscopy, in order to assess the concentration and presence of individual tubes or bundles [16, 17, 175, 187, 226, 255, 293, 303, 304, 319, 321, 342, 355, 421, 456, 463]. As an example, Fig. 9.9 plots absorption spectra of SWNTs in different solvent-surfactant systems. The peaks in these spectra represent excitonic transitions, while their sharpness is an indication of the presence of isolated SWNTs [151, 308, 319]. For example, the features from 400 to 550 nm, 550 to 900 nm and 1,100 to 1,430 nm in the spectra represent eh_{11} of m-SWNTs, eh_{22} of s-SWNTs and eh_{11} of s-SWNTs, respectively [116, 187, 293, 319, 421, 456]. Bundling results in broadening of the absorption peaks [116, 321, 352, 421, 422, 457] and reduction of the transition energies. This causes a red-shift in the absorption spectra [116, 257, 321, 352, 422, 457]. However, changes in excitonic transition energies can also be caused by increase in the dielectric constant (ε) of the surrounding environment, the so-called "dielectric screening effect" [60, 116, 124, 335, 452, 457]. Due to this, the absorption spectra of SWNTs dispersed in aqueous solvents exhibit small shifts in transition energies depending on the dispersants used [299,

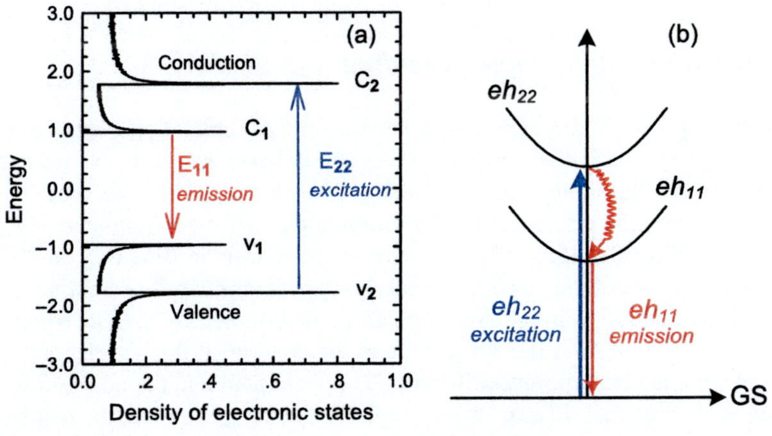

Fig. 9.8 Schemes of the optical excitation and emission of s-SWNTs based on the (**a**) band-to-band model and (**b**) exciton picture. GS represents ground state

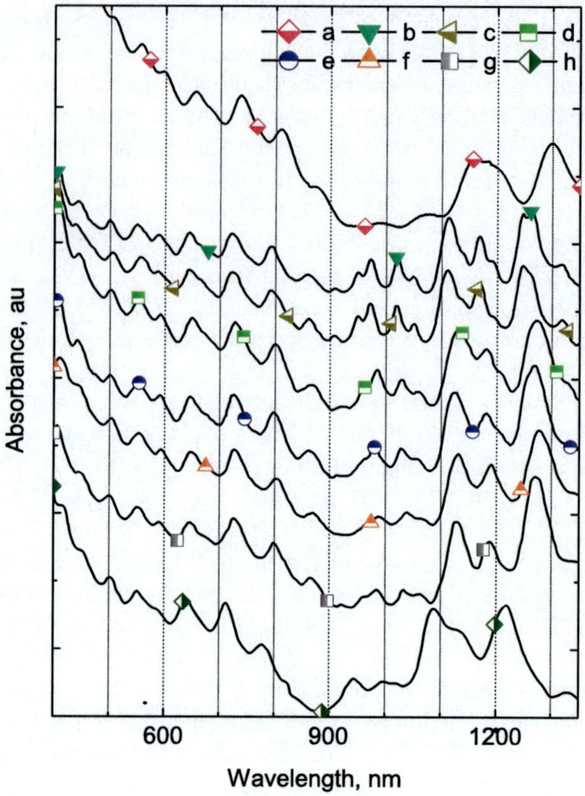

Fig. 9.9 Absorption spectra of SWNTs in different solvent-surfactant systems, (**a**) Pure NMP [158]; (**b**) Water/SDBS [308]; (**c**) Water/SDS [308]; (**d**) Water/CTAB [308]; (**e**) Water/Brij [308]; (**f**) Water/Na-CMC [299]; (**g**) Water/DNA [495]; (**h**) Water/Nano-1 [498]. These spectra illustrate the shift in absorption peaks due to the different dielectric environments surrounding the SWNTs

308, 495, 498]. In the case of organic solvents, a larger red-shift of 30–50 meV is usually observed, and attributed to large increase in dielectric screening [138, 158, 279, 384].

9.7.1 Estimation of Nanotube Loading

When preparing SWNT-polymer composites for optical applications, estimation of the SWNT aggregation in the dispersions (i.e. bundle size) is usually carried out by comparing SWNT dispersions prepared using the same dispersant/solvent combination, to eliminate the effects caused by different dielectric environments [60, 79, 124, 227, 305, 335, 352, 452, 457]. However, more specific information on SWNT bundling can be obtained using PLE spectroscopy, as discussed in Section 9.7.2. The SWNT concentration in a dispersion can be determined by the Beer-Lambert law $A_\lambda = \alpha_\lambda lc$, where A_λ is the absorbance of the material at the wavelength λ, α_λ is the corresponding absorption coefficient, l is optical path length and c is the concentration of the material. In case of very high SWNT concentration, the dispersion

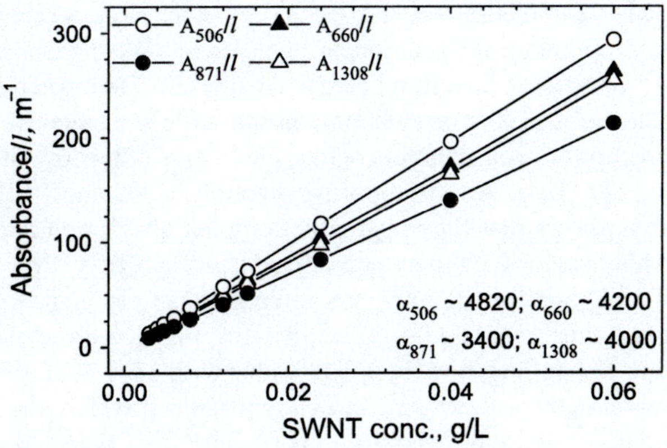

Fig. 9.10 Absorption coefficients of HiPco SWNTs in NMP at four wavelengths. Adapted from Ref. [158]

needs to be diluted to avoid scattering losses [39, 158, 159, 299]. In order to get a reliable result, it is necessary to determine α_λ at several wavelengths from a set of solutions with known SWNT concentration [138, 158, 245]. These wavelengths are chosen to match well-defined peaks in the absorption spectra of SWNTs in dispersions, e.g. eh_{11} of m-SWNTs (at 506 nm), eh_{22} of s-SWNTs (at 660 and 871 nm), and eh_{11} of s-SWNTs (at 1,308 nm) for SWNTs dispersed in NMP. Figure 9.10 shows an example used to estimate α_λ of HiPco SWNTs dispersed in pure NMP at four different wavelengths [158]. The values of α_λ thus obtained can then be used to assess the SWNT concentrations in unknown samples [138, 158, 245].

9.7.2 Detection of Nanotube Bundles

Bundling can decrease the gap in s-SWNTs even if the surrounding dielectric environment remains unchanged [321, 421, 422, 457]. Indeed, red-shift and broadening of excitonic transitions as a consequence of bundling was observed by resonant Raman profiles of radial breathing modes [321], Rayleigh scattering [457], absorption and PL spectroscopy [421, 422]. This is attributed to the modification of Coulomb interactions by dielectric screening induced by the adjacent nanotubes [452, 457]. In comparison to Raman scattering and Rayleigh scattering, absorption and PL spectroscopy are faster techniques to probe the red-shift of excitonic transitions in SWNT ensembles [321, 421, 422, 457]. However, because the optical transitions of SWNTs are strongly modulated by the dielectric environment [60, 79, 124, 227, 305, 335, 352, 452, 457], the comparison between different SWNT dispersions only works between dispersions prepared with the same combination of dispersant and solvent. It is not known how sensitively absorption or PLE spectroscopy can probe the bundle size in dispersions and composite films, because the optical transitions of SWNTs are usually very broad due to bundling inhomogeneity, packing efficiency and wide distribution of bundle sizes in the ensemble samples

[421, 422]. Therefore, it is necessary to explore a direct, simple and independent way to identify the presence of bundles in SWNT dispersions and films.

We observed exciton energy transfer (EET) between nanotubes in bundles [371, 421, 422]. EET is very common and widely studied in biological systems, conjugated polymers, quantum wires, dots, and other low-dimensional systems [1, 27, 36, 127, 194]. A thorough investigation of PL and PLE spectra of SWNT dispersions shows that the apparently complex absorption and emission features can be explained by EET between adjacent s-SWNTs within a bundle [421]. Detection of EET does not require any reference sample; hence, is an independent method to monitor bundles. For example, Ref. [421] showed that 2 months after preparation, CoMoCAT dispersions in water with 1 wt% SDBS form small bundles. This was hinted by a red-shift in eh_{11} emission wavelengths [321, 422, 457]. However, a number of new resonance spots were also detected, not corresponding to any exciton-exciton resonances of SWNTs in the emission range from 1,150 to 1,350 nm [16, 17]. We attributed these spots to EET from large-gap s-SWNTs (donors) to small-gap s-SWNTs (acceptors). Because of the large exciton binding energies [293, 335, 456], energy transfer between s-SWNTs occurs via excitons [421], not via inter-tube electron or hole migration as suggested by Ref. [433]. Note that these optical features are distinct from the deep excitonic states (DE) reported in Ref. [248]. The intensity of the DE features is very weak and their positions are dependent on the associated excitonic transition energies, while the EET features can be very strong and are exclusively dependent on the excitonic transition energies of the donor and acceptor s-SWNTs [421]. Figure 9.11 schematically describes EET from a donor nanotube to an acceptor. The emission-absorption overlap between donor and acceptor SWNTs depends on the specific donor-acceptor couple. We proposed Förster resonance energy transfer (FRET) [127] as the EET mechanism [421] because of the high degree of orientation and small wall to wall distance between tubes in bundles, the latter also favoring multipolar contributions [127, 194]. The EET features, marked in circles and ellipses in the PLE map shown in Fig. 9.12, are summarized in Table 9.1. These can thus be used to detect the presence of bundles in SWNT dispersions.

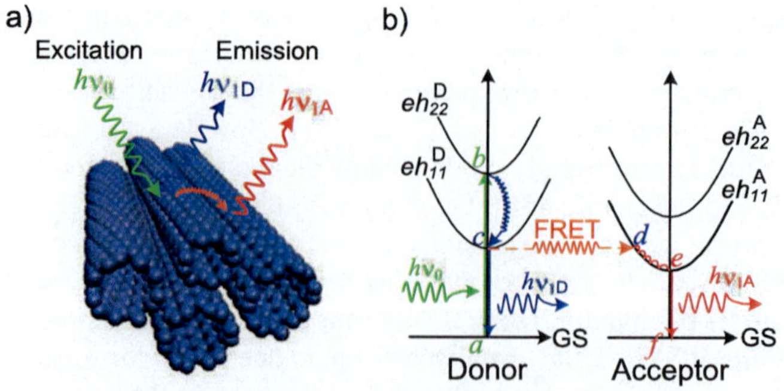

Fig. 9.11 Schematic illustration of EET from a large gap s-SWNT to a smaller gap s-SWNT [421]

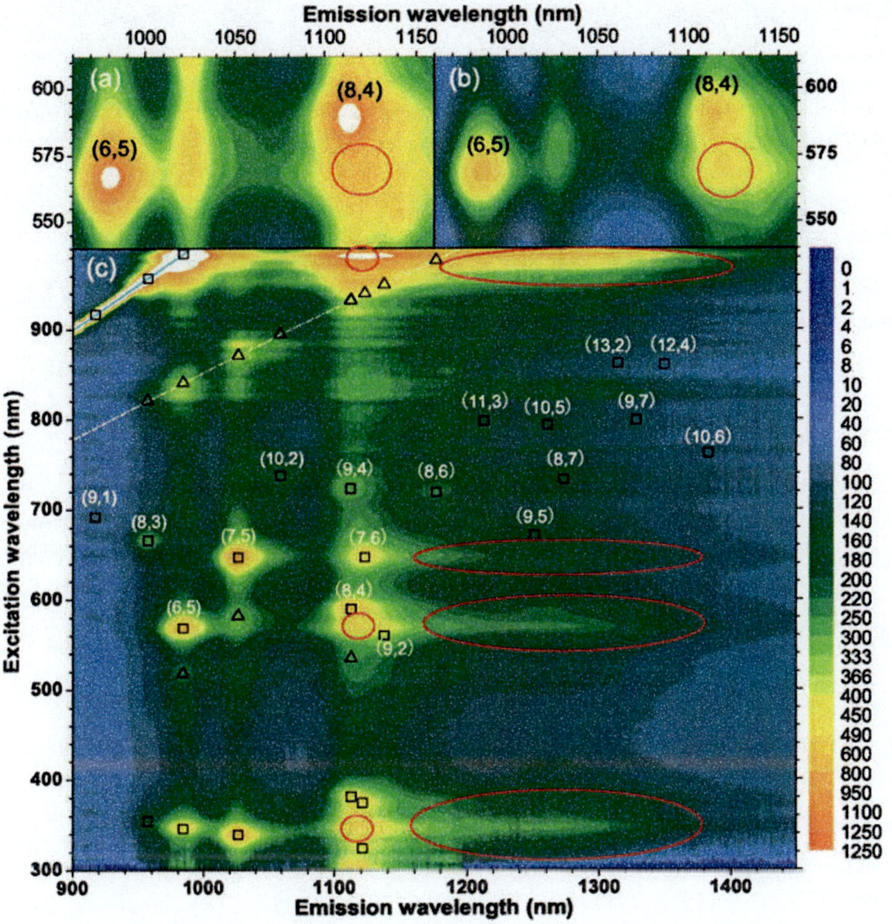

Fig. 9.12 PLE map for (**a**) as-prepared dispersions and (**b**, **c**) after 2 months of incubation, where the PLE maps in (**a**) and (**b**) are within the eh_{11} emission and the eh_{22} excitation of (6,5) and (8,4) tubes. *Solid lines* at upper left corners represent resonances with same excitation and recombination energies. The *dashed-dotted lines* are associated with the *D* sideband of eh_{11} excitons. *Open squares* represent eh_{11} emission of SWNTs for which excitation matches their eh_{11}, eh_{22}, eh_{33} and eh_{44} transitions. Each (eh_{22}, eh_{11}) resonance is labeled with the chiral index of the corresponding SWNT. *Open up-triangles* are phonon sidebands of eh_{11} and eh_{22} excitons. *Open circles* mark emission from (8, 4), (7, 6) and (9,4) SWNTs, with excitation matching eh_{11}, eh_{22} and eh_{33} of (6,5). Broad spectral features marked by *ellipses* are assigned to EET between s-SWNTs [421]

9.7.3 Optical Characterizations of Graphene in Dispersions

The linear dispersion of Dirac electrons in graphene [40, 160, 311, 411] results in flat absorption spectra from the visible to the near IR region. A UV peak is also observed which is a signature of the van Hove singularity in the graphene density of states [234]. As for SWNTs, the loading of graphene in aqueous or non-aqueous

Table 9.1 Assignment of exciton-exciton bands and corresponding EET features in the PLE map of as-prepared CoMoCAT SWNT dispersions in D_2O/SDBS. λ_{ex} and λ_{em} are excitation and emission wavelengths in nm, E_{ex} and E_{em} the excitation and emission energy in eV. (n, m) is the chirality. eh^D_{ii} ($i = 1,2,3,4$) and eh^A_{11} are excitonic transitions of donors (D) and excitonic emission of acceptors (A), respectively. Here, eh_{ii} (i=1,2,3,4) correspond to the excitonic states associated with the i-th electronic inter band transition E_{ii} (i=1,2,3,4) in the single-particle picture

		EET features		
$(\lambda_{ex},\lambda_{em})$	(E_{ex}, E_{em})	Donor	Assign.	Acceptor
(980,1025)	(1.265,1.210)	(6,5)	(eh^D_{11}, eh^A_{11})	(7,5)
(980,~1116)	(1.265,1.111)	(6,5)	(eh^D_{11}, eh^A_{11})	(8,4),(9,4),(7,6)
(980,1139)	(1.265,1.088)	(6,5)	(eh^D_{11}, eh^A_{11})	(9,2)
(980,1180)	(1.265,1.051)	(6,5)	(eh^D_{11}, eh^A_{11})	(8,6)
(980,~1260)	(1.265,0.984)	(6,5)	(eh^D_{11}, eh^A_{11})	(9,5),(10,5),(8,7)
(980,~1330)	(1.265,0.932)	(6,5)	(eh^D_{11}, eh^A_{11})	(13,2),(9,7),(12,4)
(914,~1116)	(1.357,1.111)	(9,1)	(eh^D_{11}, eh^A_{11})	(8,4),(9,4),(7,6)
(879,~1116)	(1.411,1.111)	(6,4)	(eh^D_{11}, eh^A_{11})	(8,4),(9,4),(7,6)
(828,980)	(1.498,1.265)	(5,4)	(eh^D_{11}, eh^A_{11})	(6,5)
(828,1025)	(1.498,1.210)	(5,4)	(eh^D_{11}, eh^A_{11})	(7,5)
(828,~1116)	(1.498,1.111)	(5,4)	(eh^D_{11}, eh^A_{11})	(8,4),(9,4),(7,6)
(828,1139)	(1.498,1.088)	(5,4)	(eh^D_{11}, eh^A_{11})	(9,2)
(828,1180)	(1.498,1.051)	(5,4)	(eh^D_{11}, eh^A_{11})	(8,6)
(828,~1260)	(1.498,0.984)	(5,4)	(eh^D_{11}, eh^A_{11})	(9,5),(10,5),(8,7)
(828,~1330)	(1.498,0.932)	(5,4)	(eh^D_{11}, eh^A_{11})	(13,2),(9,7),(12,4)
(645,1060)	(1.922,1.170)	(7,5)	(eh^D_{22}, eh^A_{11})	(10,2)
(646,1139)	(1.919,1.089)	(7,5)(7,6)	(eh^D_{22}, eh^A_{11})	(9,2)
(646,1180)	(1.919,1.051)	(7,5)(7,6)	(eh^D_{22}, eh^A_{11})	(8,6)
(646,~1260)	(1.919,0.984)	(7,5)(7,6)	(eh^D_{22}, eh^A_{11})	(9,5)(10,5)(8,7)
(646,~1330)	(1.919,0.932)	(7,5)(7,6)	(eh^D_{22}, eh^A_{11})	(13,2),(9,7),(12,4)
(589,1139)	(2.105,1.089)	(8,4)	(eh^D_{22}, eh^A_{11})	(9,2)
(589,1180)	(2.105,1.051)	(8,4)	(eh^D_{22}, eh^A_{11})	(8,6)
(589,~1260)	(2.105,0.984)	(8,4)	(eh^D_{22}, eh^A_{11})	(9,5)(10,5)(8,7)
(566,1025)	(2.191,1.209)	(6,5)	(eh^D_{22}, eh^A_{11})	(7,5)
(566,~1116)	(2.191,1.111)	(6,5)	(eh^D_{22}, eh^A_{11})	(8,4)(9,4)(7,6)
(566,1139)	(2.191,1.089)	(6,5)	(eh^D_{22}, eh^A_{11})	(9,2)
(566,1180)	(2.191,1.051)	(6,5)	(eh^D_{22}, eh^A_{11})	(8,6)
(566,~1260)	(2.191,0.984)	(6,5)	(eh^D_{22}, eh^A_{11})	(9,5)(10,5)(8,7)
(566,~1330)	(2.191,0.932)	(6,5)	(eh^D_{22}, eh^A_{11})	(13,2),(9,7),(12,4)
(371,1139)	(3.342,1.089)	(7,6)	(eh^D_{33}, eh^A_{11})	(9,2)

Table 9.1 (continued)

		EET features		
$(\lambda_{ex},\lambda_{em})$	(E_{ex}, E_{em})	Donor	Assign.	Acceptor
(371,1180)	(3.342,1.051)	(7,6)	(eh^D_{33}, eh^A_{11})	(8,6)
(371,~1260)	(3.342,0.984)	(7,6)	(eh^D_{33}, eh^A_{11})	(9,5)(10,5)(8,7)
(321,1139)	(3.863,1.089)	(7,6)	(eh^D_{44}, eh^A_{11})	(9,2)
(321,1180)	(3.863,1.051)	(7,6)	(eh^D_{44}, eh^A_{11})	(8,6)
(346,1060)	(3.584,1.170)	(6,5)	(eh^D_{33}, eh^A_{11})	(10,2)
(346,~1116)	(3.584,1.111)	(6,5)	(eh^D_{33}, eh^A_{11})	(8,4)(9,4)(7,6)
(346,1139)	(3.584,1.089)	(6,5)	(eh^D_{33}, eh^A_{11})	(9,2)
(346,1180)	(3.584,1.051)	(6,5)	(eh^D_{33}, eh^A_{11})	(8,6)
(346,~1260)	(3.584,0.984)	(6,5)	(eh^D_{33}, eh^A_{11})	(9,5)(10,5)(8,7)
(346,~1330)	(3.584,0.932)	(6,5)	(eh^D_{33}, eh^A_{11})	(13,2),(9,7),(12,4)
(337,1060)	(3.679,1.170)	(7,5)	(eh^D_{33}, eh^A_{11})	(10,2)
(337,~1116)	(3.679,1.111)	(7,5)	(eh^D_{33}, eh^A_{11})	(8,4)(9,4)(7,6)
(337,1139)	(3.679,1.089)	(7,5)	(eh^D_{33}, eh^A_{11})	(9,2)
(337,1180)	(3.679,1.051)	(7,5)	(eh^D_{33}, eh^A_{11})	(8,6)
(337,~1260)	(3.679,0.984)	(7,5)	(eh^D_{33}, eh^A_{11})	(9,5)(10,5)(8,7)
(337,~1330)	(3.679,0.932)	(7,5)	(eh^D_{33}, eh^A_{11})	(13,2),(9,7),(12,4)

dispersions can be estimated by UV-Vis absorption spectroscopy in conjunction with Beer-Lambert law, using experimentally determined absorption co-efficient of graphene at the desired wavelength [161, 166, 271].

9.8 Nanotube/Graphene Polymer Composites

The following subsections review generalized procedures used to prepare SWNT or graphene polymer composites and the desirable characteristics of host matrices for some selected applications, such as saturable absorbers (SAs).

9.8.1 Incorporation of Nanotube/Graphene in Host Polymer Matrices

SWNTs or graphene are dispersed in appropriate solvents, generally by ultrasonic treatment. The dispersions may initially contain aggregates or bundles, which can then be removed by centrifugation or filtration. There is a trade-off between the desired SWNT bundle sizes/unexfoliated graphene flakes and their concentration.

In case of water, high loading of both SWNTs and graphene may be achieved using surfactants, making it easier to control the optical density in the resultant composites. Organic solvents, on the other hand, cannot generally disperse a high amount of SWNTs/graphene. If allowed by the final application, nanotubes can be oxidized [2, 476] or functionalized [260, 347] to improve dispersion and loading. This also holds true for graphene, where functionalization dramatically improves the amount of material that can be dispersed [332]. For SWNTs, when covalent functionalization is not an option, different polymers are employed as dispersants, which can also act as the host in the final composite [206, 362, 476]. Commonly used organic solvents are *o*-DCB [476], chloroform [240, 338, 476], NMP [158, 159, 362] and toluene [206]. For graphene, NMP and *o*-DCB are usually the common non-aqueous solvents. After removal of aggregates or un-exfoliated material, the dispersions are mixed with the host polymer. The same protocol is followed in organic solvents if a different polymer is used as the dispersant. The mixtures are then drop-cast or spin coated, depending on the final application. Free standing or substrate-bound composites with homogeneous, submicrometer distribution of SWNTs/graphene are then obtained by evaporating the solvent [241, 357, 359, 360, 363, 370, 371].

For SWNTs, composites with individually dispersed nanotubes may also be fabricated using cellulose derivatives [299, 428]. In this instance, the dispersions are drop-cast [299, 428]. Cellulose-based composites eliminate the need for surfactants, as they are used both as the dispersant and host matrix [299, 428]. Individually dispersed SWNTs can also be prepared in gelatine films [224]. SWNT-SDS dispersions, mixed with a gelatine aqueous solution, are sonicated at 40°C and cast on a substrate to dry at room temperature to obtain such composites. Gelatine undergoes gelation at 37°C while cooling down, thus preventing re-aggregation [224].

9.8.1.1 Alignment of Nanotubes in Composites

Alignment of SWNTs is important, due to the high anisotropic interaction of SWNTs with light [6, 266, 387]. In aligned SWNTs, absorption is maximum when polarization is parallel to the alignment direction [177, 179, 182, 263, 309, 357]. Various methods, in particular mechanical stretching and Langmuir-Blodgett (LB), have been used to achieve alignment of SWNTs.

High degree of alignment within the polymer matrix can be achieved by laterally stretching the composite [179, 224, 299, 357]. A SWNT-polytstyrene-toluene dispersion cast on a teflon sheet was reported to be stretchable 10 times, yielding 56% of SWNTs aligned within $\sim$15° of the stretching axis [179]. PVA can be stretched up to 6 times by heating at 60°C in a humid environment [357], whereas gelatine films may be stretched up to three times under swelling in a water-ethanol mixture, and dried under constant elongation [224]. Hydroxyethylcellulose composites can be stretched up to three times at 100°C by adding glycerine as plasticizer [299]. Figure 9.16 shows the change of absorption of a mechanically stretched SWNT-PVA composite for incident light with varying polarization [179, 357].

LB may also be used to fabricate self-aligned SWNTs [150, 225, 385]. A graphical illustration of the LB technique used with a SWNT aqueous dispersion is shown

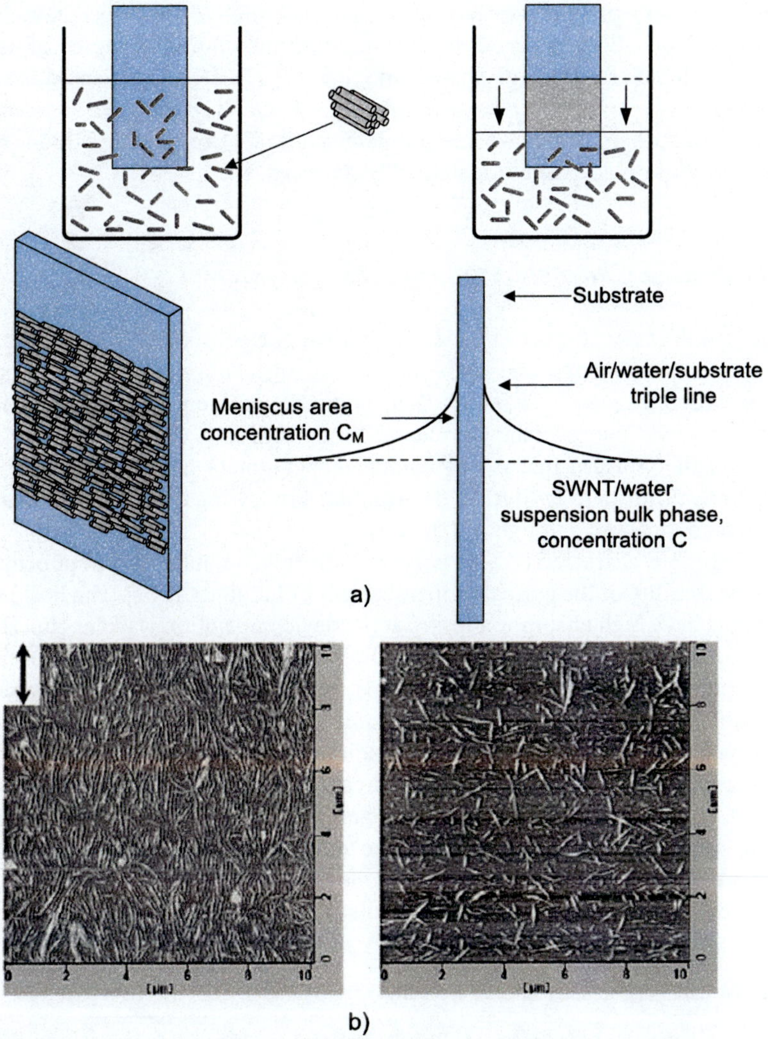

Fig. 9.13 (**a**) Schematic illustration of the SWNT self-assembly process. A hydrophilic glass slide is vertically immersed in a stable dispersion of short SWNTs. With gradual evaporation of the water, the SWNT bundles self-assemble on the glass substrate around the air/dispersion/substrate triple line. As the triple line progress downwards, a continuous SWNT film forms on the substrate. Figure adapted from [385]. (**b**) AFM images of an s-SWNT single layer on mica prepared by vertical dipping (*left*) and a drop-and-dry film prepared from the same dispersion (*right*). The *arrow* indicates the dipping direction. Adapted from Ref. [225]

in Fig. 9.13a [385]. A comparison between the AFM images of randomly oriented and aligned SWNTs can be seen in Fig. 9.13b [225]. The nanotube self-alignment process occurs at the air-substrate-solution triple line as the solvent gradually evaporates [385]. Also, in-plane compression can be used to further align the tubes

[225]. Composites with poly(N-dodecylacrylamide) (PDDA) polymer have been reported using the LB method, but no specific information on the degree of alignment was available [150]. As discussed in Section 9.3.3, LCs can also be used to align SWNTs.

LB may also be used to fabricate graphite oxide [71] or graphene [264, 425] composites with in-plane flake orientation or stacking.

9.8.2 Desirable Characteristics of Host Polymers

Important requirements for high efficiency photovoltaic cells include effective hole-transport, suitable bandgap and high stability against humidity [205]. Nanotube and graphene composites with conjugated polymers are generally used for electroluminescent and photovoltaic devices [2, 205, 206, 240, 260, 333, 347, 476]. For the case of SWNTs, the use of conjugated polymers has the added advantage of improved dispersion during the solution processing step. However, these are sensitive to moisture and light [405, 406].

Thus, desirable characteristics of polymers for telecommunication applications also include stability of the optical properties against humidity [277]. The host polymer must not have high absorption losses at the device operation wavelength. These usually arise from vibration overtones, see Table 9.2. The bonds giving higher overtone absorption intensities are C–H and O–H, while C–F overtones give the least absorption [277]. As shown in Table 9.3, fluorinated polymers are the most transparent polymers commercially available for telecommunication applications [277].

Polymers commonly used for optical applications are, e.g., polymethylmethacrylate (PMMA), polystyrene (PS), polycarbonate (PC) and epoxy resins [277]. Over the years, several new polymers have been developed, which comply with the requirement of low optical loss and environmental stability, such as deuterated or halogenated polyacrylates and fluorinated polyimides [277]. These have low losses

Table 9.2 Wavelengths and intensities of some important vibration overtones (Adapted from [66])

	Overtone order	Wavelength [nm]	Intensity (relative)
C–H	1	3,390	1
C–H	2	1,729	7.2×10^{-2}
C–H	3	1,176	6.8×10^{-3}
C–D	3	1,541	1.6×10^{-3}
C–D	4	1,174	1.3×10^{-4}
C–F	5	1,626	6.4×10^{-6}
C–F	6	1,361	1.9×10^{-7}
C–F	7	1,171	6.4×10^{-9}
C=O	3	1,836	1.2×10^{-2}
C=O	4	1,382	4.3×10^{-4}
C=O	5	1,113	1.8×10^{-5}
O–H	2	1,438	7.2×10^{-2}

Table 9.3 Characteristics of commercially available polymers for optical applications (from [277])

Company	Polymer type	Patterning techniques	Propagation loss, single-mode waveguide [db/cm] (wavelength, [nm])	Other properties (wavelength, [nm])
Optical crosslinks (formerly Dupont and Polymer photonics)	Acrylate (Polyguide)	Diffusion	0.18 (800) 0.2 (1,300) 0.6 (1,550)	Laminated sheets excimer-laser machinable
Corning (formerly AlliedSignal)	Acrylate	Photoexposure/wet etch, RIE, laser ablation	0.02 (840) 0.3 (1,300) 0.8 (1,550)	Birefringence: 0.0002 (1,550) crosslinked, Tg: 25°C Environmentally stable
	Halogenated acrylate	Photoexposure/wet etch, RIE, laser ablation	0.01 (840) 0.06 (1,300) 0.2 (1,550)	Birefringence: 0.000001 (1,550) Crosslinked, Tg: −50°C Environmentally stable
NTT	Halogenated acrylate	RIE	0.02 (830) 0.07 (1,310) 1.7 (1,550)	Birefringence: 0.000006 (1,310) Tg: 110°C
	Deuterated polysiloxane	RIE	0.17 (1,310) 0.43 (1,550)	Environmentally stable
	Fluorinated polyimide	RIE	TE:0.3, TM:0.7 (1,310) PDL: 0.4 dB/cm (1,310)	Environmentally stable
Amoco	Fluorinated polyimide (Ultradel)	Photoexposure/wet etch	0.4 (1,300) 1.0 (1,550)	Birefringence: 0.025, crosslinked, thermally stable
General electric	Polyetherimide (Ultem)	RIE, laser ablation	0.24 (830)	Thermally stable
Hitachi	Fluorinated polyimide	Photoexposure/wet etch	TE:0.5, TM:0.6 (1,300)	Birefringence: 0.009 (1,300), PDL: 0.1 dB/cm (1,300), Tg: 310°C, thermally stable
Dow chemical	Perfluorocyclobutane (XU 35121)	Photoexposure/wet etch	0.25 (1,300) 0.25 (1,550)	Tg: 400°C
	Benzocyclobutene (Cyclotene)	RIE	0.8 (1,300) 1.5 (1,550)	Tg: > 350°C
Asahi glass	Perfluorovinyl ether cyclopolymer (CYTOP)			n = 1.34 Tg = 108°C

Table 9.3 (continued)

Company	Polymer type	Patterning techniques	Propagation loss, single-mode waveguide [db/cm] (wavelength, [nm])	Other properties (wavelength, [nm])
Dupont	Tetrafluoroethylene and perfluorovinyl ether copolymer (Teflon AF)			n = 1.31 (AF 1,600) n = 1.29 (AF 2,400)
JDS uniphase (formerly Akzo Nobel) telephotonics	Polycarbonate (BEAMBOX)	RIE	0.6 (1,550)	Thermally stable
	(OASIC)	Photoexposure/wet etch, RIE, laser ablation	<0.01 (840) 0.03 (1,300) 0.1 (1,550)	Environmentally stable
Gemfire	(Gemfire)	Photoexposure, wet etch	1.0 (1,550)	Birefringence: 0.0002 (1,550) crosslinked
K-JIST	Fluorinated poly(arylene ether sulfide) (FPAESI)	RIE	TE:0.42, TM:0.4 (1,550)	Birefringence: 0.0003 (1,550) PDL: 0.02 dB/cm (1,550), crosslinked, thermally stable
Redfern	Inorganic polymer glass (IPG)	RIE		Environmentally stable
Hoechst celanese	PMMA copolymer (P2ANS)	Photobleaching	1.0 (1,330)	NLO polymer
PacificWave	Polycarbonate with CLD-1 chromophore (PC-CLD-1)	RIE	1.8 (1,550)	NLO polymer, $r_{33} = 70$ pm/V (1,310), pigtail loss = 1.5 dB/facet
Lumera	Polyurethane with FTC chromophore (PU-FTC)	RIE	2.0 (1,330)	NLO polymer, $r_{33} = 25$ pm/V (1,310), pigtail loss = 5 dB/facet
Ipitek	Poly (methylmethacrylate) with CLD-1 chromophore (PMMA-CLD-1)	RIE	5.0 (1,330)	NLO polymer, $r_{33} = 60$ pm/V (1,300), pigtail loss = 3.5 dB/facet

at the telecommunication wavelengths. However, their thermal stability is poor [277]. Fluorinated polyimides have high thermal stability and low optical losses and are, therefore, an ideal choice for telecommunication applications [277]. Polymers should also be resistant to laser-induced damage.

Nevertheless, water-soluble polymers, such as polyvinylalcohol (PVA) and cellulose derivatives, have predominantly been used for SAs due to their compatibility with high concentration SWNT/graphene aqueous solutions [160, 344, 345, 357, 359, 360, 363, 370, 371, 411, 414, 428]. High SWNT/graphene concentration allows to obtain the desired level of optical density, while minimizing unwanted non-saturable absorption losses [96, 160, 358, 363, 370]. PVA has been employed for the fabrication of freestanding films embedding both bundles and individualized SWNTs for fundamental studies of saturable absorption [357, 359], and for SWNT/graphene devices [96, 160, 344, 345, 358, 360, 363, 370, 411, 414]. Polyimides [362] and Polycarbonates (PC) [372] have been considered as alternative to PVA. In particular, PC has higher transparency and environmental stability compared to PVA and cellulose derivatives. High SWNT loading can be obtained in DCB using regioregular poly(3-hexylthiophene-2,5-diyl) (P3HT) as dispersant polymer [372].

Thermal stability is an important issue for all photonic and optical applications, as polymers lose their transparency over time due to oxidation [206, 277]. This process depends on the chemical structure, as it is induced by the formation of double bonds resulting mainly from the expulsion of H-halogen molecules [277]. Totally halogenated materials are thus the most stable due to the absence of hydrogen [277].

9.9 Characterization of Composites

9.9.1 Optical Microscopy

Composites prepared for optical applications can be characterized using different methods. Examination of the presence of voids, cracks, bubbles, particles, SWNT/graphene aggregations or other physical defects is generally carried out by optical or scanning electron microscopy. Figure 9.14 shows a set of optical microscope images of SWNT composites to be used as a SA at the telecommunication wavelength ($\sim$1.5 μm) [359]. No significant aggregation can be resolved, indicating homogeneous dispersion. Composites of similar optical quality may also be prepared from liquid phase exfoliated graphene [160, 411, 414].

9.9.2 UV-Vis-IR Spectrophotometry

The optical density of SWNTs/graphene in SA composites at the device operation wavelength is an important indicator of the expected performance [358]. UV-Vis-IR spectrophotometry can be used to determine the optical density. Figure 9.15 shows

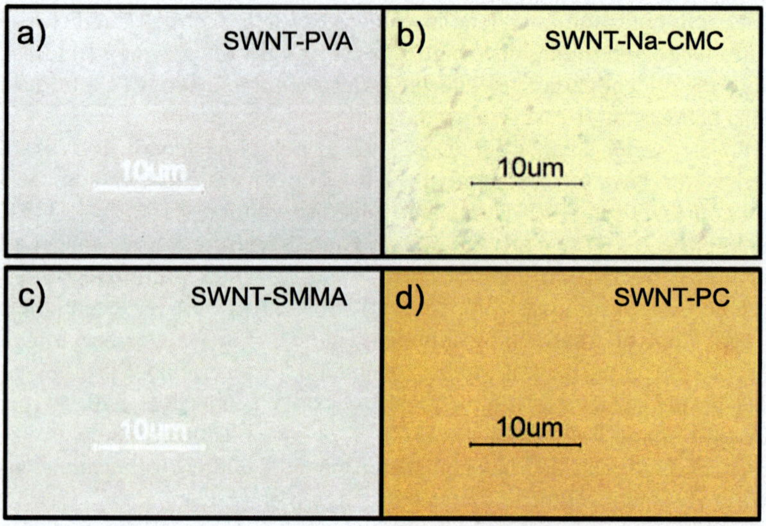

Fig. 9.14 Representative optical micrographs of SWNT-polymer composites from: aqueous dispersion (**a**) PVA, (**b**) Na-CMC and non-aqueous dispersion (**c**) SMMA and (**d**) PC [160]

the absorption spectra of a SWNT-PVA film (thickness ~80 μm), a pure PVA film and the SWNT dispersion from which the SWNT-PVA film is prepared [371]. The SA made from this composite is intended to operated at ~1.5 μm, where the SWNTs in the composite exhibit strong absorption [371]. Note the shift in absorbance for the SWNTs embedded in the host polymer compared to the dispersion. This can be attributed to change in dielectric environment [60, 116, 452, 457] as well as mechanical stress experienced by the SWNTs embedded in the host matrix [468].

Alignment of SWNTs in samples with same loading and thickness can be studied using UV-Vis-IR absorption. For example, in Figure 9.16, the three major SWNT

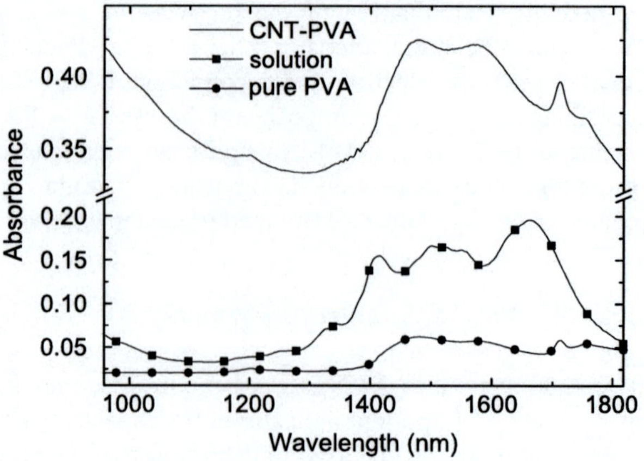

Fig. 9.15 Absorption spectra of SWNTs in a PVA-LA-SWNT composite (*top*) exhibits red-shift compared to that in the dispersion due to change in dielectric environment and mechanical stress

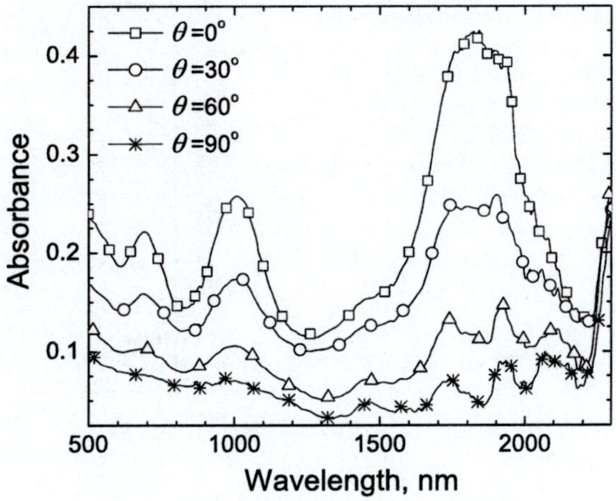

Fig. 9.16 Change in the absorption of aligned SWNTs with the polarization of incident light. Adapted from Ref. [357]

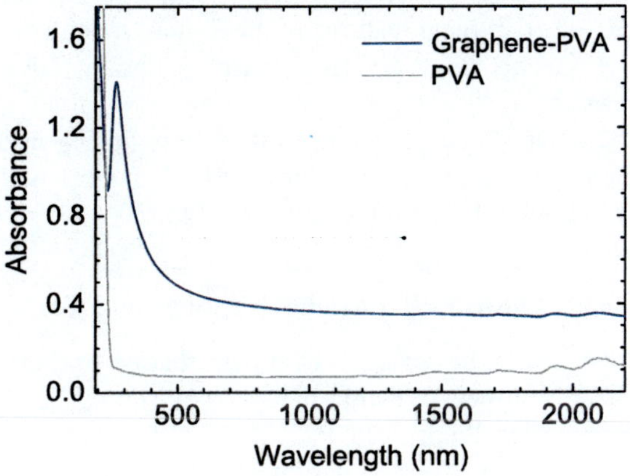

Fig. 9.17 Absorption spectra of a graphene-PVA composite and a reference PVA film. Adapted from Ref. [411]

absorption bands at 700, 1,000 and 1,800 nm decrease in intensity with increasing polarization angle, even though the SWNT loading remains unchanged [357].

Similar observations on optical density may also be made from graphene polymer composites. Figure 9.17 shows absorption spectrum of a graphene-PVA composite. The flat absorption band, due to the linear dispersion of Dirac electrons in graphene [40, 160, 311, 411], underscores the potential of graphene for wideband applications [414].

9.9.3 Raman Spectroscopy

Raman spectroscopy is a fast, powerful and non-destructive method for characterization of carbon materials [119]. A number of important information, such as

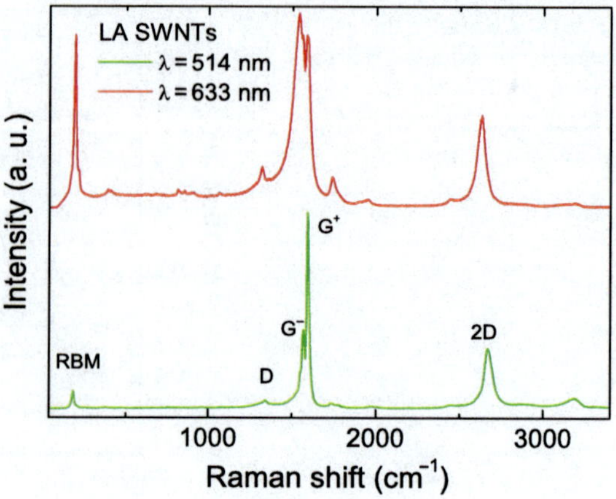

Fig. 9.18 Typical features in the Raman spectrum of SWNTs, measured at 514 and 633 nm excitation on Laser Ablation SWNTs

diameter, orientation, metallic or semiconducting character and chirality can be obtained from the Raman spectra of SWNTs [116, 189, 190, 247, 291, 340, 341, 350, 429]. Typical features in the Raman spectra of SWNTs are shown in Fig. 9.18. The spectra are taken from Laser Ablation (LA) SWNTs [249] in the form of powder.

Raman spectroscopy also allows monitoring of doping, defects, strain, disorder, chemical modifications, edges, and relative orientation of the graphene layers [26, 61, 62, 89, 90, 110, 118, 120, 122, 143, 306, 339, 340, 375, 480].

9.9.3.1 Optical Phonons in Graphene and Carbon Nanotubes

Phonons can be regarded as a perturbation of a crystal [43, 53, 111, 487]. In general, given their dynamic nature, they should be described by the time-dependent perturbation theory (TDPT) [42, 43, 53, 111, 487]. Within the adiabatic Born-Oppenheimer Approximation (BOA), they are seen as static perturbations and treated by the time-independent perturbation theory (TIPT) [42, 43, 53, 111, 487].

In materials without an electronic band-gap, like graphene, graphite and m-SWNTs, the interaction between phonons and electrons must be taken into account to calculate phonon energies [42, 43, 53, 111, 341]. In a metal, for certain phonons with a wave vector connecting two points of the Fermi surface, it is possible to have an abrupt change of the electronic screening of the atomic vibrations. This results in a sudden softening of the phonon frequencies, which is called the Kohn Anomaly (KA), and appears as a singularity in the dymamical matrix [217]. In graphene, the Fermi surface consists of the two inequivalent points **K** and **K'**. Thus, KA can occur only for phonons with $\mathbf{q} = 0$ or $\mathbf{q}$ connecting the two Fermi points (i. e. $\mathbf{q} = \mathbf{K}$). In graphene, a necessary condition for KA is a significant nonzero electron-phonon coupling (EPC) between electrons near the Fermi energy for phonons at $\mathbf{q} = 0$ or $\mathbf{q} = \mathbf{K}$ [340]. Due to their reduced dimensionality, m-SWNTs are expected to have stronger KA than graphite and graphene for the corresponding phonons [340, 341].

In graphene, KAs can be observed as sharp kinks for modes E_{2g} at Γ and A'_1 at **K** [292, 340]. The E_{2g} mode involves the in-plane bond-stretching of pairs of C

sp^2 atoms, while the A'_1 corresponds to the breathing mode of carbon rings. These two phonons are of major relevance to Raman spectroscopy of sp^2 carbon materials [120, 121, 312, 440], as the *G* peak originates from the E_{2g} phonon at Γ, while the *D* peak originates from modes around **K** [120, 340, 432].

In SWNTs, the E_{2g} mode splits in two components: the longitudinal optical (LO, polarized along the tube axis) and the transverse optical (TO, polarized along the tube circumference) modes. In s-SWNTs, the LO-TO splitting is usually explained in terms of curvature. Indeed, the σ-π mixing along the circumference results in a softening of the TO mode, which accounts for both the splitting and the diameter dependence [51, 189, 190, 338, 341]. In m-SWNTs, however, the reason for this splitting is different: the LO mode is modified by a KA, which causes a strong downshift of the LO frequency [247, 341].

Another important, low-frequency, optical mode of SWNTs is the so-called Radial Breathing Mode (RBM), which arises from the radial, in phase, vibration of all the carbon atoms in the SWNT unit cell [186, 350].

In materials without an electronic gap, the BOA is not easily justifiable, although experience proves that in most cases this accurately reproduces the phonon dispersion of metals. While this approximation still holds in intrinsic graphene, it fails in doped graphene and SWNTs [339, 341]. In these cases, phonons must be considered as time-dependent perturbations. It can be shown that KAs occur, within the static approach of the BOA, at $q = 0$ and $q = 2\,k_F$. The description of KAs is modified by the time-dependent approach, for which KAs occur at $q = \pm\hbar\omega_q/\beta$ and $q = k_F \pm \hbar\omega_q/\beta$ resulting in a shift of the position of the KAs [341].

9.9.3.2 Radial Breathing Mode

RBMs are important Raman fingerprints of SWNTs. A diameter dependence of this mode is expected [103, 165, 186, 188, 238, 280, 346, 365], with its frequency increasing as the diameter decreases [186]:

$$\omega_{RBM} = \frac{C_1}{d} + C_2 \tag{9.5}$$

Several groups have derived a variety of C_1 and C_2, as summarized in Table 9.4. More recently, the literature has converged to the values of $C_1 = 214.4$ cm^{-1} nm, $C_2 = 18.7$ cm^{-1} [11, 116, 297, 429]. References [11, 116, 297, 429] report tables where for each *(n,m)* the corresponding RBM frequency and transition energies are assigned. However, if we are just interested in an estimation of the band gap, any parameter can be used, as the difference in the calculated diameter is negligible. For example, let us consider two well-separated RBM frequencies (100 and 350 cm^{-1}) and calculate the diameter using the two most different values for C_1 (218 and 261) from Table 9.4. For the higher RBM frequency (i. e. lower diameters) the calculated diameters (0.62 and 0.75 nm) differ from their average by ~10%; an average bandgap of ~1.3 eV can be inferred from [200], from which the two extremes deviate by ~ 8%. For the lower RBM frequency (higher diameters) the

Table 9.4 Values of C_1 and C_2 proposed in the literature

Authors	C1	C2
Jishi et al. [186]	218	
Rao et al. [350]	221	
Bandow et al. [21]	223	
Kürti et al. [238]	236 (armchair) 232 (zig-zag)	
Jorio et al. [188]	248	
Alvarez et al. [8]	238	
Sánchez-Portal et al. [365]	233	
Popov et al. [346]	230	
Henrard et al. [165]	261 (armchair) 256 (zig-zag) 258 (averaged)	
Mahan et al. [280]	227	
Dobardžić et al. [103]	224.3	
Bachilo et al. [17]	223.5	12.5
Milnera et al. [298]	234	10
Telg et al. [429]	214.4	18.7
Meyer et al. [297]	204	27
Araujo et al. [11]	217.8	15.7

calculated diameters (2.18 and 2.61 nm) differ from their average by ~10%; an average bandgap of ~ 0.35 eV can be inferred, from which the two extremes deviate by ~ 9%. Thus the evaluation of optical absorption through the Kataura plot [200] is not deeply affected.

Matching the diameter given by RBM with excitation wavelength in the Kataura plot gives information on the semiconducting or metallic character. For example, in the spectra shown in Fig. 9.18 at 514 nm the RBM is at 185 cm^{-1}, from which a diameter of 1.29 nm is derived. From Ref. [200] we deduce that s-SWNTs are excited. At 633 nm, the RBM is at 193 cm^{-1}, from which a diameter of 1.23 nm is derived, and from Ref. [200] we deduce that m-SWNTs are excited.

If we want to probe the overall diameter distribution of a SWNT sample and whether they are metallic or semiconducting, Raman spectra have to be taken at as many wavelengths as possible, as a single excitation may probe only a limited range of diameters or only semiconducting or m-SWNTs. Usually, three well separated wavelengths, such as 514, 633 and 785 nm, can probe a wide range of tubes within a material and are enough to get significant, albeit not fully comprehensive, information on the diameter distribution.

9.9.3.3 D and 2D Peaks

The D peak is due to the breathing modes of sp^2 rings and requires a defect for its activation [120, 121, 340, 343, 432, 440, 446]. It comes from TO phonons around the **K** point of the Brillouin zone [120, 440], is active by double resonance (DR) [25, 432] and is strongly dispersive with excitation energy due the Kohn Anomaly at

K [340]. The activation process for the D peak is an inter-valley process as follows: (i) a laser induced excitation of an electron/hole pair; (ii) electron-phonon scattering with an exchanged momentum $\mathbf{q} \sim \mathbf{K}$; (iii) defect scattering; (iv) electron/hole recombination. The D peak intensity is not related to the number of graphene layers, but only to the amount of disorder [120, 440]. Indeed, when moving from graphite to nanocrystalline graphite, the ratio between the intensity of D and G peak, I(D)/I(G), varies inversely with the size of the crystalline grain or inter-defect distance [58, 120, 440]. DR can also happen as intra-valley process i.e. connecting two points belonging to the same cone around **K** (or **K**′). This gives rise to the so-called D′ peak, which can be seen around 1,620 cm^{-1} in defected graphite [312]. The 2D peak is the second order of the D peak. This is a single peak in monolayer graphene, whereas it splits in four bands in bilayer graphene, reflecting the evolution of the band structure [122]. The 2D′ peak is the second order of the D′ peak. Since 2D and 2D′ originate from a Raman scattering process where momentum conservation is obtained by the participation of two phonons with opposite wavevectors (**q** and −**q**), they do not require the presence of defects for their activation, and are thus always present.

Figure 9.19 plots the evolution of the 2D band as a function of the number of layers for 514.5 and 633 nm excitations [122]. Bi-layer graphene has a much broader and up-shifted 2D band with respect to single layer. This is also quite different from that of bulk graphite, which consists of two components $2D_1$ and $2D_2$ [312, 446] roughly 1/4 and 1/2 the height of the G peak, respectively. Indeed, the 2D peak in bi-layer graphene has 4 components, $2D_{1B}$, $2D_{1A}$, $2D_{2A}$, $2D_{2B}$, 2 of which, $2D_{1A}$ and $2D_{2A}$, have higher relative intensities than the other 2. A further increase of the number of layers leads to a significant decrease of the relative intensity of the lower frequency $2D_1$ peaks. For more than 5 layers the Raman spectrum becomes hardly distinguishable from that of bulk graphite [440]. On the other hand the shape of the G peak does not change with the number of layers.

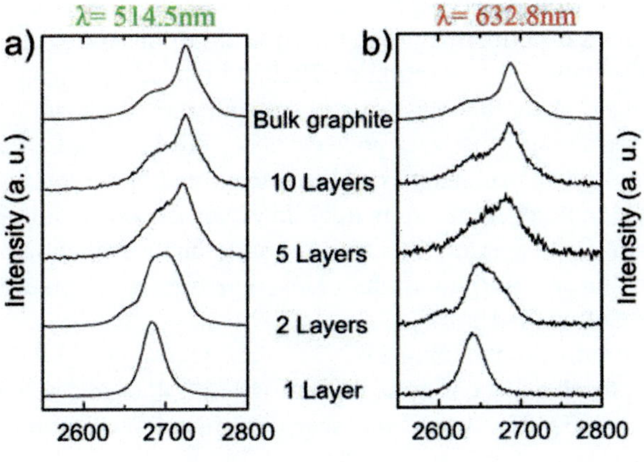

Fig. 9.19 Evolution of the 2D band in Raman spectra of graphene and few layer graphene at (**a**) 514.5 and (**b**) 632.8 nm with the number of layers. Adapted from Ref. [122]

9.9.3.4 G Peak

The G peak corresponds to the E_{2g} phonon at the Brillouin zone centre of graphene [120, 440]. The Raman spectrum of SWNTs is characterized by the presence of two distinct features around 1,550 cm^{-1}: the G^+ and G^- peaks. These arise from the splitting of the doubly degenerate, Raman-active $E_{2\,g}$ mode of graphene [120, 440]. The G^- peak position decreases with diameter, while the G^+ is almost diameter independent [190]. While in s-SWNT both peaks appear as sharp Lorentzians, the G^- peak of metallic SWNTs is very broad and its position is strongly downshifted with respect to that in s-SWNTs of the same diameter (Fig. 9.18) [51, 189, 190, 291, 338, 429].

The G^+ and G^- peaks originate from the LO and TO modes, which are polarized along the tube axis and along the tube circumference, respectively (Fig. 9.20) [341]. In s-SWNTs, the G^+ peak is assigned to the LO mode, while the G^- is assigned to the TO [51, 189, 190, 338, 341]. In m-SWNTs, however, the assignment is the opposite: the LO mode is affected by KA, which causes a strong downshift of its frequency [247, 341]. Thus the G^+ peak is due to the TO mode, while the G^- arises from the LO (Fig. 9.20) [247, 341].

9.9.4 PL Spectroscopy

PL spectroscopy can be used to determine the presence of isolated SWNTs or small bundles in the composites [224, 299]. Figure 9.21 shows the polarized PL signal from a mechanically stretched SWNT-SDS-gelatine film excited by depolarized 662 nm light, indicating isolated (8,3), (7,5), (7,6), (9,5) tubes or small bundles [224].

9.9.5 Z-Scan

Z-scan experiments probe the optical nonlinearities associated with the change in refraction and absorption coefficient induced by intense laser power. The sample is moved along the waist of a Gaussian beam as shown in Fig. 9.22a. This results in a variation of the laser power density on the sample, reaching its maximum at the focal point. An analysis of the transmitted beam through the sample as a function of sample position, Z, is done in either the open and close aperture scheme [383, 444]. Open aperture Z-scan is used for the investigation of processes associated with nonlinear absorption, while close aperture Z-scan is used to investigate nonlinear refraction [444].

Samples containing SWNTs can be investigated by Z-scan in order to understand the mechanisms of optical limiting [322, 450] and saturable absorption [278, 357, 359, 363, 427]. A typical Z-scan trace for a SWNT-polymer composite, taken in near

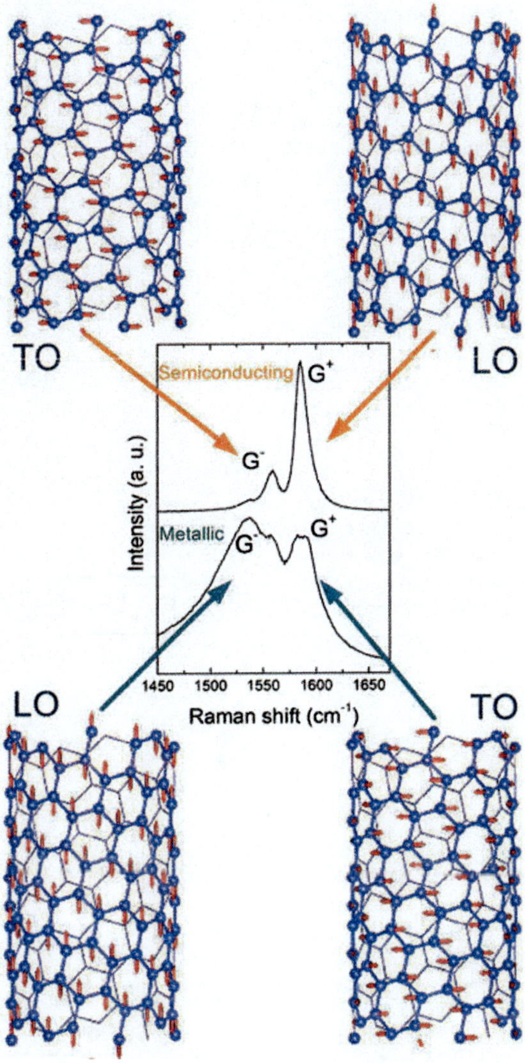

Fig. 9.20 Schematic representation of the G mode for m- and s-SWNTs. Adapted from Ref. [341]

resonant conditions, is shown in Fig. 9.22b. Here, SWNTs show strong saturable absorption, with about 30% increase in transmission when the sample passes the focal point.

9.9.6 Pump-Probe Spectroscopy

Pump-probe spectroscopy is widely in use to characterize response time of SAs [212]. The material is irradiated by a pump pulse, yielding a carrier excitation. Then,

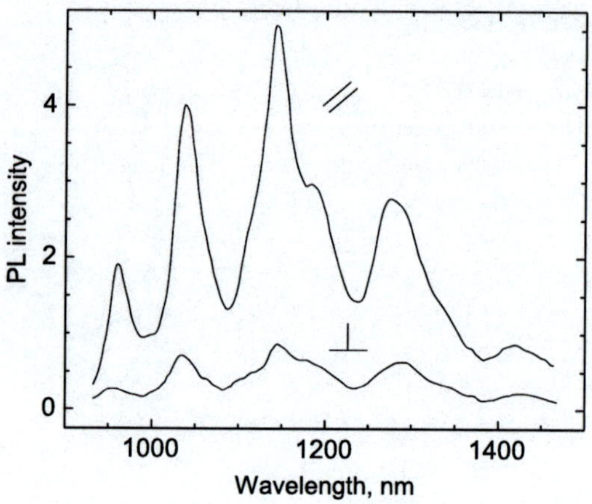

Fig. 9.21 Polarized photoluminescence from a stretch-aligned SWNT/SDS/gelatin dried film (draw ratio=3) excited by depolarized 662 nm light. Adapted from [224]

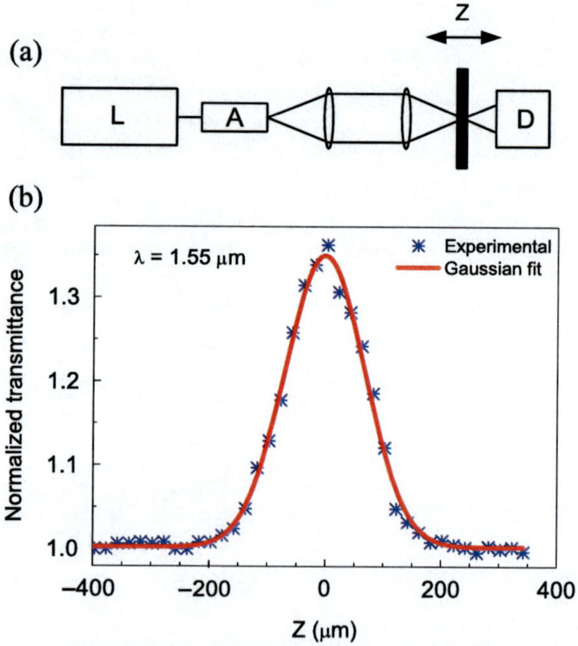

Fig. 9.22 (**a**) Z-scan setup. L: laser, A: attenuator, D: photodetector. (**b**) Z-scan measurement of a SWNT-PVA composite

after a short time, controlled by an optical delay line, the sample is irradiated by a probe pulse. By studying the transmittance or reflectance of the probe pulse it is possible to obtain information on the excitation decay (SA recovery time) caused by the pump pulse. SWNTs have an intrinsic bitemporal response, including a fast component and a slow component, which are due to different time-scale relaxation processes (interband and intraband, respectively) [215, 329]. Graphene also exhibit an ultrafast bitemporal relaxation process. This is further discussed in Section 9.12.

9.10 Nanotube/Graphene-Composites for Photonics and Optoelectronics

The desirable characteristics of the host polymer matrix for optical applications vary depending on the type of applications. The most important requirement is the compatibility with SWNTs/graphene and solvents so that a homogeneous dispersion of SWNTs/graphene in the composite can be achieved. It is also essential to reduce scattering losses in such applications, which generally arise from aggregates or particles, cracks, bubbles or voids inside the composite [277]. The optical applications so far demonstrated using SWNT/graphene-polymer composites are SA for optical signal regeneration and passive mode-locking [29, 96, 130, 160, 161, 209, 210, 220, 344, 345, 358, 360, 363, 370, 372, 380, 411, 413, 415, 416, 453–455, 479] electroluminescence [206, 476], and photovoltaic applications [2, 205, 206, 240, 241, 260, 347, 469, 476]. Significant progresses have also been made in using percolating SWNT and graphene networks and their composites as transparent, flexible conductors [40, 99, 123, 149, 221, 331, 469, 470, 472]. In addition, such composites could be used as optical modulators under strong electric field [216, 334, 493].

9.10.1 Graphene/Nanotube Networks as Transparent Conductors

Optoelectronic devices such as displays, touch-screens, light emitting diodes and solar cells require materials with low sheet resistance R_s and high transparency (T). The majority of flexible transparent optoelectronic devices utilizes Indium Thin Oxide (ITO) as transparent conductive (TC) material. ITO suffers from severe limitations: an ever-increasing cost due to indium scarcity [153], brittleness [154], difficulties in patterning [145, 153] and a sensitivity to both acidic and basic environments [154]. This demands new TC materials with improved performance. Metal grids [250], metallic nanowires [95], or other metal oxides [145] have been explored as alternative. SWNTs [136, 149, 373, 472, 497] and graphene [18, 40] also show great promise.

R_s and T need to be considered linked to analyze data for TC films. Both are determined by the response of electrons to either dynamic (light) or static (voltage) electric fields. R_s is ultimately controlled by the "dc" conductivity, σ_{dc}, via

$$R_s = (\sigma_{dc} t)^{-1} \tag{9.6}$$

where t is the film thickness.

T depends on the optical conductivity G_0,

$$T = \left(1 + \frac{G_0}{2\epsilon_0 c} t\right)^{-2}, \tag{9.7}$$

Combining Eqs. (9.6, 9.7), eliminating t, gives:

$$T = \left[1 + \frac{Z_0}{2R_s}\frac{G_0}{\sigma_{dc}}\right]^{-2} \tag{9.8}$$

where, $Z_0 = \frac{1}{\varepsilon_0 c} = 377\Omega$ is the free space impedance.

For instance, in graphene [135] we can take $\sigma_{dc} = n\mu e$, where n is the number of charge carriers. Note that for n~0, σ_{dc} does not go to zero, but assumes a constant value [135] $\sigma_{dc,min} \sim 4e^2/h$, resulting in $R_s \sim 6$ kΩ for an ideal intrinsic SLG with $T \sim 97.7\%$. Thus, ideal intrinsic monolayer graphene, would beat the best ITO only in terms of T, not R_s. However, real samples deposited on substrates, or in thin films, or embedded in polymers are never intrinsic. Usual exfoliated monolayer graphene has $n \geq 10^{12}$ cm^{-2}, and much smaller R_s. The range of T and R_s that can be realistically achieved for graphene layers of varying thickness can be estimated taking $n = 10^{12}$–10^{13} cm^{-2} and $\mu = 10^3$–2×10^4 cm^2/Vs, as typical for chemical vapor deposition (CVD) grown films. For instance, taking $n = 3.4 \times 10^{12}$ cm^{-2} and $\mu = 2\times10^4$ cm^2/Vs, typical for CVD grown films, it would be possible to get $T = 90\%$ and $R_s \leq 20\Omega\square^{-1}$.

Ultrathin SWNT/graphene films as transparent conductors are usually prepared on membranes by filtering SWNT/graphene dispersions and then transferring the resultant film from the membrane surface to the substrates (e.g. polymer, glass etc) [149, 472, 497] or directly spray coating the dispersions on substrates [373], or via roll-to-roll processing [18]. In such preparation methods, SWNTs are not incorporated in the polymer matrix, but are deposited on the substrate surface. Therefore, they are not truly composites. On the other hand, transparent and conducting SWNT-polymer composites, which are much easier to implement, have been prepared [99, 123, 221, 331, 490]. Inkjet printing of SWNT-polymer dispersions on flexible substrates has been demonstrated to yield conducting layers of percolating SWNTs with T up to ~80% in the visible range [392]. Nevertheless, the conductivity and transparency of SWNT-polymer composites do not match those of SWNT only networks. It is possible to lower the resistivity of SWNT-polymer composites by functionalization [99, 123, 490]. Reference [99] reported an increase in electrical conductivity by a factor of 5 by doping the SWNTs with $SOCl_2$ while retaining good optical transparency. At 0.1 wt% loading, the authors achieved 92.4% transmittance at 500 nm with an electrical conductivity of 290 Ω.cm for a 20 μm film. By comparison, Ref. [472] reported a 50 nm thick two dimensional SWNT network with 1.5×10^{-4} Ω.cm resistivity and ca. 75% transmittance at 500 nm.

The length of SWNTs is an important factor for conductivity enhancement of SWNT networks [164]. In addition, good dispersion of SWNTs without surfactants is key to improve SWNT-polymer conducting composites, as surfactants isolate tubes from the surrounding environment, resulting in poor inter-tube connection. An approach to solve this problem is in-situ polymerization during the sonication process, as demonstrated by Ref. [331]. With 0.1% vol. SWNT loading, ca. 70% transmittance at 500 nm for 34 μm films with a conductivity of 1×10^{-8} S.cm^{-1} was achieved. Flexible, transparent and conducting thin

SWNT-poly(3,4-ethylenedioxythiophene):poly(styrenesulfonate) composites with high optical uniformity having DC conductivities of >10^5 S/m were demonstrated [92]. For a 80 nm SWNT-composite film, a transmittance (at 550 nm) of 75% and sheet resistance of 80 $\Omega\square^{-1}$ was achieved. Further improvement might be possible using enriched long m-SWNTs, or chemically doped graphene [18, 40]. Indeed, chemical doping is a key strategy to improve performance both for SWNTs [136] and graphene [18, 37] TCs. In the latter case, Ref. [37] prepared graphene-based TC films, starting from graphene produced by micromechanical cleavage (MC), with T~98%, $R_s = 400\ \Omega\square^{-1}$, exploiting a layer of polyvinyl alcohol (PVA) to induce n-type doping. Ref. [18] achieved $R_s \sim 30\ \Omega\square^{-1}$, T ~ 90% by nitric acid treatment of graphene-based TC films derived from CVD grown flakes, one order of magnitude lower in terms of R_s than previous graphene-based TC films from wet transfer of CVD films [223].

It is important to note that graphene-based TC films derived from CVD grown flakes, combined with doping, have the potential to outperform ITO and other transparent conductive materials. On the other hand, graphene-based TC films produced by other methods, such as LPE, albeit presently with higher R_s at T=90%, have already been tested in organic light emitters [290, 470] and solar cells [269, 459]. These are a cheaper and easier scalable alternative to MC or CVD films, and need be considered in applications where cost reduction is the key factor.

9.10.2 Electroluminescent and Photovoltaic Devices

Photo-induced electron transfer between semiconducting conjugated polymers and SWNTs has attracted significant attention in recent years. Electroluminescence is a phenomenon in which a material emits light in response to an applied electric field. Studies of the electroluminescent properties of SWNT-polymer composites have also been carried out [206, 476]. Poly(3-hexyl-thiophene) (P3HT), with a bandgap of 1.8–2.1 eV has been the most widely used polymer for such applications, because of its high mobility and environmental stability [206, 333, 476]. Electroluminescence has been reported mainly from s-SWNT devices, with an emission peak in the NIR assigned to a radiative decay over the first interband transition of the π bands at the K point [300]. Electroluminescence was also observed from biased m-SWNTs [112, 282], MWNTs [112] and graphene [112]. Reference [112] assigned such light emission to phonon-assisted radiative decay from π^* states at the M point to the Fermi level at the K point.

Graphene has a work function of 4.5 eV, similar to ITO. This, combined with its promise as flexible and cheap transparent conductor, makes it an ideal candidate as organic light-emitting diode (OLED) anode. Graphene-based transparent conductor film (GTCFs) anodes enable out-coupling efficiency comparable to ITO [470]. Chemically derived graphene was also implemented as transparent cathode in a metal-free and solution-processed light-emitting electrochemical cell [290].

A photovoltaic (PV) device converts light to electricity [68]. V_{OC} is the maximum open circuit voltage, while I_{SC} is the maximum short circuit current. The fill factor (FF) is defined as: FF = ($V_{max} \times I_{max}$)/($V_{OC} \times I_{SC}$), where I_{max} and V_{max} are maximum current and voltage. The energy conversion efficiency is $\eta = P_{max}/P_{inc}$, where $P_{max} = V_{OC} \times I_{SC} \times FF$ and P_{inc} is the incident power. The fraction of absorbed photons converted to current defines the internal photocurrent efficiency (IPCE).

SWNTs and graphene have been proposed as promising materials in organic [35, 460] and dye sensitized solar cells (DSSCs) [434]. Indeed, both SWNTs and graphene can fulfill multiple functions in photovoltaic devices: (1) transparent conductor window, (2) photoactive material, (3) channel for charge transport, (4) catalyst [40]. SWNTs and GTCFs are used as window electrodes both in organic and DSSC devices [40]. Organic solar cells based on SWNTs-polymer composites have also been demonstrated, due to the excellent electron transfer with poly(p-phenylene vinylene) (PPV) [2], poly(2-methoxy-5-(2-ethylhexyloxy)-1,4-phenylenevinylene)(MEHPPV) [205], P3HT [260, 347], poly(3-ocylthiophene-2,5-diyl) (P3OT) [205, 240] as the host.

Photoinduced electron transfer in bulk heterojunctions of MEH-PPV/C_{60} and β-carotene/C_{60} was reported more than 20 years ago [366], with fullerenes used as acceptors. Combining both these ideas, using C_{60} as electron acceptor and the high mobility of SWNTs, C_{60}-SWNT-P3HT composites have been reported with up to ~390 mV open circuit voltage and 2.7 mA/cm^{-2} short circuit current density [260]. Chemically modified graphene dispersions were also used in bulk heterojunction PV devices, as electron-acceptors with poly(3-hexylthiophene) and poly(3-octylthiophene) as donors, achieving $\eta \sim 1.4\%$ [269]. Reference [485] claims that $\eta > 12\%$ should be possible with graphene as photoactive material.

SWNTs and graphene can cover an even larger number of functions in DSSCs. Indeed, other than as TC windows at the photoanode [459], SWNTs and graphene can be incorporated into the nanostructured TiO_2 photoanode, through which the photoelectrons generated from dye molecules are transported to the anodes [146]. A recombination reaction (i.e. reverse charge transfer from TiO_2 to the dye or redox couple) reduces the overall cell efficiency [314, 410, 483]. One way to promote charge transfer, reducing recombination, is to incorporate a conductive network into TiO_2. In this context, SWNTs and graphene show potential not only because of their extremely high carrier mobility, but also because of their unique one and two dimensional structures.

Reference [52] used a SWNT network as scaffolds of dye-sensitized TiO_2 nanoparticles to promote charge transport in mesoscopic semiconductor films. The authors demonstrated that, although the SWNT network in the film has no noticeable influence on the charge injection process from the excited Ru(II) trisbipyridyl complex into TiO_2 particles, it plays an important role in improving the charge separation, as the rate of back electron transfer between the oxidized sensitizer (Ru(III)) and the injected electrons becomes slower in the presence of the SWNTs scaffold [52]. Reference [483] used graphene as TiO_2 bridge, achieving faster electron transport and lower recombination, leading to $\eta \sim 6.97\%$, higher than conventional nanocrystalline TiO_2 photoanodes [481, 483].

Another option in DSSCs is to use SWNTs and graphene at the counter electrode (CE). Regeneration of dye molecules is accomplished by capturing electrons from a liquid electrolyte (Iodide/Iodine solution), sandwiched on the CE, which catalyzes the reduction of tri-iodide [328]. Another important function of the counter electrode is the back-transfer of the electrons arriving from the external circuit to the redox system [328]. The most important requirements for the counter electrode material are a high exchange current density and a low charge-transfer resistance [9]. Currently, DSSC cathodes are made of Platinum (Pt) layers deposited on transparent glass, in turn, coated by a TC such as ITO [145]. ITO suffers many limitation listed above, while Pt is rare and expensive. Furthermore Pt tends to degrade over time when in contact with an iodide/iodine liquid electrolyte, reducing the overall efficiency of DSSCs [232]. Strong efforts have been directed towards the replacement of such elements with low-cost and more versatile materials. The use of SWNTs and graphene at the CE of DSSCs is attractive for several reasons, such as high specific surface area, good catalytic properties, electronic conductivity, corrosion resistance towards iodine, high reactivity, abundance, and low cost. Thus, SWNTs and graphene films and/or composites are good candidates as CE material in DSSCs [173, 417, 434].

Reference [434] demonstrated that ozone-treated SWNTs films increase their catalytic activity due to the introduction of defects. Graphene has also great potential. Semi-transparent graphene thin films on FTO were reported with high electrocatalytic activity toward Iodide/Iodine redox couple [202]. An hybrid poly(3,4-ethylenedioxythiophene (PEDOT):poly-(styrenesulfonate) (PSS)/GO composite was used as counter electrode, getting $\eta = 4.5\%$, comparable to 6.3%, for a Pt counter electrode tested under the same conditions [173], but achieved with a cheaper material.

9.10.3 Saturable Absorbers (SAs)

9.10.3.1 Nanotube Based SAs

A material's response to an electric field can be described in terms of polarization, defined as the dipole moment per unit volume. The relationship between polarization and electric field is [45]:

$$P = \varepsilon_0 \chi E \tag{9.9}$$

where χ is the dielectric susceptibility [45]. For very high electric fields Eq. (9.9) is no longer sufficient to describe the behavior of some materials. In this case polarization can be expressed as a power series in the electric field [45]

$$P = \varepsilon_0(\chi_1 E + \chi_2 E^2 + \chi_3 E^3 + ...) \tag{9.10}$$

where χ_1 is the linear susceptibility and χ_2 and χ_3 are the second- and third-order nonlinear susceptibilities. These are responsible for phenomena such as difference frequency generations, optical parametric oscillation, self-focusing, saturable absorption, two-photon absorption etc. [45].

Strong nonlinear optical effects have been demonstrated in SWNTs [76, 286, 357, 359, 363, 427]. The very high density of states at the van Hove singularities allows strong optical absorption if the frequency of the incident electromagnetic field matches their energy spacing. As strong absorption occurs, the excited energy levels fill up and the material becomes transparent to higher power irradiation. This saturable absorption, to a first approximation, can be described as [45]:

$$\alpha(I) = \frac{\alpha_0}{1 + I(t)/I_{sat}} + \alpha_{ns}, \tag{9.11}$$

where α_0 is the linear optical absorption; $I(t)$ is the laser intensity; I_{sat} is the saturation intensity, defined by Eq. (9.11) as the intensity necessary to reduce the absorption coefficient to half of the initial value; α_{ns} is the nonsaturable absorption component [45]. The dynamic response of nonlinear absorption is specified by the recovery time (τ_A), defined as the time necessary to reduce the carriers by a factor of $1/e$, and shows how fast is the relaxation to the ground state, after excitation [211]. The modulation depth is defined as the maximum possible absorption change between low power and high power irradiation [211].

Reference [76] firstly measured saturable absorption in SWNTs at 1.55 μm by pump-probe spectroscopy, showing a sub-picosecond relaxation of excited carriers. They reported $\chi_3 \sim 10^{-10}$ esu (1 esu = 1.11 × 10^{-9}m^2/V^2 [45]), due to the non-resonant condition. A much higher χ_3 value of 10^{-7} esu, was achieved under resonant condition in Ref. [427], with a recovery time of about 600 fs. Saturable absorption can be further enhanced if SWNT are highly oriented along the light polarization direction, because of the anisotropy of their interaction with light [357].

SWNTs have a number of benefits compared to other SA materials, such as organic dyes, color filter glasses [367], dye-doped solids [139] and semiconductors [213, 215]. Since the SWNT absorption depends on the diameter, this can be fine tuned across the visible and infrared spectral range. This could allow a number of ultrafast optoelectronic applications in medicine, sensing, telecommunication and materials processing. SWNT based SAs have a high laser damage threshold, excellent environmental stability and are much easier and cheaper to assemble [357, 371].

As an example, for telecommunications at 1.55 μm, the SWNTs should have a maximum absorption around this wavelength. In laser ablation grown SWNT, diameter control can be achieved by varying the temperature in the laser oven, as shown in Fig. 9.23. Figure 9.24 plots the Raman spectra from SWNTs with absorption peak centered at 1.55 μm. From the radial breathing modes, a diameter distribution of 1–1.3 nm can be deduced [116, 429]. Therefore, the SWNTs grown at 1,000°C in Fig. 9.23 are more suitable for 1.55 μm operation than the other SWNTs presented in Fig. 9.25.

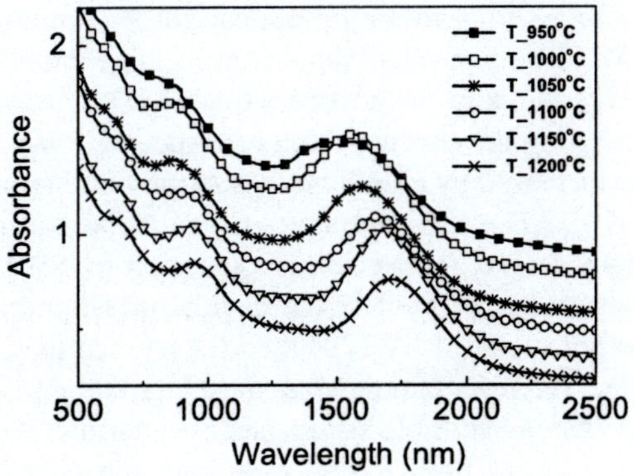

Fig. 9.23 Tuning LA-SWNT diameter by growth parameters; the absorbance reflects the change in SWNT diameter with the growth temperature

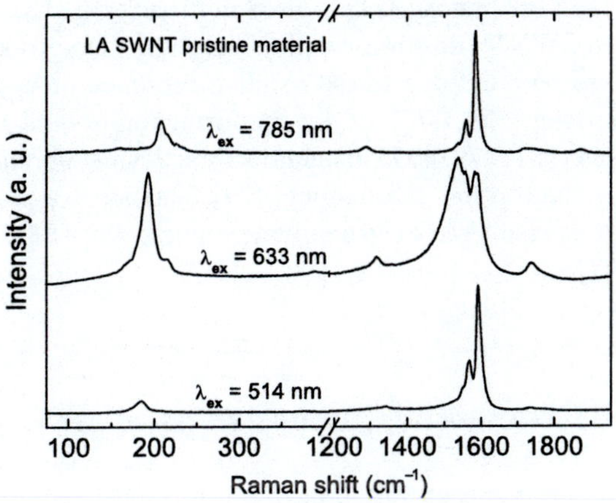

Fig. 9.24 Raman spectra of LA-SWNTs grown at 1,000°C (See Fig. 9.23)

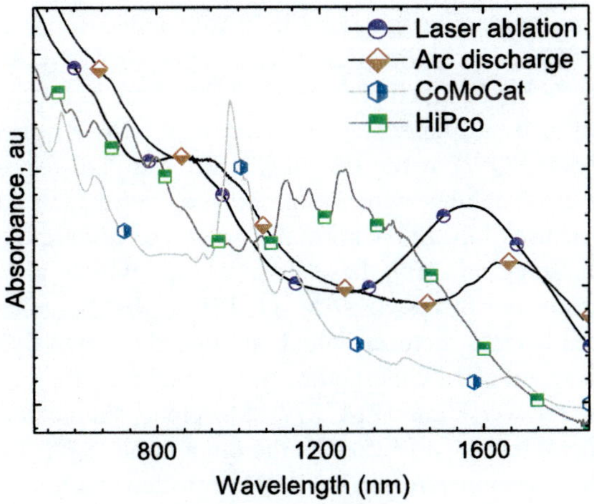

Fig. 9.25 Absorption spectra of SWNTs produced by different growth methods. The nanotubes are dispersed in D_2O with SDBS

In addition to strong optical absorption at the wavelength of interest, an ideal SWNT-polymer SAs should have a large modulation depth, small non-saturable loss [357] and ultra-fast recovery time [451]. The modulation depth is generally tuned by changing the concentration of dispersed SWNTs in the polymer matrix. It can also be increased by aligning tubes in the polymer matrix [357], as discussed in Section 9.8.1.1. The non-saturable losses can be minimized by preparing composites free from defects (e.g. cracks, voids etc.) or SWNT aggregates, using host polymers that are transparent at the device operation wavelength. However, bundles could be beneficial in SWNT-based SAs, since they allow to achieve a shorter recovery time. However bundle-sizes must be smaller than the device operation wavelength to avoid non-saturable losses due to scattering [39, 357]. It is thus important to properly characterize both SWNT solutions and composites, as discussed in Sections 9.7 and 9.9.

In earlier implementations of SWNTs as SAs, SWNTs were spray-coated on quartz substrates [381] or used in dispersion [181]. Direct synthesis of highly purified SWNT thin films on fibre-ends was also proposed [477]. However, high losses were reported due to the residual presence of large aggregates as well as catalyst particles [379, 477], or due the formation of bubbles when SWNT dispersions were used [181, 379, 477]. In addition, the device fabrication methods were time consuming and had low throughput [479]. The best way to overcome such disadvantages is to disperse SWNTs in a polymer matrix [96, 181, 358, 360, 363, 370, 371, 380, 428, 477].

9.11 Nanotube Composites as Mode Lockers for Ultrafast Lasers

The most successful application of SWNT-based SAs demonstrated thus far is as mode-lockers for ultrashort pulse lasers and noise suppression filters. The use of SAs as a mode-lockers to generate ultrashort pulses was first proposed shortly after the invention of laser itself [97, 393]. Various optical materials, such as organic dyes, color filter glasses [367], dye-doped solids [139] and semiconductors [213, 215] have so far been used for this purpose. However, it was challenging to achieve stable mode-locking operation with these conventional SAs [213, 215, 325]. The advances in molecular beam epitaxial (MBE) growth of semiconductor quantum wells (SQW) at the end of the 80s resulted in production of new semiconductor heterostructures with SA properties useful for photonics applications [215, 451]. Because of the SQW normally growth on high reflectivity mirrors (e.g. semiconductor Bragg mirrors), these structures are widely known as Semiconductor Saturable Absorber Mirrors(SESAMs) [215]. SESAMs typically are complex multi quantum well heterostructures, which are usually grown by expensive molecular beam epitaxy, and often undergo heavy-ion implantation to create defects, in order to reduce the recovery time [214, 277]. Moreover, they can only cover a narrow wavelength operation range [325]. On the other hand, SWNTs are cheap to produce, and different parameters can be well-controlled, such as modulation depth and operation wavelength range, as discussed above. SWNT composites are also mechanically robust and environmentally stable. SAs based on SWNTs have thus the potential to compete with traditional SESAMs [29, 96, 129, 160, 209, 210, 358, 360, 363, 370, 372, 413, 415, 416, 453–455, 479].

To date, SWNT-based SAs have been successfully used as mode-lockers in fibre lasers [160, 208–210, 220, 358, 360, 363, 370, 372, 380, 413, 415, 416, 453–455], waveguide lasers [29, 96], solid-state lasers [125, 376], and semiconductor lasers [396], at 0.8 [219], 1 [210], 1.1 [208], 1.3 [401], 1.55 [160, 413, 416, 453–455, 478], 1.6 [415, 478] and 2 μm [394]. Wavelength-tunable lasers using SWNTs, were demonstrated in Ref. [454] and later in Refs. [115, 219, 377, 378]. So far, the shortest reported pulse duration is 68 fs [376]. A repetition rate of 17.2 GHz was demonstrated [396]. The maximum output power reported to date is 1.6 W [416].

A large range of host polymers, e.g. polycarbonate (PC) [29, 372, 454], polyvinyl alcohol (PVA) [208–210, 413, 415, 416, 453, 455], Carboxymethyl cellulose, Polyimide (PI), Polydimethylsiloxane (PDMS), Polymethyl methacrylate (PMMA), poly(3-hexylthiophene) (P3HT) and poly(9,9-dioctyfluoreny1-2,7-diyl) (PFO) have been used [160]. Different SWNT growth technologies, e.g. laser ablation (LA), CoMoCAT, HiPCO, arc discharge (AD), chemical vapor deposition (CVD), catalytic CO disproportionation reaction [160], producing different mean diameter and diameter distribution, have been employed for mode-locking at different wavelengths.

Since ~1.55 μm is the most attractive wavelength for telecommunications, significant effort has been devoted to optimize SWNT-polymer SAs for mode-locking of Erbium-doped fiber lasers at this wavelength [160, 372, 413, 416, 453–455]. ~123 fs pulses were reported with a repetition rate of ~18 MHz [413]. Mode-locking in an active waveguide laser was reported, with transform-limited 1.6 ps [96], and 320 fs pulses [29].

To realize various functions, different configurations of SAs using SWNTs were proposed, such as evanescent field interaction in a tapered fibre [395], in a D-shaped optical fibre [398, 399], and with vertically aligned SWNTs [397]. Incorporating SWNT polymer composites into the evanescent field of the fibre taper [220] and polymer fiber [442] was also proposed. Thus far, the most common way to integrate SWNTs-devices into fiber lasers is to sandwich a SWNT polymer composite film between two fiber connectors, offering ease of integration into various lightwave systems with the flexibility of polymer photonics [160, 360, 363, 370, 372, 378, 413, 416, 453–455]. Figure 9.26a is a photograph of a FC/PC fibre patchcord with a SWNT-PVA film incorporated into it. A typical mode-locked fibre laser setup is schematized in Fig. 9.26b. The laser cavity is constructed using a mode-locker, a coupler, a fibre polarization controller (PC), an isolator (ISO), an Erbium-doped fiber (EDF), and a wavelength division multiplexer (WDM) [160]. The optical isolator ensures unidirectional light propagation. To improve the output pulse stability, a polarization controller is used, consisting of 2 spools of single-mode fiber acting as retarders. The total retardation induced by the polarization controller is a function of the fiber geometry in the spool [3]. One port of the coupler is used for a feedback

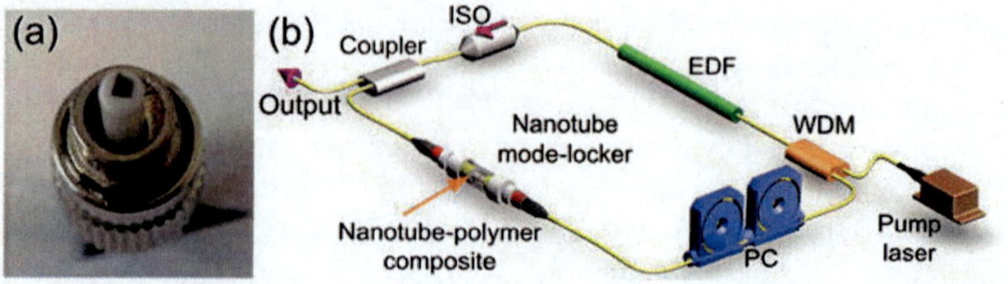

Fig. 9.26 (**a**) Fiber connector with SWNT-PVA film; (**b**) EDF laser with SWNT-mode-locker

into the cavity, while the other serves to study the cavity repetition rate, autocorrelation trace, pulse spectrum and output power [160]. The EDF is pumped by a diode laser at 960 nm to provide gain for lasing [160]. This results in the excitation of the Er^{3+} to its high-energy states, followed by a non-radiative recombination to lower-energy excited states. The subsequent radiative recombination from these levels to ground state gives an emission around ~1.5 μm [3]. The laser feedback is created using the optical fibre coupler to obtain gain [360]. The laser beam returns to the EDFA through a wavelength division multiplexer, which combines light at 1,529 and 960 nm in a single fibre. Thus, the stored pumping energy of the low-energy excited state is used to amplify the signal through stimulated emission. When the gain exceeds the loss induced by the intracavity components, the laser generation starts.

The mechanism of pulse formation can be understood as follows [211]. The SA works as a loss modulator when a short pulse circulates in the cavity (Fig. 9.27a). The peak intensity saturates the absorber more than the low intensity pulse wings. This produces an amplitude loss modulation with a frequency proportional to the cavity round trip time. The pulse circulation in the cavity gives enough gain to overcome the losses induced by the absorber. As a result, the net gain window has duration equal to the absorber recovery time [211]. The initial pulse formation can start from noise fluctuations in the laser cavity, when a high intensity spike significantly decreases its losses passing through the SA. It should be noted that it is possible to achieve pulses significantly shorter than the SA recovery time, see Fig. 9.27b [199]. In this case, soliton pulses form in the cavity as a balance between negative group velocity dispersion and self-phase-modulation [199]. Thus, the SA

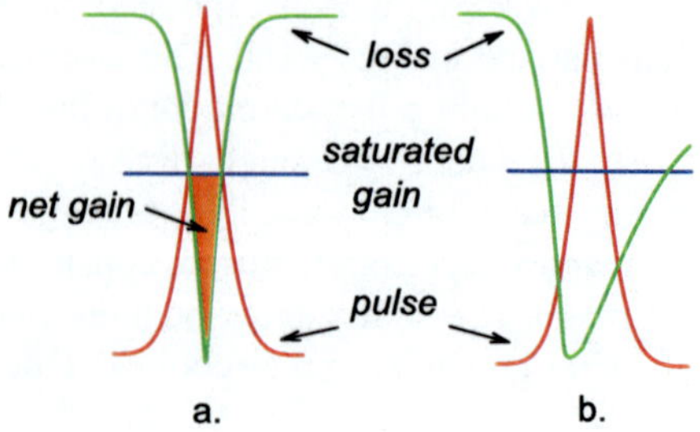

Fig. 9.27 Mode-locking mechanisms: (**a**) saturable absorber, (**b**) soliton mode-locking [360]

acts as a loss mechanism for starting the pulse formation and later on stabilizes the pulses [199].

To achieve pulse generation from the continuous-wave (CW) regime, a pig-tailed SWNT-PVA mode-locker can be placed in the cavity (Fig. 9.26b) [160]. This gives stable mode-locking with a fundamental repetition rate f_r. This can be estimated as $f_r = c/(nL)$, where c the velocity of light in vacuum, n the average refractive index of the cavity ($n \approx 1.5$ in the case of a common single mode-fiber based cavity), and L is the cavity length. Figure 9.28a plots the output pulse spectra of wavelength-tunable pulses mode-locked by SWNTs [454]. Typical soliton side-bands were observed for the laser cavity without the intracavity filter. These can usually be attributed to perturbation of pulses propagating in the cavity, caused by paths with different dispersions as well as output coupling loss [3]. Soliton pulses lose part of their energy passing through different passive cavity components (e.g. output coupler), but then regain it in EDF during the cavity round trip. The

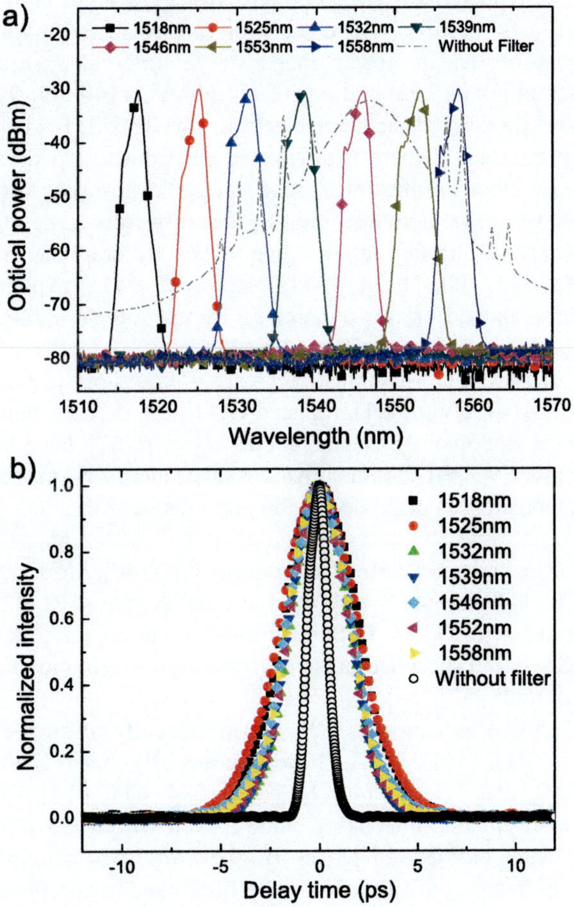

Fig. 9.28 Wavelength-tunable pulses mode-locked by nanotubes. (**a**) output optical pulse spectra, (**b**) output autocorrelation traces [454]

soliton adjusts to these perturbations by forming dispersive waves, which appear as sidebands [3]. The pulse width can be measured using second harmonic generation (SHG) auto-correlation [435]. The pulses in Fig. 9.28b are fitted with a sech^2 autocorrelation function, with a full width at half maximum (FWHM) of $\sim$3.68 ps. To get the pulse width, the autocorrelation width needs to be divided by the decorrelation factor for sech^2 (1.54) [435], giving an average pulse duration of $\sim$2.39 ps.

9.12 Graphene for Ultrafast Photonics

Various graphene based photonic and optoelectronic devices, from solar cells [269, 459] and light-emitting devices to touch screens [290, 470] and photo-detectors [473], have already been demonstrated.

In graphene, interband excitation by ultrafast optical pulses produces a non-equilibrium carrier population in the valence and conduction bands, Fig. 9.29a. Time-resolved experiments give two typical relaxation timescales: a faster one $\leq$100 fs, usually associated with carrier-carrier intraband collisions and phonon emission, and a slower one, on a ps timescale, which corresponds to interband recombination and cooling of hot phonons [40, 49, 91, 411]. In addition, the linear dispersion of the Dirac electrons [40, 160, 311, 411] in graphene implies that for any excitation, there is always an electron-hole pair in resonance. Thus graphene is an ideal ultrafast wideband saturable absorber for ultrafast pulse generation. After the first demonstration of ultrafast pulse generation using graphene [160], a variety of ultrafast lasers mode-locked by graphene have been reported [40, 160, 288, 344, 400, 411, 414, 492]. Reference [411] explained the fundamentals of the photo-excited carrier dynamics in graphene saturable absorbers (GSAs), in good agreement with experimental results.

Different fabrication strategies, e.g. liquid phase exfoliation [160, 161, 288, 344, 411, 414], CVD [491], reduced GO [400], carbon segregation from silicon carbide [445], and micro-mechanical cleavage [40, 67], have been used to fabricate GSAs. A widely used method involves wet chemistry processing of graphene-polymer composites, as discussed in the previous sections.

Saturable absorbers using functionalized graphene (e.g. graphene oxide) have also been demonstrated for mode-locking [40]. Mode-locking of fiber [40, 160, 161, 288, 344, 400, 411, 414, 491] and solid-state [412, 424] lasers have been demonstrated with GSAs. Reference [344] reported sub-200 fs pulse generation using a GSA-based stretched-pulse cavity design. High output power (>1 W) has also been reported [412].

Unlike nanotubes, GSAs are intrinsically ultrawideband (Fig. 9.29b) [40, 160, 161, 411, 414]. GSAs have successfully been used to mode-lock lasers at 1 [424], and 1.5 μm [40, 160, 161, 344, 411, 414]. Recently, we demonstrated a widely tunable fiber laser mode-locked with a GSA (Fig. 9.29c) [40, 414]. It can produce picosecond pulses (Fig. 9.29d) in a tuning range of 1,525–1,559 nm (Fig. 9.29c), only limited by the filter used inside the cavity [40, 414]. In addition

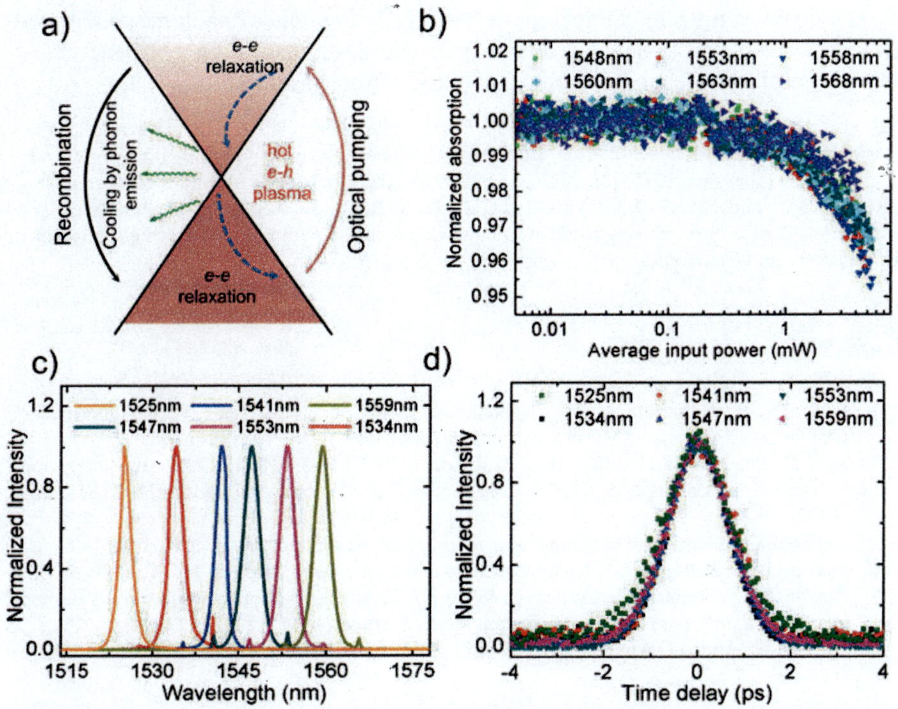

Fig. 9.29 Graphene based saturable absorbers: (**a**) Schematic of photo-excited electron kinetics in graphene, with possible relaxation mechanisms for the non-equilibrium electron population, (**b**) wideband saturable absorption of graphene, (**c**) output spectrum and (**d**) pulse duration measurements of wide-band tunable pulses mode-locked with graphene [411, 414]

to mode-locking, GSAs have also been used in Q-switching, for both wavelength-tunable [345] and dual-wavelength [274] lasers. These demonstrations underscore the potential of graphene for wideband ultrafast lasers, in principle, covering ultraviolet to THz spectral range [40, 311, 411].

9.13 Conclusions

Both nanotube and graphene are very promising for near term optoelectronic applications. Research on SWNT based optoelectronics, in particular ultrafast pulse generation, has soared over the past few years. This is primarily because of their optoelectronic properties and ease of device fabrication. In this chapter, we presented an overview of their polymer composites, starting from solution processing of the raw nanotubes, their sorting, characterization and incorporation into polymers, device fabrication and testing. We have also discussed some of the main applications, with particular focus on saturable absorbers for ultrafast lasers. Graphene has also emerged as a strong competitor of SWNTs in the field of photonics and optoelectronics. We thus concluded this chapter with a discussion on graphene based devices for ultra wideband photonic and optoelectronic applications.

Acknowledgments We thank D. Popa, F. Torrisi, F. Wang, W. B. Cho for useful discussions. TH acknowledges funding from King's College, Cambridge, FB from a Newton International Fellowship, PHT from NSF of China (No. 10874177). ACF from EPSRC (Grant Nos. GR/S97613/01 and EP/E500935/1) ERC NANOPOTS, Royal Society Brian Mercer Award for Innovation, The Cambridge Integrated Knowledge Centre in Advanced Manufacturing Technology for Photonics and Electronics, the EU grants GENIUS and RODIN.

References

1. Adams, S. R., Harootunian, A. T., Buechler, Y. J., Taylor, S. S. & Tsien, R. Y. Fluorescence ratio imaging of cyclic AMP in single cells. *Nature* **349**, 694–697 (1991).
2. Ago, H., Petritsch, K., Shaffer, M. S. P., Windle, A. H. & Friend, R. H. Composites of carbon nanotubes and conjugated polymers for photovoltaic devices. *Adv. Mater.* **11**, 1281–1285 (1999).
3. Agrawal, G. P. *Application of nonlinear fiber optics*. Academic, San Diego (2001).
4. Ajayan, P. M. & Banhart, F. Nanotubes: Strong bündles. *Nat. Mater.* **3**, 135–136 (2004).
5. Ajayan, P. M., Stephan, O., Colliex, C. & Trauth, D. Aligned carbon nanotube arrays formed by cutting a polymer resin-nanotube composite. *Science* **265**, 1212–1214 (1994).
6. Ajiki, H. & Ando, T. Aharonov-Bohm effect in carbon nanotubes. *Physica B* **201**, 349–352 (1994).
7. Allington, R. W., Brakke, M. K., Nelson, J. W., Aron, C. G. & Larkins, B. A. Optimum conditions for high-resolution gradient analysis. *Anal. Biochem.* **73**, 78–92 (1976).
8. Alvarez, L., Righi, A., Rols, S., Anglaret, E., Sauvajol, J. L., Munoz, E., Maser, W. K., Benito, A. M., Martinez, M. T. & de la Fuente, G. F. Diameter dependence of Raman intensities for single-wall carbon nanotubes. *Phys. Rev. B* **63**, 153401 (2001).
9. Anandan, S. Recent improvements and arising challenges in dye-sensitized solar cells. *Solar Energy Mater. Solar Cells* **91**, 843–846 (2007).
10. Ando, T. Excitons in carbon nanotubes. *J. Phys. Soc. Jpn.* **66**, 1066–1073 (1997).
11. Araujo, P. T., Doorn, S. K., Kilina, S., Tretiak, S., Einarsson, E., Maruyama, S., Chacham, H., Pimenta, M. A. & Jorio, A. Third and fourth optical transitions in semiconducting carbon nanotubes. *Phys. Rev. Lett.* **98**, 067401 (2007).
12. Arnold, M. S., Green, A. A., Hulvat, J. F., Stupp, S. I. & Hersam, M. C. Sorting carbon nanotubes by electronic structure using density differentiation. *Nat. Nanotech.* **1**, 60–65 (2006).
13. Arnold, M. S., Stupp, S. I. & Hersam, M. C. Enrichment of single-walled carbon nanotubes by diameter in density gradients. *Nano Lett.* **5**, 713–718 (2005).
14. Artukovic, E., Kaempgen, M., Hecht, D. S., Roth, S. & Gruner, G. Transparent and flexible carbon nanotube transistors. *Nano Lett.* **5**, 757–760 (2005).
15. Ausman, K. D., Piner, R., Lourie, O., Ruoff, R. S. & Korobov, M. Organic solvent dispersions of single-walled carbon nanotubes: Toward solutions of pristine nanotubes. *J. Phys. Chem. B* **104**, 8911–8915 (2000).
16. Bachilo, S. M., Balzano, L., Herrera, J. E., Pompeo, F., Resasco, D. E. & Weisman, R. B. Narrow (n,m)-distribution of single-walled carbon nanotubes grown using a solid supported catalyst. *J. Am. Chem. Soc.* **125**, 11186–11187 (2003).
17. Bachilo, S. M., Strano, M. S., Kittrell, C., Hauge, R. H., Smalley, R. E. & Weisman, R. B. Structure-assigned optical spectra of single-walled carbon nanotubes. *Science* **298**, 2361–2366 (2002).

18. Bae, S., Kim, H., Lee, Y., Xu, X., Park, J.-S., Zheng, Y., Balakrishnan, J., Lei, T., Ri Kim, H., Song, Y. I., Kim, Y.-J., Kim, K. S., Ozyilmaz, B., Ahn, J.-H., Hong, B. H. & Iijima, S. Roll-to- roll production of 30-inch graphene films for transparent electrodes. *Nat. Nano* **5**, 574–578 (2010).
19. Bahr, J. L., Mickelson, E. T., Bronikowski, M. J., Smalley, R. E. & Tour, J. M. Dissolution of small diameter single-wall carbon nanotubes in organic solvents? *Chem. Commun.* 193–194 (2001).
20. Bahr, J. L., Yang, J., Kosynkin, D. V., Bronikowski, M. J., Smalley, R. E. & Tour, J. M. Functionalization of carbon nanotubes by electrochemical reduction of aryl diazonium salts: A bucky paper electrode. *J. Am. Chem. Soc.* **123**, 6536–6542 (2001).
21. Bandow, S., Asaka, S., Saito, Y., Rao, A. M., Grigorian, L., Richter, E. & Eklund, P. C. Effect of the growth temperature on the diameter distribution and chirality of single-wall carbon nanotubes. *Phys. Rev. Lett.* **80**, 3779 (1998).
22. Bandyopadhyaya, R., Nativ-Roth, E., Regev, O. & Yerushalmi-Rozen, R. Stabilization of individual carbon nanotubes in aqueous solutions. *Nano Lett.* **2**, 25–28 (2002).
23. Banerjee, S., Hemraj-Benny, T. & Wong, S. S. Covalent surface chemistry of single-walled carbon nanotubes. *Adv. Mater.* **17**, 17–29 (2005).
24. Banerjee, S., Kahn, M. G. C. & Wong, S. S. Rational chemical strategies for carbon nanotube functionalization. *Chem. Eur. J.* **9**, 1898–1908 (2003).
25. Baranov, A. V., Bekhterev, A. N., Bobovich, Y. S. & Petrov, V. I. Interpretation of some singularities in raman spectra of graphite and glass carbon. *Opt Spektrosk.* **62**, 1036–1042 (1987).
26. Basko, D. M., Piscanec, S. & Ferrari, A. C. Electron-electron interactions and doping dependence of the two-phonon Raman intensity in graphene. *Phys. Rev. B* **80**, 165413 (2009).
27. Becker, K., Lupton, J. M., Muller, J., Rogach, A. L., Talapin, D. V., Weller, H. & Feldmann, J. Electrical control of Förster energy transfer. *Nat. Mater.* **5**, 777–781 (2006).
28. Beecher, P., Servati, P., Rozhin, A., Colli, A., Scardaci, V., Pisana, S., Hasan, T., Flewitt, A. J., Robertson, J., Hsieh, G. W., Li, F. M., Nathan, A., Ferrari, A. C. & Milne, W. I. Ink-jet printing of carbon nanotube thin film transistors. *J. Appl. Phys.* **102**, 043710 (2007).
29. Beecher, S. J., Thomson, R. R., Psaila, N. D., Sun, Z., Hasan, T., Rozhin, A. G., Ferrari, A. C. & Kar, A. K. 320 fs pulse generation from an ultrafast laser inscribed waveguide laser mode-locked by a nanotube saturable absorber. *Appl. Phys. Lett.* **97**, 111114 (2010).
30. Behrens, M. In Abdelhalden's Handbuch der Biologischen Arbeitsmethoden. Abt. 5, Teil 10, Part 2, pp. 1363–1392. Urban and Schwarzenberg, Berlin (1938).
31. Berber, S., Kwon, Y.-K. & Tománek, D. Unusually high thermal conductivity of carbon nanotubes. *Phys. Rev. Lett.* **84**, 4613–4616 (2000).
32. Bergin, S. D., Nicolosi, V., Streich, P. V., Giordani, S., Sun, Z., Windle, A. H., Ryan, P., Niraj, N. P. P., Wang, Z.-T., Carpenter, L., Blau, W. J., Boland, J. J., Hamilton, J. P. & Coleman, J. N. Towards solutions of single-walled carbon nanotubes in common solvents. *Adv. Mater.* **20**, 1876 (2008).
33. Bergin, S. D., Sun, Z., Streich, P., Hamilton, J. & Coleman, J. N. New solvents for nanotubes: Approaching the dispersibility of surfactants. *J. Phys. Chem. C* **114**, 231–237 (2010).
34. Bharathan, J. & Yang, Y. Polymer electroluminescent devices processed by inkjet printing: I. Polymer light-emitting logo. *Appl. Phys. Lett.* **72**, 2660–2662 (1998).
35. Bhattacharyya, S., Kymakis, E. & Amaratunga, G. A. J. Photovoltaic properties of dye functionalized single-wall carbon nanotube/conjugated polymer devices. *Chem. Mater.* **16**, 4819–4823 (2004).
36. Biju, V., Itoh, T., Baba, Y. & Ishikawa, M. Quenching of photoluminescence in conjugates of quantum dots and single-walled carbon nanotube. *J. Phys. Chem. B* **110**, 26068–26074 (2006).
37. Blake, P., Brimicombe, P. D., Nair, R. R., Booth, T. J., Jiang, D., Schedin, F., Ponomarenko, L. A., Morozov, S. V., Gleeson, H. F., Hill, E. W., Geim, A. K. & Novoselov, K. S. Graphene-based liquid crystal device. *Nano Lett.* **8**, 1704–1708 (2008).

38. Bo, X. Z., Lee, C. Y., Strano, M. S., Goldfinger, M., Nuckolls, C. & Blanchet, G. B. Carbon nanotubes-semiconductor networks for organic electronics: The pickup stick transistor. *Appl. Phys. Lett.* **86**, 182102 (2005).
39. Bohren, C. F. & Huffman, D. R. *Absorption and scattering of light by small particles*. Wiley, New York 1998.
40. Bonaccorso, F., Sun, Z., Hasan, T. & Ferrari, A. C. Graphene photonics and optoelectronics. *Nat. Photonics* **4**, 611–622 (2010).
41. Bonaccorso, F., Hasan, T., Tan, P. H., Sciascia, C., Privitera, G., Di Marco, G., Gucciardi, P. G. & Ferrari, A. C. Density gradient ultracentrifugation of nanotubes: Interplay of bundling and surfactants encapsulation. *J. Phys. Chem. C* **114**, 17267–17285 (2010).
42. Born, M. & Oppenheimer, R. Zur Quantentheorie der Molekeln. *Ann. Phys.* **84**, 457 (1927).
43. Born, M. & Huang, K. *Dynamical theory of crystal lattices*. Clarendon Press, Oxford (1954).
44. Botti, S., Ciardi, R., De Dominicis, L., Asilyan, L. S., Fantoni, R. & Marolo, T. DFWM measurements of third-order susceptibility of single-wall carbon nanotubes grown without catalyst. *Chem. Phys. Lett.* **378**, 117–121 (2003).
45. Boyd, R. W. *Nonlinear optics*. Academic, San Diego (2003).
46. Brakke, M. K. Zonal separations by density-gradient centrifugation. *Arch. Biochem.* **45**, 275–290 (1953).
47. Brakke, M. K. Photometric scanning of centrifuged density gradient columns. *Arch. Biochem.* **5**, 271–283 (1963).
48. Brakke, M. K. Density gradient centrifugation and its application to plant viruses. *Advan. Virus Res.* **7**, 193–224 (1960).
49. Breusing, M., Ropers, C. & Elsaesser, T. Ultrafast carrier dynamics in graphite. *Phys. Rev. Lett.* **102**, 086809 (2009).
50. Bronikowski, M. J., Willis, P. A., Colbert, D. T., Smith, K. A. & Smalley, R. E. Gas-phase production of carbon single-walled nanotubes from carbon monoxide via the HiPco process: A parametric study. *J. Vacuum Sci. Technol. A* **19**, 1800–1805 (2001).
51. Brown, S. D. M., Jorio, A., Corio, P., Dresselhaus, M. S., Dresselhaus, G., Saito, R. & Kneipp, K. Origin of the Breit-Wigner-Fano lineshape of the tangential G-band feature of metallic carbon nanotubes. *Phys. Rev. B* **63**, 155414 (2001).
52. Brown, P., Takechi, K. & Kamat, P. V. Single-walled carbon nanotube scaffolds for dyesensitized solar cells. *J. Phys. Chem. C* **112**, 4776–4782 (2008).
53. Bruesch, P. *Phonons: Theory and experiments I, lattice dynamics and models of interatomic forces*. Springer, Berlin, Heidelberg, New York (1982).
54. Butt, H., Graf, K. & Kappl, M. *Physics and chemistry of interfaces*. Wiley VCH, Berlin (2006).
55. Cacace, M. G., Landau, E. M. & Ramsden, J. J. The Hofmeister series: Salt and solvent effects on interfacial phenomena. *Quart. Rev. Bioph.* **3**, 241–277 (1997).
56. Calvert, P. Inkjet printing for materials and devices. *Chem. Mater.* **13**, 3299–3305 (2001).
57. Calzolari, A., Marzari, N., Souza, I. & Buongiorno Nardelli, M. Ab initio transport properties of nanostructures from maximally localized Wannier functions. *Phys. Rev. B* **69**, 035108 (2004).
58. Cançado, L. G., Jorio, A., Martins Ferreira, E. H., Stavale, F., Achete, C. A., Capaz, R. B., Moutinho, M. V. O., Lombardo, A., Kulmala, T. & Ferrari, A. C. Quantifying defects in grapheme via Raman spectroscopy at different excitation energies. arXiv:1105.0175 (2011).
59. Cancado, L. G., Pimenta, M. A., Neves, B. R. A., Dantas, M. S. S. & Jorio, A. *Phys. Rev. Lett.* **93**, 247401 (2004).
60. Capaz, R. B., Spataru, C. D., Ismail-Beigi, S. & Louie, S. G. Diameter and chirality dependence of exciton properties in carbon nanotubes. *Phys. Rev. B* **74**, 121401(R) (2006).
61. Casiraghi, C., Hartschuh, A., Qian, H., Piscanec, S., Georgi, C., Fasoli, A., Novoselov, K. S., Basko, D. M. & Ferrari, A. C. *Nano Lett.* **9**, 1433 (2009).
62. Casiraghi, C., Pisana, S., Novoselov, K. S., Geim, A. K. & Ferrari, A. C. *Appl. Phys. Lett.* 91, 233108 (2007).

63. Carlson, L. J., Maccagnano, S. E., Zheng, M., Silcox, J. & Krauss, T. D. Fluorescence efficiency of individual carbon nanotubes. *Nano Lett.* **7**, 3698–3703 (2007).
64. Cathcart, H., Quinn, S., Nicolosi, V., Kelly, J. M., Blau, W. J. & Coleman, J. N. Spontaneous debundling of single-walled carbon nanotubes in DNA-based dispersions. *J. Phys. Chem. C* **111**, 66–74 (2007).
65. Chandrasekhar, R., Charles, S. W. & O'Grady, K. Cobalt ferrite fluids and their application to magnetic ink-jet printing. *J. Imaging Technol.* **13**, 55–59 (1987).
66. Chang, E., Bussi, G., Ruini, A. & Molinari, E. Excitons in carbon nanotubes: An Ab Initio symmetry-based approach. *Phys. Rev. Lett.* **92**, 196401 (2004).
67. Chang, Y. M., Kim, H., Lee, J. H. & Song, Y.-W. Multilayered graphene efficiently formed by mechanical exfoliation for nonlinear saturable absorbers in fiber mode-locked lasers. *Appl. Phys. Lett.* **97**, 211102–3 (2010).
68. Chapin, D. M., Fuller, C. S. & Pearson, G. L. A new silicon p-n junction photocell for converting solar radiation into electrical power. *J. Appl. Phys.* **25**, 676–677 (1954).
69. Charlier, J.-C., Lambin, Ph. & Ebbesen, T. W. Electronic properties of carbon nanotubes with polygonized cross sections. *Phys. Rev. B* **54**, R8377–R8380 (1996).
70. Collins, K. & Washabaugh, M. The Hofmeister effect and the behaviour of water at interfaces. *Q. Rev. Biophys.* **18**, 323–422 (1985).
71. Cote, L. J., Kim, F. & Huang, J. Langmuir-Blodgett assembly of graphite oxide single layers. *J. Am. Chem. Soc.* **131**, 1043–1049 (2008).
72. Chen, G. Z., Shaffer, M. S. P., Coleby, D., Dixon, G., Zhou, W., Fray, D. J. & Windle, A. H. Carbon nanotube and polypyrrole composites: Coating and doping. *Adv. Mater.* **12**, 522–526 (2000).
73. Chen, H.-Y. & Lee, W. Suppression of field screening in nematic liquid crystals by carbon nanotubes. *Appl. Phys. Lett.* **88**, 222105 (2006).
74. Chen, H.-Y., Lee, W. & Clark, N. A. Faster electro-optical response characteristics of a carbonnanotube-nematic suspension. *Appl. Phys. Lett.* **90**, 033510-3 (2007).
75. Chen, J., Rao, A. M., Lyuksyutov, S., Itkis, M. E., Hamon, M. A., Hu, H., Cohn, R. W., Eklund, P. C., Colbert, D. T., Smalley, R. E. & Haddon, R. C. Dissolution of full-length single-walled carbon nanotubes. *J. Phys. Chem. B* **105**, 2525–2528 (2001).
76. Chen, Y.-C., Raravikar, N. R., Schadler, L. S., Ajayan, P. M., Zhao, Y.-P., Lu, T.-M., Wang, G.-C. & Zhang, X.-C. Ultrafast optical switching properties of single-wall carbon nanotube polymer composites at 1.55 um. *Appl. Phys. Lett.* **81**, 975–977 (2002).
77. Cheng, H. M., Li, F., Sun, X., Brown, S. D. M., Pimenta, M. A., Marucci, A., Dresselhaus, G. & Dresselhaus, M. S. Bulk morphology and diameter distribution of single-walled carbon nanotubes synthesized by catalytic decomposition of hydrocarbons. *Chem. Phys. Lett.* **289**, 602–610 (1998).
78. Cheng, W. D., Wu, D. S., Zhang, H., Li, X. D., Lan, Y. Z., Chen, D. G. & Wang, H. X. Enhancements of third-order nonlinear optical response in the triplet excited state of finite open single-walled carbon nanotubes. *J. Chem. Phys.* **119**, 13100–13105 (2003).
79. Choi, J. H. Solvatochromism in single-walled carbon nanotubes. *Appl. Phys. Lett.* **90**, 223114 (2007).
80. Cohen, R. & Claverie, J. M. Sedimentation of generalized systems of interacting particles. II. Active enzyme centrifugation – theory and extensions of its validity range. *Biopolymers* **14**, 1701–1716 (1975).
81. Coleman, J. N., Khan, U., Blau, W. J. & Gun'ko, Y. K. Small but strong: A review of the mechanical properties of carbon nanotube-polymer composites. *Carbon* **44**, 1624–1652 (2006).
82. Coleman, J. N., Khan, U. & Gun'ko, Y. K. Mechanical reinforcement of polymers using carbon nanotubes. *Adv. Mater.* **18**, 689–706 (2006).
83. Coleman, J. N. Liquid-phase exfoliation of nanotubes and graphene. *Adv. Funct. Mater.* **19**, 3680–3695 (2009).
84. Crawford, L. V. & Waring, M. J. Supercoiling of polyoma virus DNA measured by its interaction with ethium bromide. *J. Mol. Biol.* **25**, 23–30 (1967).

85. Creeth, J. H. & Pain, R. H. The determination of molecular weights of biological macromolecules by ultracentrifuge methods. *Prog. Biophys. Mol. Biol.* **17**, 217–287 (1967).
86. Crochet, J., Clemens, M. & Hertel, T. Quantum yield heterogeneities of aqueous single-wall carbon nanotube suspensions. *J. Am. Chem. Soc.* **129**, 8058–8059 (2007).
87. Dai, H. Carbon nanotubes: Synthesis, integration, and properties. *Acc. Chem. Res.* **35**, 1035–1044 (2002).
88. Dalton, A. B., Collins, S., Muñoz, E., Razal, J. M., Ebron, V. H., Ferraris, J. P., Coleman, J. N., Kim, B. G. & Baughman, R. H. Super-tough carbon-nanotube fibres. *Nature* **423**, 703 (2003).
89. Das, A., Chakraborty, B., Piscanec, S., Pisana, S., Sood, A. K. & Ferrari, A. C. Phonon renormalization in doped bilayer graphene. *Phys. Rev. B* **79**, 155417 (2009).
90. Das, A., Pisana, S., Chakraborty, B., Piscanec, S., Saha, S. K., Waghmare, U. V., Novoselov, K. S., Krishnamurthy, H. R., Geim, A. K., Ferrari, A. C. & Sood, A. K. *Nat. Nanotechnol.* **3**, 210 (2008).
91. Dawlaty, J. M., Shivaraman, S., Strait, J., George, P., Chandrashekhar, M., Rana, F., Spencer, M. G., Veksler, D. & Chen, Y. Measurement of the optical absorption spectra of epitaxial graphene from terahertz to visible. *Appl. Phys. Lett.* **93**, 131905 (2008).
92. De, S., Lyons, P. E., Sorel, S., Doherty, E. M., King, P. J., Blau, W. J., Nirmalraj, P. N., Boland, J. J., Scardaci, V., Joimel, J. & Coleman, J. N. Transparent, flexible, and highly conductive thin films based on polymer-nanotube composites. *ACS Nano* **3**, 714–720 (2009).
93. De Dominicis, L., Botti, S., Asilyan, L. S., Ciardi, R., Fantoni, R., Terranova, M. L., Fiori, A., Orlanducci, S. & Appolloni, R. Second- and third- harmonic generation in single-walled carbon nanotubes at nanosecond time scale. *Appl. Phys. Lett.* **85**, 1418–1420 (2004).
94. De, S., King, P. J., Lotya, M., O'Neill, A., Doherty, E. M., Hernandez, Y., Duesberg, G. S. & Coleman, J. N. Flexible, transparent, conducting films of randomly stacked graphene from surfactant-stabilized, oxide-free graphene dispersions. *Small* **6**, 458–464 (2010).
95. De, S., Higgins, T. M., Lyons, P. E., Doherty, E. M., Nirmalraj, P. N., Blau, W. J., Boland, J. J. & Coleman, J. N. Silver nanowire networks as flexible, transparent, conducting films: Extremely high DC to optical conductivity ratios. *ACS Nano* **3**, 1767–1774 (2009).
96. Della Valle, G., Osellame, R., Galzerano, G., Chiodo, N., Cerullo, G., Laporta, P., O., S., Morgner, U., Rozhin, A. G., Scardaci, V. & Ferrari, A. C. Passive mode locking by carbon nanotubes in a femtosecond laser written waveguide laser. *Appl. Phys. Lett.* **89**, 231115 (2006).
97. DeMaria, A. J., Stetser, D. A. & Heynau, H. Self mode-locking of lasers with saturable absorbers. *Appl. Phys. Lett.* **8**, 174–176 (1966).
98. Denneulin, A., Bras, J., Blayo, A., Khelifi, B., Roussel-Dherbey, F. & Neuman, C. The influence of carbon nanotubes in inkjet printing of conductive polymer suspensions. *Nanotechnology* **20**, 385701 (2009).
99. Dettlaff-Weglikowska, U., Kaempgen, M., Hornbostel, B., Skakalova, V., Wang, J., Liang, J. & Roth, S. Conducting and transparent SWNT/polymer composites. *Phys. Stat. Sol. (b)* **243**, 3440–3444 (2006).
100. Dieckmann, G. R., Dalton, A. B., Johnson, P. A., Razal, J., Chen, J., Giordano, G. M., Muñoz, E., Musselman, I. H., Baughman, R. H. & Draper, R. K. Controlled assembly of carbon nanotubes by designed amphiphilic peptide helices. *J. Am. Chem. Soc.* **125**, 1770–1777 (2003).
101. Dierking, I., Scalia, G. & Morales, P. Liquid crystal–carbon nanotube dispersions. *J. Appl. Phys.* **97**, 044309 (2005).
102. Dierking, I., Scalia, G., Morales, P. & LeClere, D. Aligning and reorienting carbon nanotubes with nematic liquid crystals. *Adv. Mater.* **16**, 865–869 (2004).
103. Dobardžić, E., Maultzsch, J., Milošević, I., Thomsen, C. & Damjanović, M. The radial breathing mode frequency in double-walled carbon nanotubes: An analytical approximation. *Phys. Stat. Sol. (b)* **237**, R7 (2003).

104. Doorn, S. K., Strano, M. S., O'Connell, M. J., Haroz, E. H., Rialon, K. L., Hauge, R. H. & Smalley, R. E. Capillary electrophoresis separations of bundled and individual carbon nanotubes. *J. Phys. Chem. B* **107**, 6063–6069 (2003).
105. Dyke, C. A. & Tour, J. M. Solvent-free functionalization of carbon nanotubes. *J. Am. Chem. Soc.* **125**, 1156–1157 (2003).
106. Dyke, C. A. & Tour, J. M. Covalent functionalization of single-walled carbon nanotubes for materials applications. *J. Phys. Chem. A* **108**, 11151–11159 (2004).
107. Dyke, C. A. & Tour, J. M. Overcoming the insolubility of carbon nanotubes through high degrees of sidewall functionalization. *Chem. Eur. J.* **10**, 812–817 (2004).
108. Eda, G. & Chhowalla, M. Chemically derived graphene oxide: Towards large-area thin-film electronics and optoelectronics. *Adv. Mater.* **22**, 2392–2415 (2010).
109. Eivindvik, K. & Sjogren, C. E. Physicochemical properties of iodixanol. *Acta Radiologica* **36**, 32–38 (1995).
110. Elias, D. C., Nair, R. R., Mohiuddin, T. M. G., Morozov, S. V., Blake, P., Halsall, M. P., Ferrari, A. C., Boukhvalov, D. W., Katsnelson, M. I., Geim, A. K. & Novoselov, K. S. Control of graphene's properties by reversible hydrogenation: Evidence for graphane. *Science* **323**, 610 (2009).
111. Eschrig, H. Unified perturbation treatment for phonons in metallic covalent and ionic crystals. *Phys. Stat. Sol. (b)* **56**, 197 (1973).
112. Essig, S., Marquardt, C. W., Vijayaraghavan, A., Ganzhorn, M., Dehm, S., Hennrich, F., Ou, F., Green, A. A., Sciascia, C., Bonaccorso, F., Bohnen, K.-P., v Lonhneysen, H., Kappes, M. M., Ajayan, P. M., Hersam, M. C., Ferrari, A. C. & Krupke, R. Phonon-assisted electroluminescence from metallic carbon nanotubes and graphene. *Nano lett.* **10**, 1589–1594 (2010).
113. Fagan, J. A., Becker, M. L., Chun, J., Nie, P., Bauer, B. J., Simpson, J. R., Hight-Walker, A. & Hobbie, E. K. Centrifugal length separation of carbon nanotubes. *Langmuir* **24**, 13880–13389 (2008).
114. Fairhurst, C. E., Fuller, S., Gray, J., Holmes, M. C. & Tiddy, G. J. T. In *Lyotropic surfactant liquid crystals*, Vol. 3. High molecular mass liquid crystals (Eds.: Demus, D., Goodby, J., Gray, G. W., Spiess, H. W., and Vill, V.), New York, Wiley-VCH (1998).
115. Fang, Q., Kieu, K. & Peyghambarian, N. An all-fiber 2-um wavelength-tunable mode-locked laser. *IEEE Photonics Technol. Lett.* **22**, 1656–1658 (2010).
116. Fantini, C., Jorio, A., Souza, M., Strano, M. S., Dresselhaus, M. S. & Pimenta, M. A. Optical transition energies for carbon nanotubes from resonant Raman spectroscopy: Environment and temperature effects. *Phys. Rev. Lett.* **93**, 147406 (2004).
117. Feng, W., Bai, X. D., Lian, Y. Q., Liang, J., Wang, X. G. & Yoshino, K. Well-aligned polyaniline/carbon-nanotube composite films grown by in-situ aniline polymerization. *Carbon* **41**, 1551–1557 (2003).
118. Ferrari, A. C. Raman spectroscopy of graphene and graphite: Disorder, electron-phonon coupling, doping and nonadiabatic effects. *Solid State Comm.* **143**, 47 (2007).
119. Ferrari, A. C., Robertson, J. *Raman spectroscopy in carbons: From nanotubes to diamond.* Philos. Trans. R. Soc. A, **362**, 2267–2565 (2004)
120. Ferrari, A. C. & Robertson, J. Interpretation of Raman spectra of disordered and amorphous carbon. *Phys. Rev. B* **61**, 14095–14107 (2000).
121. Ferrari, A. C. & Robertson, J. Resonant Raman spectroscopy of disordered, amorphous, and diamondlike carbon. *Phys. Rev. B* **64**, 075414 (2001).
122. Ferrari, A. C., Meyer, J. C., Scardaci, V., Casiraghi, C., Lazzeri, M., Mauri, F., Piscanec, S., Jiang, D., Novoselov, K. S., Roth, S. & Geim, A. K. Raman spectrum of graphene and graphene layers. *Phys. Rev. Lett.* **97**, 187401 (2006).
123. Ferrer-Anglada, N., Kaempgen, M., Skakalova, V., Dettlaf-Weglikowska, U. & Roth, S. Synthesis and characterization of carbon nanotube-conducting polymer thin films. *Diamond Relat. Mater.* **13**, 256–260 (2004).
124. Finnie, P., Homma, Y. & Lefebvre, J. Band-gap shift transition in the photoluminescence of single-walled carbon nanotubes. *Phys. Rev. Lett.* **94**, 247401 (2005).

125. Fong, K. H., Kikuchi, K., Goh, C. H., Set, S. Y., Grange, R., Haiml, M., Schlatter, A. & Keller, U. Solid-state Er:Yb:glass laser mode-locked by using single-wall carbon nanotube thin film. *Opt. Lett.* **32**, 38–40 (2007).
126. Ford, T., Graham, J. & Rickwood, D. Iodixanol-a non ionic isosmotic centrifugation medium for the formation of self-generated gradients. *Anal. Biochem.* **220**, 360–366 (1994).
127. Förster, T. Transfer mechanisms of electronic excitation. *Discuss. Faraday Soc.* **27**, 7–17 (1959).
128. Fox, N. A., Youh, M. J., Steeds, J. W. & Wang, W. N. Patterned diamond particle films. *J. Appl. Phys.* **87**, 8187–8191 (2000).
129. Freifelder, D. Molecular weights of coliphages and coliphages DNA. Molecular weights of DNA from bacteriophages T4, T5, and general problem of determination of M. *J. Mol. Bio.* **54**, 567–577 (1970).
130. Gambetta, A., Galzerano, G., Rozhin, A. G., Ferrari, A. C., Ramponi, R., Laporta, P. & Marangoni, M. Sub-100 fs two-color pump-probe spectroscopy of single wall carbon nanotubes with a 100 MHz Er-fiber laser system. *Opt. Express* **16**, 11727–11734 (2008).
131. Gamerith, S., Klug, A., Scheiber, H., Scherf, U., Moderegger, E., & List, E. J. W. Direct ink-jet printing of Ag-Cu nanoparticle and Ag-Precursor based electrodes for OFET applications. *Adv. Funct. Mater.* **17**, 3111–3118 (2007).
132. Gans, B.-J. de, Duineveld, P. C. & Schubert, U. S. Inkjet printing of polymers: State of the art and future developments. *Adv. Mater.* **16**, 203–213 (2004).
133. Garber, R. C. & Yoder, O. C. Isolation of DNA from filamentous fungi and separation into nuclear, mitochondrial, ribosomal, and plasmid components. *Anal. Biochem.* **135**, 416–422 (1983).
134. Garcia-Vidal, F. J. & Pitarke, J. M. Optical absorption and energy-loss spectra of aligned carbon nanotubes. *Eur. Phys. J. B* **22**, 257–265 (2001).
135. Geim, A. K. & Novoselov, K. S. The rise of graphene. *Nat. Mater.* **6**, 183–191 (2007).
136. Geng, H. Z., Kim, K. K., So, K. P., Lee, Y. S., Chang, Y. & Lee, Y. H. Effect of acid treatment on carbon nanotube-based flexible transparent conducting films. *J. Am. Chem. Soc.* **129**, 7758–7759 (2007).
137. Ghatee, M. H. & Pakdel, L. Surface tension regularity of non-polar, polar, and weak electrolyte liquid hydrocarbons. *Fluid Phase Equilib.* **234**, 101–107 (2005).
138. Giordani, S., Bergin, S. D., Nicolosi, V., Lebedkin, S., Kappes, M. M., Blau, W. J. & Coleman, J. N. Debundling of single-walled nanotubes by dilution: Observation of large populations of individual nanotubes in amide solvent dispersions. *J. Phys. Chem. B* **110**, 15708–15718 (2006).
139. Giorgetti, E., Margheri, G., Sottini, S., Casalboni, M., Senesi, R., Scarselli, M. & Pizzoferrato, R. Dye-doped zirconia-based Ormosil planar waveguides: Optical properties and surface morphology. *J. Non-Cryst. Solids* **255**, 193–198 (1999).
140. Girifalco, L. A., Hodak, M. & Lee, R. S. Carbon nanotubes, buckyballs, ropes, and a universal graphitic potential. *Phys. Rev. B* **62**, 13104–13110 (2000).
141. Girifalco, L. A. & Lad, R. A. Energy of cohesion, compressibility, and the potential energy functions of the graphite system. *J. Chem. Phys.* **25**, 693–697 (1956).
142. Gracia-Espino, E., Sala, G., Pino, F., Halonen, N., Luomahaara, J., Maklin, J., Toth, G., Kordas, K., Jantunen, H., Terrones, M., Helisto, P., Seppa, H., Ajayan, P. M. & Vajtai, R. Electrical transport and field-effect transistors using inkjet-printed SWCNT films having different functional side groups. *ACS Nano* **4**, 3318–3324 (2010).
143. Graf, D., Molitor, F., Ensslin, K., Stampfer, C., Jungen, A., Hierold, C. & Wirtz, L. Spatially resolved Raman spectroscopy of single- and few-layer graphene. *Nano Lett.* **7**, 238 (2007).
144. Graham, J. M. *Biological centrifugation.* BIOS Scientific Publisher Limited, Oxford (2001).
145. Granqvist, C. G. Transparent conductors as solar energy materials: A panoramic review. *Sol. Energ. Mat. Sol. C.* **91**, 1529–1598 (2007).
146. Gratzel, M. Dye-sensitized solar cells. *J. Photochem. Photobiol. C* **4**, 145–153 (2003).
147. Green, A. A. & Hersam, M. C. Processing and properties of highly enriched double-wall carbon nanotubes. *Nat. Nanotechnol.* **4**, 64–70 (2009).

148. Green, A. A. & Hersam, M. C. Solution phase production of graphene with controlled thickness via density differentiation. *Nano Lett.* **9**, 4031–4036 (2009).
149. Gruner, G. Carbon nanotube films for transparent and plastic electronics. *J. Mater. Chem.* **16**, 3533–3539 (2006).
150. Guo, Y., Minami, N., Kazaoui, S., Peng, J., Yoshida, M. & Miyashita, T. Multi-layer LB films of single-wall carbon nanotubes. *Physica B* **323**, 235–236 (2002).
151. Hagen, A. & Hertel, T. Quantitative analysis of optical spectra from individual single-wall carbon nanotubes. *Nano Lett.* **3**, 383–388 (2003).
152. Hagen, A., Steiner, M., Raschke, M. B., Lienau, C., Hertel, T., Qian, H. H., Meixner, A. J. & Hartschuh, A. Exponential decay lifetimes of excitons in individual single-walled carbon nanotubes. *Phys. Rev. Lett.* **95**, 197401 (2005).
153. Hamberg, I. & Granqvist, C. G. Evaporated Sn-doped In2O3 films: Basic optical properties and applications to energy-efficient windows. *J. Appl. Phys.* **60**, R123–R160 (1986).
154. Han, J.-I. Stability of externally deformed ITO films. In *Flexible and flat panel dispolays* (Ed.: Crawford, G. P.). Sussex, England (2005).
155. Hansen, C. M. *Hansen solubility parameters: A user's handbook.* CRC Press Inc., Boca Raton, FL (2007).
156. Harris, P. J. F. Carbon nanotube composites. *Int. Mat. Rev.* **49**, 31–43 (2004).
157. Hasan, T., Tan, P. H., Bonaccorso, F., Rozhin, A., Scardaci, V., Milne, W. I. & Ferrari, A. C. Polymer-assisted isolation of singlewall carbon nanotubes in organic solvents for optical-quality nanotube-polymer composites. *J. Phys. Chem. C* **112**, 20227–20232 (2008).
158. Hasan, T., Scardaci, V., Tan, P. H., Rozhin, A. G., Milne, W. I. & Ferrari, A. C. Stabilization and 'de-bundling' of single-wall carbon nanotube dispersions in N-Methyl-2-Pyrrolidone (NMP) by polyvinylpyrrolidone (PVP). *J. Phys. Chem. C* **111**, 12594–12602 (2007).
159. Hasan, T., Scardaci, V., Tan, P. H., Rozhin, A. G., Milne, W. I. & Ferrari, A. C. Dispersibility and stability improvement of unfunctionalized nanotubes in amide solvents by polymer wrapping. *Physica E* **40**, 2414–2418 (2008).
160. Hasan, T., Sun, Z., Wang, F., Bonaccorso, F., Tan, P. H., Rozhin, A. G. & Ferrari, A. C. Nanotube-polymer composites for ultrafast photonics. *Adv. Mater.* **21**, 3874–3899 (2009).
161. Hasan, T., Torrisi, F., Sun, Z., Popa, D., Nicolosi, V., Privitera, G., Bonaccorso, F. & Ferrari, A. C. Solution-phase exfoliation of graphite for ultrafast photonics. *Physica Status Solidi (b)* **247**, 2953–2957 (2010).
162. Haschemeyer, R. H. & Bowers, W. F. Exponential analysis of concentration or concentration difference data for discrete molecular weight distributions in sedimentation equilibrium. *Biochemistry* **9**, 435–445 (1970).
163. Hearst, J. E. & Vinograd, J. A Three-Component Theory of Sedimentation Equilibrium in a Density Gradient. *J. Natl. Acad. Sci. USA* **47**, 999–1004 (1961).
164. Hecht, D, Hu, L. & Gruner, G. Conductivity scaling with bundle length and diameter in single walled carbon nanotube networks. *Appl. Phys. Lett.* **89**, 133112 (2006).
165. Henrard, L., Hernández, E., Bernier, P. & Rubio, A. Van der Waals interaction in nanotube bundles: Consequences on vibrational modes. *Phys. Rev. B* **60**, R8521 (1999).
166. Hernandez, Y., Nicolosi, V., Lotya, M., Blighe, F. M., Sun, Z., De, S., McGovern, I. T., Holland, B., Byrne, M., Gun'Ko, Y. K., Boland, J. J., Niraj, P., Duesberg, G., Krishnamurthy, S., Goodhue, R., Hutchison, J., Scardaci, V., Ferrari, A. C. & Coleman, J. N. High-yield production of graphene by liquid-phase exfoliation of graphite. *Nat. Nano* **3**, 563–568 (2008).
167. Hersam, M. Progress towards monodisperse single-walled carbon nanotubes. *Nat. Nanotech.* **3**, 387 (2008).
168. Hertel, T., Fasel, R. & Moos, G. Charge-carrier dynamics in single-wall carbon nanotube bundles: A time-domain study. *Appl. Phys. A* **75**, 449–465 (2002).
169. Hertel, T. & Moos, G. Electron-phonon interaction in single-wall carbon nanotubes: A time-domain study. *Phys. Rev. Lett.* **84**, 5002–5005 (2000).

170. Hertel, T., Perebeinos, V., Crochet, J., Katharina, A., Kappes, M. & Phaedon, A. Intersubband decay of 1-D exciton resonances in carbon nanotubes. *Nano Lett.* **8**, 87–91 (2008).
171. Hertel, T., Walkup, R. E. & Avouris, Ph. Deformation of carbon nanotubes by surface van der Waals forces. *Phys. Rev. B* **58**, 13870–13873 (1998).
172. Hirsch, A. Functionalization of single-walled carbon nanotubes. *Angew. Chem. Int. Ed.* **41**, 1853–1859 (2002).
173. Hong, W., Xu, Y., Lu, G., Li, C. & Shi, G. Transparent graphene/PEDOT-PSS composite films as counter electrodes of dye-sensitized solar cells. *Electrochem. Commun.* **10**, 1555–1558 (2008).
174. Hsieh, G. W., Beecher, P., Li, F. M., Servati, P., Colli, A., Fasoli, A., Chu, D., Nathan, A., Ong, B., Robertson, J., Ferrari, A. C. & Milne, W. I. Formation of composite organic thin film transistors with nanotubes and nanowires. *Physica E* **40**, 2406–2413 (2008).
175. Htoon, H., O'Connell, M. J., Doorn, S. K. & Klimov, V. I. Single carbon nanotubes probed by photoluminescence excitation spectroscopy: The role of phonon-assisted transitions. *Phys. Rev. Lett.* **94**, 127403 (2005).
176. Huang, C.-Y., Hu, C.-Y., Pan, H.-C. & Lo, K.-Y. Electrooptical responses of carbon nanotube-doped liquid crystal devices. *Jpn. J. Appl. Phys.* **44**, 8077–8081 (2005).
177. Hwang, J., Gommans, H. H., Ugawa, A., Tashiro, H., Haggenmueller, R., Winey, K. I., Fisher, J. E., Tanner, D. B. & Rinzler, A. G. Polarized spectroscopy of aligned single-wall carbon nanotubes. *Phys. Rev. B* **62**, 13310–13313 (2000).
178. Ichida, M., Hamanaka, Y., Kataura, H., Achiba, Y. & Nakamura, A. Ultrafast relaxation dynamics of photoexcited carriers in metallic and semiconducting single-walled carbon nanotubes. *J. Phys. Soc. Jpn.* **73**, 3479–3483 (2004).
179. Ichida, M., Mizuno, S., Kataura, H., Achiba, Y. & Nakamura, A. Anisotropic optical properties of mechanically aligned single-walled carbon nanotubes in polymer. *Appl. Phys. A* **78**, 1117–1120 (2004).
180. Ifft, J. B. & Vinograd, J. The Buoyant behavior of Bovine serum mercaptalbumin in salt solutions at equilibrium in the ultracentrifuge. II. Net hydration, ion binding, and solvated molecular weight in various salt solutions. *J. Phys. Chem.* **70**, 2814–2822 (1966).
181. Il'ichev, N. N., Garnov, S. V. & Obraztsova, E. D. Single-wall carbon nanotube suspension as a passive Q-switch for self mode-locking of solid state lasers. *AIP Conf. Proc. Electron. Prop. Novel Nanostruct.* **786**, 611–615 (2005).
182. Islam, M. F., Milkie, D. E., Kane, C. L., Yodh, A. G. & Mikkawa, J. M. Direct measurement of the polarized optical absorption cross section of single-wall carbon nanotubes. *Phys. Rev. Lett.* **93**, 037404 (2004).
183. Islam, M. F., Rojas, E., Bergey, D. M., Johnson, A. T. & Yodh, A. G. High weight fraction surfactant solubilization of single-wall carbon nanotubes in water. *Nano Lett.* **3**, 269–273 (2003).
184. Jiang, J., Dong, J. M., Wan, X. G. & Xing, D. Y. A new kind of nonlinear optical material: The fullerene tube. *J. Phys. B* **31**, 3079–3086 (1998).
185. Jiang, J., Saito, R., Gruneis, A., Dresselhaus, G. & Dresselhaus, M. S. Optical absorption matrix elements in single-wall carbon nanotubes. *Carbon* **42**, 3169–3176 (2004).
186. Jishi, R. A., Venkataraman, L., Dresselhaus, M. S. & Dresselhaus, G. Phonon modes in carbon nanotubes. *Chem. Phys. Lett.* **209**, 77 (1993)
187. Jorio, A., Fantini, C., Pimenta, M. A., Heller, D. A., Strano, M. S., Dresselhaus, M. S., Oyama, Y., Jiang, J. & Saito, R. Carbon nanotube population analysis from Raman and photoluminescence intensities. *Appl. Phys. Lett.* **88**, 023109 (2006).
188. Jorio, A., Saito, R., Hafner, J. H., Lieber, C. M., Hunter, M., McClure, T., Dresselhaus, G. & Dresselhaus, M. S. Structural (n,m) determination of isolated single-wall carbon nanotubes by resonant Raman scattering. *Phys. Rev. Lett.* **86**, 1118–1121 (2001).
189. Jorio, A., Fantini, C., Dantas, M. S. S., Pimenta, M. A., Souza, A. G., Samsonidze, G. G., Brar, V. W., Dresselhaus, G., Dresselhaus, M. S., Swan, A. K., Ünlü, M. S., Goldberg,

B. B. & Saito, R. Linewidth of the Raman features of individual single-wall carbon nanotubes. *Phys. Rev. B* **66**, 115411 (2002).

190. Jorio, A., Souza, A. G., Dresselhaus, G., Dresselhaus, M. S., Swan, A. K., Unlu, M. S., Goldberg, B. B., Pimenta, M. A., Hafner, J. H., Lieber, C. M. & Saito, R. G-band resonant Raman study of 62 isolated single-wall carbon nanotubes. *Phys. Rev. B* **65**, 155412 (2002).
191. Jost, O., Gorbunov, A. A., Pompe, W., Pichler, T., Friedlein, R., Knupfer, M., Reibold, M., Bauer, H. D., Dunsch, L., Golden, M. S. & Fink, J. Diameter grouping in bulk samples of single-walled carbon nanotubes from optical absorption spectroscopy. *Appl. Phys. Lett.* **75**, 2217–2219 (1999).
192. Journet, C., Maser, W. K., Bernier, P., Loiseau, A., Lamy de la Chapelle, M., Lefrant, S., Deniard, P., Lee, R. & Fischer, J. E. Large-scale production of single-walled carbon nanotubes by the electric-arc technique. *Nature* **388**, 756–758 (1997).
193. Kaempgen, M., Dettlaff, U. & Roth, S. Characterization of carbon nanotubes by optical spectra. *Synth. Met.* **135–136**, 755–756 (2003).
194. Kagan, C. R., Murray, C. B., Nirmal, M. & Bawendi, M. G. Electronic energy transfer in CdSe quantum dot solids. *Phys. Rev. Lett.* **76**, 1517–1520 (1996).
195. Kamaraju, N., Kumar, S., Sood, A. K., Guha, S., Krishnamurthy, S. & Rao, C. N. R. Large nonlinear absorption and refraction coefficients of carbon nanotubes estimated from femtosecond z-scan measurements. *Appl. Phys. Lett.* **91**, 251103 (2007).
196. Kamyshny, A., Ben-Moshe, M., Aviezer, S. & Magdassi, S. Ink-jet printing of metallic nanoparticles and microemulsions. *Macromol. Rapid Commun.* **26**, 281–288 (2005).
197. Kane, C. L. & Mele, E. J. Ratio problem in single carbon nanotube fluorescence spectroscopy. *Phys. Rev. Lett.* **90**, 207401 (2003).
198. Kang, Y. & Taton, T. A. Micelle-encapsulated carbon nanotubes: A route to nanotube composites. *J. Am. Chem. Soc.* **125**, 5650–5651 (2003).
199. Kartner, F. H., Brovelli, L. R., Kopf, D., Kamp, M., Calasso, I. & Keller, U. Control of solid-state laser dynamics by semiconductor devices. *Opt. Eng.* **34**, 2024–2036 (1995).
200. Kataura, H., Kumazawa, Y., Maniwa, Y., Ohtsuka, Y., Sen, R., Suzuki, S. & Achiba, Y. Diameter control of single-walled carbon nanotubes. *Carbon* **38**, 1691–1697 (2000).
201. Kataura, H., Kumazawa, Y., Maniwa, Y., Umezu, I., Suzuki, S., Ohtsuka, Y. & Achiba, Y. Optical properties of single-wall carbon nanotubes. *Synth. Met.* **103**, 2555–2558 (1999).
202. Kavan, L., Ho Yum, J. & Gratzel, M. Optically transparent cathode for dye-sensitized solar cells based on graphene nanoplatelets. *ACS Nano*, **5**, 165–172 (2011).
203. Kawase, T., Shimoda, T., Newsome, C., Sirringhaus, H. & Friend, R. H. Inkjet printing of polymer thin film transistors. *Thin Solid Films* **438–439**, 279–287 (2003).
204. Kawase, T., Sirringhaus, H., Friend, R. H. & Shimoda, T. Inkjet printed via-hole interconnections and resistors for all-polymer transistor circuits. *Adv. Mater.* **13**, 1601–1605 (2001).
205. Kazaoui, S., Minami, N., Nalini, B., Kim, Y. & Hara, K. Near-infrared photoconductive and photovoltaic devices using single-wall carbon nanotubes in conductive polymer films. *J. Appl. Phys.* **98**, 084314 (2005).
206. Kazaoui, S., Minami, N., Nalini, B., Kim, Y., Takada, N. & Hara, K. Near-infrared electroluminescent devices using single-wall carbon nanotubes thin flms. *Appl. Phys. Lett.* **87**, 211914 (2005).
207. Kazaoui, S., Minami, N., Yamawaki, H., Aoki, K., Kataura, H. & Achiba, Y. Pressure dependence of the optical absorption spectra of single-walled carbon nanotube films. *Phys. Rev. B* **62**, 1643 (2000).
208. Kelleher, E. J. R., Travers, J. C., Sun, Z., Ferrari, A. C., Golant, K. M., Popov, S. V. & Taylor, J. R. Bismuth fiber integrated laser mode-locked by carbon nanotubes. *Laser Phys. Lett.* **7**, 790–794 (2010).
209. Kelleher, E. J. R., Travers, J. C., Ippen, E. P., Sun, Z., Ferrari, A. C., Popov, S. V. & Taylor, J. R. Generation and direct measurement of giant chirp in a passively mode-locked laser. *Opt. Lett.* **34**, 3526–3528 (2009).

210. Kelleher, E. J. R., Travers, J. C., Sun, Z., Rozhin, A. G., Ferrari, A. C., Popov, S. V. & Taylor, J. R. Nanosecond-pulse fiber lasers mode-locked with nanotubes. *Appl. Phys. Lett.* **95**, 111108 (2009).
211. Keller, U. Recent developments in compact ultrafast lasers. *Nature* **424**, 831–838 (2003).
212. Keller, U. In *Ultrafast solid-state lasers,* Vol. 46 (Ed.: Wolf, E.), Elsevier, 2004.
213. Keller, U., Miller, D. A. B., Boyd, G. D., Chiu, T. H., Ferguson, J. F. & Asom, M. T. Solidstate low-loss intracavity saturable absorber for Nd:YLF lasers: An antiresonant semiconductor Fabry – Perot saturable absorber. *Opt. Lett.* **17**, 505–507 (1992).
214. Keller, U. & Tropper, A. C. Passively modelocked surface-emitting semiconductor lasers. *Phys. Rep.* **429**, 67–120 (2006).
215. Keller, U., Weingarten, K. J., Kartner, F. X., Kopf, D., Braun, B., Jung, I. D., Fluck, R., Hönninger, C., Matuschek, N. & Aus der Au, J. Semiconductor Saturable Absorber Mirrors (SESAM's) for Femtosecond to Nanosecond pulse generation in solid-state lasers. *IEEE J. Sel. Top. Quant. Electron.* **2**, 435–453 (1996).
216. Kennedy, J. W., Vardeny, Z. V., Collins, S., Baughman, R. H., Zhao, H. & Mazumdar, S. Electroabsorption spectroscopy of single walled nanotubes. *arXiv:cond-mat/0505071v1* (2005).
217. Kohn, W. Image of the Fermi surface in the vibration spectrum of a metal. *Phys. Rev. Lett* **2**, 393 (1959).
218. Khoo, I. C., Ding, J., Zhang, Y., Chen, K. & Diaz, A. Supra-nonlinear photorefractive response of single-walled carbon nanotube- and C60-doped nematic liquid crystal. *Appl. Phys. Lett.* **82**, 3587–3589 (2003).
219. Khudyakov, D. V., Lobach, A. S. & Nadtochenko, V. A. Passive mode locking in a Ti:sapphire laser using a single-walled carbon nanotube saturable absorber at a wavelength of 810 nm. *Opt. Lett.* **35**, 2675–2677 (2010).
220. Kieu, K. & Mansuripur, M. Femtosecond laser pulse generation with a fiber taper embedded in carbon nanotube/polymer composite. *Opt. Lett.* **32**, 2242–2244 (2007).
221. Kilbride, B. E., Coleman, J. N., Fraysse, J., Fournet, P., Cadek, M., Drury, A., Hutzler, S., Roth, S. & Blau, W. J. Experimental observation of scaling laws for alternating current and direct current conductivity in polymer-carbon nanotube composite thin films. *J. Appl. Phys.* **92**, 4024–4030 (2002).
222. Kim, D. S., Nepal, D. & Geckeler, K. E. Individualization of single-walled carbon nanotubes: Is the solvent important? *Small* **1**, 1117–1124 (2005).
223. Kim, K. S., Zhao, Y., Jang, H., Lee, S. Y., Kim, J. M., Kim, K. S., Ahn, J. H., Kim, P., Choi, J. Y. & Hong, B. H. Large-scale pattern growth of graphene films for stretchable transparent electrodes. *Nature* **457**, 706–710 (2009).
224. Kim, Y., Minami, N. & Kazaoui, S. Highly polarized absorption and photoluminescence of stretch-aligned single-wall carbon nanotubes dispersed in gelatin films. *Appl. Phys. Lett.* **86**, 073103 (2005).
225. Kim, Y., Minami, N., Zhu, W., Kazaoui, S., Azumi, R. & Matsumoto, M. Langmuir-Blodgett films of single-wall carbon nanotubes: Layer-by-layer deposition and in-plane orientation of tubes. *Jpn. J. Appl. Phys.* **42**, 7629–7634 (2003).
226. Kiowski, O., Lebedkin, S., Hennrich, F. & Kappes, M. M. Single-walled carbon nanotubes show stable emission and simple photoluminescence spectra with weak excitation sidebands at cryogenic temperatures. *Phys. Rev. B* **76**, 075422 (2007).
227. Kiowski, O., Lebedkin, S., Hennrich, F., Malik, S., Rösner, H., Arnold, K., Sürgers, C. & Kappes, M. M. Photoluminescence microscopy of carbon nanotubes grown by chemical vapor deposition: Influence of external dielectric screening on optical transition energies. *Phys. Rev. B* **75**, 075421 (2007).
228. Kirschner, M. W. & Schachman, H. K. Conformational changes in proteins as measured by difference sedimentation studies. II. Effect of stereospecific ligands on the catalytic subunit of aspartate transcarbamylase. *Biochemistry* **10**, 1919–1926 (1971).

229. Kis, A., Csanyi, G., Salvetat, J. P., Lee, T.-N., Couteau, E., Kulik, A. J., Benoit, W., Brugger, J. & Forro, L. Reinforcement of single-walled carbon nanotube bundles by intertube bridging. *Nat. Mater.* **3**, 153–157 (2004).
230. Kono, J., Ostojic, G. N., Zaric, S., Strano, M. S., Moore, V. C., Shaver, J., Hauge, R. H. & Smalley, R. E. Ultra-fast optical spectroscopy of micelle-suspended single-walled carbon nanotubes. *Appl. Phys. A* **78**, 1093–1098 (2004).
231. Koo, H. S., Chen, M., Pan, P. C., Chou, L. T., Wu, F. M., Chang, S. J. & Kawai, T. Fabrication and chromatic characteristics of the greenish LCD colour-filter layer with nano-particle ink using inkjet printing technique. *Displays* **27**, 124–129 (2006).
232. Koo, B.-K., Lee, D.-Y., Kim, H.-J., Lee, W.-J., Song, J.-S. & Kim, H.-J. Seasoning effect of dye-sensitized solar cells with different counter electrodes. *J. Electroceram.* **17**, 79–82 (2006).
233. Kordas, K., Mustonen, T., Toth, G., Jantunen, H., Lajunen, M., Soldano, C., Talapatra, S., Kar, S., Vajtai, R. & Ajayan, Pulickel M. Inkjet printing of electrically conductive patterns of carbon nanotubes. *Small* **2**, 1021–1025 (2006).
234. Kravets, V. G., Grigorenko, A. N., Nair, R. R., Blake, P., Anissimova, S., Novoselov, K. S. & Geim, A. K. Spectroscopic ellipsometry of graphene and an exciton-shifted van Hove peak in absorption. *Phys. Rev. B* **81**, 155413 (2010).
235. Krupke, R., Hennrich, F., von Lohneysen, H. & Kappes, M. M. Separation of metallic from semiconducting single-walled carbon nanotubes. *Science* **301**, 344–347 (2003).
236. Krupke, R., Hennrich, F., Hampe, O. & Kappes, M. M. Near-infrared absorbance of single-walled carbon nanotubes dispersed in dimethylformamide. *J. Phys. Chem. B* **107**, 5667–5669 (2003).
237. Kumar, S., Dang, T. D., Arnold, F. E., Bhattacharyya, A. R., Min, B. G., Zhang, X., Vaia, R. A., Park, C., Adams, W. W., Hauge, R. H., Smalley, R. E., Ramesh, R. & Willis, P. A. Synthesis, structure, and properties of PBO/SWNT composites. *Macromolecules* **35**, 9039–9043 (2002).
238. Kurti, J., Kresse, G. & Kuzmany, H. First-principles calculations of the radial breathing mode of single-wall carbon nanotubes. *Phys. Rev. B* **58**, R8869–R8872 (1998).
239. Kwon, Y.-K., Saito, S. & Tomanek, D. Effect of intertube coupling on the electronic structure of carbon nanotube ropes. *Phys. Rev. B* **58**, R13314–R13317 (1998).
240. Kymakis, E., Alexandrou, I. & Amaratunga, G. A. J. High open-circuit voltage photovoltaic devices from carbon-nanotube-polymer composites. *J. Appl. Phys.* **93**, 1764–1768 (2003).
241. Kymakis, E. & Amaratunga, G. A. J. Single-wall carbon nanotube/conjugated polymer photovoltaic devices. *Appl. Phys. Lett.* **80**, 112–114 (2002).
242. Labban, A. S. & Marcus, Y. Solvatochromic parameters of ethanolamines. *J. Chem. Soc. Faraday Trans.* **93**, 77–79 (1997).
243. Lagerwall, J. P. F., Scalia, G., Haluska, M., Dettlaff-Weglikowska, U., Giesselmann, F. & Roth, S. Simultaneous alignment and dispersion of carbon nanotubes with lyotropic liquid crystals. *Phys. Stat. Sol. (b)* **243**, 3046–3049 (2006).
244. Lagerwall, J. P. F., Scalia, G., Haluska, M., Dettlaff-Weglikowska, U., Roth, S. & Giesselmann, F. Nanotube alignment using lyotropic liquid crystals. *Adv. Mater.* **19**, 359–364 (2007).
245. Landi, B. J., Ruf, H. J., Worman, J. J. & Raffaelle, R. P. Effects of alkyl amide solvents on the dispersion of single-wall carbon nanotubes. *J. Phys. Chem. B* **108**, 17089–17095 (2004).
246. Lauret, J. S., Voisin, C., Cassabois, G., Delalande, C., Roussignol, P., Jost, O. & Capes, L. Ultrafast carrier dynamics in single-wall carbon nanotubes. *Phys. Rev. Lett.* **90**, 057404 (2003).
247. Lazzeri, M., Piscanec, S., Mauri, F., Ferrari, A. C. & Robertson, J. Phonon linewidths and electron-phonon coupling in graphite and nanotubes. *Phys. Rev. B* **73**, 155426 (2006).
248. Lebedkin, S., Hennrich, F., Kiowski, O. & Kappes, M. M. Photophysics of carbon nanotubes in organic polymer-toluene dispersions: Emission and excitation satellites and relaxation pathways. *Phys. Rev. B* **77**, 165429 (2008).

249. Lebedkin, S., Schweiss, P., Renker, B., Malik, S., Hennrich, F., Neumaier, M., Stoermer, C. & Kappes, M. M. Single-wall carbon nanotubes with diameters approaching 6 nm obtained by laser vaporization. *Carbon* **40**, 417–423 (2002).
250. Lee, J. Y., Connor, S. T., Cui, Y. & Peumans, P. Solution-processed metal nanowire mesh transparent electrodes. *Nano Lett.* **8**, 689–692 (2008).
251. Lee, W. & Chiu, C.-S. Observation of self-diffraction by gratings in nematic liquid crystals doped with carbon nanotubes. *Opt. Lett.* **26**, 521–523 (2001).
252. Lee, W., Gau, J.-S. & Chen, H.-Y. Electro-optical properties of planar nematic cells impregnated with carbon nanosolids. *Appl. Phys. B* **81**, 171–175 (2005).
253. Lee, W., Wang, C.-Y. & Shih, Y.-C. Effects of carbon nanosolids on the electro-optical properties of a twisted nematic liquid-crystal host. *Appl. Phys. Lett.* **85**, 513–515 (2004).
254. Lee, Y.-S., Buongiorno Nardelli, M. & Marzari, N. Band structure and quantum conductance of nanostructures from maximally localized wannier functions: The case of functionalized carbon nanotubes. *Phys. Rev. Lett.* **95**, 076804 (2005).
255. Lefebvre, J. & Finnie, P. Polarized photoluminescence excitation spectroscopy of single-walled carbon nanotubes. *Phys. Rev. Lett.* **98**, 167406 (2007).
256. Lefebvre, J., Fraser, J. M., Finnie, P. & Homma, Y. Photoluminescence from an individual single-walled carbon nanotube. *Phys. Rev. B* **69**, 075403 (2004).
257. Lefebvre, J., Fraser, J. M., Homma, Y. & Finnie, P. Photoluminescence from single walled carbon nanotubes: A comparison between suspended and micelle-encapsulated nanotubes. *Appl. Phys. A* **78**, 1107–1110 (2004).
258. Lefebvre, J., Homma, Y. & Finnie, P. Bright band gap photoluminescence from unprocessed single-walled carbon nanotubes. *Phys. Rev. Lett.* **90**, 217401 (2003).
259. Heller, D. A., Mayrhofer, R. M., Baik, S., Grinkova, Y. V., Usrey, M. L. & Strano, M. S. Concomitant Length and Diameter Separation of Single-Walled Carbon Nanotubes. *J. Am. Chem. Soc.* **126**, 14567–14573 (2004).
260. Li, C., Chen, Y., Wang, Y., I., Z., Chhowalla, M. & Mitra, S. A fullerene-single wall carbon nanotube complex for polymer bulk heterojunction photovoltaic cells. *J. Mater. Chem.* **17**, 2406–2411 (2007).
261. Li, F., Cheng, H. M., Bai, S., Su, G. & Dresselhaus, M. S. Tensile strength of single-walled carbon nanotubes directly measured from their macroscopic ropes. *Appl. Phys. Lett.* **77**, 3161–3163 (2000).
262. Li, Y.-L., Kinloch, I. A. & Windle, A. H. Direct spinning of carbon nanotube fibers from chemical vapor deposition synthesis. *Science* **304**, 276–278 (2004).
263. Li, Z. M., Tang, Z. K., Liu, H. J., Wang, N., Chan, C. T., Saito, R., Okada, S., Li, G. D., Chen, J. S., Nagasawa, N. & Tsuda, S. Polarized absorption spectra of single-walled 4 angstrom carbon nanotubes aligned in channels of an AlPO4-5 single crystal. *Phys. Rev. Lett.* **8712**, 127401 (2001).
264. Li, X. L., Zhang, G. Y., Bai, X. D., Sun, X. M., Wang, X. R., Wang, E. & Dai, H. J. Highly conducting graphene sheets and Langmuir-Blodgett films. *Nat. Nano* **3**, 538–542 (2008).
265. Lin, M. F. Optical spectra of single-wall carbon nanotube bundles. *Phys. Rev. B* **62**, 13153–13159 (2000).
266. Lin, M. F. & Chuu, D. S. π plasmons in carbon nanotube bundles. *Phys. Rev. B* **57**, 10183–10187 (1998).
267. Liu, P. Modifications of carbon nanotubes with polymers. *Eur. Polym. J.* **41**, 2693–2703 (2005).
268. Liu, X., Si, J., Chang, B., Xu, G., Yang, Q., Pan, Z., Xie, S., Ye, P., Fan, J. & Wan, M. Thirdorder optical nonlinearity of the carbon nanotubes. *Appl. Phys. Lett.* **74**, 164–166 (1999).
269. Liu, Z., Liu, Q., Huang, Y., Ma, Y., Yin, S., Zhang, X., Sun, W. & Chen, Y. Organic photovoltaic devices based on a novel acceptor material: Graphene. *Adv. Mater.* **20**, 3924–3930 (2008).
270. Lopez, M. J., Rubio, A., Alonso, J. A., Qin, L. C. & Iijima, S. Novel polygonized single-wall carbon nanotube bundles. *Phys. Rev. Lett.* **86**, 3056 (2001).

271. Lotya, M., Hernandez, Y., King, P. J., Smith, R. J., Nicolosi, V., Karlsson, L. S., Blighe, F. M., De, S., Wang, Z., McGovern, I. T., Duesberg, G. S. & Coleman, J. N. Liquid phase production of graphene by exfoliation of graphite in surfactant/water solutions. *J. Am. Chem. Soc.* **131**, 3611–3620 (2009).
272. Lu, J., Nagase, S., Zhang, X. W., Wang, D., Ni, M., Maeda, Y., Wakahara, T., Nakahodo, T., Tsuchiya, T., Akasaka, T., Gao, Z. X., Yu, D. P., Ye, H. Q., Mei, W. N. & Zhou, Y. S. Selective interaction of large or charge-transfer aromatic molecules with metallic single-wall carbon nanotubes: Critical role of the molecular size and orientation. *J. Am. Chem. Soc.* **128**, 5114–5118 (2006).
273. Lu, J. P. Elastic properties of carbon nanotubes and nanoropes. *Phys. Rev. Lett.* **79**, 1297 (1997).
274. Luo, Z., Zhou, M., Weng, J., Huang, G., Xu, H., Ye, C. & Cai, Z. Graphene-based passively Q-switched dual-wavelength erbium-doped fiber laser. *Opt. Lett.* **35**, 3709–3711 (2010).
275. Lyklema, J. The surface tension of pure liquids: Thermodynamic components and corresponding states. *Colloids Surf. A* **156**, 413–421 (1999).
276. Lynch, M. D. & Patrick, D. L. Organizing carbon nanotubes with liquid crystals. *Nano Lett.* **2**, 1197–1201 (2002).
277. Ma, H, Jen, A. K.-H. & Dalton, A. R. Polymer-based optical waveguides: Materials, processing and devices. *Adv. Mater.* **14**, 1339–1365 (2002).
278. Maeda, A., Matsumoto, S., Kishda, H., Takenobu, T., Iwasa, Y., Shraishi, M., Ata, M. & Okamoto, H. Large optical nonlinearity of semiconducting single-walled carbon nanotubes under resonant excitations. *Phys. Rev. Lett.* **94**, 047404 (2005).
279. Maeda, Y., Kimura, S.-i., Hirashima, Y., Kanda, M., Lian, Y., Wakahara, T., Akasaka, T., Hasegawa, T., Tokumoto, H., Shimizu, T., Kataura, H., Miyauchi, Y., Maruyama, S., Kobayashi, K. & Nagase, S. Dispersion of single-walled carbon nanotube bundles in nonaqueous solution. *J. Phys. Chem. B* **108**, 18395–18397 (2004).
280. Mahan, G. D. Oscillations of a thin hollow cylinder: Carbon nanotubes. *Phys. Rev. B* **65**, 235402 (2002).
281. Mamedov, A. A., Kotov, N. A., Prato, M., Guldi, D. M., Wicksted, J. P. & Hirsch, A. Molecular design of strong single-wall carbon nanotube/polyelectrolyte multilayer composites. *Nat. Mater.* **1**, 190–194 (2002).
282. Mann, D., Kato, Y. K., Kinkhabwala, A., Pop, E., Cao, J., Wang, X., Zhang, L., Wang, Q., Guo, J. & Dai, H. Electrically driven thermal light emission from individual single-walled carbon nanotubes *Nat. Nano.* **2**, 32–38 (2007).
283. Manzoni, C., Gambetta, A., Menna, E., Meneghetti, M., Lanzani, G. & Cerullo, G. Intersubband exciton relaxation dynamics in single-walled carbon nanotubes. *Phys. Rev. Lett.* **94**, 207401 (2005).
284. Marago, O. M., Bonaccorso, F., Saija, R., Privitera, G., Gucciardi, P. G., Iati, M. A., Calogero, G., Jones, P. H., Borghese, F., Denti, P., Nicolosi, V. & Ferrari, A. C. Brownian motion of graphene. *ACS Nano* **4**, 7515–7523 (2010).
285. Marcus, Y. The properties of organic liquids that are relevant to their use as solvating solvents. *Chem. Soc. Rev.* **22**, 409–416 (1993).
286. Margulis, V. A. Theoretical estimations of third-order optical nonlinearities for semiconductor carbon nanotubes. *J. Phys. Condens. Mat.* **11**, 3065 (1999).
287. Margulis, V. A. & Sizikova, T. A. Theoretical study of third-order nonlinear optical response of semiconductor carbon nanotubes. *Physica B* **245**, 173–189 (1998).
288. Martinez, A., Fuse, K., Xu, B. & Yamashita, S. Optical deposition of graphene and carbon nanotubes in a fiber ferrule for passive mode-locked lasing. *Opt. Express* **18**, 23054–23061 (2010).
289. Kozinsky, B. & Marzari, N. Static dielectric properties of carbon nanotubes from firstprinciples. *Phys. Rev. Lett.* **96**, 166801-4 (2006).
290. Matyba, P., Yamaguchi, H., Eda, G., Chhowalla, M., Edman, L. & Robinson, N. D. Graphene and mobile ions: The key to all-plastic, solution-processed light-emitting devices. *ACS Nano* **4**, 637–642 (2010).

291. Maultzsch, J., Reich, S., Schlecht, U. & Thomsen, C. High-energy phonon branches of an individual metallic carbon nanotube. *Phys. Rev. Lett.* **91**, 087402 (2003).
292. Maultzsch, J., Reich, S., Thomsen, C., Requardt, H. & Ordejón, P. Phonon dispersion of graphite. *Phys. Rev. Lett.* **92**, 075501 (2004).
293. Maultzsch, J., Pomraenke, R., Reich, S., Chang, E., Prezzi, D., Ruini, A., Molinari, E., Strano, M. S., Thomsen, C. & Lienau, C. Exciton binding energies in carbon nanotubes from two-photon photoluminescence. *Phys. Rev. B* **72**, 241402(R) (2005).
294. Maultzsch, J., Telg, H., Reich, S. & Thomsen, C. Radial breathing mode of single-walled carbon nanotubes: Optical transition energies and chiral-index assignment. *Phys. Rev. B* **72**, 205438 (2005).
295. Mazer, N. A., Carey, M. C., Kwasnick, R. F. & Benedeck, G. B. Quasielastic light scattering studies of aqueous biliary lipid systems. Size, shape, and thermodynamics of bile salt micelles. *Biochemistry* **18**, 3064–3075 (1979).
296. McDonald, T. J., Engtrakul, C., Jones, M., Rumbles, G. & Heben, M. J. Kinetics of PL quenching during single-walled carbon nanotube rebundling and diameter-dependent surfactant interactions. *J. Phys. Chem. B* **110**, 25339 (2006).
297. Meyer, J. C., Paillet, M., Michel, T., Moreac, A., Neumann, A., Duesberg, G. S., Roth, S. & Sauvajol, J. L. Raman modes of index-identified freestanding single-walled carbon nanotubes. *Phys. Rev. Lett.* **95**, 217401 (2005).
298. Milnera, M., Kürti, J., Hulman, M. & Kuzmany, H. Periodic resonance excitation and intertube interaction from quasicontinuous distributed helicities in single-wall carbon nanotubes. *Phys. Rev. Lett.* **84**, 1324 (2000).
299. Minami, N., Kim, Y., Miyashita, K., Kazaoui, S. & Nalini, B. Cellulose derivatives as excellent dispersants for single-wall carbon nanotubes as demonstrated by absorption and photoluminescence spectroscopy. *Appl. Phys. Lett.* **88**, 093123 (2006).
300. Misewich, J. A., Martel, R., Avouris, Ph., Tsang, J. C., Heinze, S. & Tersoff, J. Electrically induced optical emission from a carbon nanotube FET. *Science* **300**, 783–786 (2003).
301. Miyajima, K. Correlation between the hydrophobic nature of monosaccharides and cholates, and their hydrophobic indices. *J. Chem. Soc. Faraday Trans.* **84**, 2537–2544 (1988).
302. Miyauchi, Y., Chiashi, S., Murakami, Y., Hayashida, Y. & Maruyama, S. Fluorescence spectroscopy of single-walled carbon nanotubes synthesized from alcohol. *Chem. Phys. Lett.* **387**, 198–203 (2004).
303. Miyauchi, Y. & Maruyama, S. Identification of an excitonic phonon sideband by photoluminescence spectroscopy of single-walled carbon-13 nanotubes. *Phys. Rev. B* **74**, 035415 (2006).
304. Miyauchi, Y., Oba, M. & Maruyama, S. Cross-polarized optical absorption of single-walled nanotubes by polarized photoluminescence excitation spectroscopy. *Phys. Rev. B* **74**, 205440 (2006).
305. Miyauchi, Y., Saito, R., Sato, K., Ohno, Y., Iwasaki, S., Mizutani, T., Jiang, J. & Maruyama, S. Dependence of exciton transition energy of single-walled carbon nanotubes on surrounding dielectric materials. *Chem. Phys. Lett.* **442**, 394–399 (2007).
306. Mohiuddin, T. M. G., Lombardo, A., Nair, R. R., Bonetti, A., Savini, G., Jalil, R., Bonini, N., Basko, D. M., Galiotis, C., Marzari, N., Novoselov, K. S., Geim, A. K. & Ferrari, A. C. *Phys. Rev. B* **79**, 205433 (2009).
307. Moniruzzaman, M. & Winey, K. I. Polymer nanocomposites containing carbon nanotubes. *Macromolecules* **39**, 5194–5205 (2006).
308. Moore, V. C., Strano, M. S., Haroz, E. H., Hauge, R. H. & Smalley, R. E. Individually suspended single-walled carbon nanotubes in various surfactants. *Nano Lett.* **3**, 1379–1382 (2003).
309. Murakami, Y., Einarsson, E., Edamura, T. & Maruyama, S. Polarization dependence of the optical absorption of single-walled carbon nanotubes. *Phys. Rev. Lett.* **94**, 087402 (2005).
310. Nakazawa, M., Nakahara, S., Hirooka, T., Yoshida, M., Kaino, T. & Komatsu, K. Polymer saturable absorber materials in the 1.5 mm band using poly-methyl-methacrylate and

polystyrene with single-wall carbon nanotubes and their application to a femtosecond laser. *Opt. Lett.* **31**, 915–917 (2006).

311. Nair, R. R., Blake, P., Grigorenko, A. N., Novoselov, K. S., Booth, T. J., Stauber, T., Peres, N. M. R. & Geim, A. K. Fine structure constant defines visual transparency of graphene. *Science* **320**, 1308 (2008).
312. Nemanich, R. J. & Solin, S. A. First- and second-order Raman scattering from finite-size crystals of graphite. *Phys. Rev. B* **20**, 392 (1979).
313. Newell, J. O. & Schachman, H. K. Amino acid substitution which stabilize aspirate transcarbamoylase in the R state disrupt both homotropic and heterotropic effects. *Biophys. Chem.* **37**, 183–196 (1960).
314. Ng, Y. H., Lightcap, I. V., Goodwin, K., Matsumura, M. & Kamat, P. V. To what extent do graphene Scaffolds improve the photovoltaic and photocatalytic response of TiO2 nanostructured films? *J. Phys. Chem. Lett.* **1**, 2222–2227 (2010).
315. Nguyen, B. T., Gautrot, J. E., Nguyen, M. T. & Zhu, X. X. Nitrocellulose-stabilized silver nanoparticles as low conversion temperature precursors useful for inkjet printed electronics. *J. Mater. Chem.* **17**, 1725–1730 (2007).
316. Niyogi, S., Boukhalfa, S., Chikkannanavar, S. B., Mcdonald, T. J., Heben, M. J. & Doorn, S. K. Selective aggregation of single-walled carbon nanotubes via salt addition. *J. Am. Chem. Soc.* **129**, 1898 (2007).
317. Niyogi, S., Hamon, M. A., Hu, H., Zhao, B., Bhowmik, P., Sen, R., Itkis, M. E. & Haddon, R. C. Chemistry of single-walled carbon nanotubes. *Acc. Chem. Res.* **35**, 1105 (2002).
318. Noh, Y. Y., Zhao, N., Caironi, M. & Sirringhaus, H. Downscaling of self-aligned, all-printed polymer thin-film transistors. *Nat. Nanotech.* **2**, 784–789 (2007).
319. O'Connell, M. J., Bachilo, S. M., Huffman, C. B., Moore, V. C., Strano, M. S., Haroz, E. H., Rialon, K. L., Boul, P. J., Noon, W. H., Kittrell, C., Ma, J., Hauge, R. H., Weisman, R. B. & Smalley, R. E. Band gap fluorescence from individual single-walled carbon nanotubes. *Science* **297**, 593–596 (2002).
320. O'Connell, M. J., Boul, P., Ericson, L. M., Huffman, C., Wang, Y., Haroz, E., Kuper, C., Tour, J., Ausman, K. D. & Smalley, R. E. Reversible water-solubilization of single-walled carbon nanotubes by polymer wrapping. *Chem. Phys. Lett.* **342**, 265–271 (2001).
321. O'Connell, M. J., Sivaram, S. & Doorn, S. K. Near-infrared resonance Raman excitation profile studies of single-walled carbon nanotube intertube interactions: A direct comparison of bundled and individually dispersed HiPco nanotubes. *Phys. Rev. B* **69**, 235415 (2004).
322. O'Flaherty, S. A., Murphy, R., Hold, S. V., Cadek, M., Coleman, J. N. & Blau, W. J. Material investigation and optical limiting properties of carbon nanotube and nanoparticle dispersions. *J. Phys. Chem. B* **107**, 958–964 (2003).
323. Okazaki, T., Saito, T., Matsuura, K., Ohshima, S., Yumura, M., Oyama, Y., Saito, R. & Iijima, S. Photoluminescence and population analysis of single-walled carbon nanotubes produced by CVD and pulsed-laser vaporization methods. *Chem. Phys. Lett.* **420**, 286–290 (2006).
324. Okazaki, T., Saito, T., Matsuura, K., Ohshima, S., Yumura, M. & Iijima, S. Photoluminescence mapping of "As-Grown" single-walled carbon nanotubes: A comparison with Micelle-encapsulated nanotube solutions. *Nano Lett.* **5**, 2618–2623 (2005).
325. Okhotnikov, O., Grudinin, A. & Pessa, M. Ultra-fast fibre laser systems based on SESAM technology: New horizons and applications. *New J. Phys.* **6**, 177 (2004).
326. Okimoto, H., Takenobu, T., Yanagi, K., Miyata, Y., Shimotani, H., Kataura, H. & Iwasa, Y. Tunable carbon nanotube thin-film transistors produced exclusively via inkjet printing. *Adv. Mater.* **22**, 3981–3986 (2010).
327. Olek, M., Ostrander, J., Jurga, S., Möhwald, H., Kotov, N., Kempa, K. & Giersig, M. Layerby-layer assembled composites from multiwall carbon nanotubes with different morphologies. *Nano Lett.* **4**, 1889–1895 (2004).
328. O'Regan, B. & Gratzel, M. A low-cost, high-efficiency solar cell based on dye-sensitized colloidal TiO2 films. *Nature*, **353**, 737–740 (1991).

329. Ostojic, G. N., Zaric, S., Kono, J., Strano, M. S., Moore, V. C., Hauge, R. H. & Smalley, R. E. Interband recombination dynamics in resonantly excited single-walled carbon nanotubes. *Phys. Rev. Lett.* **92**, 117402 (2004).
330. Oyama, Y., Saito, R., Sato, K., Jiang, J., Samsonidze, G. G., Gruneis, A., Miyauchi, Y., Maruyama, S., Jorio, A., Dresselhaus, G. & Dresselhaus, M. S. Photoluminescence intensity of single-wall carbon nanotubes. *Carbon* **44**, 873–879 (2006).
331. Park, C., Ounaies, Z., Watson, K. A., Crooks, R. E., Smith, J., Lowther, S. E., Connell, J. W., Siochi, E. J., Harrison, J. S. & Clair, T. L. S. Dispersion of single wall carbon nanotubes by in situ polymerization under sonication. *Chem. Phys. Lett.* **364**, 303–308 (2002).
332. Park, S. & Ruoff, R. S. Chemical methods for the production of graphenes. *Nat. Nano* **4**, 217–224 (2009).
333. Pasquier, A.-Du, Unalan, H. E., Kanwal, A., Miller, S. & Chhowalla, M. Conducting and transparent single-wall carbon nanotube electrodes for polymer-fullerene solar cells. *Appl. Phys. Lett.* **87**, 203511-3 (2005).
334. Perebeinos, V. & Avouris, P. Exciton ionization, Franz-Keldysh, and Stark effects in carbon nanotubes. *Nano Lett.* **7**, 609–613 (2007).
335. Perebeinos, V., Tersoff, J. & Avouris, P. Scaling of excitons in carbon nanotubes. *Phys. Rev. Lett.* **92**, 257402 (2004).
336. Peumans, P., Uchida, S. & Forrest, S. R. Efficient bulk heterojunction photovoltaic cells using small-molecular-weight organic thin films. *Nature* **425**, 158–162 (2003).
337. Pickels, E. G. & Smadel, J. E. Ultracentrifugation studies on the elementary bodies of vaccine virues: I. General methods and determination of particle size *J. Exp. Med.* **68**, 583–606 (1938).
338. Pimenta, M. A., Marucci, A., Empedocles, S. A., Bawendi, M. G., Hanlon, E. B., Rao, A. M., Eklund, P. C., Smalley, R. E., Dresselhaus, G. & Dresselhaus, M. S. Raman modes of metallic carbon nanotubes. *Phys. Rev. B* **58**, 16016 (1998).
339. Pisana, S., Lazzeri, M., Casiraghi, C., Novoselov, K. S., Geim, A. K., Ferrari, A. C. & Mauri, F. Breakdown of the adiabatic Born-Oppenheimer approximation in graphene. *Nat. Mater.* **6**, 198 (2007).
340. Piscanec, S., Lazzeri, M., Mauri, F., Ferrari, A. C. & Robertson, J. Kohn anomalies and electron-phonon interactions in graphite. *Phys. Rev. Lett.* **93**, 185503 (2004).
341. Piscanec, S., Lazzeri, M., Robertson, J., Ferrari, A. C. & Mauri, F. Optical phonons in carbon nanotubes: Kohn anomalies, Peierls distortions, and dynamic effects. *Phys. Rev. B* **75**, 035427 (2007).
342. Plentz, F., Ribeiro, H. B., Jorio, A., Strano, M. S. & Pimenta, M. A. Direct experimental evidence of exciton-phonon bound states in carbon nanotubes. *Phys. Rev. Lett.* **95**, 247401 (2005).
343. Pocsik, I., Hundhausen, M., Koos, M. & Ley, L. Origin of the D peak in the Raman spectrum of microcrystalline graphite. *J. Non-cryst. Solids* **230**, 1083–1086 (1998).
344. Popa, D., Sun, Z., Torrisi, F., Hasan, T., Wang, F. & Ferrari, A. C. Sub 200 fs pulse generation from a graphene mode-locked fiber laser. *Appl. Phys. Lett.* **97**, 203106-3 (2010).
345. Popa, D., Sun, Z., Hasan, T., Torrisi, F., Wang, F. & Ferrari, A. C. Graphene Q-switched, tunable fiber laser. *Appl. Phys. Lett.* **98**, 073106-3 (2011).
346. Popov, V. N., Van Doren, V. E. & Balkanski, M. Lattice dynamics of single-walled carbon nanotubes. *Phys. Rev. B* **59**, 8355–8358 (1999).
347. Pradhan, B., Batabyal, S. K. & Pal, A. J. Functionalized carbon nanotubes in donor/acceptortype photovoltaic devices. *Appl. Phys. Lett.* **88**, 093106 (2006).
348. Priya, B. R. & Byrne, H. J. Investigation of sodium dodecyl benzene sulfonate assisted dispersion and debundling of single-wall carbon nanotubes. *J. Phys. Chem. C* **112**, 332–337 (2008).
349. Qu, L., Lin, Y., Hill, D. E., Zhou, B., Wang, W., Sun, X., Kitaygorodskiy, A., Suarez, M., Connell, J. W., Allard, L. F. & Sun, Y.-P. Polyimide-functionalized carbon nanotubes: Synthesis and dispersion in nanocomposite films. *Macromolecules* **37**, 6055–6060 (2004).

350. Rao, A. M., Richter, E., Bandow, S., Chase, B., Eklund, P. C., Williams, K. A., Fang, S., Subbaswamy, K. R., Menon, M., Thess, A., Smalley, R. E., Dresselhaus, G. & Dresselhaus, M. S. Diameter-selective Raman scattering from vibrational modes in carbon nanotubes. *Science* **275**, 187–191 (1997).
351. Reich, S., Dworzac, M., Hoffmann, A., Thomsen, C. & Strano, M. S. Excited-state carrier lifetime in single-walled carbon nanotubes. *Phys. Rev. B* **71**, 033402 (2005).
352. Reich, S., Thomsen, C. & Ordejón, P. Electronic band structure of isolated and bundled carbon nanotubes. *Phys. Rev. B* **65**, 155411 (2002).
353. Reich, S., Thomsen, C. & Robertson, J. Exciton resonances Quench the photoluminescence of zigzag carbon nanotubes. *Phys. Rev. Lett.* **95**, 077402-4 (2005).
354. Riddik, J. A., Bunger, W. B. & Sakano, T. K. *Organic solvents, physical properties and methods of purification.* Wiley, New York (1986).
355. Rinzler, A. G., Liu, J., Dai, H., Nikolaev, P., Huffman, C. B., Rodríguez-Macías, F. J., Boul, P. J., Lu, A. H., Heymann, D., Colbert, D. T., Lee, R. S., Fischer, J. E., Rao, A. M., Eklund, P. C. & Smalley, R. E. Large-scale purification of single-wall carbon nanotubes: Process, product, and characterization. *Appl. Phys. A* **67**, 29 (1998).
356. Roda, A., Hofmann, A. F. & Mysels, K. J. The influence of bile salt structure on selfassociation in aqueous solutions. *J. Bio. Chem.* **258**, 6362 (1983).
357. Rozhin, A. G., Sakakibara, Y., Kataura, H., Matsuzaki, S., Ishida, K., Achiba, Y. & Tokumoto, M. Anisotropic saturable absorption of single-wall carbon nanotubes aligned in polyvinyl alcohol. *Chem. Phys. Lett.* **405**, 288–293 (2005).
358. Rozhin, A. G., Sakakibara, Y., Namiki, S., Tokumoto, M. & Kataura, H. Sub-200-fs pulsed erbium-doped fiber laser using a carbon nanotube-polyvinylalcohol mode locker. *Appl. Phys. Lett.* **88**, 051118 (2006).
359. Rozhin, A. G., Sakakibara, Y., Tokumoto, M., Kataura, H. & Achiba, Y. Near-infrared nonlinear optical properties of single-wall carbon nanotubes embedded in polymer film. *Thin Solid Films* **464–465**, 368–372 (2004).
360. Rozhin, A. G., Scardaci, V., Wang, F., Hennrich, F., White, I. H., Milne, W. I. & Ferrari, A. C. Generation of ultra-fast laser pulses using nanotube mode-lockers. *Phys. Stat. Sol. (b)* **243**, 3551 (2006).
361. Sackett, D. L. & Lippoldt, R. E. Thermodynamics of reversible monomer-dimer association of tubulin. *Biochemistry* **30**, 3511–3517 (1991).
362. Sakakibara, Y., Kintaka, K., Rozhin, A. G., ltatani, T., Soe, W. M., ltatani, H., Tokumoto, M. & Kataura, H. Optically uniform carbon nanotube-polyimide nanocomposite: Application to 165 fs mode-locked fiber laser and waveguide, presented at *31st European Conference on Optical Communication*, Glasgow, Scotland (2005).
363. Sakakibara, Y., Rozhin, A. G., Kataura, H., Achiba, Y. & Tokumoto, H. Carbon nanotube-poly(vinylalcohol) nanocomposite film devices: Applications for femtosecond fiber laser mode lockers and optical amplifier noise suppressors. *Jpn. J. Appl. Phys.* **44**, 1621–1625 (2005).
364. Salvetat, J.-P., Briggs, G. A. D., Bonard, J.-M., Bacsa, R. i. R., Kulik, A. J., Stöckli, T., Burnham, N. A. & Forró, L. Elastic and shear moduli of single-walled carbon nanotube ropes. *Phys. Rev. Lett.* **82**, 944 (1999).
365. Sánchez-Portal, D., Artacho, E., Soler, J. M., Rubio, A. & Ordejón, P. Ab initio structural, elastic, and vibrational properties of carbon nanotubes. *Phys. Rev. B* **59**, 12678–12688 (1999).
366. Sariciftci, N. S., Smilowitz, L., Heeger, A. J. & Wudl, F. Photoinduced electron transfer from a conducting polymer to buckminsterfullerene. *Science* **258**, 1474–1476 (1992).
367. Sarukura, N., Ishida, Y., Yanagawa, T. & Nakano, H. All solid-state cw passively modelocked Ti:sapphire laser using a colored glass filter. *Appl. Phys. Lett.* **57**, 229–230 (1990).
368. Sasaki, Y., Igura, T., Miyassu, Y. I., Lee, S., Nagadome, S., Takiguchi, H. & Sugihara, G. The adsorption behavior of four bile salt species on graphite in water – Evaluation of effective hydrophobicity of bile acids. *Colloids Surf. B* **5**, 241–247 (1995).

369. Scalia, G., Lagerwall, J. P. F., Schymura, S., Haluska, M., Giesselmann, F. & Roth, S. Carbon nanotubes in liquid crystals as versatile functional materials. *Phys. Stat. Sol. (b)* **244**, 4212–4217 (2007).
370. Scardaci, V., Rozhin, A. G., Hennrich, F., Milne, W. I. & Ferrari, A. C. Carbon nanotube-polymer composites for photonic devices. *Physica E* **37**, 115 (2007).
371. Scardaci, V., Rozhin, A. G., Tan, P. H., Wang, F., White, I. H., Milne, W. I. & Ferrari, A. C. Carbon nanotubes for ultrafast photonics. *Phys. Stat. Sol. (b)* **244**, 4303–4307 (2007).
372. Scardaci, V., Sun, Z., Wang, F., Rozhin, A. G., Hasan, T., Hennrich, F., White, I. H., Milne, W. I. & Ferrari, A. C. Carbon nanotube-polycarbonate composites for ultrafast lasers. *Adv. Mater.* **20**, 4040 (2008).
373. Scardaci, V., Coull, R. & Coleman, J. N. Very thin transparent, conductive carbon nanotube films on flexible substrates *Appl. Phys. Lett.* **97**, 023114 (2010).
374. Schabel, M. C. & Martins, J. L. Energetics of interplanar binding in graphite. *Phys. Rev. B* **46**, 7185 (1992).
375. Schedin, F., Lidorikis, E., Lombardo, A., Kravets, V. G., Geim, A. K., Grigorenko, A. N., Novoselov K. S. & Ferrari A. C. Surface-enhanced Raman spectroscopy of graphene. *ACS Nano* **4**, 5617 (2010).
376. Schibli, T. R., Minoshima, K., Kataura, H., Itoga, E., Minami, N., Kazaoui, S., Miyashita, K., Tokumoto, M. & Sakakibara, Y. Ultrashort pulse-generation by saturable absorber mirrors based on polymer-embedded carbon nanotubes. *Opt. Express* **13**, 8025–8031 (2005).
377. Schmidt, A., Rivier, S., Cho, W. B., Yim, J. H., Choi, S. Y., Lee, S., Rotermund, F., Rytz, D., Steinmeyer, G., Petrov, V. & Griebner, U. Sub-100 fs single-walled carbon nanotube saturable absorber mode-locked Yb-laser operation near 1 μm. *Opt. Express* **17**, 20109–20116 (2009).
378. Senoo, Y., Nishizawa, N., Sakakibara, Y., Sumimura, K., Itoga, E., Kataura, H. & Itoh, K. Polarization-maintaining, high-energy, wavelength-tunable, Er-doped ultrashort pulse fiber laser using carbon-nanotube polyimide film. *Opt. Express* **17**, 20233–20241 (2009).
379. Set, S., Yaguchi, H., Jablonski, M., Tanaka, Y., Sakakibara, Y., Rozhin, A., Tokumoto, M., Kataura, H., Achiba, Y. & Kikuchi, K. A noise suppressing saturable absorber at 1550 nm based on carbon nanotube technology", presented at *Optical Fiber Communications Conference (OFC)*, Washington, DC (2003).
380. Set, S. Y., Yaguchi, H., Tanaka, Y. & Jablonski, M. Ultrafast fiber pulsed lasers incorporating carbon nanotubes. *IEEE J. Sel. Top. Quant. Electron.* **10**, 137–146 (2004).
381. Set, S. Y., Yaguchi, H., Tanaka, Y., Jablonski, M., Sakakibara, Y., Rozhin, A., Tokumoto, M., Kataura, H., Achiba, Y. & Kikuchi, K. Mode-locked fiber lasers based on a saturable absorber incorporating carbon nanotubes", presented at *Optical Fiber Communication Conference (OFC)*, Washington, DC (2003).
382. Shaffer, M. S. P. & Windle, A. H. Fabrication and characterization of carbon nanotube/poly(vinyl alcohol) composites. *Adv. Mater.* **11**, 937–941 (1999).
383. Sheik-Bahae, M., Said, A. A. & Van Stryland, E. W. High-sensitivity, single-beam n2 measurements. *Opt. Lett.* **14**, 955–957 (1989).
384. Shigeta, M., Komatsu, M. & Nakashima, N. Individual solubilization of single-walled carbon nanotubes using totally aromatic polyimide. *Chem. Phys. Lett.* **418**, 115–118 (2006).
385. Shimoda, H., Oh, S. J., Geng, H. Z., Walker, R. J., Zhang, X. B., McNeil, L. E. & Zhou, O. Self-assembly of carbon nanotubes. *Adv. Mater.* **14**, 899–901 (2002).
386. Shimoda, T., Morii, K., Seki, S. & Kiguchi, H. Inkjet printing of light-emitting polymer displays. *Mater. Res. Soc. Bull.* **28**, 821–827 (2003).
387. Shyu, F. L. & Lin, M. F. π plasmons in two-dimensional arrays of aligned carbon nanotubes. *Phys. Rev. B* **60**, 14434–14440 (1999).
388. Singh, M., Haverinen, H. M., Dhagat, P. & Jabbour, G. E. Inkjet printing–process and its applications. *Adv. Mater.* **22**, 673–685 (2010).
389. Sirringhaus, H., Kawase, T., Friend, R. H., Shimoda, T., Inbasekaran, M., Wu, W. & Woo, E. P. High-resolution inkjet printing of all-polymer transistor circuits. *Science* **290**, 2123–2126 (2000).

390. Smale, G. & Sasse, J. RNA isolation form cartilage using density gradient centrifugation in cesium trifluoroacetate: An RNA preparation technique effective in the presence of high proteoglycan content. *Anal. Biochem.* **203**, 352–356 (1992).
391. Small, D. M., Penkett, S. A. & Chapman, D. Studies on simple and mixed bile salt micelles by nuclear magnetic resonance spectroscopy. *Biochim. Biophys. Acta* **176**, 178–189 (1969).
392. Small, W. R. & Panhuis, M. In het Inkjet printing of transparent, electrically conducting single-walled carbon-nanotube composites. *Small* **3**, 1500–1503 (2007).
393. Smith, Peter W. Mode-locking of lasers. *Proc. IEEE* **58**, 1342–1357 (1970).
394. Solodyankin, M. A., Obraztsova, E. D., Lobach, A. S., Chernov, A. I., Tausenev, A. V., Konov, V. I. & Dianov, E. M. Mode-locked 1.93 μm thulium fiber laser with a carbon nanotube absorber. *Opt. Lett.* **33**, 1336–1338 (2008).
395. Song, Y.-W., Morimune, K., Set, S. Y. & Yamashita, S. Polarization insensitive all-fiber mode-lockers functioned by carbon nanotubes deposited onto tapered fibers. *Appl. Phys. Lett.* **90**, 021101 (2007).
396. Song, Y.-W. & Yamashita, S. Passively mode-locked lasers with 17.2-GHz fundamentalmode repetition rate pulsed by carbon nanotubes. *Opt. Lett.* **32**, 430–432 (2007).
397. Song, Y.-W., Yamashita, S., Einarsson, E. & Maruyama, S. All-fiber pulsed lasers passively mode locked by transferable vertically aligned carbon nanotube film. *Opt. Lett.* **32**, 1399–1401 (2007).
398. Song, Y.-W., Yamashita, S., Goh, C. S. & Set, S. Y. Carbon nanotube mode lockers with enhanced nonlinearity via evanescent field interaction in D-shaped fibers. *Opt. Lett.* **32**, 148–150 (2007).
399. Song, Y.-W., Yamashita, S. & Maruyama, S. Single-walled carbon nanotubes for high-energy optical pulse formation. *Appl. Phys. Lett.* **92**, 021115 (2008).
400. Song, Y.-W., Jang, S.-Y., Han, W.-S. & Bae, M.-K. Graphene mode-lockers for fiber lasers functioned with evanescent field interaction. *Appl. Phys. Lett.* **96**, 051122-3 (2010).
401. Song, Y. W., Set, S. Y., Yamashita, S., Goh, C. S. & Kotake, T. 1300-nm pulsed fiber lasers mode-locked by purified carbon nanotubes. *IEEE Photonics Technol. Lett.* **17**, 1623–1625 (2005).
402. Spataru, C. D., Ismail-Beigi, S., Benedict, L. X. & Louie, S. G. Excitonic effects and optical spectra of single-walled carbon nanotubes. *Phys. Rev. Lett.* **92**, 077402 (2004).
403. Star, A., Liu, Y., Grant, K., Ridvan, L., Stoddart, J. F., Steuerman, D. W., Diehl, M. R., Boukai, A. & Heath, J. R. Noncovalent side-wall functionalization of single-walled carbon nanotubes. *Macromolecules* **36**, 553–560 (2003).
404. Star, A. & Stoddart, J. F. Dispersion and solubilization of single-walled carbon nanotubes with a hyperbranched polymer. *Macromolecules* **35**, 7516–7520 (2002).
405. Star, A., Stoddart, J. F., Steuerman, D., Diehl, M., Boukai, A., Wong, E. W., Yang, X., Chung, S.-W., Choi, H. & Heath, J. R. Preparation and properties of polymer-wrapped single-walled carbon nanotubes. *Angew. Chem. Int. Ed.* **40**, 1721–1725 (2001).
406. Steuerman, D. W., Star, A., Narizzano, R., Choi, H., Ries, R. S., Nicolini, C., Stoddart, J. F. & Heath, J. R. Interactions between conjugated polymers and single-walled carbon nanotubes. *J. Phys. Chem. B* **106**, 3124–3130 (2002).
407. Stillemark, P., Boren, J., Andersson, M., Larsson, T., Rustaeus, S., Karlsson, K. A. & Olofsson, S. O. The assembly and secretion of apolipoprotein B-48-containing very low density lipoproteins in McA-RH7777 cells. *J. Biol. Chem.* **275**, 10506–10513 (2000).
408. Stutzmann, N., Friend, R. H. & Sirringhaus, H. Self-aligned, vertical-channel, polymer field-effect transistors. *Science* **299**, 1881–1884 (2003).
409. Sugihara, G., Shigematsu, D.-S., Nagadome, S., Lee, S., Sasaki, Y. & Igimi, H. Thermodynamic study on the langmuir adsorption of various bile salts including taurine and glycine conjugates onto graphite in water. *Langmuir* **16**, 1825–1833 (2000).
410. Sun, Y., Wu, Q. & Shi, G. Graphene based new energy materials. *Energy Environ. Sci.* doi:10.1039/C0EE00683A (2011).
411. Sun, Z., Hasan, T., Torrisi, F., Popa, D., Privitera, G., Wang, F., Bonaccorso, F., Basko, D. M. & Ferrari, A. C. Graphene mode-locked ultrafast laser. *ACS Nano* **4**, 803–810 (2010).

412. Sun, Z., Lin, X., Yu, H., Hasan, T., Torrisi, F., Zhang, L., Sun, L., Guo, L., Hou, W., Li, J. & Ferrari, A. C. High-power ultrafast solid-state laser using graphene based saturable absorber. In *The Conference on Lasers and Electro-Optics* (Baltimore, US), JWA79 (2010).
413. Sun, Z., Hasan, T., Wang, F., Rozhin, A., White, I. & Ferrari, A. Ultrafast stretched-pulse fiber laser mode-locked by carbon nanotubes. *Nano Res.* **3**, 404–411 (2010).
414. Sun, Z., Popa, D., Hasan, T., Torrisi, F., Wang, F., Kelleher, E. J. R., Travers, J. C., Nicolosi, V. & Ferrari, A. C. A Stable, wideband tunable, near transform-limited, graphene-mode-locked, ultrafast laser. *Nano Res.* **3**, 653–660 (2010).
415. Sun, Z., Rozhin, A. G., Wang, F., Scardaci, V., Milne, W. I., White, I. H., Hennrich & F. Ferrari, A. C. L-band ultrafast fiber laser mode locked by carbon nanotubes. *Appl. Phys. Lett.* **93**, 061114 (2008).
416. Sun, Z., Rozhin, A. G., Wang, F., Hasan, T., Popa, D., O'Neill, W. & Ferrari, A. C. A compact, high power, ultrafast laser mode-locked by carbon nanotubes. *Appl. Phys. Lett.* **95**, 253102 (2009).
417. Suzuki, K., Yamaguchi, M., Kumagai, M. & Yanagida, S. Application of carbon nanotubes to counter electrodes of dye-sensitized solar cells. *Chem. Lett.* **32**, 28–29 (2003).
418. Svedberg, T. & Pedersen, K. O. *The ultracentrifuge.* Clarendon Press, Oxford (1940).
419. Takenobu, T., Miura, N., Lu, S.-Y., Okimoto, H., Asano, T., Shiraishi, M. & Iwasa, Y. Ink-jet printing of carbon nanotube thin-film transistors on flexible plastic substrates. *Appl. Phys. Express* **2**, 025005 (2009).
420. Takahashi, T., Tsunoda, K., Yajima, H. & Ishii, T. Dispersion and purification of single-wall carbon nanotubes using carboxymethylcellulose. *Jpn. J. Appl. Phys.* **43**, 3636–3639 (2004).
421. Tan, P. H., Rozhin, A. G., Hasan, T., Hu, P., Scardaci, V., Milne, W. I. & Ferrari, A. C. Photoluminescence spectroscopy of carbon nanotube bundles: Evidence for exciton energy transfer. *Phys. Rev. Lett.* **99**, 137402 (2007).
422. Tan, P. H., Hasan, T., Bonaccorso, F., Scardaci, V., Rozhin, A. G., Milne, W. I. & Ferrari, A. C. Optical properties of nanotube bundles by photoluminescence excitation and absorption spectroscopy. *Physica E* **40**, 2352–2359 (2008).
423. Tan, Y. & Resasco, D. E. Dispersion of single-walled carbon nanotubes of narrow diameter distribution. *J. Phys. Chem. B* **109**, 14454–14460 (2005).
424. Tan, W. D., Su, C. Y., Knize, R. J., Xie, G. Q., Li, L. J. & Tang, D. Y. Mode locking of ceramic Nd:yttrium aluminum garnet with graphene as a saturable absorber. *Appl. Phys. Lett.* **96**, 031106 (2010).
425. Tang, Z., Zhuang, J. & Wang, X. Exfoliation of graphene from graphite and their self-assembly at the oil-water interface. *Langmuir* **26**, 9045–9049 (2010).
426. Tasis, D., Tagmatarchis, N., Georgakilas, V. & Prato, M. Soluble carbon nanotubes. *Chem. Eur. J.* **9**, 4000–4008 (2003).
427. Tatsuura, S., Furuki, M., Sato, Y., Iwasa, I., Tian, M. & Mitsu, H. Semiconductor carbon nanotubes as ultrafast switching materials for optical telecommunications. *Adv. Mater.* **15**, 534–537 (2003).
428. Tausenev, A. V., Obraztsova, E. D., Lobach, A. S., Chernov, A. I., Konov, V. I., Konyashchenko, A. V., Kryukov, P. G. & Dianov, E. M. Self-mode-Iocking in erbium-doped fibre lasers with saturable polymer film absorbers containing single-wall carbon nanotubes synthesised by the arc discharge method. *Quantum Electron.* **37**, 205–208 (2007).
429. Telg, H., Maultzsch, J., Reich, S., Hennrich, F. & Thomsen, C. Chirality distribution and transition energies of carbon nanotubes. *Phys. Rev. Lett.* **93**, 177401 (2004).
430. Tersoff, J. & Ruoff, R. S. Structural properties of a carbon-nanotube crystal. *Phys. Rev. Lett.* **73**, 676 (1994).
431. Thess, A., Lee, R., Nikolaev, P., Dai, H., Petit, P., Robert, J., Xu, C., Lee, Y. H., Kim, S. G., Rinzler, A. G., Colbert, D. T., Scuseria, G. E., Tománek, D., Fischer, J. E. & Smalley, R. E. Crystalline ropes of metallic carbon nanotubes. *Science* **273**, 483–487 (1996).
432. Thomsen, C. & Reich, S. Double resonant Raman scattering in graphite. *Phys. Rev. Lett.* **85**, 5214–5217 (2000).

433. Torrens, O. N., Milkie, D. E., Zheng, M. & Kikkawa, J. M. Photoluminescence from intertube carrier migration in single-walled carbon nanotube bundles. *Nano Lett.* **6**, 2864–2867 (2006).
434. Trancik, J. E., Barton, S. C. & Hone, J. Transparent and catalytic carbon nanotube films, *Nano Lett.* **8**, 982–987 (2008).
435. Trebino, R. *Frequency-resolved optical gating: The measurement of ultrashort laser pulses.* Kluwer Academic Publishers, Norwell, MA (2002).
436. Trickey, S. B., Müller-Plathe, F., Diercksen, G. H. F. & Boettger, J. C. Interplanar binding and lattice relaxation in a graphite dilayer. *Phys. Rev. B* **45**, 4460 (1992).
437. Trushkevych, O., Collings, N., Crossland, W. A., Wilkinson, T. D., Georgiou, A. & Milne, W. I. Projection of holograms from photorefractive optically addressed spatial light modulators. *J. Nonlinear Opt. Phys.* **16**, 307–316 (2007).
438. Trushkevych, O., Collings, N., Hasan, T., Scardaci, V., Ferrari, A. C., Wilkinson, T. D., Crossland, W. A., Milne, W. I., Geng, J., Johnson, B. F. G. & Macaulay, S. Characterization of carbon nanotube-thermotropic nematic liquid crystal composites. *J. Phys. D* **41**, 125106 (2008).
439. Tsyboulsky, D., Rocha, J. D. R., Bachilo, S. M., Cognet, L. & Weisman, R. B. Structuredependent fluorescence efficiencies of individual single-walled cardon nanotubes. *Nano Lett.* **7**, 3080–3085 (2007).
440. Tuinstra, F. & Koenig, J. L. Raman spectrum of graphite. *J. Chem. Phys.* **53**, 1126–1130 (1970).
441. Tu, X., Manohar, S., Jagota, A. & Zheng, M. DNA sequence motifs for structure-specific recognition and separation of carbon nanotubes. *Nature* **460**, 250–253 (2009).
442. Uchida, S., Martinez, A., Song, Y.-W., Ishigure, T. & Yamashita, S. Carbon nanotube-doped polymer optical fiber. *Opt. Lett.* **34**, 3077–3079 (2009).
443. Umeyama, T., Kadota, N., Tezuka, N., Matano, Y. & Imahori, H. Photoinduced energy transfer in composites of poly[(p-phenylene-1,2-vinylene)-co-(p-phenylene-1,1-vinylidene)] and single-walled carbon nanotubes. *Chem. Phys. Lett.* **444**, 263–267 (2007).
444. Van Stryland, E. W. & Sheik-Bahae, M. In *Z-scan* (Eds.: Kuzyk, M. G. and Dirk, C. W.), New York, Marcel Dekker Inc., 1998.
445. Yu, H., Chen, X., Zhang, H., Xu, X., Hu, X., Wang, Z., Wang, J., Zhuang, S. & Jiang, M. Large energy pulse generation modulated by Graphene epitaxially grown on silicon carbide. *ACS Nano* **4**, 7582–7586 (2010).
446. Vidano, R. P., Fischbach, D. B., Willis, L. J. & Loehr, T. M. Observation of Raman band shifting with excitation wavelength for carbons and graphites. *Solid State Commun.* **39**, 341–344 (1981).
447. Vieira, O. V., Laranjinha, J. A. N., Madeira, V. M. C. & Almeida, L. M. Rapid isolation of low density lipoproteins in a concentrated fraction free from water-soluble plasma antioxidants. *J. Lipid Res.* **37**, 2715–2721 (1996).
448. Vigolo, B., Penicaud, A., Coulon, C., Sauder, C., Pailler, R., Journet, C., Bernier, P. & Poulin, P. Macroscopic fibers and ribbons of oriented carbon nanotubes. *Science* **290**, 1331–1334 (2000).
449. Viswanathan, G., Chakrapani, N., Yang, H. C., Wei, B. Q., Chung, H. S., Cho, K. W., Ryu, C. Y. & Ajayan, P. M. Single-step in situ synthesis of polymer-grafted single-wall nanotube composites. *J. Am. Chem. Soc.* **125**, 9258–9259 (2003).
450. Vivien, L., Anglaret, E., Riehl, D., Hache, F., Bacou, F., Andrieux, M., Lafonta, F., Journet, C., Goze, C., Brunet, M. & Bernier, P. Optical limiting properties of singlewall carbon nanotubes. *Opt. Commun.* **174**, 271–275 (2000).
451. Wada, O. Femtosecond all-optical devices for ultrafast communication and signal processing. *New J. Phys.* **6**, 183 (2004).
452. Walsh, A. G., Vamivakas, A. N., Yin, Y., Ünlü, M. S., Goldberg, B. B., Swan, A. K. & Cronin, S. B. Screening of excitons in single, suspended carbon nanotubes. *Nano Lett.* **7**, 1485–1488 (2007).

453. Wang, F., Rozhin, A., Sun, Z., Scardaci, V., Penty, R., White, I. & Ferrari, A. C. Fabrication, characterization and mode locking application of single-walled carbon nanotube/polymer composite saturable absorbers. *Int. J. Mater. Form.* **1**, 107–112 (2008).
454. Wang, F., Rozhin, A. G., Scardaci, V., Sun, Z., Hennrich, F., White, I. H., Milne, W. I. & Ferrari, A. C. Wideband-tuneable, nanotube mode-locked, fibre laser. *Nat. Nanotechnol.* **3**, 738–742 (2008).
455. Wang, F., Rozhin, A. G., Sun, Z., Scardaci, V., White, I. H. & Ferrari, A. C. Soliton fiber laser mode-locked by a single-wall carbon nanotube-polymer composite. *Phys. Stat. Sol. (b)* **245**, 2319–2322 (2008).
456. Wang, F., Dukovic, G., Brus, L. E. & Heinz, T. F. The optical resonances in carbon nanotubes arise from excitons. *Science* **308**, 838–841 (2005).
457. Wang, F., Sfeir, M. Y., Huang, L., Huang, X. M. H., Wu, Y., Kim, J., Hone, J., O'Brien, S., Brus, L. E. & Heinz, T. F. Interactions between individual carbon nanotubes studied by Rayleigh scattering spectroscopy. *Phys. Rev. Lett.* **96**, 167401 (2006).
458. Wang, J. Z., Zheng, Z. H., Li, H. W., Huck, W. T. S. & Sirringhaus, H. Dewetting of conducting polymer inkjet droplets on patterned surfaces. *Nat Mater* **3**, 171–176 (2004).
459. Wang, X., Zhi, L. & Mullen, K. Transparent, conductive graphene electrodes for dye-sensitized solar cells. *Nano Lett.* **8**, 323–327 (2008).
460. Wang, X., Zhi, L., Tsao, N., Tomovic, Z., Li, J. & Müllen, K. Transparent carbon films as electrodes in organic solar cells. *Angew. Chem.* **47**, 2990 (2008).
461. Wei, L., Wang, B., Goh, T. H., Li, L. J., Yang, Y. H., Chan-Park, M. B. & Chen, Y. Selective enrichment of (6,5) and (8,3) single-walled carbon nanotubes via cosurfactant extraction from narrow (n,m) distribution samples. *J. Phys. Chem. B* **112**, 2771–2774 (2008).
462. Weisman, R. B. & Bachilo, S. M. Dependence of optical transition energies on structure for single-walled carbon nanotubes in aqueous suspension: An empirical Kataura plot. *Nano Lett.* **3**, 1235 (2003).
463. Weisman, R. B., Bachilo, S. M. & Tsyboulsky, D. Fluorescence spectroscopy of singlewalled carbon nanotubes in aqueous suspension. *Appl. Phys. A* **78**, 1111–1116 (2004).
464. Weiss, V., Thiruvengadathan, R. & Regev, O. Preparation and characterization of a carbon nanotube-lyotropic liquid crystal composite. *Langmuir* **22**, 854–856 (2006).
465. Wenseleers, W., Vlasov, I. I., Goovaerts, E., Obraztsova, E. D., Lobach, A. S. & Bouwen, A. Efficient isolation and solubilization of pristine single-walled nanotubes in bile salt micelles. *Adv. Funct. Mater.* **14**, 1105–1112 (2004).
466. White, B., Banerjee, S., O'Brien, S., Turro, N. J. & Herman, I. P. Zeta-potential measurements of surfactant-wrapped individual single-walled carbon nanotubes. *J. Phys. Chem. C* **111**, 13684–13690 (2007).
467. Williams, J. W., Van Holde, K. E., Baldwin, R. L. & Fujita, H. The theory of sedimentation analysis. *Chem Rev.* **58**, 715–806 (1958).
468. Wu, J., Walukiewicz, W., Shan, W., Bourret-Courchesne, E., Ager III, J. W., Yu, K. M., Haller, E. E., Kissell, K., Bachilo, S. M., Weisman, R. B. & Smalley, R. E. Structure-dependent hydrostatic deformation potentials of individual single-walled carbon nanotubes. *Phys. Rev. Lett.* **93**, 017404 (2004).
469. Wu, J., Becerril, H. A., Bao, Z., Liu, Z., Chen, Y. & Peumans, P. Organic solar cells with solution-processed graphene transparent electrodes. *Appl. Phys. Lett.* **92**, 263302-3 (2008).
470. Wu, J., Agrawal, M., Becerril, H. C. A., Bao, Z., Liu, Z., Chen, Y. & Peumans, P. Organic light-emitting diodes on solution-processed graphene transparent electrodes. *ACS Nano* **4**, 43–48 (2009).
471. Wu, Y., Li, Y., Ong, B. S., Liu, P., Gardner, S. & Chiang, B. High-performance organic thin-film transistors with solution-printed gold contacts. *Adv. Mater.* **17**, 184–187 (2005).
472. Wu, Z., Chen, Z., Du, X., Logan, J. M., Sippel, J., Nikolou, M., Kamaras, K., Reynolds, J. R., Tanner, D. B., Hebard, A. F. & Rinzler, A. G. Transparent, conductive carbon nanotube films. *Science* **305**, 1273–1276 (2004).
473. Xia, F., Mueller, T., Lin, Y.-M., Valdes-Garcia, A. & Avouris, P. Ultrafast graphene photodetector. *Nat. Nano* **4**, 839–843 (2009).

474. Xie, H., Ortiz-Acevedo, A., Zorbas, V., Baughman, R. H., Draper, R. K., Musselman, I. H., Dalton, A. B. & Dieckmann, G. R. Peptide cross-linking modulated stability and assembly of peptide-wrapped single-walled carbon nanotubes. *J. Mater. Chem.* **15**, 1734–1741 (2005).
475. Xie, Xiao-Lin, Mai, Yiu-Wing & Zhou, Xing-Ping. Dispersion and alignment of carbon nanotubes in polymer matrix: A review. *Mater. Sci. Eng. R* **49**, 89–112 (2005).
476. Xu, Z., Wu, Y., Hu, B., Ivanov, I. N. & Geohegan, D. B. Carbon nanotube effects on electroluminescence and photovoltaic response in conjugated polymers. *Appl. Phys. Lett.* **87**, 263118 (2005).
477. Yamashita, S., Inoue, Y., Maruyama, S., Murakami, Y., Yaguchi, H., Jablonski, M. & Set, S. Y. Saturable absorbers incorporating carbon nanotubes directly synthesized onto substrates and fibers and their application to mode-locked fiber lasers. *Opt. Lett.* **29**, 1581–1583 (2004).
478. Yamashita, S., Inoue, Y., Yaguchi, H., Jablonski, M. & Set, S. Y. S-,C-,L-Band picosecond fiber pulse sources using a broadband carbon-nanotube-based mode-locker. *Euro. Conf. on Opt. Comm.* (ECOC), (Stockholm, 2004). pp. No.Th.1.3.4.
479. Yamashita, S., Set, S. Y., Goh, C. S. & Kikuchi, K. Ultrafast saturable absorbers based on carbon nanotubes and their applications to passively mode-locked fiber lasers. *Electron. Comm. Jpn.* **90**, 17–24 (2007).
480. Yan, J., Zhang, Y. B., Kim, P. & Pinczuk, A. *Phys. Rev. Lett.* **98**, 166802 (2007).
481. Yan, X., Cui, X., Li, B. & Li, L.-S. Large, solution-processable graphene quantum dots as light absorbers for photovoltaics. *Nano Lett.*, **10**, 1869–1873 (2010).
482. Yanagi, K., Miyata, Y. & Kataura, H. Optical and conductive characteristics of metallic single-wall carbon nanotubes with three basic colors; Cyan, magenta, and yellow. *Appl. Phys. Express* **1**, 034003 (2008).
483. Yang, N., Zhai, J., Wang, D., Chen, Y. & Jiang, L. Two-dimensional graphene bridges enhanced photoinduced charge transport in dye-sensitized solar cells. *ACS Nano* **4**, 887–894 (2010).
484. Yaws, C. L. *Thermophysical properties of chemicals and hydrocarbons.* William Andrew Inc. Norwich, NY (2008).
485. Yong, V. & Tour, J. M. Theoretical efficiency of nanostructured graphene-based photovoltaics. *Small* **6**, 313–318 (2009).
486. Yu, A., Hu, H., Bekyarova, E., Itkis, M. E., Gao, J., Zhao, B. & Haddon, R. C. Incorporation of highly dispersed single-walled carbon nanotubes in a polyimide matrix. *Compos. Sci. Tech.* **66**, 1187 (2006).
487. Yu, P. & Cardona, M. *Fundamentals of semiconductors.* Springer, Berlin (2005).
488. Zacharia, R., Ulbricht, H. & Hertel, T. Interlayer cohesive energy of graphite from thermal desorption of polyaromatic hydrocarbons. *Phys. Rev. B* **69**, 155406 (2004).
489. Zhang, J., Kang, D. E., Xia, W., Okochi, M., Mori, H., Selkoe, D. J. & Koo, E. H. Subcellular distribution and turnover of presenilins in transfected cells. *J. Biolog. Chem.* **273**, 12436–12442 (1998).
490. Zhang, D., Ryu, K., Liu, X., Polikarpov, E., Ly, J., Tompson, M. E. & Zhou, C. Transparent, conductive, and flexible carbon nanotube films and their application in organic light-emitting diodes. *Nano Lett.* **6**, 1880–1886 (2006).
491. Zhang, H., Tang, D. Y., Zhao, L. M., Bao, Q. L. & Loh, K. P. Large energy mode locking of an erbium-doped fiber laser with atomic layer graphene. *Opt. Express* **17**, 17630–17635 (2009).
492. Zhao, H. & Mazumdar, S. Electron-electron interaction effects on the optical excitations of semiconducting single-walled carbon nanotubes. *Phys. Rev. Lett.* **93**, 157402 (2004).
493. Zhao, H. & Mazumdar, S. Elucidation of the electronic structure of semiconducting single-walled carbon nanotubes by electroabsorption spectroscopy. *Phys. Rev. Lett.* **98**, 166805 (2007).
494. Zhao, N., Chiesa, M., Sirringhaus, H., Li, Y., Wu, Y. & Ong, B. Self-aligned inkjet printing of highly conducting gold electrodes with submicron resolution. *J. Appl. Phys.* **101**, 064513-6 (2007).

495. Zheng, M., Jagota, A., Semke, E. D., Diner, B. A., Mclean, R. S., Lustig, S. R., Richardson, R. E. & Tassi, N. G. DNA-assisted dispersion and separation of carbon nanotubes. *Nat. Mater.* **2**, 338–342 (2003).
496. Zheng, M., Jagota, A., Strano, M. S., Santos, A. P., Barone, P., Chou, S. G., Diner, B. A., Dresselhaus, M. S., McLean, R. S., Onoa, G. B., Samsonidze, G. G., Semke, E. D., Usrey, M. & Walls, D. J. Structure-based carbon nanotube sorting by sequence-dependent DNA assembly. *Science* **302**, 1545–1548 (2003).
497. Zhou, Y., Hu, L. & Gruner, G. A method of printing carbon nanotube thin films. *Appl. Phys. Lett.* **88**, 123109 (2006).
498. Zorbas, V., Ortiz-Acevedo, A., Dalton, A. B., Yoshida, M. M., Dieckmann, G. R., Draper, R. K., Baughman, R. H., Jose-Yacama, M. & Musselman, I. H. Preparation and characterization of individual peptide-wrapped single-walled carbon nanotubes. *J. Am. Chem. Soc.* **126**, 7222–7227 (2004).

Chapter 10
Electronic Transport in Carbon Nanotube Field-Effect Transistors

J. Knoch and J. Appenzeller

Abstract In the present chapter we will discuss the electronic transport properties of carbon nanotube field-effect transistors (CNFETs). Three different device concepts will be studied in more detail: Schottky-barrier CNFETs with metallic source and drain contacts, conventional-type CNFETs with doped nanotube segments as source and drain electrodes and finally a new concept, the tunneling CNFET. As it turns out, the main factors determining the electrical behavior of CNFETs are the geometry, the one-dimensionality of the electronic transport and the way of making contacts to the nanotube. Analytical as well as simulation results will be given and compared with each other and with experimental data in order to explain the different influences on the electronic transport in CNFETs and thus on the device behavior.

10.1 Introduction

The enormous evolution of information technology has been made possible to a large extend by modern CMOS technology. In particular, the continued downscaling of the metal-oxide-semiconductor field-effect transistors (MOSFET) has led to tremendous performance improvements over the past three decades. The gain of performance when scaling down the transistor dimensions has two origins: (i) an increase of the number of MOSFET devices per chip and (ii) an improvement of the electrical performance of each individual device. The very successful route of scaling, however, runs into severe problems in the very near future. This is the reason why researchers have been looking intensively at alternatives to the conventional bulk silicon MOSFET by exploring new materials as well as new device architectures as replacements and/or add-ons to current CMOS technology.

In recent years, carbon nanotubes have attracted a great deal of attention as building blocks of future nanoelectronics circuits. As will be discussed in detail below, the main reasons for this are the inherent small dimensions of carbon nanotubes

J. Knoch (✉)
Faculty of Electrical Engineering and Information Technology, RWTH Aachen University, Templergraben 55, 52056 Aachen, Germany
e-mail: knoch@iht.rwth-aachen.de

O. Hayden, K. Nielsch (eds.), *Molecular- and Nano-Tubes*,
DOI 10.1007/978-1-4419-9443-1_10, © Springer Science+Business Media, LLC 2011

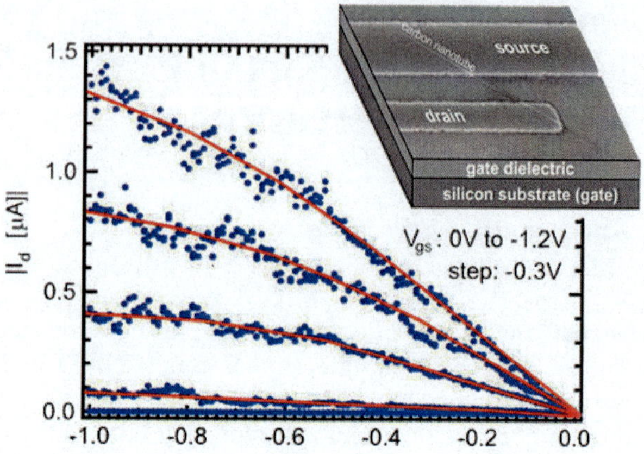

Fig. 10.1 Typical output characteristics of a p-type CNFET; the lines are guides to the eye. The *inset* shows an electron micrograph of a CNFET with metallic contacts and large area back-gate

yielding optimal electrostatics gate control in ultimately scaled FET devices as well as a large carrier mobility even at room temperature [1–3].

The inset of Fig. 10.1 shows an SEM of a typical research device with metallic source/drain contacts and a large area back-gate. Representative output characteristics of a CNFET are shown in the main panel of Fig. 10.1. The device exhibits regular transistor characteristics with a linear increase of current for small and current saturation for large drain-source bias V_{ds} showing that carbon nanotubes are a suitable class of material for realizing field-effect transistors. Substantial progress has been made recently in terms of understanding the transport properties as well as the role of the contacts on the behavior of nanotube-based FET devices [4–10]. Furthermore, researchers were able to fabricate and study novel device concepts including band-to-band tunnel FETs [11–13]. Significant progress has also been made regarding the growth of nanotubes in general, at pre-defined locations and in ordered arrays [14–17], as well as regarding the placement and alignment of nanotubes [18–21] and the integration into electronic circuits [22, 23]. Recently, a ring-oscillator fabricated on a single nanotube has been demonstrated for the first time that exhibits a strongly improved frequency response compared to previous realizations [24]. Figure 10.2 shows an SEM of such a ring-oscillator which allows studying the high-frequency capability of carbon nanotube devices representing a major step forward towards the realization of integrated circuits.

Despite this very impressive progress it should be noted that a future nanoelectronics based on nanotubes would require fabricating billions of carbon nanotube transistors at certain positions with each tube having the same electronic properties on a small chip. Present technology is not yet at this stage. Independent of all technology related aspects it is fair to say however, that carbon nanotubes are an ideal work-horse to study and explore possible advantages of one-dimensional systems as active channel material for ultimately scaled transistor devices.

The aim of the present chapter is to illuminate the specific advantages of carbon nanotubes for nanoelectronics applications which were already mentioned above:

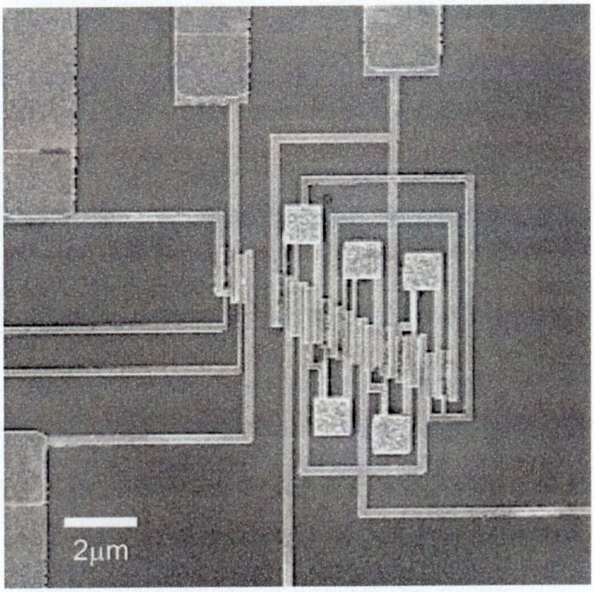

Fig. 10.2 SEM image of a ring oscillator fabricated on a single carbon nanotube [24]

the geometrical smallness enabling ultimately scaled FET devices, unique electronic transport properties and one-dimensional transport. After exploring these aspects in greater detail we will discuss specific examples that illustrate how the unique properties of nanotubes play an important role and enable a level of device optimization that is difficult if not impossible to obtain using other materials.

10.2 MOSFETs, Scaling and Short Channel Effects

Figure 10.3 shows a schematics of an n-type MOSFET consisting of two n-doped regions called source and drain in a p-doped host substrate. A gate electrode of length L and width W resides on top of the substrate surface insulated from it by a gate dielectric of thickness d_{ox}. In operation a drain-source voltage V_{ds} is applied and a drain current I_d is mediated by applying a gate-source voltage V_{gs} at the same time. The electrical behavior of a MOSFET can be understood to a large extend by looking at the potential landscape inside the MOSFET at the substrate-gate dielectric interface as shown in Fig. 10.3 (light gray line). The two back-to-back n-p junctions at the source-channel and channel-drain interfaces lead to a potential variation in-between the source-drain electrodes with a potential barrier maximum denoted Φ_f^0. It is this potential barrier that determines the injection of carriers from the source and drain contacts into the channel giving rise to a certain charge density in and current through the channel.

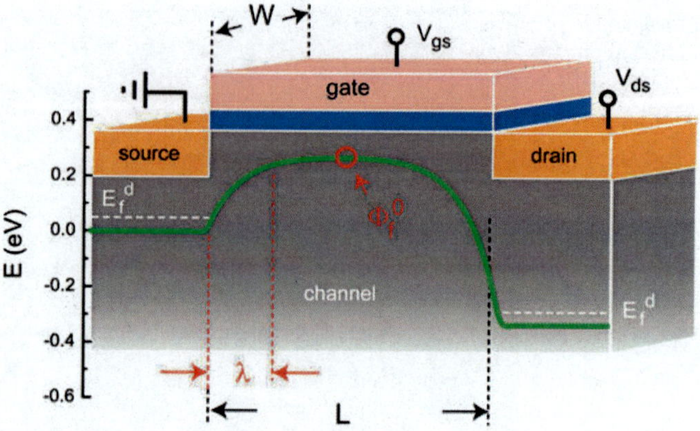

Fig. 10.3 Schematics of a conventional n-type bulk silicon MOSFET with channel length L and width W. The *light gray line* illustrates the conduction band profile within the device. The p-n-junctions at the source-channel and channel-drain interfaces have a spatial extend of λ

For low gate voltages, the potential barrier (in the channel) is rather high and only carriers in the exponential tail of the source and drain Fermi distribution functions with energies larger than the potential maximum Φ_f^0 are injected into the channel yielding an off-state current $I_d^{off} \approx exp\left(-\Phi_f^0/k_BT\right)$ [25]. Changing the gate voltage yields a change of Φ_f^0 and consequently the off-state current increases exponentially since an increasing fraction of the source Fermi distribution function contributes to the current; Fig. 10.4 shows this scenario schematically. Plotting the drain current I_d as a function of V_{gs} on a logarithmic scale (see Fig. 10.4, right panel) yields an increasing current in the device's off-state characterized by the so-called inverse subthreshold slope S given by $S = ln(10)\left(\partial I_d/\partial V_{gs} \cdot 1/I_d\right)^{-1} = ln(10)\left(\partial I/\partial \Phi_f^0 \cdot \partial \Phi_f^0/\partial \Phi_g \cdot (-q)/I_d\right)^{-1}$. In the ideal case of perfect gate control the change of gate voltage leads to the same change of Φ_f^0, i.e. $\partial \Phi_f^0/\partial \Phi_g = 1$ and therefore $S = k_BT/q \cdot ln(10) \approx 60\,\mathrm{mV/dec}$ at room temperature. This means that for a change of current by one order of magnitude at least 60 mV gate voltage change is required. It is important that S should be as steep as possible in order to reach low off-state currents within a certain, fixed voltage range. The thermal broadening of the source Fermi distribution function ultimately is responsible for the minimum inverse subthreshold slope of 60 mV/dec which is true for any conventional FET independent of dimension or material in use.

For gate voltages larger than the threshold voltage V_{th} (see Fig. 10.4), Φ_f^0 is moved close to the Fermi level and the transistor reaches its on-state. The electrical behavior and in particular the dependence of drain current on V_{ds} can again be understood by looking at the conduction band shown in Fig. 5a. The total drain current I_d is essentially the difference between the current flow from source to drain and vice versa which is determined by the difference of the source and drain Fermi distribution functions. Thus, for low V_{ds} (black curve in Fig. 10.5a) this difference

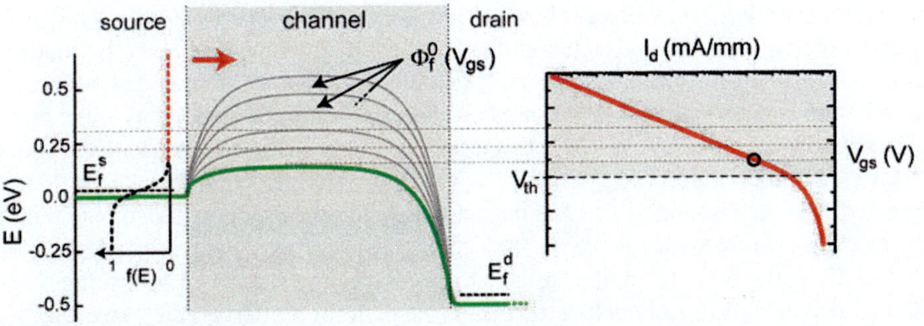

Fig. 10.4 *Left*: Conduction band profile along current transport for various gate voltages in a conventional FET. *Right*: Transfer characteristic showing an exponential increase of current in the device's off-state

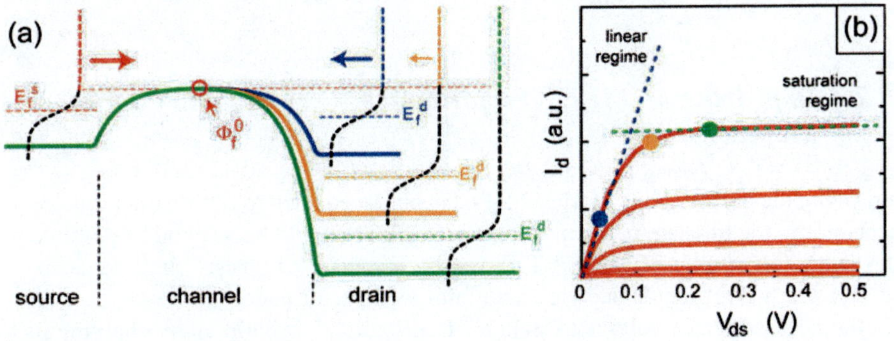

Fig. 10.5 (**a**) Conduction band along current transport direction in the on-state of a MOSFET for three different bias voltages. (**b**) Displays the output characteristics indicating the current that belongs to the respective band profiles

yields a linear relationship $I_d \propto V_{ds}$. However, with increasing bias, the drain Fermi level is moved away from the source Fermi level and eventually I_d consist solely of carriers with energies larger than Φ_f^0 injected from source. As a result, the drain current saturates and output characteristics as displayed in Fig. 10.5b are obtained. It is important to note, that saturation is only achieved if Φ_f^0 is not affected by V_{ds} which is the case in a long-channel MOSFET where the source-channel and channel-drain p-n-junctions are well separated from each other (see e.g. Fig. 10.3).

The saturation current can be calculated using the so-called gradual channel approximation leading to [26]

$$I_d^{sat} \approx W C_{ox} \frac{\mu_{eff}}{L} \frac{\left(V_{gs} - V_{th}\right)^2}{2} \tag{10.1}$$

where $C_{ox} = \varepsilon_0 \varepsilon_{ox} / d_{ox}$ is the geometrical gate oxide capacitance per area, μ_{eff} is the effective carrier mobility and W and L are the width and channel length of the device (see Fig. 10.3). Note, however, that expression (10.1) is only valid in long channel devices. In transistors with a short channel length, high electric fields occur that lead to a saturation of carrier velocity v_{sat} due to increased scattering mainly with optical phonons. As a result, the saturation current deviates from the $1/L$ dependence stated above [27]. In the limit of very short channel lengths, the velocity-saturation current $I_d^{sat} \approx WC_{ox}v_{sat}(V_{gs} - V_{th})$ is substantially smaller than expected from expression (10.1) and does not depend on L anymore. Also note that I_d^{sat} exhibits a linear dependence on $V_{gs} - V_{th}$ consistent with experimental observations. If, on the other hand, the ballistic transport regime is reached by scaling the channel length significantly below the mean free path for scattering, the current I_d^{sat} also becomes independent of the channel length L. However, in the ballistic case the saturation current is significantly larger than the current given by Eq. (10.1) as well as the velocity-saturation current. In this case, current saturation occurs due to an exhaustion of carriers injected from source over the maximum potential barrier Φ_f^0 as schematically shown in Fig. 10.5.

10.2.1 Improving MOSFET Performance – Scaling and Carrier Mobility

Improving the performance of MOSFETs can be traced back to two distinct requirements. First, the on-state performance of MOSFETs should be as good as possible. The saturation current of MOSFETs given by Eq. (10.1) suggests, that increasing e.g. the width W of the device yields an improved performance. However, in a circuit the necessary gate voltage to switch a transistor "2" is build up by charging its gate capacitance with the drain current of a device "1". Therefore, since the drain current of device "1" and the gate capacitance of "2" both scale with W increasing the width does not improve the switching speed of MOSFETs. An appropriate figure of merit to measure the on-state performance is the device delay time $\tau = C_g V_{dd}/I_d$ where V_{dd} is the supply voltage and $C_g \approx W \cdot L \cdot C_{ox}$ in a conventional bulk-MOSFET [27]. τ is a measure of how much charge on the gate is needed in order to realize a certain on-state current and therefore provides a figure of merit that is independent of geometrical factors such as the gate width and also the gate oxide thickness. Inserting the saturation current of a long channel MOSFET given by Eq. (10.1) or the velocity-saturation current into the expression for τ yields

$$\tau_{long\ channel} \propto \frac{L^2}{\mu_{eff}} \cdot \frac{V_{dd}}{(V_{dd} - V_{th})^2}, \quad \tau_{velocity\ saturation} \propto \frac{L}{v_{sat}} \cdot \frac{V_{dd}}{(V_{dd} - V_{th})} \tag{10.2}$$

respectively, where V_{dd} is the supply voltage. Increasing V_{dd} and/or the difference $V_{dd} - V_{th}$ reduces τ. However, the second performance requirement states that the total power consumption given approximately by

$$P \approx AC_{tot}(V_{dd})^2 f + V_{dd} I_{leak} \tag{10.3}$$

should be as small as possible. Here, the first term is the dynamic part of the power consumption due to continues charging and discharging of a total load capacitance C_{tot} with frequency f and the second term is the static power consumption due to leakage currents. Obviously, Eq. (10.3) suggests reducing the supply voltage due to the quadratic dependence on V_{dd} which, however, will degrade the on-state performance. Moreover, increasing $V_{dd} - V_{th}$ by reducing V_{th} leads to exponentially increasing leakage currents in the device's off-state thereby drastically increasing the static part of the power consumption. The reason for this is the limitation of any FET to an inverse subthreshold slope of 60 mV/dec that requires a minimal gate voltage range for a proper switching behavior. As a result, there is a lower limit for V_{dd} and $V_{dd} - V_{th}$ scaling and therefore, a further performance gain results from a down-scaling of the channel length L and an improvement of the carrier transport properties; ultimately, ballistic transport is most desirable yielding the smallest device delay times (note that although in the ballistic case the delay time scales only proportional to L it is always smaller than in the case of scattering limited transport [12]). As will become clear below, carbon nanotubes are ideally suited for future nanoelectronics devices since they allow combining ultimate scaling of FET device with superior electronic transport properties.

10.2.2 Short Channel Effects

Scaling is a very powerful concept that enables significant performance improvements. However, it potentially leads to so-called short-channel effects (SCE) if the channel length becomes comparable to the spatial extent of the source-channel and channel-drain p-n junctions. Figure 10.6 shows the conduction band profile in the direction of current transport in the case of a long-channel FET and a device with short L. The spatial extent of the p-n-junctions is denoted with λ which is the relevant length scale for potential variations. Obviously, in the case of a short channel length, the p-n-junctions strongly overlap such that the maximum potential barrier Φ_f^0 is substantially reduced compared to the long-channel case. As a result, short channel devices will exhibit significantly larger off-state currents at the same gate voltages. In addition, the overlapping p-n-junctions yield Φ_f^0 to be dependent on V_{ds} and moreover result in a loss of gate control, i.e. $\partial\Phi_f^0/\partial\Phi_g < 1$. Hence, the switching behavior of FETs suffering from SCE is deteriorated and off-state leakage is strongly increased which in turn drastically increases the power consumption. To avoid the appearance of SCE one has to make sure that during scaling $\lambda \ll L$ is preserved.

It has been shown that the potential profile within MOSFETs can be described by a one-dimensional, modified Poisson equation for the surface potential Φ_f at the channel-gate dielectric interface of the following form [13, 28]

$$\frac{\partial^2\Phi_f(x)}{\partial x^2} - \frac{\Phi_f(x) - \Phi_g - \Phi_{bi}}{\lambda^2} = -\frac{e\rho(x)}{\varepsilon_0\varepsilon_r} \tag{10.4}$$

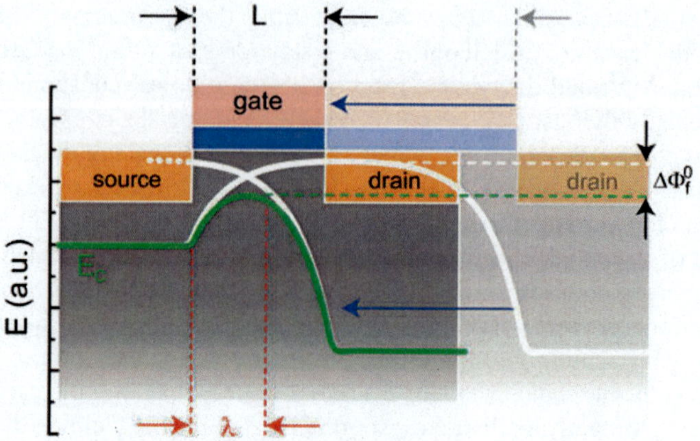

Fig. 10.6 Down-scaling of the channel length. If $\lambda \approx L$, the source-channel and channel-drain p-n-junctions overlap and lead to "drain-induced-barrier-lowering"

Here, Φ_g and Φ_{bi} are the gate and built-in potentials, ρ is the carrier density and ε_r is the relative dielectric constant. Equation (10.4) implies that the presence of the gate (enforcing a constant potential) leads to an exponential screening of potential variations on the length scale λ, where λ depends on the device structure, the gate oxide thickness d_{ox} and channel layer thickness d_{ch}. If, for instance, instead of a single gate a double-gate device structure is realized the potential within the channel will be restricted to the area in-between the two gates and as a result, the source-channel and channel-drain p-n-junctions will exhibit a smaller spatial extent, i.e. λ is smaller. Eventually, multi-gate devices and in particular gate-all-around (GAA)-FETs provide optimal scalability. To be specific, in a cylindrical GAA-FET as illustrated in Fig. 10.7a the screening length is given by $\lambda = \sqrt{(\varepsilon_{nt} d_{nt}^2 ln(1 + 2d_{ox}/d_{nt}))/(8\varepsilon_{ox})}$; scaling down the gate oxide thickness and the channel diameter yields smaller λ

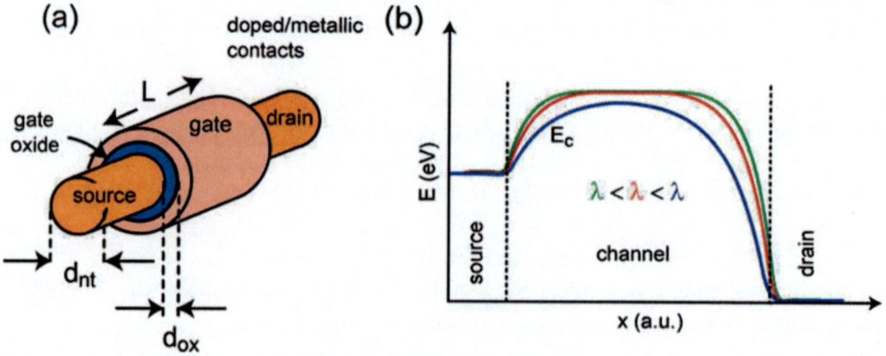

Fig. 10.7 (**a**) Illustration of the cylindrical device geometry under consideration. (**b**) Shows the conduction band profile for different screening lengths λ

and thus steeper p-n-junctions effectively suppressing SCE as schematically shown in Fig. 10.7b (compare green and blue conduction band profile). For instance, in the case of a diameter of the cylindrical channel of $d_{ch} \approx 2\,\text{nm}$ and an oxide thickness of $d_{ox} = 1\,\text{nm}$ one obtains $\lambda = 1.2\,\text{nm}$. In order to avoid SCE, the channel length should be at least $L \geq 3.4 \times \lambda$ which in turn means that a minimal channel length of approximately 5 nm can be obtained.

10.3 Why Carbon Nanotubes?

Consider a sheet of graphene as schematically shown in Fig. 10.8a. In graphene, sp2 hybridization yields three of the four valence electrons of carbon to form in-plane, covalent σ bonds with their neighboring carbon atoms while the fourth valence electron is located in the remaining p-orbitals that result in bonding π and antibonding π^* bands. Figure 10.8c shows schematically the hexagonal 1st Brillouin zone. The π and π^* bands, i.e. valence and conduction bands meet at the K and K′ points and form cones with linear dispersion relation. Since each carbon atom provides one electron the π bands are occupied whereas the π^* bands are empty and hence the Fermi energy is energetically located at the band crossing. In a two-dimensional system, however, the density of states vanishes at the crossing point and thus, graphene is a zero-gap semiconductor.

A so-called (n,m) carbon nanotube is formed by cutting out a graphene stripe perpendicular to the vector $\vec{C} = n\vec{a_1} + m\vec{a_2}$ (gray shaded area in Fig. 10.8a) and rolling up the sheet to a nanotube with circumference $|\vec{C}|$ (depicted in Fig. 10.8b). Quantization along $\vec{C}$ leads to the requirement $k_\perp \cdot |\vec{C}| = n \cdot 2\pi$ due to periodic boundary conditions. The intersection of planes with constant $k_\perp^n = 2\pi n/|\vec{C}|$ with the cones of the graphene dispersion relation results in a sequence of one-dimensional subbands (see Fig. 10.8c, d). If the quantization condition is such that a constant $k_\perp$-plane crosses the K/K' points, a constant density of states at the Fermi energy and hence a metallic nanotube results (due to the linear dispersion relation and one-dimensionality). If, on the other hand, the K/K'-points do not fulfill the quantization condition, the nanotube will be semiconducting with an energy gap E_g that scales with $\Delta k_\perp = 2\pi/|\vec{C}|$, meaning that E_g is inversely proportional to the circumference and hence diameter of the nanotube (see Fig. 10.8c, d). Furthermore, the energetic spacing between two subsequent one-dimensional subbands (depicted in Fig. 10.8d) also increases with decreasing nanotube circumference and it has been shown that the spacing between first and second subband is in the range of a few 100 meV for nanotube diameters of 1.2 nm (see e.g. [5]). Hence, electronic transport can be truly one-dimensional at room temperature in carbon nanotubes of sufficiently small diameter. As will be discussed in detail below, the one-dimensionality plays a decisive role for the performance of novel device designs such as tunneling FETs.

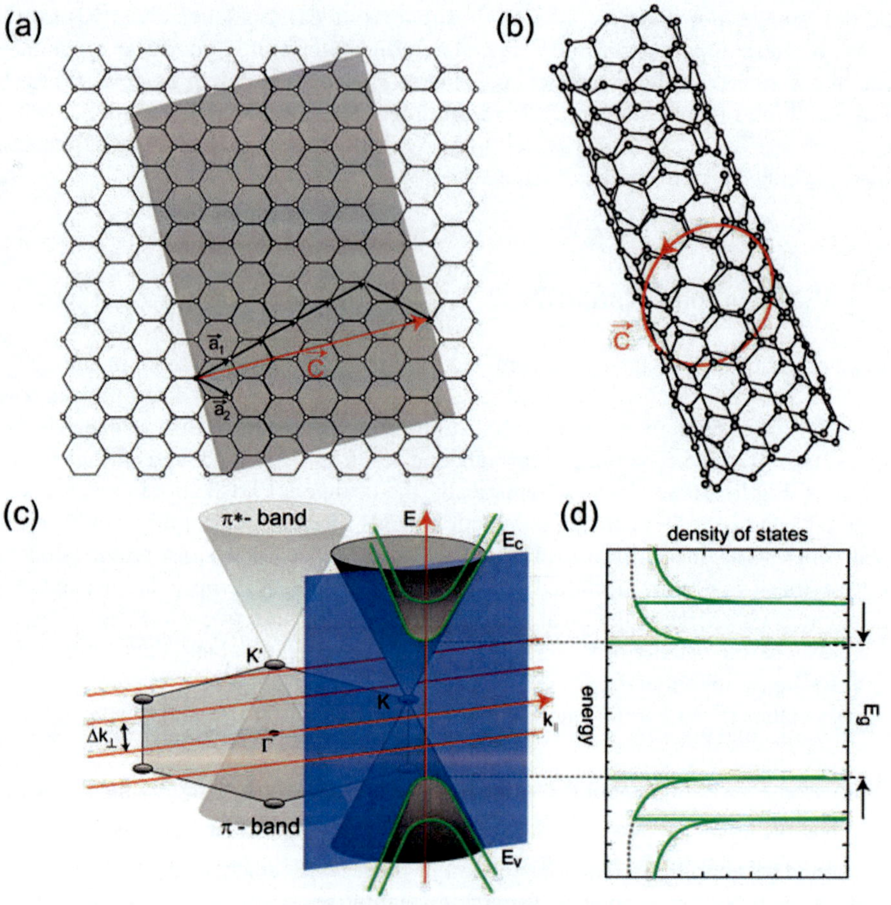

Fig. 10.8 (**a**) Cutting a stripe (*gray shaded*) out of a graphene layer and rolling up along the circumference yields a carbon nanotube (**b**). (**c**) displays the 1st Brillouin zone in graphene. The quantization condition along $\vec{C}$ yields one-dimensional subbands. If the quantization planes do not cross the K/K'-points the nanotube is semiconducting (**d**)

The fact that a nanotube consists of a stripe of graphene rolled up to a seamless tube with perfectly periodic boundary conditions for the electronic waver functions perpendicular to the nanotube axis is the main reason why carbon nanotubes are such ideal one-dimensional systems. While 1D nanowires can be fabricated either by e.g. a bottom-up method such as vapour-liquid-solid growth or by a top-down approach, in particular from III-V semiconductors, these systems always exhibit a disturbance of the periodic lattice at the boundaries leading to interface states, surface roughness etc., all of which lead to charge carrier scattering and hence limit the effective carrier mobility. In carbon nanotubes exceptionally high carrier mobilities up to 100,000 cm^2/Vs at room temperature [1] have been measured and nanotubes therefore can operate as ballistic FET devices [2, 29]. Moreover, when scaling down the channel length of an FET, λ has to be scaled accordingly which essentially means that, both, the gate oxide thickness and the channel layer thickness (body thickness) have to be scaled down as was discussed in Section 10.2. In this respect, one of the major benefits of carbon nanotubes is their inherent small diameter in the few nanometer range making them ideally suited for ultimately scaled FET devices. In addition, due to their cylindrical shape, nanotubes allow realizing a wrap-gate architecture that enables the tightest gate control over the channel layer, i.e. they allow realizing the smallest possible values for λ which can be on the order of 1–2 nm for thin gate oxides [30].

As a result, carbon nanotubes allow obtaining an optimum on-state performance by providing ultimate scalability in combination with superior electronic transport properties making nanotubes a premier choice for future nanoelectronics devices.

10.4 CNFETs – Ultimate Ultrathin-Body Schottky-Barrier MOSFETs

The easiest way of fabricating a CNFET is to disperse a carbon nanotube on an oxidized piece of silicon and contact it with metal electrodes (see Fig. 10.9). The silicon wafer then serves as a large area back gate and the oxide on top plays the role of the gate dielectric. At the metal-nanotube contact a Schottky-contact can be expected with a Schottky-barrier of height Φ_{SB} as illustrated in the right panel of Fig. 10.9. Although it has been shown that the picture of a simple metal-semiconductor contact is insufficient to describe all aspects of the electrical behavior of CNFETs with metallic contacts [31], it is able to explain the transport phenomena relevant to the present analysis. Doped source and drain contacts are incorporated by taking a part of the electrodes in the calculation into account.

Figure 10.10a shows representative transfer characteristics for various drain-source voltages of a SB-CNFET. The device exhibits ambipolar behavior, typical of Schottky-barrier FET devices. Moreover, if one extracts the inverse subthreshold

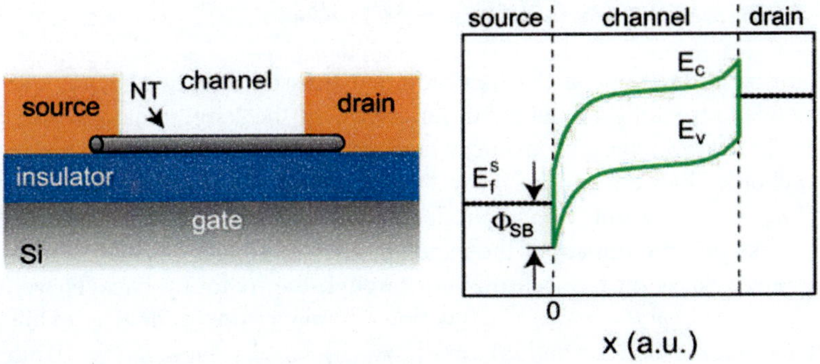

Fig. 10.9 *Left*: Schematics of a back-gated SB-CNFET. The *right panel* shows the conduction and valence bands of a p-type SB-CNFET in the device's on-state

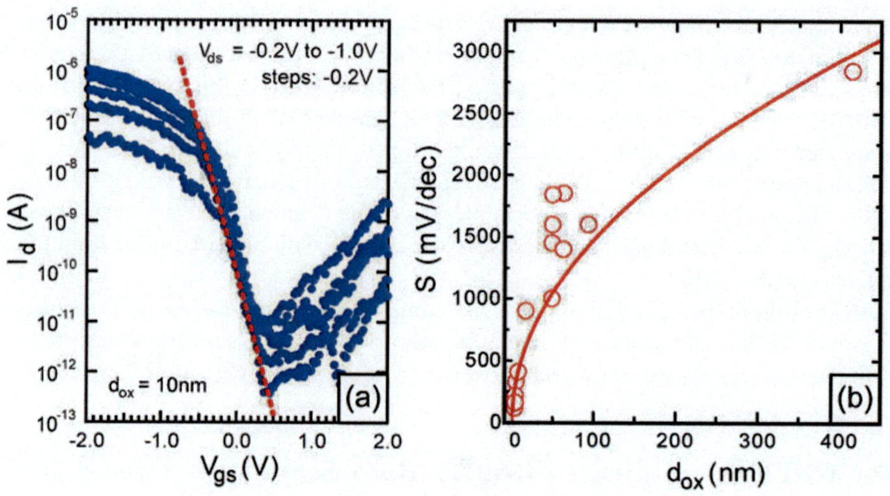

Fig. 10.10 (**a**) Transfer characteristics of a SB-CNFET. The inverse subthreshold slope is extracted as inidicated by the *red dashed line*. (**b**) Inverse subthreshold slope as a function of gate oxide thickness extracted from a number of experimental devices ([7] and references therein)

slope S (indicated by the red dashed line) for devices with varying gate oxide thickness one observes a strong dependence of S on the gate oxide thickness as shown in Fig. 10.10b [7]. Using a simple analytical model for the potential profile at the source Schottky diode it will be shown in the next section, that this peculiar behavior (not expected in conventional-type FETs as long as electrostatic integrity is preserved) is a consequence of the dependence of the carrier injection through the Schottky barrier on the screening length λ [32]. The solid red line in Fig. 10.10b is a result of this model showing excellent agreement with the experimental data in the device's off-state.

10.4.1 Transport in SB-CNFETs – Off-State

It has been mentioned above that the modified Poisson equation (10.4) leads to an exponential screening of potential variations on the length scale λ which in turn depends on the nanotube diameter and gate oxide thicknesses. This means that the Schottky barriers can be made "thinner" and therefore more transparent if d_{nt} and d_{ox} are made thin which is reflected in a different electrical behavior. In order to investigate the impact of the nanotube diameter, gate oxide thickness and effective carrier mass on the transmission through the Schottky barriers we have simulated transfer characteristics as a function of temperature for devices exhibiting (i) $d_{nt} = 1.4\,\text{nm}$, $d_{ox} = 2\,\text{nm}$ and $m^* = 0.1\,\text{m}_0$, (ii) $d_{nt} = 1.4\,\text{nm}$, $d_{ox} = 10\,\text{nm}$ and $m^* = 1.0\,\text{m}_0$ (iii) $d_{nt} = 5\,\text{nm}$, $d_{ox} = 10\,\text{nm}$ and $m^* = 1.0\,\text{m}_0$. Typical Arrhenius plots were employed to extract the effective Schottky barrier height Φ_{SB}^{eff} using

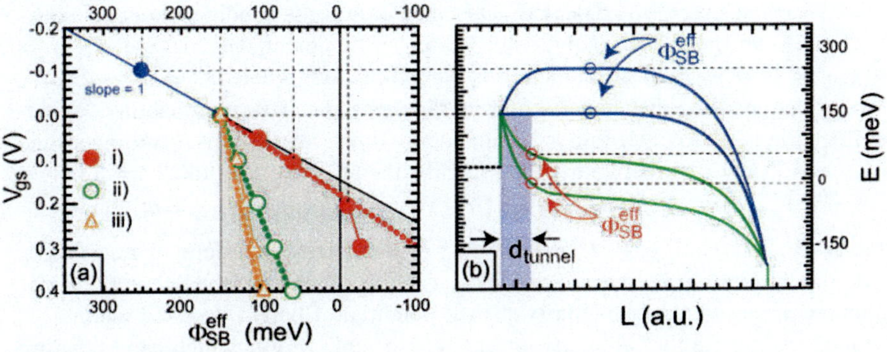

Fig. 10.11 (**a**) Extracted effective SB heights (data points, see text for details) versus gate voltage. The *lines* belong to the analytical calculation. (**b**) shows the conduction band for four different gate voltages. If the extracted SB is larger than the actual barrier height a one-to-one change of Φ_{SB}^{eff} with gate voltage is observed as indicated by the *horizontal dashed lines*

$I_d \propto T^2 exp(-q\Phi_{SB}^{eff}/k_B T)$ in our analysis [33]. If the gate voltage is such that the potential Φ_f^0 lies above the Schottky barrier at the source contact (dark gray curves in Fig. 10.11b) then Φ_f^0 is the maximum potential barrier that determines the current flow instead of the Schottky barrier. Furthermore, this potential barrier changes one-to-one with changing gate voltage (see discussion above) and consequently, one expects that in this regime $\Phi_f^0 = \Phi_{SB}^{eff}$ and that Φ_{SB}^{eff} changes one-to-one with gate voltage. This behavior is indeed observed indicated by the horizontal lines in Fig. 10.11a. For larger gate voltages, on the other hand, the bands are pushed below the source side Schottky barrier (light gray curves in Fig. 10.11b) and Φ_{SB}^{eff} as a function of V_{gs} deviates from the one-to-one behavior. In fact, in bulk-like, large Schottky contacts, increasing the gate voltage should only lead to a small decrease of Φ_{SB}^{eff} (due to the Schottky effect [26]). However, in ultrathin-body devices such as CNFETs one would expect a stronger reduction of the effective Schottky-barrier due to the tighter gate control over the potential distribution of the barrier. This is exactly what is observed when Φ_{SB}^{eff} is extracted in the cases (ii) and in particular (i). Whereas in the case (iii) with large carrier mass, thick gate oxide and nanotube diameter, the effective barrier is only lowered slightly with increasing gate voltage. The impact of V_{gs} on Φ_{SB}^{eff} increases continuously when all three parameters (m^*, d_{ox}, d_{nt}) are decreased. In the case (i) the gate control over the effective Schottky-barrier is so strong that Φ_{SB}^{eff} changes almost one-to-one even if Φ_f^0 is significantly less than Φ_{SB}. For large gate voltages negative effective barriers can be achieved in this case, meaning that an excellent injection of carriers from the source Schottky contact into the channel can be expected. A CNFET with such a "thin" Schottky-barrier should accordingly exhibit an excellent off- as well as on-state.

We will now quantify the impact of d_{nt} and d_{ox} on the effective Schottky barrier Φ_{SB}^{eff}. In the off-state of a transistor the charge in the channel is small and can to first order be neglected. This means that the modified Poisson equation (10.4) can be solved analytically leading to a potential landscape $\Phi_f(x) = (\Phi_{SB} - \Phi_f^0)exp(-x/\lambda) + \Phi_f^0$ at the source Schottky diode where $\Phi_f^0 = \Phi_g + \Phi_{bi}$. We introduce a tunneling distance d_{tunnel} such that the transmission probability for carrier injection is set to unity if at a certain energy the potential barrier is thinner than d_{tunnel} and $T(E) = 0$ otherwise. Consequently, the effective Schottky barrier height is simply $\Phi_{SB}^{eff}(\lambda, \Phi_g) = \Phi_f(d_{tunnel}) = (\Phi_{SB} - \Phi_f^0)exp(-d_{tunnel}/\lambda) + \Phi_f^0$. This scenario is illustrated in Fig. 10.11b where Φ_{SB}^{eff} is shown for two different gate voltages (dark gray circles). An estimation for d_{tunnel} can be obtained if the Schottky barrier is approximated with a triangularly shaped potential of height Φ_{SB} and width λ. It is important to note that d_{tunnel} *is not* the length scale λ over which band bending occurs. Using the transmission probability calculated with the WKB approximation an expression for d_{tunnel} can be obtained that shows a weak dependence on Φ_{SB} as well as d_{nt} and d_{ox}; however, d_{tunnel} scales as $1/\sqrt{m^*}$ [34]. Due to the exponential dependence of $\Phi_{SB}^{eff}(\lambda, \Phi_g)$ on λ the effective Schottky barrier height can be lowered more efficiently with increasing gate voltage in the case of ultrathin d_{nt} and d_{ox} (and small effective masses). Hence, the same behavior of Φ_{SB}^{eff} is obtained with this simple model as in the case discussed above where the effective Schottky barrier height was extracted from temperature dependent simulations of the transfer characteristics. Indeed, the dotted lines in Fig. 10.11a are calculated using the model (all calculated with the same, constant d_{tunnel}) and show excellent agreement with the extracted data points. Only in the case of (i) and large gate voltages a deviation is seen which is expected since for such low barriers the charge within the channel plays an important role in determining the potential landscape of the source side Schottky diode. This, however, has not been taken into account in this analysis (see the next section for details).

A closed expression for the inverse subthreshold slope can be calculated using the effective Schottky barrier height as follows. For large V_{ds} and $\Phi_{SB}^{eff}(\lambda, \Phi_g) > k_B T$ the current in the off-state is proportional to $I_d \propto exp\left(-\left(\Phi_{SB}^{eff} - E_f^s\right)/k_B T\right)$ and with $S = ln(10)\left(\left(\partial I_d/\partial \Phi_{SB}^{eff}\right)\left(\partial \Phi_{SB}^{eff}/\partial V_{gs}\right)(1/I_d)\right)^{-1}$ it is easy to show that

$$S = \frac{k_B T}{q} ln(10) \frac{1}{1 - exp(-d_{tunnel}/\lambda)} \overset{d_{tunnel}>\lambda}{\approx} \frac{k_B T}{q} ln(10) \left(\frac{1}{2} + \frac{\lambda}{d_{tunnel}}\right). \quad (10.5)$$

Equation (10.1) implies that in an SB-FET, even in the long channel case, S strongly depends on λ which is not the case in conventional-type devices. The solid line in Fig. 10.10b is calculated using Eq. (10.5) showing excellent agreement between experimental data and model. Second, S in an SB-FET is always larger than in a conventional FET but can be made small using an ultrathin channel layer and thin gate oxides. Hence, the small diameter of nanotubes explains why in SB-CNFETs steep inverse subthreshold slopes of ~100 mV/dec can be observed even in devices exhibiting a gate oxide thickness as large as 10 nm [7, 35]. (Note that the discussion

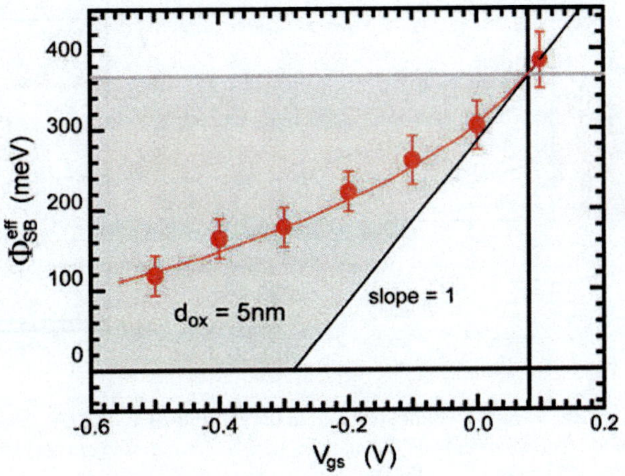

Fig. 10.12 Effective Schottky barrier height as a function of gate voltage extracted from temperature dependent measurements of experimental SB-CNFETs (taken from [33])

above is not restricted to carbon nanotubes but holds true also for other UTB devices such as SOI SB-MOSFETs [36]).

The ability to significantly lower the effective Schottky barrier height with increasing gate voltage in SB-CNFETs has also been observed experimentally. Figure 10.12 shows Φ_{SB}^{eff} as a function of gate voltage. The experimental curve exhibits qualitatively the same dependence as the simulated data.

10.4.2 Transport in SB-CNFETs – On-State

In the device's on-state the charge in the channel cannot be neglected anymore since it will have a significant impact on the potential distribution within the channel and in turn also on the potential landscape of the source and drain Schottky diodes. In order to account for the charge in the channel and also for scattering in the channel we subdivide the channel into two segments. The first segment (called "1" in the following) comprises the Schottky diode and has a spatial extend on the order of λ, as illustrated in the left panel of Fig. 10.13. The second segment ("2") extends up to the drain contact (where the particular potential landscape at the drain Schottky diodes is disregarded for simplicity). Depending on the tunneling probability through the source Schottky barrier and the scattering within the channel, part of the drain-source voltage will drop across segment 1 and the rest drops along the remainder of the channel.

In order to calculate an approximate expression for the on-state current one has to compute Φ_f^0 which allows computing the effective Schottky barrier height. However, to do so we need to know the charge density at the point $x = \lambda$ (see Fig. 10.13). If we know the quasi-Fermi level in $x = \lambda$, E_f^0, we can approximately calculate the charge at this position by simply integrating over the product of the one-dimensional density of states at this position with an equilibrium Fermi

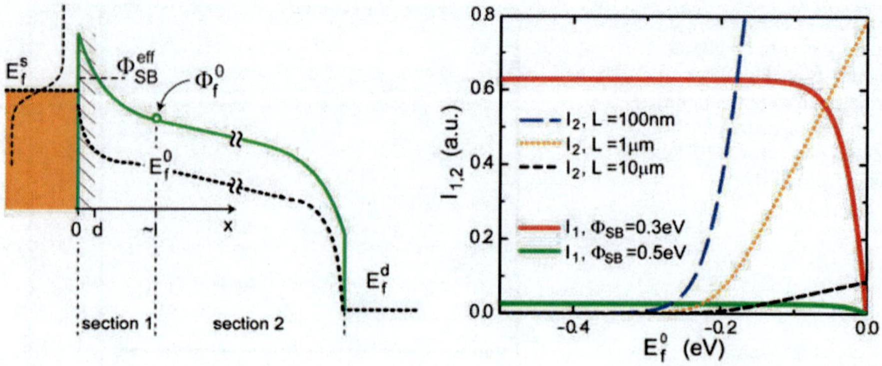

Fig. 10.13 *Left*: Conduction band in the on-state of a SB-FET. *Right*: Current through the Schottky diode (*solid curves*) and the channel (*dashed and dotted curves*) as a function of the Fermi level at point "0" (taken from [37])

function with E_f^0. The quasi-Fermi level E_f^0 on the other hand can be calculated by equating the individual current components through segment "1" and "2", $I_1 = I_2$, yielding

$$\frac{l_{scat}}{l_{scat}+\lambda}\int_{\Phi_{SB}^{eff}}^{\infty} dE\left(f(E_f^s)-f(E_f^0)\right) \propto I_1 \stackrel{!}{=} I_2 \propto \frac{l_{scat}}{l_{scat}+L}\int_{\Phi_f^0}^{\infty} dE\left(f(E_f^0)-f(E_f^d)\right) \tag{10.6}$$

where we have accounted for scattering in the channel with a simple energy-independent transmission function $T(E) = l_{scat}/(l_{scat}+L)$ [38] as was also done above. For carriers that scatter within the steep potential variation region within segment one it is unlikey to be scattered back into the source contact once they have lost k_BT in energy since the Schottky diode rapidly becomes "thicker" [39, 36] preventing the carriers to tunnel back. Thus, one has to modify the transmission function and replace the channel length L with λ in the I_1-term [36]. Equation (10.6) yields a transcendent equation for E_f^0 that can be solved numerically or graphically. If we plot both current contributions in the same graph, then E_f^0 is given by the intersect of the two curves. Note that E_f^0/q is the voltage that has dropped across segment one, i.e. at the Schottky barrier. This means, if $E_f^0 \approx 0$ then almost all of the drain-source bias drops across the channel whereas for $|E_f^0| \rightarrow qV_{ds}$ all voltage drops at the source Schottky diode. Since the tunneling probability through the Schottky barrier depends on d_{ox}, d_{nt} but also on the gate voltage, E_f^0 will be a function of these three quantities and the transcendent equation has to be solved for each set of parameters individually. Exemplarily, the right panel of Fig. 10.13 shows I_1 (gray solid line) for a back-gated SB-CNFET with $d_{ox} = 10\,\text{nm}$, $d_{nt} = 1\,\text{nm}$, a Schottky barrier of 0.3 eV and a mean free path of $l_{scat} = 100\,\text{nm}$; a bias of $V_{ds} = 0.5\,\text{V}$ and $V_{gs} = 0.5\,\text{V}$, i.e. in the device's on-state, were assumed in the present case. I_2 is plotted for three different channel lengths (see the figure for details). In the present case (which is typical of back-gated CNFETs) a significant part of the potential drops across the Schottky barrier for $L = 100$ nm and $L = 1$ μm. This means that the current through SB-CNFETs is determined by the Schottky barrier rather than the transport properties of the channel (note that this is also true for any other SB-FET [36]). As such, SB-FETs can be denoted as "contact-switching" devices. The current is dominated by scattering within the channel only for very long channels (10 μm in the present case; see the black dashed line in the right panel of Fig. 10.13). The channel length for which scattering in the channel dominates over the scattering across the Schottky barrier depends exponentially on the SB height and λ; the resistance of the channel on the other hand increases only linearly with channel length L. Whereas in the case of $L = 10$ μm and the parameters given above (gray solid curve) E_f^0 is close to zero, i.e. only a small fraction of the bias drops at the Schottky diode, this is very different in the case of increasing the actual Schottky height to 0.5 eV (green solid curve): again a substantial part of the bias drops across the Schottky diode making even longer channel lengths necessary for the electronic transport to be dominated by scattering within the channel.

As a result, scattering in the channel only plays a role if (i) Φ_{SB} is rather low and/or the Schottky barrier is very thin and thus highly transmissive or (ii) the channel is appropriately long. This has an important implication, namely that care has to be taken when one extracts the mobility from SB-CNFETs. This is only possible if the channel length is long enough such that the electrical behavior of the device is determined by the scattering in the channel and not by the tunneling through the Schottky diodes at the contacts. The required length can be quite long particularly in case of carbon nanotubes which exhibit rather long mean free paths.

Having determined the quasi-Fermi level the charge at position "0" can be computed and hence Φ_f^0 can be determined. To do so, we assume that around the position "0" the curvature term in the modified Poisson equation can be neglected and Φ_f^0 is employed to compute Φ_{SB}^{eff} as a function of gate voltage. This task has to be done either numerically or with some approximation for the Fermi-Dirac intregral. The dependence of current on λ, however, can be stated explicitly as follows

$$I_d \approx \frac{2e}{h}\frac{l_{scat}}{l_{scat}+\lambda}k_BT\left(ln\left(exp\left(\left(E_f^s - \Phi_{SB}^{eff}\right)/k_BT\right)+1\right) - ln\left(exp\left(\left(E_f^0 - \Phi_{SB}^{eff}\right)/k_BT\right)+1\right)\right). \tag{10.7}$$

Hence, the drain current strongly improves when decreasing λ even in the ballistic case due to the exponential dependence of the effective Schottky-barrier height on λ (see for instance [40]). The reason for this is again that the carrier injection into the channel is strongly improved in UTB SB-FETs with thin gate oxides.

10.4.3 *Transport in SB-CNFETs – Ambipolar Behavior*

Contacting carbon nanotubes with metallic electrodes is an easy way to investigate the electronic transport in nanotubes in an FET device configuration. In the preceding sections we saw that the geometrical smallness of the nanotubes leads to a peculiar device behavior, particularly to a strong dependence of the on- as well as the off-state on the nanotube diameter and the gate oxide thickness. Both, the on-state as well as the off-state in terms of the inverse subthreshold slope improve when λ is made as small as possible. This means in particular, that in a contact switching device such as SB-CNFETs the gate oxide thickness should be as thin as possible in order to improve the device performance. However, an improvement of the carrier injection by making the Schottky barriers thinner leads at the same time to a significant increase of the off-state leakage current due to the ambipolar operation of SB-FETs [36]. For instance, suppose the Fermi level is pinned in a midgap position. Then the minimum drain current flows when $\Phi_f^0 = \Phi_d/2 + \Phi_{SB} = 1/2 \cdot (\Phi_d + E_g)$ where Φ_d is again the drain potential. Since the minimum current situation is characterized by an equal electron and hole current as shown in the left panel of Fig. 10.14, I_d^{min} (for large enough source-drain bias) is given by

$$I_d^{min} \approx 2 \times \frac{4e}{h} \int_{\Phi_{SB}^{eff}}^{\infty} dE\, f(E_f^s) = \frac{8e}{h} k_B T\, exp\left(-\frac{\Phi_d(1 - exp(-d/\lambda) + E_g)}{2k_B T}\right) \quad (10.8)$$

where we have set $E_f^s = 0$ for simplicity. It is apparent that the minimum off-state leakage obtainable in SB-CNFETs exponentially depends on the size of the energy gap, the source-drain bias *and on* λ, making SB-CNFETs particularly vulnerable to high off-state leakage currents.

Figure 10.14 (left panel) shows transfer characteristics of a SB-CNFET exhibiting the discussed strong ambipolar behavior with significantly increased off-state leakage. In fact, the on/off-ratio shrinks to less than three orders of magnitude in the case of $V_{ds} = 0.7$ V which is attributed to the rather small band gap of the nanotube used for the present device [35]. The high off-state leakage which increases as λ is

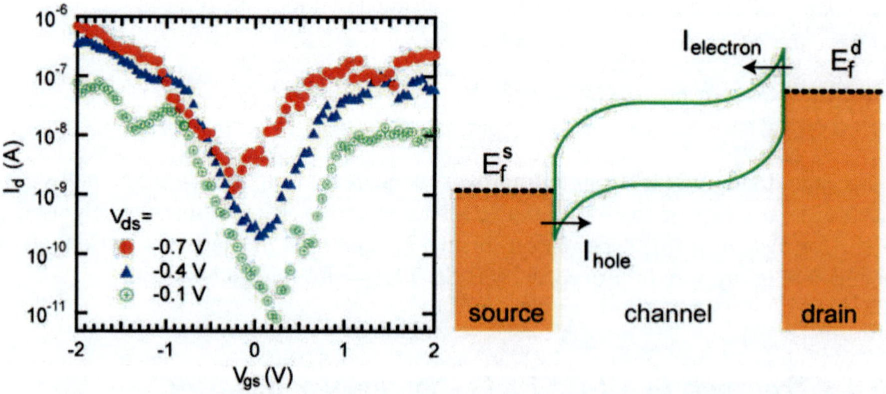

Fig. 10.14 Experimental transfer characteristics of a SB-CNFET with titanium electrodes (taken from [35]). The device exhibits a strong ambipolar behavior with bias dependent off-state leakage. The *right panel* shows the conduction and valence bands at V_{gs} yielding the minimum current

made smaller and the small on/off-ratio are unfavorable from an application point of view. In the next section we will therefore discuss conventional-type CNFETs and explore their electronic transport and applicability as FET devices.

10.5 Conventional-Type CNFETs

In a conventional-type FET the source and drain electrodes consist of doped semiconductor portions, i.e. doped nanotube segments in the case of CNFETs. Since within the band gap of a semiconductor there are no states that can be occupied, doped semiconducting electrodes lead to a strong suppression of the ambipolar behavior and unipolar characteristics can be expected resulting in much smaller off-state leakage currents.

Doping the nanotube on the other hand is not as simple as in silicon based devices. However, using a dual-gate architecture, doping can be realized electrostatically hence facilitating the investigation of transport in conventional-type carbon nanotube FETs (C-CNFETs). Such a dual-gate device has recently been demonstrated by Lin and coworkers which allowed a controlled transition from ambipolar to unipolar device behavior [35, 41].

Figure 10.15 shows a schematics of such a dual-gate transistor. A highly doped silicon substrate serves as a large area back-gate separated from the nanotube by a thermally grown SiO_2 (10 nm in thickness in the actual experiment). This back-gate is used to electrostatically "dope" the source/drain extensions of length L_{bg} (typically on the order of 200 nm). In order to exclude any influence of the back-gate on the channel the actual gate is deposited on top of the SiO_2, effectively screening the electrical field of the back-gate. Afterwards, a thin oxide is grown on top of the actual gate (approximately 4 nm thick Al_2O_3 is formed in the case of the devices presented in Ref. [35]; for more details on the fabrication see Ref. [41]).

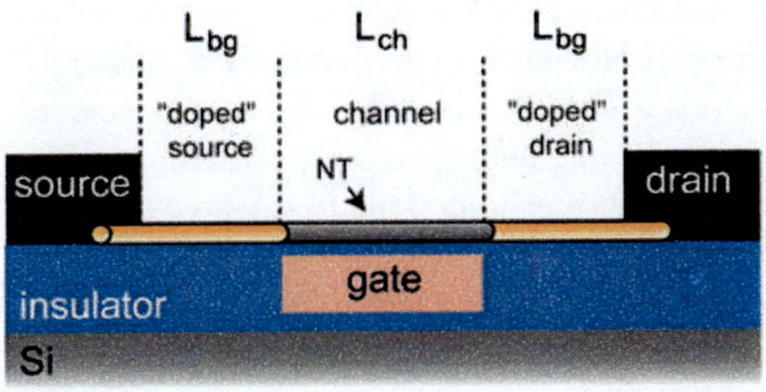

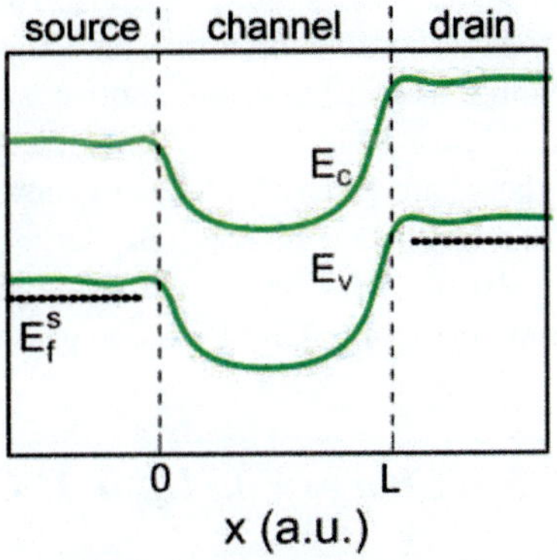

Fig. 10.15 Schematics of a dual-gate CNFET structure. The source/drain extensions are electrostatically "doped" by applying a large back-gate voltage. The *right panel* shows the conduction and valence bands in case of a p-type C-CNFET in the "doped" source/drain region and the channel

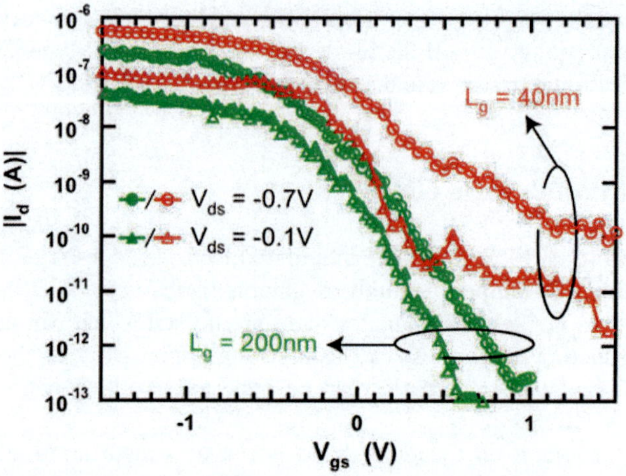

Fig. 10.16 Experimental transfer characteristics of dual-gate C-CNFETs (taken from [42]) in the case of a long channel with $L = 200$ nm (*green curves*) and for a scaled device with $L = 40$ nm (*red curves*)

Finally, the nanotube is dispersed on top of this dual-gate structure and contacted with titanium electrodes. For negative back-gate voltages, the device basically represents a conventional p-i-p FET structure. In operation, the back-gate is kept at a constant, negative gate voltage while the actual gate is swept from negative to positive voltages. For negative V_{gs} the device operates as conventional p-type FET with a unipolar behavior.

Figure 10.16 shows transfer characteristics of a dual-gate C-CNFET with a channel length of 200 nm (green curves). The device shows regular FET transfer characteristics with an almost ideal inverse subthreshold slope of 65 mV/dec reinforcing that the device acts as a conventional-type transistor [35]. However, scaling down the channel length of such a transistor leads to an unexpected behavior in the device's off-state: If the electrostatic integrity is preserved during scaling, the off-state should exhibit the same almost ideal inverse subthreshold slope in the long channel as well as in the short channel case. However, experimentally – although it is ensured that the device is electrostatically well-behaved – the scaled device with a channel length of 40 nm shows an increasing leakage current with increasing bias as well as a substantially larger inverse subthreshold slope [42] (red curves in Fig. 10.16). In order to investigate and explain this unusual behavior we performed simulations of C-CNFETs, discussed in the next section.

10.5.1 Charge Pile-Up in C-CNFETs

Instead of calculating the entire device structure with back-gate we assumed a high doping level in the source/drain contacts giving rise to a certain Fermi energy below the valence band edge (see right panel of Fig. 10.15). Again, the non-equilibrium

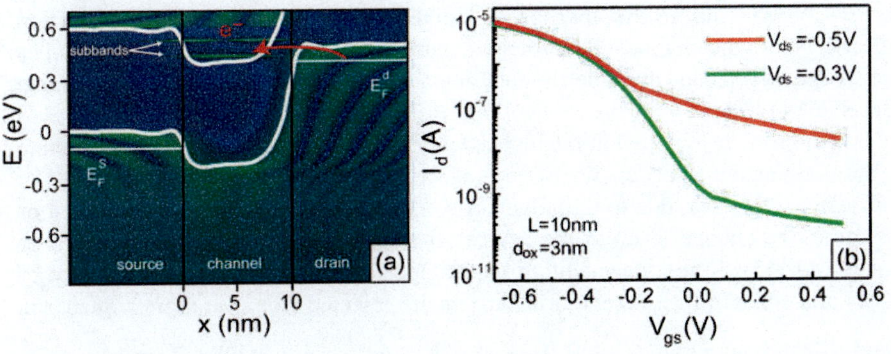

Fig. 10.17 (**a**) Local density of states in a C-CNFET. The large band-to-band tunneling probability gives rise to a charge pileup in the channel even in the device's off-state [31]. (**b**) Shows transfer characteristics exhibiting the same behavior as experimentally observed

Green's function formalism on a finite difference grid is used to compute the charge in and current through the transistor. We have simulated C-CNFETs with a channel length of 10 nm and gate oxide thickness of 3 nm for $V_{ds} = 0.3$ and 0.5 V; the Fermi energy in the source/drain contacts of the C-CNFET is set to 0.1 eV and the band gap of the nanotube under consideration was 0.6 eV (see [43] for more details).

Looking at the transfer characteristics of the simulated device, as displayed in Fig. 10.17b, it is apparent that they show the same, unusual behavior in the off-state as the experimental devices: below a certain gate voltage the current levels off at an inverse subthreshold slope far in excess of 60 mV/dec. The reason for this can be inferred by a closer look at Fig. 10.17a which is a log-scale plot of the local density of states in the C-CNFET at zero gate voltage. Since the maximum $V_{ds} = 0.5$ V is smaller than the band gap and because the band gap truncates the Fermi distribution in the source/drain contacts of the C-CNFET, no electron leakage current can flow from drain to source. However, Fig. 10.17a reveals bound states with a high DOS in the conduction band within the channel area. Electrons are likely to be injected into the channel via band-to-band tunneling due to the low transport effective mass in nanotubes (around $0.1 \times m_0$) and the steep n-p junction at the drain side of the device. This steep n-p junction is in turn a result of the small screening length λ, i.e. stems from the smallness of the nanotube diameter and the thin gate oxide. As a result, electrons are injected into the channel and get trapped since their flow is blocked by the energy gap of the source contact. Consequently, this leads to a charge pileup preventing the gate from effectively moving the bands to turn the device off. The reduced potential barrier on the other hand leads to a large hole leakage current in the off-state. However, as can be seen from the transfer characteristics in Fig. 10.17b, the charge pileup significantly deteriorates device performance only if $eV_{ds} > E_g/2$. For smaller bias the charge pileup is much less pronounced and allows for a proper off-state performance of the device. In this case, the C-CNFET provides higher on-state currents and steeper inverse subthreshold slopes compared to SB-CNFETs due to the absence of a potential barrier that limits the injection of carriers into the channel. Nevertheless, particularly for small band gap nanotubes the charge pileup can have a detrimental impact on the device performance of scaled C-CNFETs [43].

It is interesting to note that the increasing leakage current in the device's off-state is not due to usual short channel effects since electrostatic integrity is preserved in the device. However, due to the pileup of electrons, there is a significant amount of charge in the channel even in the device off-state giving rise to a quantum capacitance [44, 45]. Since the quantum capacitance is proportional to the density of states and since in the present case carriers are injected from the drain contact via band-to-band tunneling, $C_q \approx 2/h\sqrt{8m^*/(\Phi_f^0 - (E_f^s - eV_{ds}))}$ with $E_f^d = E_f^s - eV_{ds}$. Hence, the charge in the channel is given by $Q = -\Phi_f^0/e \cdot (C_{ox} + C_q)$ where we have neglected a source/drain capacitance for simplicity since a long channel device exhibiting electrostatic integrity is assumed. At the same time $Q = C_{ox}V_{gs} + C_qV_{ds}$ and therefore we obtain

$$\Phi_f^0 = \frac{C_{ox}}{C_{ox} + C_q}\Phi_g + \frac{C_q}{C_{ox} + C_q}\Phi_d \tag{10.9}$$

For a proper device functionality, i.e. without short channel effects, it is required that in the off-state $\partial\Phi_f^0/\partial\Phi_g = 1$ for constant Φ_d since this leads to a one-to-one change of the surface potential with changing gate voltage and hence to $S = 60\,\text{mV/dec}$ as discussed in Section 10.1. In addition, to suppress drain-induced barrier lowering, i.e. SCE, $\partial\Phi_f^0/\partial\Phi_d$ should be as small as possible (see Section 10.2). In a conventional and electrostatically well-behaved device $\partial\Phi_f^0/\partial\Phi_d \approx 0$ and in the *off-state* also $C_{ox} \gg C_q$ and consequently the device will be free of short channel effects. However, in the present, peculiar case $C_q > C_{ox}$ even in the off-state because of the charge pileup. This means that if the device does not operate in the quantum capacitance limit, i.e. $C_q > C_{ox}$ short channel effects appear (as indeed observed) although the device – from a pure geometrical point of view – should electrostatically be well-behaved. Again, the reason for this situation is the substantial band-to-band tunneling that leads to a large carrier density in the channel although the device is in the off-state.

In order to avoid the charge pileup in devices based on low-band gap nanotubes to have a significant impact on the device characteristics the gate oxide has to be scaled towards the quantum capacitance limit where $C_{ox} \gg C_q$. In this case the large off-state leakage and its dependence on V_{ds} vanishes as has been confirmed with simulations (not shown here). As a result, in scaled C-CNFETs based on small band gap nanotubes the gate oxide needs to be scaled much more than expected to avoid short channel effects contrary to conventional bulk-like MOSFETs.

10.6 Tunneling CNFETs

We have seen above that in terms of applicability the conventional CNFET shows a superior on- as well as off-state performance compared to the SB-CNFET if the drain voltages are restricted to about half of the band gap of the carbon nanotube. In addition, the small diameter together with a wrap-gate architecture allows to scale conventional-type transistors to smallest dimensions. However, even if an ideal conventional-type CNFET can be manufactured, this device architecture *in general* suffers from a fundamental problem: any conventional-type FET is limited to a minimum inverse subthreshold slope of 60 mV/dec at room temperature as has been discussed in Section 10.1. This limitation is a major obstacle to further reduce the supply voltage and hence the power consumption of integrated circuits. Provided a certain ratio between the off-state and the on-state current of approximately 3 orders of magnitude is required and if we assume that two thirds of the maximum applied gate voltage are needed to obtain a high on-state current, one needs at least a gate voltage range of $3 \times (3 \cdot 60) = 540\,\text{mV}$ to properly operate the device. In turn this means that scaling down the supply voltage of devices limited to an $S = 60$ mV/dec leaves only two options: either the off-state leakage is increased or the on-state performance deteriorated. Therefore, transistor devices that show an inverse subthreshold slope significantly steeper than 60 mV/dec and still provide a high on-state performance are highly desirable.

It has been discussed above that the reason for the limit of S to a minimal value of 60 mV/dec is the fact that the switching mechanism of conventional-type devices relies on the modulation of the injection of carriers from a thermally broadened Fermi function. Hence, in order to achieve subthreshold swings below 60 mV/dec, the current injection from the source contact has to be modified in a way that it becomes independent of a thermally broadened Fermi distribution function. Recently, band-to-band tunneling (BTBT) has been proposed as an effective means to accomplish this [11, 21, 46–48]. A device consisting of a p-doped source(drain), an intrinsic channel and an n-doped drain(source) (p-i-n) is ideally suited for this purpose. In the following we will investigate the electronic transport in such p-i-n- or tunneling FETs and show that carbon nanotubes enable the realization of high-performance tunneling FETs with steep inverse subthreshold slopes and good on-state performance due to their small diameter and in particular due to the one-dimensional transport.

Consider a device structure as depicted in Fig. 10.18. In order to obtain an optimal gate control over the bands in the channel the transistor exhibits a wrap-gate and a thin gate oxide. The entire device is embedded into SiO_2 and sits on a grounded silicon wafer. We have simulated the electronic transport in such tunneling nanotube transistors using the non-equilibrium Green's function formalism together with the modified Poisson equation (10.4) presented above. The right panel of Fig. 10.18 shows the conduction and valence bands in a tunneling FET in the on-state. At the source-side n-p junction BTBT leads to injection of holes from the source contact into the channel constituting the on-state current.

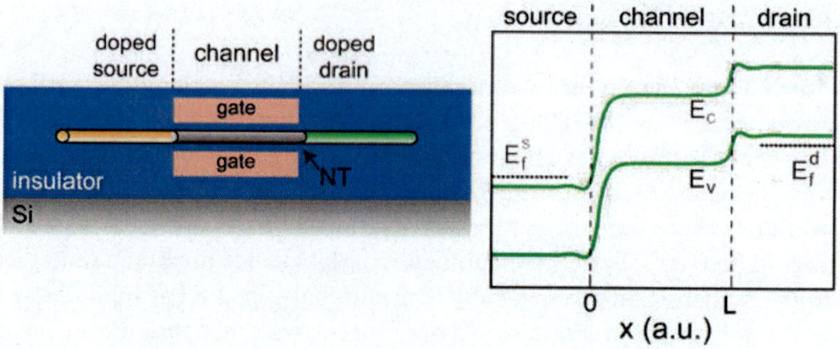

Fig. 10.18 *Left*: Schematics of the device structure of the tunneling CNFET under consideration. The *right panel* shows the conduction and valence bands in the device's on-state

In the succeeding sections we will first discuss the working principle of a tunneling FET and investigate under what circumstance an S<60 mV/dec can be achieved. Subsequently, the performance of tunneling CNFETs and their peculiarities are discussed in more detail.

10.6.1 Working Principle of the Tunneling FET

Before we will investigate the electronic transport in tunneling CNFETs with simulations we will discuss the working principle of the tunneling FET and under what circumstances an *S* smaller than 60 mV/dec can be achieved. Figure 10.19a shows the conduction and valence band in the off-state of the tunneling FET. Only electrons residing in the tail of the drain Fermi function and holes in the tail of the source Fermi function can contribute to the current and hence very low leakage currents – depending on the size of the energy gap – can be expected. As soon as the gate pulls the valence band in the channel above the conduction band in the source contact a

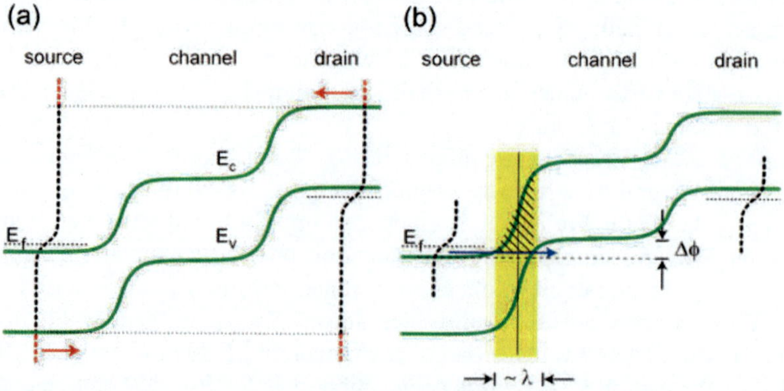

Fig. 10.19 Conduction and valence bands along current transport in a tunneling FET. (**a**) shows the off-state and (**b**) the on-state

channel for band-to-band tunneling is opened up. For a sufficiently "thin" BTBT barrier the current through the device is then determined by the tunneling of carriers within the energetic window $\Delta\Phi$ (see Fig. 10.19b). An analytic expression for the current can be obtained using the WKB approximation for the BTB tunneling probability [49]

$$I_d = \frac{2e}{h}\int_0^{\Delta\Phi} dE\; T_{WKB}\left(f(E_f^d) - f(E_f^s)\right)$$
$$\approx \frac{2e}{h} T_{WKB} k_B T \left\{ ln\left(exp\left(\frac{\Delta\Phi - E_f^s}{k_B T}\right) + 1\right) - ln\left(exp\left(\frac{-E_f^s}{k_B T}\right) + 1\right)\right\} \tag{10.10}$$

where we assumed a drain-source bias large enough to ensure that within $\Delta\Phi$ the drain Fermi function is approximately one. The tunneling probability in WKB approximation becomes energy-independent if one approximates the source-channel n-p junction with a triangular potential barrier as illustrated with the dashed area in Fig. 10.19b [26]. In addition, we know from the discussion above that p-n junctions in nanotube transistors (and also any other ultrathin body device) have a spatial extend on the order of λ. As a result, T_{WKB} can be written as [49]

$$T_{WKB} \approx exp\left(-\frac{4\lambda\sqrt{2m^*}E_g^{3/2}}{3\hbar(E_g + \Delta\Phi)}\right). \tag{10.11}$$

A number of insights can be obtained from this simple expression together with Eq. (10.10): Obviously, the tunneling probability and hence the drain current can be made large, if λ is made small. Furthermore, the inverse subthreshold slope can either be easily calculated by noting that $S = ln(10)\left(\partial I_d/\partial V_{gs} \cdot 1/I_d\right)^{-1}$ or extracted from log-scale plots of $I_d - V_{gs}$. Figure 10.20 shows transfer characteristics calculated using expression (10.10) for four different cases specified in the box of Fig. 10.20 (note that for simplicity we did not take the impact of the charge in the channel on $\Delta\Phi$ into account which is justified in the off-state and also in the on-state as long as the tunneling probability is not too close to unity). When extracting the inverse subthreshold slope it is important to note that a steep subthreshold swing at some gate voltage is not sufficient: an averaged $\overline{S} < 60\,\text{mV/dec}$ over several orders of magnitude in drain current is needed for an FET based on BTBT to exhibit an off-state superior to a conventional device. In order to extract an average S we defined four orders of magnitude to be sufficient between the off-state current and the current at threshold (black horizontal line in Fig. 10.20). We will first concentrate on the curves belonging to devices with $E_f^s = 0.05\,\text{eV}$. It can then be seen that an average $S < 60$ mV/dec can only be obtained in the case of small effective mass and small gate oxide thickness (red solid line). In the other two cases S is smaller 60 mV/dec only in a limited range of $V_{gs} = -\Delta\Phi/q$ (green and blue curves). For proper

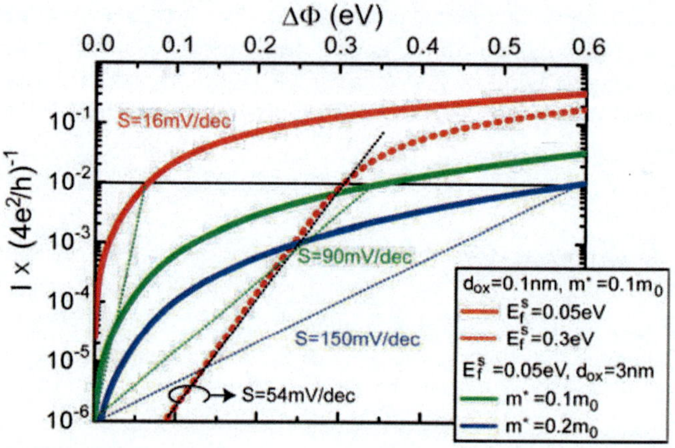

Fig. 10.20 Normalized drain current in a one-dimensional tunneling FET. A constant current of 10^{-2} was taken to determine the potential at threshold. The average inverse subthreshold slopes for the four different cases displayed are indicated by the *dashed lines* (taken from [37])

functioning of the tunneling FET the position of the Fermi energy in the source contact also plays an important role. If E_f^s becomes too large, the curly bracket term in Eq. (10.10) is approximately equal to $exp\left(-(\Delta\Phi - E_f^s)/k_BT\right)$ for small $\Delta\Phi$. Hence, an inverse subthreshold slope of approximately 60 mV/dec is obtained in the case of a large tunneling probability like in a conventional-type FET (see red dashed line in Fig. 10.20). The reason for this is that the low energetic exponential tail is not cut-off efficiently enough and the energetic window $\Delta\Phi$ in which carriers are injected into the channel now resides within the exponential tail of the source Fermi function [50].

In case of a large tunneling probability and rather small source Fermi energy S approximately equals $ln(10)/q \cdot \Delta\Phi$ meaning that in this case the particular conduction and valence band profiles represent a band-pass-filter effectively cutting off the high and low energy tails of the Fermi function. This band pass filter behavior can be seen in Fig. 10.21 which shows simulated conduction and valence band profiles together with the current spectrum in a tunneling FET for three different gate voltages. In the off-state (a) the band gap of source, channel and drain effectively block current transport leading to a low off-state leakage. However, once the valence band in the channel is lifted above the conduction band in source a conducting channel via BTBT is opened up. Looking at the current spectra in the device's on-state (red lines in Fig. 10.21b, c) it becomes apparent that a significant current contribution stems exclusively from the energetic window between the valence band edge in the channel and the conduction band edge in the source contact, i.e. the band gap in source and channel act as the bandpass filter mentioned above. Thus, in the tunneling FET, the injection of carriers occurs from an effectively "cooled" Fermi function and therefore inverse subthreshold slopes much steeper than 60 mV/dec are feasible.

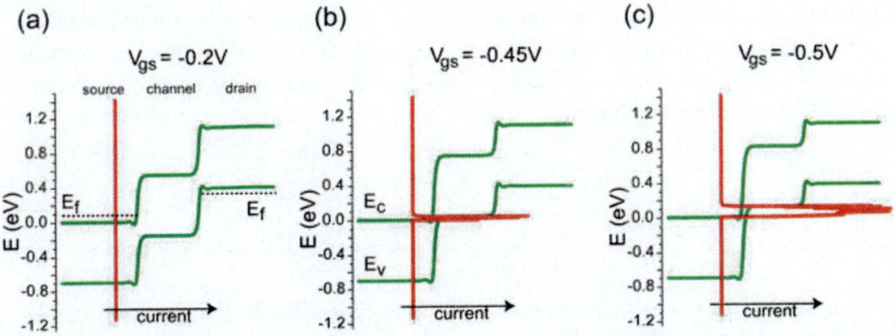

Fig. 10.21 Band profile in a tunneling FET with large BTB tunneling probability; (**a**) off-state, (**b**) slightly above threshold and (**c**) on-state. The *red line* is the current spectrum showing that the p-i-n structure represents a band-pass filter effectively cooling the source Fermi function [50]

The one-dimensionality of the nanotube tunneling FET has an interesting consequence which makes a 1D system very attractive for the realization of tunneling devices. For realistic systems the BTBT probability will always be smaller than unity and hence the on-state current of a tunneling FET is deteriorated compared to a conventional-type transistor. However, the device delay $\tau = C_g V_{dd}/I_d$ is a much more appropriate measure to quantify device performance as has been discussed above. Assuming the long-channel case (i.e. $\lambda \gg L$) the gate capacitance is given by $C_g = C_{ox}C_q/(C_{ox} + C_q)$ due to the serial combination of quantum and geometrical oxide capacitance [25, 44, 45]. For large enough drain-source bias the quantum capacitance can be calculated using the WKB approximation. Since T_{WKB} is an exponential, after differentiation it turns out that $C_q \propto T_{WKB}$:

$$
\begin{aligned}
C_q \approx \frac{\partial}{\partial \Phi_f^0} Q &= e \frac{\partial}{\partial \Phi_f^0} T_{WKB} \int dE\, D(E) \left(1 - f(E_f^s)\right) \\
&= \frac{4\lambda \sqrt{2m^*} E_g^{3/2}}{3\hbar (E_g + \Delta\Phi)^2} T_{WKB} \int dE\, D(E) \left(1 - f(E_f^s)\right) \\
&\quad + e T_{WKB} \frac{\partial}{\partial \Phi_f^0} \int dE\, D(E) \left(1 - f(E_f^s)\right)
\end{aligned}
\tag{10.12}
$$

where $D(E)$ is the one-dimensional density of states. In 1D systems the quantum capacitance limit can be reached where $C_{ox} >> C_q$ due to the decreasing density of states. In a tunneling FET C_q will be even smaller than in a conventional-type FET due to the proportionality to T_{WKB} (this is true, as long as E_f^d is larger than the valence band in the channel). This means that in a wrap-gate tunneling FET with thin gate dielectric it is very likely that the quantum capacitance limit is reached and thus $C_g \approx C_q$. At the same time, $I_d \propto T_{WKB}$ (cf. Eq. (10.10)) and therefore, the device delay $C_g V_{dd}/I_d \approx C_q V_{dd}/I_d$ becomes independent of T_{WKB}. This means that to first order the presence of the BTBT barrier does not deteriorate the on-state of 1D tunneling FETs in terms of the device delay. Furthermore, in case of a small λ, i.e. small nanotube diameter and thin gate oxide thickness, and well in the on-state (large $\Delta\Phi > 0$) the first term of Eq. (10.12) (lower part) can be neglected. In this case the device delay is given by

$$\frac{C_g V_{dd}}{I_d} \approx \frac{e\partial/\partial\Delta\Phi \int dE\, D(E)\left(1 - f(E_f^s)\right) V_{dd}}{2e/h \int dE \left(1 - f(E_f^s)\right)} \qquad (10.13)$$

which is the result one obtains in a conventional-type FET in the quantum capacitance limit. As a result, one-dimensional systems such as carbon nanotubes are ideally suited for the realization of tunneling FETs since one-dimensional tunneling FETs allow combining a high on-state performance with steep subthreshold swings.

10.6.2 Vertical Scaling of Tunneling CNFETs

In order to get a large BTBT probability the exponent in Eq. (10.11) must be as small as possible. At the point where the tunneling FET starts to switch (i.e. where $\Delta\Phi \approx 0$) the tunneling probability scales as $\sqrt{E_g m^*}\lambda$ and thus a large tunneling probability can be obtained in principle by making all three quantities small. However, in reality certain trade-offs and dependencies have to be taken into account. Firstly, decreasing E_g improves the tunneling probability but also leads to a rapid increase of the off-state leakage since the latter depends exponentially on the size of the energy gap. Furthermore, the tunneling FET also shows ambipolar behavior and only yields low off-state currents when the semiconductor gap in the channel area blocks electron injection from the valence band of the drain contact into the conduction band of the channel as well as hole injection from the conduction band in source into the valence band in the channel area at the same time. Consequently, a reduction of the energy gap lowers the applicable drain-source bias range accordingly. Secondly, when scaling down the diameter of the active channel material one has to take into account the dependence of the effective mass and the energy gap on the diameter. The particular dependencies on d_{nt} are advantageous in carbon nanotubes: E_g as well as m^* scale as $1/d_{nt}$ [3, 51]. On the other hand, in a wrap-gate architecture $\lambda = \sqrt{\varepsilon_{nt} d_{nt}^2 ln(1 + 2d_{ox}/d_{nt})/8\varepsilon_{ox}}$ and as a result, the product $\sqrt{E_g m^*}\lambda$ shows only a weak dependence on the diameter (it only strongly increases if $d_{nt} < 1\,\text{nm}$) with increasing the diameter leading to a slight improvement of the tunneling probability. For a desired size of energy gap and hence nanotube diameter, scaling the gate dielectric thickness and in particular employing high-k gate dielectrics are the most effective means to decrease λ, leaving E_g well as m^* unaffected [49, 52].

Figure 10.22 shows a simulated transfer curve of a tunneling CNFET with a wrap-gate. Rather aggressively scaled geometrical parameters with $d_{nt} = 1\,\text{nm}$ and $d_{ox} = 1\,\text{nm}$ are chosen in the present case (see Fig. 10.22a). The device exhibits

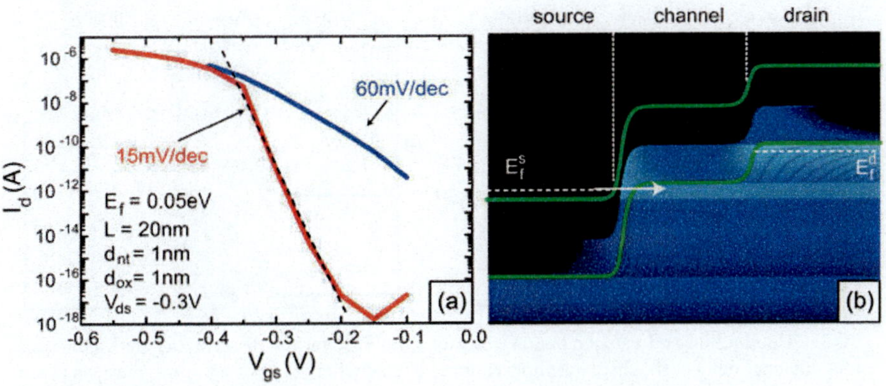

Fig. 10.22 Transfer characteristics of a 1D tunneling CNFET (**a**). (**b**) displays the local hole distribution showing that holes are only injected within the energetic window $\Delta\Phi$

an inverse subthreshold slope of 15 mV/dec over several orders of magnitude. For comparison, the inverse subthreshold slope of a conventional-type device is shown as well (black line). Figure 10.22b displays a log-scale plot of the local hole distribution, again showing that holes are injected into the channel only within the energetic window between the conduction band in source and the valence band in the channel.

Similar to SB-FETs, in tunneling FETs the gate oxide should be as small as possible in order to guarantee a thin BTBT barrier and therefore a sufficiently high on-state current as well as steep inverses subthreshold slopes.

10.6.3 Lateral Scaling of Tunneling CNFETs

In the present section we want to address two important points related to the scaling of the channel length in tunneling FETs. First, it is often said that tunneling FETs exhibit less short channel effects compared to conventional devices since the tunneling process only happens at the source-channel interface. However, similarly to conventional-type FETs the n-p junctions at the contact-channel interfaces have a spatial extend on the order of λ (see gray-shaded area in Fig. 10.23a). For a tunneling FET to show steep inverse subthreshold slopes it is necessary that the gate has a good electrostatic control to manipulate the bands one-to-one with changing gate voltage. This requires that the channel length must be significantly larger than λ which is the same requirement as for suppressing short channel effects in a conventional-type MOSFET.

Figure 10.23a, b show conduction and valence band profiles in the case of a properly designed tunneling FET (a) and a device that exhibits short channel effects (b). In the case (a), the channel length is larger than the two n-p and p-n junctions (light gray-shaded area) such that a good gate control of the bands in the channel area can be expected. In the device (b) on the other hand the n-p and the p-n junctions significantly overlap (dark gray-shaded area) leading to a potential profile in

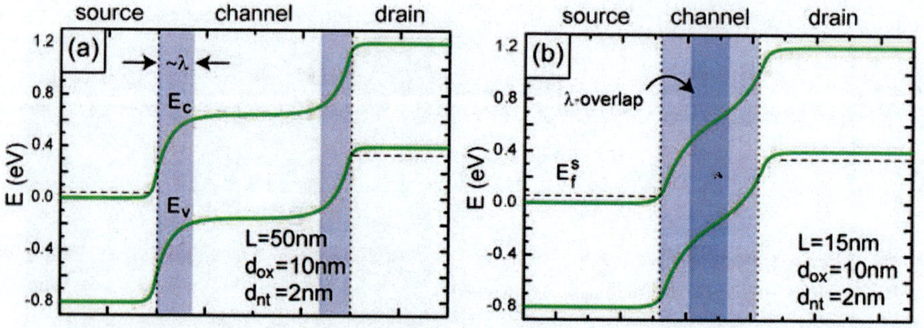

Fig. 10.23 Conduction and valence bands in a tunneling FET in the off-state. (**a**) shows the case of a long channel device; (**b**) displays short channel device where the n-p and p-n junctions at the contact channel interfaces overlap significantly

the channel that is to a large extend determined by the drain potential rather than the gate. As a result, device (b) exhibits short channel effects leading to a much larger inverse subthreshold slope and drain-induced-barrier-thinning. Due to the n-i-p structure, however, a tunneling transistor that suffers from short channel effects can still exhibit small off-state currents (if the device is not scaled to such small channel lengths that direct source to drain tunneling leads to an increase of the off-state current). The averaged inverse subthreshold slope on the other hand will in the case of short channel effects be significantly larger than 60 mV/dec. Figure 10.24 displays transfer characteristics of tunneling FETs for different channel lengths; the smaller the channel lengths the more pronounced do short channel effects appear. Obviously, tunneling FETs that are not properly scaled show an increase of the inverse subthreshold slope diminishing the major benefit of the tunneling FET architecture.

The second important point that we want to address in this section is the insensitivity of the drain current to changes in the channel length. It is often argued

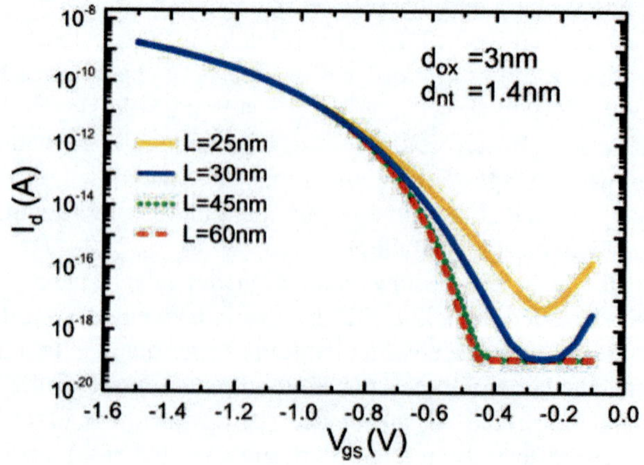

Fig. 10.24 Transfer characteristics of tunneling CNFETs with different channel lengths. Devices with small channel length exhibit short channel effects

that this is a major advantage of a tunneling FET. However, a tunneling FET is a contact-switching device very similar to a SB-FET. Hence, the same argument about the relevance of scattering applies in the present case: If the BTB tunneling is the main scattering event then scattering within the channel does not play a role anymore unless the channel is made sufficiently long. Again, the smaller the BTBT probability the longer the channel has to be in order for the scattering in the channel to dominate. As a result, the price for the insensitivity of the tunneling FET to channel length variations is a low BTBT probability and hence a low on-state current. On the other hand, if the BTBT probability is made large (i.e. λ is made small) much smaller channel lengths are sufficient to obtain scattering dominated transport through the device. Equation (10.11) allows obtaining an estimate of the channel length required for the scattering transport in the channel to become dominant. The transmission function due to scattering for carriers flowing through the channel can be estimated to be $T_{scat} \approx l_{scat} / (l_{scat} + L)$. Then the overall transmission function is given as [38]

$$T_{tot} = \frac{T_{WKB} T_{scat}}{T_{WKB} + T_{scat} - T_{WKB} T_{scat}} = \frac{1}{exp\left(\frac{4\lambda\sqrt{2m^* E_g^{3/2}}}{3q\hbar(E_g + \Delta\Phi)}\right) + \frac{L}{l_{scat}}}. \quad (10.14)$$

This means, scattering in the channel becomes significant only for minimum channel lengths $L_{min} \geq l_{scat}\, exp\left((4\lambda\sqrt{2m^* E_g^{3/2}}) / (3q\hbar(E_g + \Delta\Phi))\right)$. The exponential dependence of L_{min} on λ on the other hand shows that for a realistic device geometry the minimum channel length can be rather long. In particular, if one considers planar tunneling FETs in silicon-on-insulator, for instance, the minimum channel length can easily be in the few micrometer range. Therefore, scattering in tunneling FETs usually can be neglected. However, one has to keep in mind that this implies a deteriorated on-state current due to a small BTB tunneling probability.

10.6.4 Experimental Realization

As a first attempt to realize a tunneling FET experimentally we have measured the dual-gate nanotube FET presented in Section 10.4. The device essentially is a p-i-p structure where the p-doped segments are realized with electrostatic "doping" of the source/drain extensions between the actual gate and metallic contact electrodes [11]. For negative gate voltages the device acts as a conventional-type transistor as has been discussed above: Negative gate voltages pull up the conduction and valence bands in the channel such that the thermal emission of holes increases. This operation mode and the respective band diagrams are schematically shown on the left of the main panel of Fig. 10.25. However, if large positive gate voltages are applied, the conduction and valence bands are pushed down far enough such that a channel for BTBT opens up between the valence band in source and the conduction band in

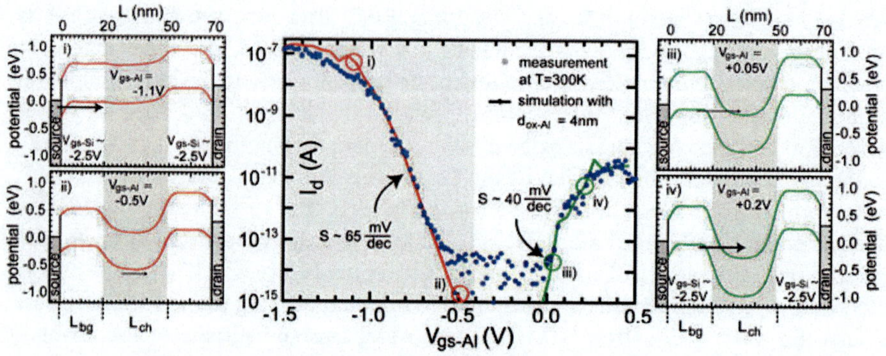

Fig. 10.25 Transfer characteristic of a dual-gate CNFET. For negative gate voltages the device acts as a C-CNFET, for positive gate voltages as a T-CNFET with an inverse subthreshold slope of 40 mV/dec. The band profiles to the *left* and *right* show the main path for current flow in the respective operation mode (taken from [11])

the channel as indicated by the arrows in the band diagrams shown in Fig. 10.25 (to the right of the main panel). For increasing gate voltage the barrier becomes thinner and consequently the current increases. In the present case, we observe an inverse subthreshold slope in the BTB tunneling regime of $S = 40\,\text{mV/dec}$ which is to the best of our knowledge the first experimental demonstration of transistor operation with a slope better than 60 mV/dec due to controlled BTB tunneling in any material system [11].

In Fig. 10.25 we also plot the result of a simulation using the model presented above (solid lines). In order to keep the computational burden as small as possible a length $L_{bg} = 20\,\text{nm}$ and $L_{ch} = 30\,\text{nm}$ were found to be well suited to describe the experimental device structure. The measured and the simulated transfer curves are in excellent agreement showing that the model captures well the essential transport mechanism in our dual-gate CNFET and more importantly it shows that the interpretation of the experimental data in terms of BTB tunneling is appropriate. Note that the rather low on-state current in the BTBT branch of the characteristics shown in Fig. 10.25 is due to the fact that the dual-gate structure is not ideal for a tunneling FET since the carriers have to tunnel through two potential barriers. Optimizing the gate oxide thicknesses and in particular realizing a p-i-n structure will greatly increase the current level in the device's on-state (see for instance [21, 42, 47]).

10.7 Conclusion

Carbon nanotube field effect transistors represent a relatively new class of field-effect transistor devices exhibiting very promising electrical characteristics as well as new and interesting device physics aspects. In the present chapter we studied the electronic transport in three different CNFET architectures. As it turned out, the inherently small diameter of carbon nanotubes and their cylindrical shape make them ideal objects for ultimately scaled devices. In addition, the one-dimensionality of the electronic transport in carbon nanotubes leads to distinctly different behavior if compared to bulk-like MOSFET devices. For instance, a charge pileup in the off-state of conventional-type CNFETs leads to short channel effects although the device is electrostically well behaved. Furthermore, due to the one-dimensional transport device operation in the quantum capacitance limit becomes attainable which is beneficial for the transistor performance of e.g. tunneling nanotube FETs.

From an application point of view, the SB-CNFET is the easiest structure to fabricate but exhibits some drawbacks: the presence of the Schottky barrier yields either inverse subthreshold slopes significantly larger than 60 mV/dec if nanotubes with larger diameter are employed. Or improving the injection of carriers by using small diameter tubes and thin gate oxides leads to a rather large off-state leakage current. The C-CNFET shows characteristics superior to the SB-CNFETs. However, due to the charge pileup in these structures, the drain source-voltage should not exceed half of the band gap since otherwise short channel effect appear leading to large off-state leakage currents. The third transistor structure, the T-CNFET, shows very promising characteristics since this device architecture allows combining a superior off-state with inverse subthreshold slopes significantly smaller than 60 mV/dec together with a high on-state performance.

References

1. Duerkop, T.; Kim, B.M. and Fuhrer, M.S.; Properties and application of high-mobility semiconducting nanotubes, *J. Phys.: Condens. Matter*, **16**, R553–R580 (2004).
2. Javey, A.; Guo, J.; Wang, Q.; Lundstrom, M. and Dai, H.; Ballistic carbon nanotube field-effect transistors, *Nature*, **424**, 654–657 (2003).
3. Saito, R.; Dresselhaus, G. and Dresselhaus, M.S.; Physical properties of carbon nanotubes, Imperial College Press, London, 1998.
4. Appenzeller, J.; Lin, Y.-M.; Knoch, J.; Chen, Z. and Avouris, Ph.; 1/f noise in carbon nanotube devices – on the impact of contacts and device geometry, *IEEE Trans. Nanotechnol.*, **6**(3), 368–373 (2007).
5. Appenzeller, J.; Knoch, J.; Radosavljevic, M. and Avouris, Ph.; Multi-mode transport in Schottky barrier carbon nanotube field-effect transistors, *Phys. Rev. Lett.*, **92**, 226802-1-4 (2004).
6. Appenzeller, J.; Knoch, J.; Martel, R.; Derycke, V.; Wind, S. and Avouris, Ph.; Short-channel like effects in Schottky barrier carbon nanotube field-effect transistors, *Internat. Electron Dev. Meeting 2002, Tech. Dig.*, 285–288 (2002).
7. Appenzeller, J.; Knoch, J.; Derycke, V.; Wind, S. and Avouris, Ph.; Field-modulated carrier transport in carbon nanotube transistors, *Phys. Rev. Lett.*, **89**, 126801-1-4 (2002).
8. Chen, Z.; Appenzeller, J.; Knoch, J.; Lin, Y.-M. and Avouris, Ph.; The role of metal-nanotube contact in the performance of carbon nanotube field-effect transistors, *Nano Lett.*, **5**(7), 1497–1502 (2005).
9. Guo, J.; Datta, S. and Lundstrom, M.; Assesment of silicon MOS and carbon nanotube FET performance limits using a general theory of ballistic transistors, *Internat. Electron Dev. Meeting, Tech. Dig.*, 711–714 (2002).
10. Zhou, Ch.; Kong, J. and Dai, H.; Electrical measurements of individual semiconducting single-walled carbon nanotubes of various diameters, *Appl. Phys. Lett.*, **76**, 1597–1599 (2000).
11. Appenzeller, J.; Lin, Y.-M.; Knoch, J. and Avouris, Ph.; Band-to-band tunneling in carbon nanotube field-effect transistors, *Phys. Rev. Lett.*, **93**(19), 196805 (2004).

12. Knoch, J.; Riess, W. and Appenzeller, J.; Outperforming the conventional scaling rules in the quantum-capacitance limit, *IEEE Electron Dev. Lett.*, **29**(4), 372–374 (2008).
13. Young, K.K.; Short-channel effect in fully-depleted SOI MOSFET's, *IEEE Trans. Electron Dev.*, **36**, 399–402 (1989).
14. Franklin, N.R.; Wang, Q.; Tombler, T.W.; Javey, A.; Shim, M. and Dai, H.; Integration of suspended carbon nanotube arrays into electronic devices and electromechanical systems, *Appl. Phys. Lett.*, **81**, 913–915 (2002).
15. Huang, Sh.; Woodson, M.; Smalley, R. and Liu, J.; Growth mechanism of oriented long single walled carbon nanoubes using fast heating chemical vapor deposition process, *Nano Lett.*, **4**, 1025–1028 (2004).
16. Wang, Y.; Kim, M.J.; Shan, H.; Kittrell, C.; Fan, H.; Ericson, L.M.; Hwang, W.-F.; Arepalli, S.; Hauge, R.H. and Smalley, R.; Continued growth of single-walled carbon nanotubes, *Nano Lett.*, **5**, 997–1002 (2004).
17. Yin, A.; Tzolov, M.; Cardimona, D.A. and Xu, J.; Template-growth of highly ordered carbon nanotube arrays on silicon, *IEEE Trans Nanotechnol.*, **5**, 564–567 (2006).
18. Hannon, J.B.; Afzali, A.; Klinke, Ch. and Avouris, Ph.; Selective placement of carbon nanotubes on metal-oxide surfaces, *Langmuir*, **21**, 8569–8671 (2005).
19. Hedberg, J.; Dong, L. and Jiao, J.; Air flow technique for large scale dispersion and alignment of carbon nanotubes on various substrates, *Appl. Phys. Lett.*, **86**, 143111-1-4 (2005).
20. Huang, X.M.H.; Caldwell, R.; Huang, L.; Jun, S.C.; Huang, M.; Sfeir, M.Y.; O'Brien, S.P. and Hone, J.; Controlled placement of individual carbon nanotubes, *Nano Lett.*, **5**, 1515–1518 (2005).
21. Zhang, G.; Wang, X.; Li, X.; Lu, Y.; Javey, A. and Dai, H.; Carbon nanotubes: From growth, placement, and assembly control to 60 mV/decade and sub-60 mV/decade tunnel transistors, *IEEE Internat. Electron Dev. Meeting, Technical Digest* (2006).
22. Bachtold, A.; Hadley, P.; Nakanishi, T. and Dekker, C.; Logic circuits with carbon nanotube transistors, *Science*, **294**, 1317 (2001).
23. Javey, A.; Wang, Q.; Ural, A.; Li, Y. and Dai, H.; Carbon nanotube transistor arrays for multistage complementary logic and ring oscillators, *Nano Lett.*, **2**, 929 (2002).
24. Chen, Z.; Appenzeller, J.; Lin, Y.-M.; Sippel-Oakley, J.; Rinzler, A.G.; Tang, J.; Wind, S.J.; Solomon, P.M. and Avouris, Ph.; An integrated logic circuit assembled on a single carbon nanotube, *Science*, **311**(5768), 1735 (2006).
25. Knoch, J. and Appenzeller, J.; Tunneling phenomena in carbon nanotube field-effect transistors, *Phys. Stat. Solidi A*, **205**, 679–694 (2008).
26. Sze, S.M.; Physics of semiconductor device, Wiley, New York, 1981.
27. Taur, Y. and Ning, T.H.; Fundamentals of modern VLSI devices, Cambridge University Press, Cambridge, 1998.
28. Yan, R.-H.; Ourmazd, A. and Lee, K.F.; Scaling the Si MOSFET: From bulk to SOI to bulk, *IEEE Trans. Electron Dev.*, **39**, 1704–1710 (1992).
29. Wind, S.J.; Appenzeller, J. and Avouris, Ph.; Lateral scaling in carbon – nanotube transistors, *Phys. Rev. Lett.*, **91**, 058301-1-4 (2003).
30. Auth, Ch. and Plummer, J.D.; Scaling theory for cylindrical, fully depleted, surrounding gate MOSFET's, *IEEE Electron Dev. Lett.*, **18**, 74–76 (1997).
31. Knoch, J.; Mantl, S.; Lin, Y.-M.; Chen, Z.; Avouris, Ph. and Appenzeller, J.; An extended model for carbon nanotube field-effect transistors, *Dev. Res. Conf. 2004, Conf. Dig.*, 135–136 (2004).
32. Knoch, J.; Zhang, M.; Mantl, S. and Appenzeller, J.; On the performance of single-gated ultrathin-body SOI Schottky-barrier MOSFETs, *IEEE Trans. Electron Dev.*, **53**(7), 1669–1674 (2006).
33. Appenzeller, J.; Radosavljevic, M.; Knoch, J. and Avouris, Ph.; Tunneling versus thermionic emission in one-dimensional semiconductors, *Phys. Rev. Lett.*, **92**, 048301-1-4 (2004).
34. Knoch, J. and Appenzeller, J.; in Carbon nanotube field-effect transistors – the importance of being small. Hardware technology drivers of ambient intelligence, Springer, Berlin, 2006.

35. Lin, Y.-M.; Appenzeller, J.; Knoch, J. and Avouris, Ph.; High-performance carbon nanotube field-effect transistor with runable polaritites, *IEEE Trans. Nanotechnol.*, **4**(5), 481–489 (2005).
36. Knoch, J.; Zhang, M.; Appenzeller, J. and Mantl, S.; Physics of ultrathin-body silicon-on-insulator Schottky-barrier field-effect transistors, *Appl. Phys. A*, **87**(3), 351–357 (2007).
37. Appenzeller, J.; Knoch, J.; Björk, M.T.; Riel, H.; Schmid, H. and Riess, W.; Toward nanowire electronics, *IEEE Trans. Electron Dev.*, **55**, 2827–2845 (2008).
38. Datta, S.; Electronic Transport in Mesoscopic Systems, *Cambridge University Press*, Cambridge and New York, 1995.
39. Guo, J. and Lundstrom, M.; Role of phonon scattering in carbon nanotube field-effect transistors, *Appl. Phys. Lett.*, **86**, 193103 (2005).
40. Knoch, J. and Appenzeller, J.; Impact of the channel thickness on the performance of Schottky barrier metal-oxide-semiconductor field-effect transistors, *Appl. Phys. Lett.*, **81**, 3082–3084 (2002).
41. Lin, Y.-M.; Appenzeller, J. and Avouris, Ph.; Novel structures enabling bulk switching in carbon nanotube FETs, *Dev. Res. Conf., Conf. Dig.*, 133–134 (2004).
42. Appenzeller, J.; Lin, Y.-M.; Knoch, J.; Chen, Z. and Avouris, Ph.; Comparing carbon nanotube transistors – the ideal choice: a novel tunneling device design, *IEEE Trans. Electron Dev.*, **52**(12), 2568–2576 (2005).
43. Knoch, J.; Mantl, S. and Appenzeller, J.; Comparison of transport properties in carbon nanotube field-effect transistors with Schottky contacts and doped source/drain contacts, *Solid-State Electron*, **49**, 73–76 (2005).
44. John, D.L.; Castro, L.C. and Pulfrey, D.L.; Quantum capacitance in nanoscale device modeling, *J. Appl. Phys.*, **96**, 5180–5184 (2004).
45. Luryi, S.; Quantum capacitance devices, *Appl. Phys. Lett.*, **52**, 501–503 (1988).
46. Bhuwalka, K.K.; Novel tunneling devices for future CMOS technologies, PhD thesis, University of the German Armed Forces, Munich, 2005.
47. Koswatta, S.O.; Nikonov, D.E. and Lundstrom, M.S.; Computational study of carbon nanotube p-i-n Tunnel FET, *IEEE Internat. Electron Dev. Meeting, Tech. Dig.* (2005).
48. Zhang, Q.; Zhao, W. and Seabaugh, A.; Low-subthreshold-swing tunnel transistors, *IEEE Electron Dev. Lett.*, **27**(4), 297–300 (2006).
49. Knoch, J. and Appenzeller, J.; A novel concept for field-effect transistors – the tunneling carbon nanotube FET, *Dev. Res. Conf., Conf. Digest*, 153–156 (2005).
50. Knoch, J.; Mantl, S. and Appenzeller, J.; Impact of the dimensionality on the performance of tunneling FETs: bulk versus one-dimensional devices, *Solid-State Electron*, **51**(4), 572–578 (2007).
51. Pennington, G. and Goldsman, N.; Low-field semiclassical carrier transport in semiconducting carbon nanotubes, *Phys. Rev. B*, **71**, 205318 (2005).
52. Boucart, K. and Ionescu, A.; Double-gate tunnel FET with high-k gate dielectric, *IEEE Trans. Electron Dev.*, **54**(7), 1725–1733 (2007).

Chapter 11
Inorganic Nanotubes

Maja Remskar

Abstract Since the first report on synthesis of the WS_2 nanotubes in 1992, the number of articles on successful growth of different inorganic nanotubes increases rapidly revealing the importance of this field for nanotechnology. Although some geometrical similarities with carbon nanotubes, inorganic nanotubes distinguish themselves by important peculiarities, from the growth mechanisms to the physical and chemical properties attractive for possible applications. Their structural properties, important for potential applications for biosensors, drug delivery, safe containers, nanoreactors and strengthening fibres are discussed in this chapter.

11.1 Introduction

Trend of miniaturization of electronic components by top down process is hindered by the current fabrication technology and challenged by increased power consumption and dissipation of heat. New visions are needed using low-dimensional phenomena and self-assembly. The particular physical and chemical properties of quasi one- and two-dimensional solids promise new inventions, new products and fresh contributions to human knowledge. Single molecules, single atoms or even single electrons coupled with strong quantum size effects constitute the limit stage in miniaturization. In addition to the size effects, the geometry of low dimensional materials is an important conceptual novelty in materials research. Creation of porous structures and channels is a natural way of thermodynamic stabilization of complex materials. The complementary forms to voids and pores are spheres and nanotubes, preferentially grown by spontaneous minimization of free energy by saturation of dangling bonds in self-closed nanoobjects. Mismatch strain at heterogeneous nucleation, local atmosphere in confined geometry as well as defect concentration and their distribution during growth play critical role in crystal structure and morphology of one dimensional materials, like nanowires and nanotubes. While in the emerging nanotechnology the nanowires are useful for interconnections between device elements, the nanotubes can serve as passive or/and active

M. Remskar (✉)
Jozef Stefan Institute, SI-1000 Ljubljana, Slovenia
e-mail: Maja.Remskar@ijs.si

O. Hayden, K. Nielsch (eds.), *Molecular- and Nano-Tubes*,
DOI 10.1007/978-1-4419-9443-1_11,

electronic component depending on their electric, magnetic and size-effect properties. Their passive roles do matter as charge carriers' media or as containers of encapsulated active components, while in the active role they represent a central carrier of a particular electronic property, like an electromagnetic object, transistor, field emitter of electrons, photoactive element, etc. The ultimate goal of reduction of the device size to nano scale meets both active and passive applications of nanotubes.

Since 1991, when S. Ijima [1] reported on carbon nanotubes, and since 1992 with the first report on inorganic nanotubes by R. Tenne [2], nanotubes have become a symbol of the new and fast developing nanotechnology. Intensive enthusiasm for the new form of carbon has enabled fast growth of carbon nanotube community and intensive research in the field. Inorganic nanotubes remained in a shadow of carbon counterparts nearly for a decade. The chronological list of the very first reports of different kinds of inorganic nanotubes or rolled-up structures shows a level of activity in this field in the first decade: 1992 – WS_2 [2]; 1993 – MoS_2 [3]; 1995 – BN [4], SiO_2 [5]; 1998 – TiO_2 [6], VO_x [7], $NiCl_2$ [8]; 2000 – $NbSe_2$ [9], Au [10], Co and Fe [11]; 2001– CdS [12], CdSe [13], ZnS [14], NiS [15], $Cu_{5.5}FeS_{6.5}$ [16], Al_2O_3 [17], In_2O_3 and Ga_2O_3 [18], GaN [19]; 2002 – ZrS_2 and HfS_2 [20], NbS_2 and TaS_2 [21], (Er, Tm, Yb, Lu) oxide [22], ZnO [23], $BaTiO_3$ and $PbTiO_3$ [24], Cu and Ni [25], Te [26], ReS_2 [27], silicon nanotubes [28], etc.

At present, nanotubular and fullerene-like allotropes of other chalcogenides as well as halogenides and oxides have been discovered and characterized [29–32]. Structural, optical and electrochemical properties of inorganic nanotubes, particularly MoS_2, WS_2, and of fullerene-like particles have been reviewed by Tenne [33, 34]. Methods for synthesizing inorganic nanotubes, together with simulations of their structures and predictions of their properties, have been reviewed by Ivanovskii [35]. Morphology properties were reviewed in 2004 [36]. The number of compounds synthesized in cylindrical shapes is continuously growing as well as variety of preparation techniques [31, 32]. While carbon nanotubes are already in a stage of applications, unusual properties of nanotubes made from inorganic materials offer intriguing possibilities for applications [37]. The MoS_2 and WS_2 nanotubes and fullerene-like particles, which still remain the most investigated non-carbon nanostructures, show superior catalytic and tribological properties compared with the bulk allotropes. They can be intercalated with alkali metals in the same way as the parent bulk forming new materials for micro and nano-electronics and for electrochemistry [38]. Electronic properties of MoS_2 and WS_2 are similar to silicon (an indirect band gap of 1.2 versus 1.12 eV for silicon). However, S-Mo-S or S-W-S layers with saturated surfaces are much more resistant to oxidation and humidity effects, what constitutes a considerable advantage of MoS_2 compared with semiconductors of the groups IV (e.g. Si), III-V (e.g. GaAs), II-VI (e.g. CdS) and can make MoS_2 and WS_2 appropriate in nanoscale applications [39]. The only available data are on transport properties of fullerene-like WS_2 nanoparticles published recently [40] revealing a comparable carrier density $p(300\ K) = 4 \times 10^{16}\ cm^{-3}$, while mobility was found to be very small ($0.2\ cm^2V^{-1}s^{-1}$) comparing to bulk WS_2 ($30\ cm^2V^{-1}s^{-1}$). The results are blurred by an influence of water and hydrogen in the sample.

Several applications of inorganic nanotubes can be foreseen in different industrial branches, like as nanoelectronic components, high-frequency modulators, polymer-composite materials, lubricants, shock resistance components, solar cell application, environmental purification or disinfection media, catalysis in green chemistry, biosensors, drug delivery, safe containers, nanoreactors, strengthening fibres, and others. Although there are some geometrical similarities to carbon nanotubes, inorganic nanotubes are distinguished by important peculiarities, from their growth mechanisms to the various physical and chemical properties that are attractive for potential applications. Their structural and morphological properties determining chemical stability and activity, transport properties, mechanics and interaction with other building blocks in composites, are discussed in this chapter.

11.2 Stability of Inorganic Nanotubes

The proposed basis of the growth mechanism of inorganic nanotubes is the lack of resistance to bending of thin quasi two-dimensional crystal flakes [41]. This bending can be spontaneous, as in transition metal dichalcogenide [42] nanotubes (NTs) and in misfit layer compounds [43] grown from the vapour phase, or geometrically conducted by a template growth in channels of alumina membrane [18, 28] and as coatings of nanofibres [44, 45]. Cylindrical structure as an intrinsic property of some solids appears spontaneously from vapor or liquid phase only in a few compounds, like MoS_2, WS_2, $K_4Nb_6O_{17}$-$(C_nH_{2n+1}NH_3)^+$ [46], while ternary or quarterly compounds usually need template method.

Theoretical considerations suggest that, due to the strain energy coupled with bending, the narrow tubular structure might be less energetically favoured than a finite strip. But when the diameter passes a critical value, the strain in the tubes becomes smaller than the energy associated with the edges (dangling bonds) in the layered strips and the cylindrical geometry, with self-closed layers, becomes clearly the most stable structure [47].

The simple rolling of crystal flake does not usually lead to nanotube longitudinal growth if the stacking order and orientation relationship between two adjacent turns is not satisfied. It is true that the geometry of the folded flake is cylindrical, but such a formation can only grow in the radial direction until the strain energy no longer exceeds the energy gained by van der Waals interactions between layers. The interaction between the layers is usually weak and interlayer distances are arbitrary.

On the contrary, if the conditions for stacking order and orientation relationship are satisfied, the interaction between the adjacent turns is stronger, but two limit cases still exist from the macroscopic point of view. In the case where van der Waals interaction energies between the planes are much smaller than those associated with covalent bonds, dislocations will be generated on bending. In other cases, dislocations will not form on bending and the system will remain coherent [48, 49]. Microscopically, the matching of adjacent layers at the atomic level, with consequent extension and/or contraction effects, contributes to the elastic energy caused by the enlarged circumference and by the bending of the molecular layers. The total elastic energy per primitive cell is larger in nanotubes than in microtubes, where the elastic energy is distributed over a larger number of atoms.

Quantum-mechanical calculations for tubes with radii of 4–13 nm have shown that the strain energy for the smallest MoS_2 tubes is at least one order of magnitude larger than that of carbon nanotubes with a similar diameter [50]. The rolling of a triple-atomic layer S-Mo-S into a narrow tube leads to increased steric crowding and depletion effects which result in a higher energy penalty than the one of rolling a single-atomic graphite monolayer. It also explains a preference for the formation of uncapped MoS_2 tubes with open ends. As an example, MoS_2 tubes, with a diameter larger than 6.2 nm, have been shown to be more stable than stripes $\pi \times 6.2$ nm wide. In accordance with these calculations, the smallest inner diameter for single-wall NT should not be smaller than 6 or 7 nm. The smallest stable multi-wall NT has an inner diameter about 5 nm and a corresponding outer diameter of about 12 nm or larger [47].

Taking into account that the number of atoms N in a tube is proportional to R, the total energy per atom of a single-walled tube may be written as [51]:

$$\frac{E_{tot}}{N} = \varepsilon_\infty + \frac{\beta}{N^2}.$$

ε_∞ is the energy/atom in a 2D infinite monolayer. The second term is the strain energy which occurs as a consequence of the bending, when a monolayer is rolled up to a tube. The parameter $\beta = 900$ eV·atom can be obtained directly from calculations of the total energy for single-walled nanotubes [52]. For multi-walled nanotubes, the stabilizing inter-wall van der Waals interaction has to be added to the sum of the total energies for separate single-walled tubes. The expression for the total energy of a multi-walled nanotube consisting of *k* single-walled tubes may be written as:

$$\frac{E_{tot}}{N} = \varepsilon_\infty + \frac{\beta}{N}\sum_{i=1}^{k}\frac{1}{N_i} + \frac{k-1}{k}\varepsilon_{vdW}; \qquad N = \sum_{i=1}^{k} N_i.$$

ε_{vdW} is the interlayer vander-Waals energy/atom. It is evident, that every MoS_2 single-walled nanotube is energetically less preferable than a monolayer, because of the high strain energy. The additional van der-Waals interaction between the walls rapidly stabilizes multi-walled nanotubes compared with a monolayer already for about 200 and more atoms in the unit cell.

These findings explain a preferential multiwall structure of inorganic nanotubes. There are some indications that symmetry rules increase stability if number of molecular layers matches the period of a unit cell in the corresponding direction. Even number of molecular layers in accordance with 2Hb polytypic stacking was found in narrow MoS_2 and WS_2 nanotubes. As example, number of molecular layers in a WS_2 nanotube prepared by sulphurization of W_5O_{14} nanowire decreases by two molecular layered steps from 8, 6, to 4 (Fig. 11.1).

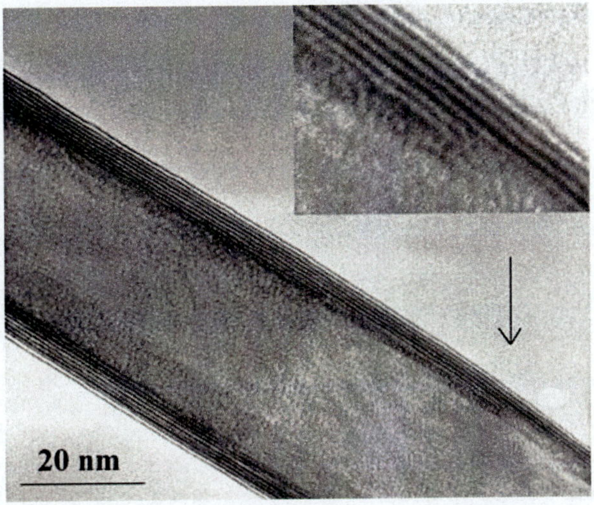

Fig. 11.1 A TEM image of a WS_2 nanotube with 8, 6, and 4 molecular layers in a wall thickness and an enlarged view of a two-molecular layer step at the inner nanotube's surface (*inset*). Figure originally published in Nano Letters [91]

11.3 Variety of Inorganic Nanotubes

Several families of inorganic nanotubes and fullerene-like particles have been synthesized up to now. The current list of some representative compounds for each of these families is as follows:

(a) Transition metal chalcogenide NTs: MoS_2 [3], $MoSe_2$ [53], WS_2 [2], WSe_2 [53], NbS_2 [21], $NbSe_2$ [9], TaS_2 [21], ZrS_2 [20], HfS_2 [20], TiS_2 [54], ZnS [14], NiS [15], CdSe [13], CdS [12], VS_2 [55];

(b) Oxide NTs:

- transition metal oxides: TiO_2 [6], ZnO [23], GaO/ZnO [56], VO_x [57], $W_{18}O_{49}$ [58], V_2O_5 [59], Al_2O_3 [17], AlOOH [60], In_2O_3 [18], $InVO_4$ [61], Ga_2O_3 [18], $BaTiO_3$ [24], $PbZr_{0.52}Ti_{0.48}O_3$ [62], $PbTiO_3$ [24], SnO_2 [63], $MgAl_2O_4$ [64]; MoO_3 [65]; RuO_2 [65]
- silicon oxide: SiO_2 [5, 59]; Ag_2SiO_3/SiO_2 [28], Fe-doped chrysotile [66];
- rare earth oxides: (Er, Tm, Yb, Lu) oxide [22]; Y_2O_3:(Eu,Tb,Dy) [67], (Y, Gd, Tb, Dy, Ho, Er, Tm, Yb, Lu) oxide [68]

(c) transition metal halogenous NTs: $NiCl_2$ [8]

(d) mixed phase and metal doped NTs: $PbNb_nS_{2n+1}$ [69], $Mo_{1-x}WS_2$ [70], $W_xMo_yC_zS_z$ [71]; Nb-WS_2 [72], WS_2-carbon NTs [44], NbS_2-carbon NTs [73]; Au-MoS_2 [74], Ag-WS_2 [74], Ag-MoS_2 [74]; $BaWO_4$ [75], $SbPS_{4-x}Se_x$ [76], InGaAs/GaAs [77]

(e) Boron and silicon based NTs: BN [59], BCN [78], Si [28];

(f) metal and metal oxide nanotubes: Au [10], Co [11], Fe [11], Cu [79], Ni [79], Te [26], Bi [80], $Ni(OH)_2$ [81], $LaNiO_3$ [82]

(g) rare earth fluorides: $NaHoF_4$ [83], $NaSmF_4$ [83]
(h) organic/inorganic nanotubes: PVP-TiO_2 [84], $K_4Nb_6O_{17}$-$(C_nH_{2n+1}NH_3)^+$ [46], $(C_4H_{12}N)_{14}[(UO_2)_{10}(SeO_4)_{17}(H_2O)]$ [85], Eu_2O_3.carbon NTs [86]
(i) Transition metal nitrides and carbides: AlN [87], WC [88], GaN [89]

The most important methods for growing inorganic nanotubes can be divided broadly into: (a) sulphurization, (b) decomposition of precursor crystals, (c) template growth, (d) precursor assisted pyrolysis, (e) misfit rolling, (f) direct synthesis from the vapour phase, (g) electrospinning, (h) precipitation, (i) self-assembly, etc. Some nanotubes grow only by the combination of several processes. For example, the first inorganic nanotubes synthesized in Tenne's group were grown by sulphurization of the transition metal oxides [2]. The needle-like WO_{3-x} served as precursor crystals for sulphurization and, at the same time, they acted as a template to dictate the nanotube size [90]. Recently it was found that the stoichiometry and surface structure of the WO_{3-x} precursor crystals influences the efficiency of such transformation process as well as the morphology of resulted products. While the most reduced tungsten oxide ($W_{18}O_{49}$) transforms to WS_2 nanotubes, only slightly less reduces W_5O_{14} nanowires have been transformed to WS_2 nanobuds [91]. Degree of reduction in transition metal substoichiometric oxides is very sensitive on local atmosphere. The process of sulphurization has to be considered as a complex and dynamical way of producing inorganic nanotubes, but simultaneously very challenging for a rich family of transition metal dichalcogenide nanotubes.

11.4 Morphology of Single Nanotubes and Hybrid Nanostructures

Inorganic NTs exist in different states of crystallinity. Relating to a standard classification of crystals with unit cell repeating in three perpendicular directions of Euclidean space, nanotubes can be described as a semi-single crystal lattice with translational periodicity along two straight directions, while the trajectory of the third translation is curved. The term of semi-single crystallinity can be valid for non-helical or mono-helical NTs with a low density of defects, while polycrystallinity appears either in the structure of the nanotube wall, composed of small thin crystal flakes, or in the radial direction as multi-helicity. Many of the inorganic NTs prepared by the decomposition process appear as an assembly of nanocrystallites forming the nanotube wall (e.g. HfS_2 [20], NbS_2 [92]). The crystallinity can be improved by additional annealing of tubes, as for example, CdSe NTs [13], or by increased in situ temperature during the growth [92]. Although the defects are a kind of relaxation and stabilization of the real system, they increase the free energy with regard to the perfect structure. It is assumed, and experimentally evidenced, that the structures produced by gas phase synthesis are more likely to be at equilibrium and more perfect than those produced by diffusion transformations [49, 92, 93].

The most accurately studied lattice structures of inorganic nanotubes were those of MoS_2 and WS_2 tubes. Two growth mechanisms have been observed [41]. Thin folded flakes can directly roll up and adopt the cylindrical shape. The folds of strongly undulated layered crystals at a micro level size can also serve as an origin of turbulent gas flow of transported molecules, which promotes a tube-like growth mode. Instability of weakly bonded molecular layers at plate-like crystal edges and at surface growth steps against bending causes the growth of very narrow nanotubes. Several tubes can nucleate by such a way at the same crystal edge, where they combine and continue in growth like a rope. The nanotube in a rope still grows in longitudinal direction, up to a few millimeters in length, while its radius is limited by co-grown nanotubes. The second kind of the nanotube nucleation sites are the helical extremities of external and inner spirally rolled-up molecular layer building the already growing tube [42]. They are subject of bending like surface steps or edges of flat crystals. The nanotubes in so created ropes are usually rolled around or inside the central tube. Figure 11.2a shows a pair of identical nanotubes, 12 nm in diameter, which are spirally rolled around a thicker nanotube of 45 nm in

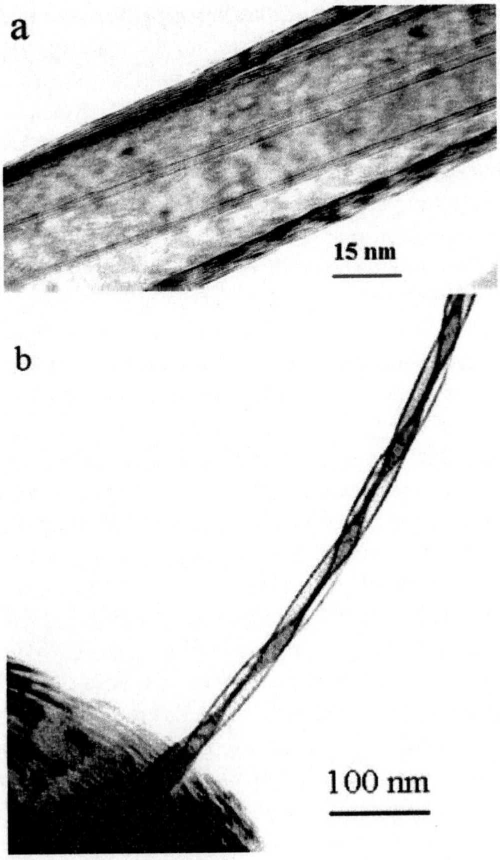

Fig. 11.2 (**a**) A pair of WS_2 nanotubes, four molecular layers in wall thickness, rolled around a central nanotube, 46 in diameter. (**b**) The WS_2 nanorope composed of two mutually rolled-up nanotubes with diameters of 15 nm and a spiral period of 110 nm. The pair of nanotubes nucleated at the inner surface step of the nanotube, 1.2 μm in diameter. Figure originally published in Applied Physics Letters [42]

diameter. The thicker nanotube with a step of helicity of 1.7 μm calculated from the corresponding electron diffraction pattern represents a central tube in this rope. Both smaller nanotubes with wall thickness of four molecular layers are wound round the central tube with the 1.8 μm period of winding. This agreement of both values in the range of accuracy supports the assumption that the growth of mutually rolled nanotube groups represents a special case of growth by syntactic coalescence, where the wound tubes follow the crystal structure of the supporting tube. In a case that several nanotubes nucleate at the same inner extremity, they are mutually rolled-up (Fig. 11.2b) [93].

The interlayer distance in nanotube walls was found extended by up to 3% relative to the plate-like crystals of the same compounds [41]. Only right-hand chirality was found using transmission electron diffraction [94]. The polytypic stacking of molecular layers in nanotubes with diameters above 2 μm (rhombohedral 3R stacking) was found to differ from that in nanotubes with diameters below 100 nm (hexagonal 2Hb stacking) [95]. The selected area diffraction on the microtube wall revealing the rhombohedral (3R) stacking, otherwise stable at elevated pressure above 4 GPa, provides indirect evidence of the presence of strain incorporated into the microtube wall. The intensity of the strain increases towards the tube axis, causing the contraction of interlayer distance and instability of the inner molecular layers at the place where the incorporated strain is relaxed with the creation of edge dislocations or stacking faults. In nanotubes with diameter below 200 nm, the strain is relaxed, stabilizing the hexagonal (2H) stacking. The incorporated strain has been confirmed by resonance Raman spectroscopy taken on single MoS_2 and WS_2 microtubes and nanotubes synthesized by chemical transport reaction [96]. The up-shift of Raman peaks compared to the bulk has been observed. The additional support to the structural reason for the peak shift is the fact that such a shift was not observed in the nanotubes produced by sulphurization process of tungsten oxides [97].

Recently it was found that MoS_2 fullerene-like particles encapsulated in thin walled MoS_2 nanotubes grow in the rhombohedral 3R stacking [98], while WS_2 fullerene-like particles grown on the surface WS_2 nanotubes forming nanobuds [91] appear in hexagonal 2Hb stacking. These observations reveal importance of local atmosphere, particularly a ratio between transition metal and sulfur on the stacking order of adjacent curved layers. Consequently, the rate of chemical reaction can change stoichiometry, defect concentration and crystal structure, in a confined geometry of chemical reactor even of a particular nanotube [92]. Diffusion of adatoms during the growth process on inner or outer surface is important especially in non-template growth from vapor phase. Incorporated strain in the nanotube walls is the largest in the most inner turns and increases with number of wrapping molecular layers. Beyond the stability conditions, the layers start to break to flakes promoting into the inner tube hole [93]. The diffusion at inner tube surface is hindered permanently if temperature for redistribution of material is not sufficient.

Nanotubes prepared by decomposition of precursor crystals at sufficient temperature for evaporation of $WO_{3-x}.nH_2O$ hydrate (H_2O is a side product of the sulphurization process) reveal a strong exhaust of the vapor at open ends, which evidences a dense crystallinity of the tube wall, which limits penetration of gasses

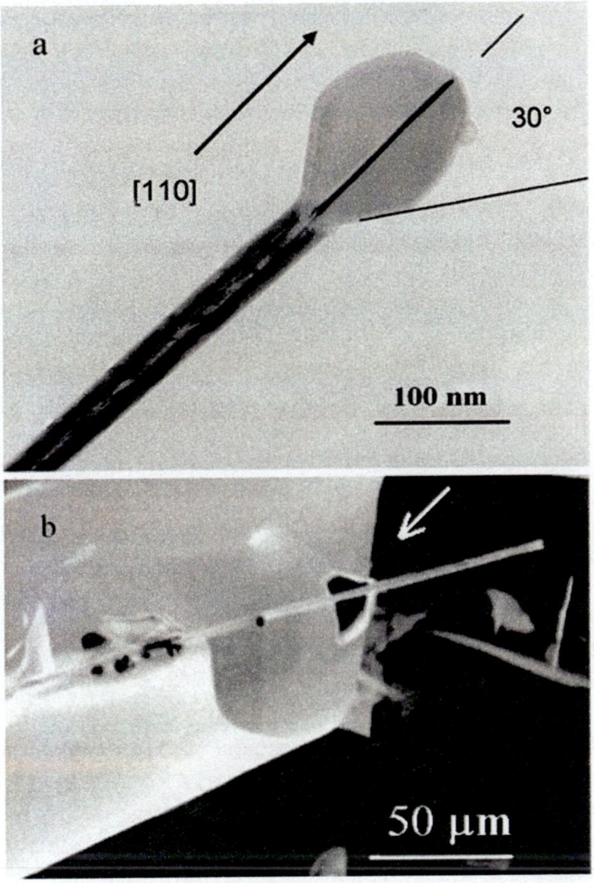

Fig. 11.3 Translocation of material: (**a**) An open WS_2 nanotube produced by sulphurization of W_5O_{14} precursor nanowires. Material which was blown out of the nanotube, formed symmetrical platelet with edged parallel with <100> MoS_2 lattice directions. The only curved part outside of the tube is the nano-fold sustainably protected against termination by the gas flow from this nano-"chimney". (**b**) A nucleation of MoS_2 nanotube in a microfold of an undulating MoS_2 crystal. The *arrow* marks a perforation of the crystal flake caused by preferential growth of nanotube with respect to a flat geometry. Material was translocated by diffusion through a contact area of the crystal flake with the growing nanotube. Figure originally published in Applied Physics Letters [41]

through the nanotube wall (Fig. 11.3a). The gas flow originates an appearance of open ended nanotubes, while redistribution of non-volatile components along the inner tube surface contributes to the closing process of the tubes. In a moment when the local gas flow of exhaust is stopped, the inside of a nanotube end represents a place with negative curvature, where vapor pressure is lowered with respect to sites with positive curvature according to the Kelvin equation and the remaining WO_{3-x} molecules therein condensate and build the nanotube cap termination.

Cylindrical geometry is stable in a relatively narrow range of thermodynamical conditions: temperature, gas flow, stoichiometry. There are some indications that nanotubes or fullerene-like particles grow at a balance of two possible states, but usually far away from equilibrium defined as minimum of total free energy. Therefore comparison with a theory is difficult especially if nanotubes of fullerene-like particles grow in a layer-by-layer mode [99]. MoS_2 nanotubes, as an example, appear in a narrow range of conditions (temperature, partial pressures, gas flow), where amorphous MoS_3 compound starts to slowly transform to nanocrystallites of MoS_2 [100], or at small sulfur excess where two polytypes (hexagonal 2Hb and rhombohedral 3R) can coexist [42]. Small deviation from appropriate conditions already leads to a "normal" (i.e. non-curved) growth of a compound in accordance with their crystal structure. As example, material which was blown out of the nanotube from Fig. 11.3a, formed symmetrical platelet with edged parallel with <100> MoS_2 lattice directions. The only curved part outside of the tube is the nano-fold sustainably protected against termination by the gas flow. Conditions outside of the confined geometry of a nanotube were obviously preferential for the flat geometry. A preferential growth mode (plate-like or cylindrical one) depends on local conditions. The different example is a translocation of material, which previously already composed a plate-like crystal flake, to obviously energetically more attractive position contributing to the growth of adjacent nanotube (Fig. 11.3b). Diffusion of material to more stable stage, in this case from flat to cylindrical geometry, formed a hole in a platelet at the contact area of the platelet with the nanotube.

The confined geometry, incorporated strain and the partial pressure of sulphur strongly influence the crystal structure of inorganic nanotubes and fullerene-like particles grown from vapor phase. It is possible to assume that also stacking in radial direction is not homogeneous, especially because stacking faults are common defects in cylindrical and spherical geometry. Nanotubes with radial dependent interlayer distances, which were found experimentally [101], have not yet been described theoretically.

Material science sometimes dictates terms as "normal" or "unusual" in accordance with a period of a particular discovery. Knowing bulk, three-dimensional compounds with flat geometrical shapes for decades or centuries, we try to describe an existence of hollow nanotubes or fullerene-like particles with some peculiar reasons leading to these "unusual" features. It is possible that cylindrical [102] or spherical geometry of materials is an intrinsic property of some compounds at some dimension range. As an example, it was evidenced that WS_2 fullerene-like particles grow spontaneously with diffusion process along longitudinal surface corrugations of W_5O_{14} precursor crystals [91]. In the early stage of growth the thin-walled buds round up on one side and narrow at the contact area, growing elongated in the direction of the supporting nanotube (Fig. 11.4a). These formations reveal that the material needed for growth is really provided by diffusion. The next stage is growth of faceted nanoparticles (Fig. 11.4b) with thicker walls, which are still narrowed at the contact area, so that further diffusion is still possible. In some cases a kind of secondary nucleation has been observed, i.e. some fullerenes grow on the surface of twinned fullerenes, where the diffusion paths are developing at the interface folds. In this stage the walls are thicker and the particles are more spherically shaped in nearly ball-like morphology (Fig. 11.4c). The fullerene-like particles can grow up to several hundreds nanometres, which exceeds the dimensions of those produced by template techniques from WO_x nanoparticles.

Next example of nanotubes grown by diffusion process are so-called "mama"tubes produced by decomposition and recrystallisation of $Mo_6S_2I_8$ nanowires in the process of sulphurization at high (1,150 K) temperature (Fig. 11.5).

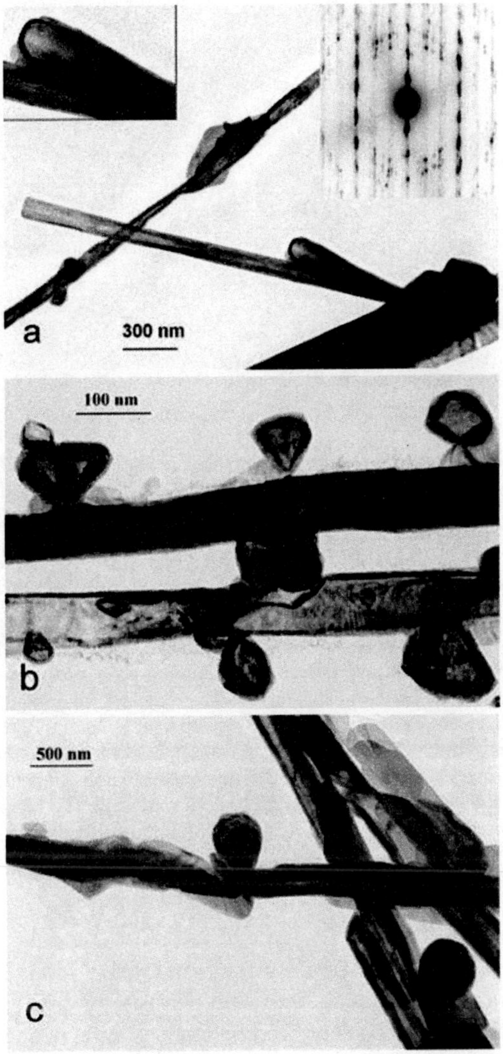

Fig. 11.4 WS_2 nanobuds: (**a**) an early stage of growth with typically elongated thin walled buds along the nanotube support, which serves as a reservoir of material. The nanotubes grow in the relaxed hexagonal 2Hb stacking (*inset*); (**b**) some faceted nanoparticles joined to the supported thin-walled nanotube with their V-shaped parts; (**c**) the final stage of growth with nearly spherical thick-walled fullerenes. Figure originally published in Nano Letters [91]

MoS_2 nanotubes with crystalline walls of 10 nm encapsulated MoS_2 fullerene-like particles, appears to originate from the following sequence of events: (a) formation of the MoS_2 nanotube as a kind of envelope, by decomposition of the topmost layers of a $Mo_6S_2I_8$ nanowire (it is possible that the transformation takes place through local appearance of other $Mo_xS_yI_z$ phases with low iodine content, including different Chevrel phases), (b) subsequent decomposition of the inner $Mo_6S_2I_8$ phase causing a large reduction of mass after complete removal of the released iodine, and (c) further sulphurization of the remaining internal material and formation of MoS_2 fullerenes by diffusion of molecules along the inner nanotube surface.

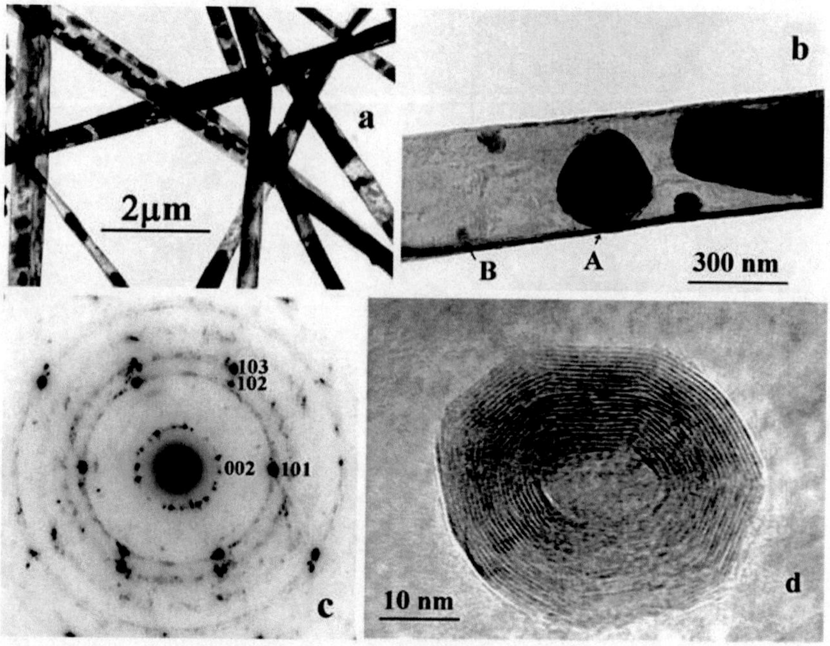

Fig. 11.5 Transmission electron micrographs of the final product: (**a**) a general view of the MoS_2 nanotubes with encapsulated MoS_2 fullerene-like nanoparticles, (**b**) single MoS_2 fullerenes and their aggregates inside a thin-walled MoS_2 nanotube, (**c**) an electron diffraction pattern taken on the particle A shown in (**b**), an overlap of electrons scattered by both walls of the tube (hexagonally arranged sharp peaks) and by the encapsulated nanoparticle (diffraction rings), and (**d**) a hollow MoS_2 fullerene with a slightly deformed spherical shape situated inside MoS_2 nanotube. Figure originally published in Advanced Materials [98]

Formation of only a few nanometers thick MoS_2 envelope in the first stage of transformation together with nearly simultaneous creation of an empty inner space due to large loose of iodine enables an appearance of special conditions inside the nanowires. The diffusion is space limited to cylindrical caves, while the surface diffusion is made difficult particularly at the inner surface, where the first curved crystal flakes appear, which are not part of the tube wall but start to self-terminate. The nucleation of these flakes could be triggered by the defects in the nanotube walls. Subsequent sulphurization with a formation of cage structures inside tubes localizes the diffusion even more, until the transformation to MoS_2 nanopods is complete. Due to very thin walls, which break under short ultra sound agitation, the fullerene-like particles can be released in a controled way.

Similar hybrid structures can be predicted in other "cluster"-materials [103]. High porosity at subsequent easy manipulation of long nanowires raise expectations on applications as hydrogen storage, catalyst, as solar cells or as additive to polymers.

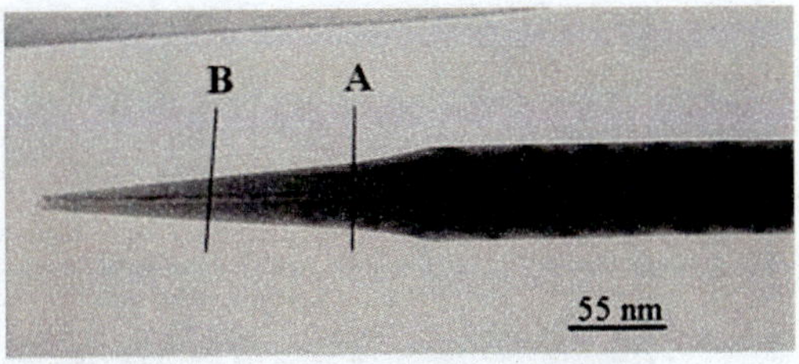

Fig. 11.6 Chirality changes in a Au-WS_2 nanotube as a function of its wall thickness. The tip reveals the transition from coaxial growth to the spiral growth mode. The detailed measurements show that the tube begins to narrow by termination of top molecular layers by steps of length of 2.2 nm, which still show coaxial growth mode. When the step length increases to 4.4 nm, the angle of chirality is $\approx 2°$ (**a**). The tip part close to the end narrows by steps of 7 nm in length with the corresponding angle of $\approx 4.5°$ (**b**). Figure originally published in Surface Review Letters [101]

The contraction of interlayer distances caused by internal strain incorporated in the tube wall influences the helicity of the nanotubes. Each winding of the molecular layer of thickness t, i.e. 0.6147 nm for MoS_2 [104], enlarges the tube circumference by πt. Since this value is not an integer multiple of lattice parameter in the basal plane (0.316 nm), the molecular layers have to be strained and/or helical, assuming that a regular crystal lattice is formed at least in narrow strips parallel with the nanotube axis. Increasing strain intensity toward the nanotube central hole causes the co-existence of several chiralities in the same nanotube (Fig. 11.6). In thick-wall nanotubes the incorporated strain causes reduction of the interlayer distances, in some tubes even below the value typical of plate-like crystals. In such cases the longitudinal growth of the nanotubes is stopped, with the typical crater-tip terminations [105].

Complete suppression of helicity is typical of MoS_2 and WS_2 nanotubes alloyed with gold and silver [74]. The noble metal species built into the tube walls increase the interaction between molecular layers and cause a double-layer by double-layer growth mode. The interlayer distances are expanded by 5–6%, causing the change in tube morphology. The tube walls are composed of molecular layers grown in a form of coaxial cylinders. The concentration of intercalated noble metal is negligible in the central part of the tubes where the layer by layer growth mode prevails. These completely stable nanotubes represent new compounds otherwise unknown in a plane geometry. Inhomogeneous distribution of alloyed gold makes the tubes interesting for nanodevice applications.

Contrary to noble metal alloyed nanotubes, the Nb-doped WS_2 nanotubes did not show any extension of interlayer distance [106]. The niobium was found homogeneously distributed in the nanotube walls, most probably by replacing the W atoms. The 2Hb stacking order typical of WS_2 is also preserved in Nb-alloyed nanotubes.

The largest and "oldest" inorganic tube was found in natural MoS_2-molybdenite from quarries at Mont Saint-Hilaire, Quebec, Canada. The tube, 0.5 mm-diameter by 6 mm long, is housed in the Canadian Museum of Nature as specimen No. 48756 [107] (Fig. 11.7). The largest synthetic MoS_2 tube grown without using

Fig. 11.7 A SEM micrograph of a natural molybdenite (MoS_2) tube, 4 mm long and 0.5 mm in diameter. Photo taken by E. Vadas of a specimen in the Canadian Museum of Nature (#48756) found in a miarolitic cavity at the large quarry at Mont Saint-Hilaire, Quebec, Canada [123]

a template was 11 μm in diameter and 2 mm in length [41]. Optimization of the growth process or use of nanosized templates enables the production of tubes with nanosized diameter. As an example, the change of stoichiometry from transition metal di-chalcogenide towards three-chalcogenides considerably reduces the average diameter of MoS_2 and WS_2 NTs.

Nanotubes with diameters below 20 nm have been synthesized quite rarely. BN NTs with inner diameters from 1 to 3 nm and outer diameters from 6 to 8 nm were made by the arc-discharge process [4]; 15 nm diameter WS_2 NTs were grown by sulphurization of WO_3 [108]; NbS_2 NTs with inner diameters in the range ~4–15 nm were made by decomposition of NbS_3 [21]; CdSe NTs with outer diameters in the range 15–20 nm have been made from cadmium oxide using surfactant assisted synthesis [12]. Titania nanotubes with diameters of about 8 nm have been produced by hydrothermal treatment of nano-size TiO_2 powder in NaOH solution [109]. A long term discussion of their composition, balancing between the anatase and rutile phases of TiO_2, came to the conclusion that the titania nanotubes are actually $H_2Ti_3O_7$ single sheets rolled up into cylindrical geometry [110, 111].

11.5 Mechanical Properties and Thermal Stability

The mechanical properties of inorganic nanotubes are of great interest both for the sake of fundamental science and for their applications. Nonetheless, the mechanical properties of inorganic nanotubes have been investigated to a relatively small extent so far.

Polarized resonance Raman measurements of individual multiwall WS_2 nanotubes were recently described [112]. A strong Raman scattering signal was obtained when the light was polarized along the nanotube axis. Using a fit to a theoretical model, an estimate of the ratio of the perpendicular to parallel polarizabilities $\alpha_{XX}/\alpha_{ZZ} = 0.16$ was obtained, value of which is comparable to that obtained for single-wall carbon nanotubes (0.09). Symmetry analysis of the Raman and infrared

(IR) active modes of MoS_2 single-wall nanotubes was also recently published [113]. The tubular structure is found to be characterized by two Raman active modes. A new high-energy breathing mode transition, which is characterized by the in-phase breathing of the sulfur shells and out of phase relative to the molybdenum atoms, was identified. Unfortunately, theory for multi-wall MoS_2 or WS_2 nanotubes is still missing.

The mechanical properties of inorganic nanotubes have been reviewed in some details in recent years [114, 115]. BN nanotubes were shown to exhibit Young's modulus almost as high as that of carbon nanotubes. The axial Young's modulus of 1.22 ± 0.24 TPa was found from the thermal vibration amplitude of a cantilevered BN nanotube observed in a transmission electron microscope [116]. Free-standing, vertically aligned GaN nanotubes were exposed to uniaxial compression of 150 μm using a nanoindenter. The tubes were 500 and 300 nm in length, with an outer radius of 40 nm and an inner radius of 20 nm. The critical buckling strain was found around 7.4–19.0%, the experimentally obtained values of Young's modulus for lengths of 500 and 300 nm were 483.9 and 223.4 GPa, respectively [117].

The mechanical properties of individual WS_2 nanotubes have been investigated in detail using both experiment and theory [118, 119]. These studies and others show that despite their smaller Young's modulus and larger specific weight compared to carbon nanotubes, inorganic nanotubes may find numerous applications in ultrahigh strength nanocomposites, mostly due to their high compression strength. As example, the Young's modulus of the WS_2 nanotubes exposed to buckling was estimated at 150–170 GPa, while tensile tests gave a value of 152 GPa. The combination of high tensile strength and ca. 14% elongation is a unique property for all the nanotubes, which were measured so far [120, 121]. Furthermore, the strength of the strongest nanotubes is about 11% of its Young's modulus, corresponding approximately to the theoretical value of the strength of the material. This value is appreciably larger than typical high-strength engineering materials. These findings indicate that the WS_2 nanotubes are remarkably free of critical defects. Furthermore, the stress-strain plots and the ab-initio calculations show that the nanotubes deformed elastically, almost until failure. This study shows that WS_2 nanotubes like other nanotubes are both ultra-strong and elastic, distinguishing them from other known materials. The association of WS_2 nanotubes shown in Fig. 11.8 was made by axial rotation of tweezers held the strand of nanotubes [122]. The nanotubes, even the smallest in the strand with diameter bellow 30 nm did not break asunder during this robust process. Their mechanical stability is explained by their chiral growth mode and with high lattice perfection, which render the transport of dislocations in radial direction needed for the break of material.

Thermal stability of some inorganic nanotubes is also important. While MoS_2 and WS_2 nanotubes grow at relatively high temperature (1,010 K) in iodine and sulfur rich local atmosphere, their annealing in air results to complete oxidation at 570 K. In-situ annealing in a high vacuum inside transmission electron microscope (10^{-3} Pa) by irradiation with electron beam enables the observation of progressive decomposition of the nanotubes [122]. The WS_2 nanotube, 35 nm in diameter, 5 nm in wall thickness and composed of 8 molecular layers, was completely stable

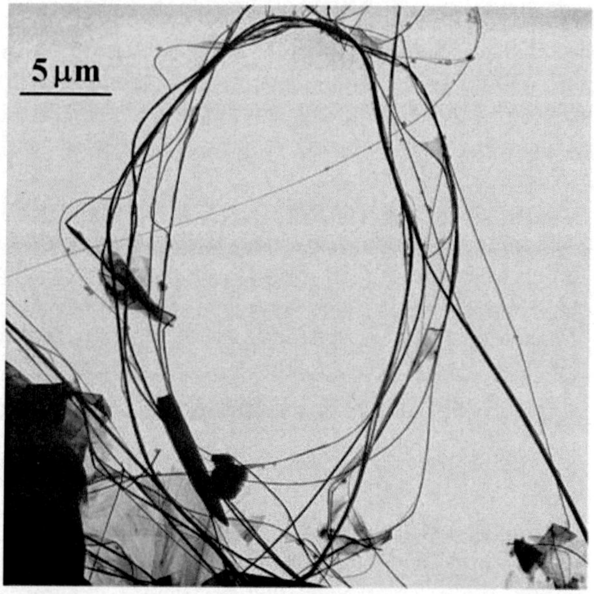

Fig. 11.8 An artificial association of WS_2 nanotubes reveals the strong stability against tensile forces. Figure originally published in Current Opinion in Solid State and Material Science [122]

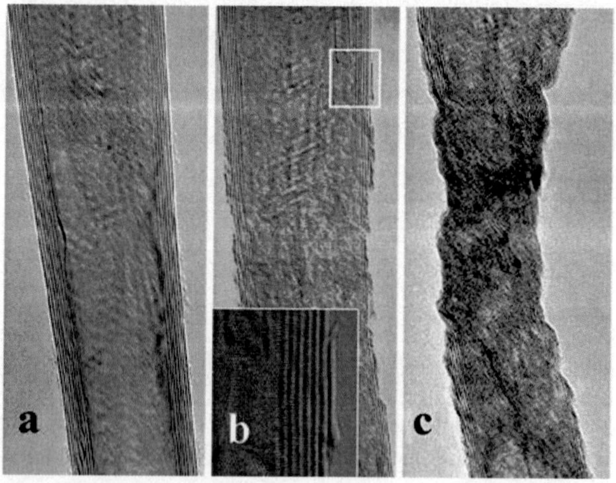

Fig. 11.9 Irradiation of the WS_2 nanotube, 35 nm in diameter, by electron beam in transmission electron microscope. Density of electrons and duration of irradiation: (**a**) 7 A/cm^2 – structure was stable; (**b**) 50 A/cm^2 – 2 min; (**c**) 180 A/cm^2 – 2 min. Figure originally published in Current Opinion in Solid State and Material Science [122]

during irradiation with a dose 7 A/cm^2 (Fig. 11.9a). Six times larger dose (50 A/cm^2) applied for two minutes started the decomposition process of molecular layers (Fig. 11.9b). The destruction of the nanotube takes place layer by layer from outside toward the inner layers. Some outer molecular layers are completely removed, what indicates indirectly the temperature above 1,120 K, where the WO_3 becomes volatile (figure inset). The increased electron dose up to 180 A/cm^2 for additional 2 min destroyed cylindrical geometry; the diameter shrunk for nearly

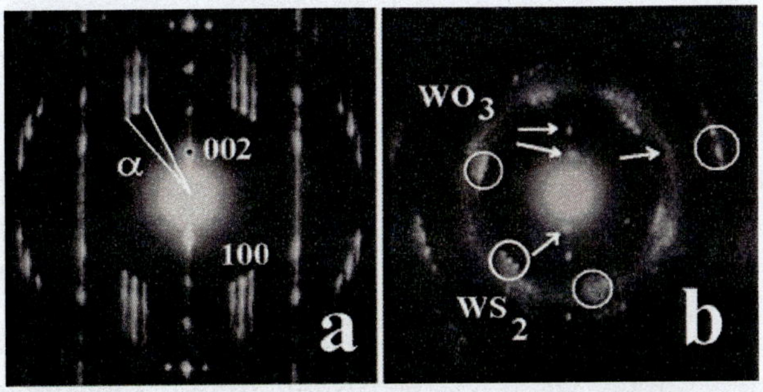

Fig. 11.10 The transmission electron diffraction pattern of electrons scattered by the WS_2 nanotube: (**a**) at low dose of electrons-7 A/cm^2; (**b**) at 180 A/cm^2. Diffraction peaks belonged to WS_2 are encircled, while WO_3 spots are marked by arrows. Figure originally published in Current Opinion in Solid State and Material Science [122]

50% and the tube became partially oxidized (Fig. 11.9c). The corresponding electron diffraction (Fig. 11.10) is a superposition of electrons scattered by both walls of the tube oriented perpendicular to the electron beam and by side parts of the wall oriented parallel with the beam. The tube wall was composed of chirally grown molecular layers with 13° chiral angle and non-chiral molecular layers (Fig. 11.10a). After the irradiation with the maximal dose the diffraction spots belonged to scattering by the side (00*l*) WS_2 planes are much weaker than before the irradiation revealing the destruction of cylindrical geometry. In addition, the position of the {100} spots are turned for 10° indicating the twisting of the nanotube during the annealing (Fig. 11.10b). The screw deformation is explained by relaxation of internal strain incorporated in the chirally grown nanotube wall.

The relatively high thermal stability of inorganic nanotubes in no oxidizing atmosphere is concluded from the observation that only 50 nm away from the irradiation spot the nanotube was nearly non-attacked (Fig. 11.9c – upper part of the image). The origin of the decomposition was obviously not only the enhanced temperature, but also the irradiation damages of the lattice by 200 keV electrons, which trigger the local decomposition of the wall.

11.6 Conclusions

Inorganic nanotubes are an important field of new nanomaterials, which expands generic structure of inorganic layered two-dimensional compounds. Although some geometrical similarities with carbon nanotubes, inorganic nanotubes distinguish themselves by important peculiarities, from the growth mechanisms to the physical and chemical properties attractive for possible applications. Study of these novel nanostructures has led to the observation of a number of interesting properties offering numerous potential applications as biosensors, drug delivery, safe containers, nanoreactors, strengthening fibres as well as in tribology, as high-energy-density batteries, sensors, in photoconversion of solar energy and in nanoelectronics. Numerous growth techniques are used for the synthesis of transition metal dichalcogenide inorganic NTs. After the first enthusiasm on the successful synthesis of cylindrical crystals from new compound, the additional demands appear with respect to their properties. The control of their dimensions is desired and crucial for some applications. The second important demand is their structural perfection, especially for their mechanical and electric properties significant in construction of nanodevices. Both demands are rarely satisfied simultaneously in inorganic nanotubes as well as in carbon nanotubes. There is still plenty of room for technological and basic phenomenological research of inorganic nanotubes.

References

1. Iijima, S. Helical microtubules of graphitic carbon. *Nature* **354**, 56–58 (1991).
2. Tenne, R., Margulis, L., Genut, M., Hodes, G. Polyhedral and cylindrical structures of tungsten disulfide. *Nature* **360**, 444–446 (1992).
3. Margulis, L., Salitra, G., Tenne, R., Talianker, M. Nested fullerene-like structures. *Nature* **365**, 113–114 (1993).
4 Chopra, N. G., et al. Boron-nitride nanotubes. *Science* **269**, 966–967 (1995).
5. Nakamura, H., Matsui, Y. Silica gel nanotubes obtained by sol-gel method. *J. Am. Chem. Soc.* **117**, 2651–2652 (1995).
6. Hoyer, P. Formation of a titanium dioxide nanotube array. *Langmuir* **12**, 1411–1413 (1996).
7. Spahr, M. E., et al. Redox-active nanotubes of vanadium oxide. *Angew. Chem., Int. Ed.* **37**, 1263–1265 (1998).
8. Hacohen, Y. R., Grunbaum, E., Tenne, R., Sloan, J., Hutchison, J. L. Cage structures and nanotubes of NiCl2. *Nature* **395**, 336–337 (1998).
9. Galvan, D. H., Kim, J. H., Maple, M. B., Avalos-Berja, M., Adem, E. Formation of NbSe2 nanotubes by electron irradiation. *Fullerene Sci. Technol.* **8**, 143–151 (2000).
10. Hutleen, J. C., Jirage, K. B., Martin, C. R. Introducing chemical transport selectivity into gold nanotubule membrane. *J. Am. Chem. Soc.* **120**, 6603–6604 (1998).
11. Tourillon, G., Pontonnier, L., Levy, J. P., Langlais, V. Electrochemically synthesized Co and Fe nanowires and nanotubes. Electrochem. *Solid-State Lett.* **3**, 20–23 (2000).
12. Rao, C. N. R., Govindaraj, A. G., Deepak, F. L., Gunari, N. A., Nath, M. Surfactant-assisted synthesis of semiconductor nanotubes and nanowires. *Appl. Phys. Lett.* **78**, 1853–1855 (2001).
13. Govindaraj, A., Deepak, F. L., Gunari, N. A., Rao, C. N. R. Semiconductor nanorods: Cu, Zn, and Cd chalcogenides. *Israel J. Chem.* **41**, 23–30 (2001).
14. Dloczik, L., et al. Hexagonal nanotubes of ZnS by chemical conversion of monochrystalline ZnO columns. *Appl. Phys. Lett.* **78**, 3687–3689 (2001).
15. Jiang, X., Xie, Y., Zhu, L., He, W., Qian, Y. Synthesis of novel nickel sulfide layer-rolled structures. *Adv. Mater.* **13**, 1278–1281 (2001).
16. Peng, Y. Y., et al. Cu5.5FeS6.5 nanotubes – a new kind of ternary sulfide nanotube. *New J. Chem.* **25**, 1359–1361 (2001).
17. Pu, L., Bao, X., Zou, J., Feng, D. Individual alumina nanotubes. *Angew. Chem., Int. Ed.* **40**, 1490–1493 (2001).
18. Cheng, B., Samulski, E. T. Fabrication and characterization of nanotubular semiconductor oxides In2O3 and Ga2O3. *J. Mater. Chem.* **11**, 2901–2902 (2001).
19. Li, J. Y., Chen, X. L., Qiao, Z. Y., Cao, Y. G., Li, H. Synthesis of GaN nanotubes. *J. Mater. Sci. Lett.* **20**, 1987–1988 (2001).

20. Nath, M., Rao, C. N. R. Nanotubes of group 4 metal disulfides. *Angew. Chem., Int. Ed.* **41**, 3451–3454 (2002).
21. Nath, M., Rao, C. N. R. New metal disulfide nanotubes. *J. Am. Chem. Soc.* **123**, 4841–4842 (2001).
22. Yada, M., Mihara, M., Mouri, S., Kuroki, M., Kijima, T. Rare earth (Er, Tm, Yb, Lu) oxide nanotubes templated by dodecylsulfate assemblies. *Adv. Mater.* **14**, 309–313 (2002).
23. Wu, J., Liu, S., Wu, C., Chen, K., Chen, L. Heterostructures of ZnO-Zn coaxial nanocables and ZnO nanotubes. *Appl. Phys. Lett.* **81**, 1312–1314 (2002).
24. Hernandez, B. A., Chang, K. S., Fisher, E. R., Dorhout, P. K. Sol-gel template synthesis and characterization of BaTiO3 and PbTiO3 nanotubes. *Chem. Mater.* **14**, 480–482 (2002).
25. Han, C. C., Bai, M. Y., Lee, J. T. A new and easy method for making Ni and Cu microtubules and their regurarly assembled structures. *Chem. Mater.* **13**, 4260–4268 (2001).
26. Mayers, B., Xia, Y. Formation of tellurium nanotubes through concentration depletion at the surfaces of seeds. *Adv. Mater.* **14**, 279–282 (2002).
27. Brorson, M., Hansen, T. W., Jacobsen, C. J. H. J. Rhenium(IV) sulfide nanotubes. *Am. Chem. Soc.* **124**, 11582–11583 (2002).
28. Sha, J., et al. Silicon nanotubes. *Adv. Mater.* **14**, 1219–1221 (2002).
29. Tenne, R., Zettl, A. K. Nanotubes from inorganic materials. *Top. Appl. Phys.* **80**, 81–112 (2001).
30. Pokropivnyi, V. V. Nanostructured materials: Non-carbon nanotubes – Properties and applications. *Powder Metallurg. Metal Ceram.* **41**, 123–135 (2002).
31. Patzke, G. R., Krumeich, F., Nesper, R. Oxidic nanotubes and nanorods-anisotropic modules for a future nanotechnology. *Angew. Chem., Int. Ed.* **41**, 2446–2461 (2002).
32. Rao Rao, C. N. R., Nath, M. Inorganic nanotubes. *Dalton Trans.* **1**, 1–24 (2003).
33. Tenne, R. Fullerene-like materials and nanotubes from inorganic compounds with a layered (2-D) structure. *Colloids and Surfaces A: Physicochem. Eng. Aspects* **208**, 83–92 (2002).
34. Tenne, R. Inorganic nanotubes and fullerene-like nanoparticles. *Nat. Nanotechnol.* **1**, 103–111 (2006).
35. Ivanovskii, A. L. Non-carbon nanotubes: synthesis and simulation. *Russ. Chem. Rev.* **71**, 175–194 (2002).
36. Remskar, M. Inorganic nanotubes. *Adv. Mater.* **16**, 1497–1504 (2004).
37. Halford, B. Inorganic menagerie. *Chem. Eng. News* **83**, 30–33 (2005).
38. Enyashin, A., Gemming, S., Seifert, G. Nanosized allotropes of molybdenum disulfide. *Eur. Phys. J. Spec. Top.* **149**, 103–125 (2007).
39. Li, Q., et al. Molybdenum Disulfide nanowires and nanoribbons by electrochemical/chemical synthesis. *J. Phys. Chem. B* **109**, 3169–3182 (2005).
40. Kopnov, F., Yoffe, A., Leitus, G., Tenne, R. Transport properties of fullerene-like WS2 nanoparticles. *Phys. Stat. Sol. B* **234**, 1229–1240 (2006).
41. Remskar, M., Skraba, Z., Cleton, F., Sanjines, R., Levy, F. MoS2 as microtubes. *Appl. Phys. Lett.* **69**, 351–353 (1996).
42. Remskar, M., Skraba, Z., Sanjines, R., Levy, F. Syntactic coalescence of WS2 nanotubes. *Appl. Phys. Lett.* **74**, 3633–3635 (1999).
43. Bernaerts, D., Amelincx, S., Van Tendeloo, G., Van Landuyt, J. Microstructure and formation mechanisms of cylindrical and conical scrolls of the misfit layer compounds PbNbnS2n+1. *J. Cryst. Growth* **172**, 433–439 (1997).
44. Whitby, R. L. D., Hsu, W. K., Boothroyd, C. B., Kroto, H. W., Walton, D. R. M. Tungsten disulphide coated multi-walled carbon nanotubes. *Chem. Phys. Lett.* **359**, 121–126 (2002).
45. Whitby, R. L. D., et al. WS2 layer formation on multi-walled carbon nanotubes. *Appl. Phys. A* **76**, 527–532 (2003).
46. Du, G., Yu, Y., Peng, L. M. Hexaniobate nanotubes with variable interlayer spacings. *Chem. Phys. Lett.* **400**, 536–540 (2004).
47. Seifert, G., Köhler, T., Tenne, R. Stability of metal chalcogenide nanotubes. *J. Phys. Chem. B* **106**, 2497–2501 (2002).

48. Srolovitz, D. J., Safran, S. A., Tenne, R. Elastic equilibrium of curved thin films. *Phys. Rev. E* **49**, 5260–5270 (1994).
49. Mendelev, M. I., Srolovitz, D. J., Safran, S. A., Tenne, R. Equilibrium structure of multilayer van der Waals films and nanotubes. *Phys. Rev. B* **65**, 075402 1–12 (2002).
50. Seifert, G., Terrones, H., Terrones, M., Jungnickel, G., Frauenheim, T. Structure and electronic properties of MoS2 nanotubes. *Phys. Rev. Lett.* **85**, 146–149 (2000).
51. Seifert, G., Köhler, T., Tenne, R. Stability of metal chalcogenide nanotubes. *J. Phys. Chem. B* **106**, 2497–2501 (2002).
52. Bar-Sadan, M., et al. Structure and stability of molybdenum sulfide fullerenes. *J. Phys. Chem. B* **110**, 25399–25410 (2006).
53. Nath, M., Rao, C. N. R. MoSe2 and WSe2 nanotubes and related structures. *Chem.Commun.* **21**, 2236–2237 (2001).
54. Chen, J., Li, S. L., Tao, Z. L., Gao, F. Low-temperature synthesis of titanium disulfide nanotubes. *Chem. Commun.* **8**, 980–981 (2003).
55. Therese, H. A., et al. VS2 nanotubes containing organic-amine templates from the NT-VOx precursors and reversible copper intercalation in NT-VS2. *Angew. Chem. Int. Ed.* **44**, 262–265 (2005).
56. Hu, J., Bando, Y., Liu, Z. Synthesis of gallium-filled gallium oxide-zinc oxide composite coaxial nanotubes. *Adv. Mater.* **15**, 1000–1003 (2003).
57. Niederberger, M., et al. Low-cost synthesis of vanadium oxide nanotubes via two novel non-alkoxide routes. *Chem. Mater.* **12**, 1995–2000 (1995).
58. Hu, W. B., et al. Generation of hollow crystalline tungsten oxide fibres. *Appl. Phys. A* **70**, 231–233 (2000).
59. Satshkumar, B. C., Govindaraj, A., Erasmus, M. V., Basumallick, L., Rao, C. N. R. Oxide nanotubes prepared using carbon nanotubes as templates. *J. Mater. Res.* **12**, 604–606 (1997).
60. Lu, J. G., Zhang, J., Ding, W. P., Shen, B., Guo, X. F. Synthesis and Characterization of Boehmite AlOOH Nanotubes. *Chinese J. Inorg. Chem.* **23**, 897–900 (2007).
61. Wang, Y., Cao, G. Z. Synthesis and electrochemical properties of InVO4 nanotube arrays. *J. Mater. Chem.* **17**, 894–899 (2007).
62. Luo, Y., et al. Nanoshell tubes of ferroelectric lead zirconate titanate and barium titanate. *Appl. Phys. Lett.* **83**, 440–442 (2003).
63. Wang, Y., Lee, J. Y., Zeng, H. C. Polycrystalline SnO2 nanotubes prepared via infiltration casting of nanocrystallites and their electrochemical applications. *Chem. Mater.* **17**, 3899–3903 (2005).
64. Fan, H., et al. Single-crystalline MgAl2O4 spinel nanotubes using a reactive and removable MgO nanowire template. *Nanotechnology* **17**, 5157–5162 (2006).
65. Satishkumar, B. C., Govindaraj, A., Nath, M., Rao, C. N. R. Synthesis of metal oxide nanorods using carbon nanotubes as templates. *J. Mater. Chem.* **10**, 2115–2119 (2000).
66. Foresti, E., et al. Morphological and chemical/physical characterization of Fe-doped synthetic chrysolite nanotubes. *Adv. Func. Mater.* **15**, 1009–1016 (2005).
67. Liu, G. X., Hong, G. Y. Synthesis and photoluminescence of Y2O3:RE3+ (RE = Eu, Tb, Dy) porous nanotubes templated by carbon nanotubes. *J. Nanosci. Nanotech.* **6**, 120–124 (2006).
68. Yada, M., et al. Hierarhical two- and three-dimensional microstructures composed of rare-earth compound nanotubes. *Adv. Mater.* **16**, 1448–1453 (2004).
69. Bernaerts, D., Amelincx, S., Van Tendeloo, G., Van Landuyt, J. Microstructure and formation mechanism of cylindrical and conical scrolls of the misfit layer compounds PbNbnS2n+1. *J. Cryst. Growth* **172**, 433–439 (1997).
70. Nath, M., Mukhopadhyay, K., Rao, C. N. R. Mo1-xWxS2 nanotubes. *Chem. Phys. Lett.* **352**, 163–168 (2002).
71. Hsu, W. K., et al. Mixed-phase WxMoyCzS2 nanotubes. *Chem. Mater.* **12**, 3541–3546 (2000).
72. Zhu, Y. Q., et al. Niobium-doped WS2 nanotubes. *Chem. Phys. Lett.* **342**, 15–21 (2001).
73. Zhu, Y. Q., Hsu, W. K., Kroto, H. W., Walton, D. R. M. An alternative route to NbS2 nanotubes. *J. Phys. Chem. B* **106**, 7623–7626 (2002).

74. Remskar, M., Skraba, Z., Stadelmann, P., Levy, F. Structural stabilization of new compounds: MoS2 and WS2 micro- and nanotubes alloyed with gold and silver. *Adv. Mater.* **12**, 814–818 (2000).
75. Li, D., et al. Multi-phase equilibrium microemulsions-based routes to synthesize nanoscale BaWO4 spheres, cylinders and rods. *Colloids Surf. A* **274**, 18–23 (2006).
76. Malliakas, C. D., Kanatzidis, M. G. Inorganic single wall nanotubes of SbPS4-xSex ($0 \leq x \leq 3$) with tunable band gap. *J. Am. Chem. Soc.* **128**, 6538–6539 (2006).
77. Prinz, A. V., Prinz, V. Ya. Application of semiconductor micro- and nanotubes in biology. *Surf. Sci.* **532–535**, 911–915 (2003).
78. Stephan, O., et al. Doping graphitic and carbon nanotube structures with boron and nitrogen. *Science* **266**, 1683–1685 (1994).
79. Han, C. C., Bai, M. Y., Lee, J. T. A new and easy method for making Ni and Cu microtubules and their regularly assembled structures. *Chem. Mater.* **13**, 4260–4268 (2001).
80. Li, Y. D., et al. Bismuth nanotubes: a rational low-temperature synthetic route. *J. Am. Chem. Soc.* **123**, 9904–9905 (2001).
81. Chou, S. L., Cheng, F. Y., Chen, J. Electrochemical deposition of Ni(OH)2 and Fe-doped Ni(OH)2 tubes. *Eur. J. Inorg. Chem.* **20**, 4035–4039 (2005).
82. Tagliazucchi, M., Sanches, R. D., Troiani, H. E., Calvo, E. J. Synthesis of lanthanum nickelate perovskite nanotubes by using a template-inorganic precursor. *Solid State Comm.* **137**, 212–215 (2006).
83. Liang, L. F., et al. Hydrothermal synthesis of prismatic NaHoF4 microtubes and NaSmF4 nanotubes. *Inorg. Chem.* **43**, 1594–1596 (2004).
84. Li, D., Xia, Y. Direct fabrication of composite and ceramic hollow nanofibers by electrospinning. *Nano Lett.* **4**, 933–938 (2004).
85. Krivovichev, S. V., et al. Highly porous uranyl selenate nanotubules. *J. Am. Chem. Soc.* **127**, 1072–1073 (2005).
86. Fu, L., et al. Coating carbon nanotubes with rare earth oxide multiwalled nanotubes. *Adv. Mater.* **16**, 350–352 (2004).
87. Wu, Q., et al. Synthesis and characterization of faceted hexagonal aluminum nitride nanotubes. *Am. Chem. Soc.* **125**, 10176–10177 (2003).
88. Pol, S. V., Pol, V. G., Gedanken, A. Synthesis of WC nanotubes. *Adv. Mater.* **18**, 2023–2027 (2006).
89. Goldberger, J., et al. Single-crystal gallium nitride nanotubes. *Nature* **422**, 599–602 (2003).
90. Rothschild, A., Popovitz-Biro, R., Lourie, O., Tenne, R. Morphology of multiwall WS2 nanotubes. *J. Phys. Chem. B* **104**, 8976–8981 (2000).
91. Remskar, M., Virsek, M., Jesih, A. WS2 nanobuds as a new hybrid nanomaterial. *Nano Lett.* **8**, 76–80 (2008).
92. Remskar, M., Mrzel, A., Jesih, A., Lévy, F. Metal-alloyed NbS2 nanotubes synthesised by the self-assembly of nanoparticles. *Adv. Mater.* **14**, 680–684 (2002).
93. Remskar, M., Skraba, Z., Ballif, C., Sanjines, R., Levy, F. New crystal structures of WS2: microtubes, ribbons and ropes. *Adv. Mater.* **10**, 246–249 (1998).
94. Margulis, L., Dluzewski, P., Feldman, Y., Tenne, R. TEM study of chirality in MoS2 nanotubes. *Journal of Microscopy* **181**, 68–71 (1996).
95. Remskar, M., Skraba, Z., Ballif, C., Sanjines, R., Levy, F. Stabilization of the rhombohedral polytype in MoS2 and WS2 microtubes: TEM and AFM study. *Surf. Sci.* **433/435**, 637–641 (1999).
96. Virsek, M., Jesih, A., Milosevic, I., Damnjanovic, M., Remskar, M. Raman scattering of the MoS2 and WS2 single nanotubes. *Surf. Sci.* **601**, 2868–2872 (2007).
97. Frey, G. L., Tenne, R., Matthews, M. J., Dresselhaus, M. S., Dresselhaus, G. Raman and resonance Raman investigation of MoS2 nanoparticles. *Phys. Rev. B* **60**, 2883–2892 (1999).
98. Remskar, M., Mrzel, A., Virsek, M., Jesih, A. Inorganic nanotubes as nanoreactors: the first MoS2 nanopods. *Adv. Mater* **19**, 4276–4278 (2007).

99. Srolovitz, D. J., Safran, S. A., Homyonfer, M., Tenne, R. Morphology of nested fullerenes. *Phys. Rev. Lett.* **74**, 1779–1782 (1995).
100. Feldman, Y., Wasserman, E., Srolovitz, D. J., Tenne, R. High-rate gas-phase growth of MoS2 nested inorganic fullerenes and nanotubes. *Science* **276**, 222–225 (1995).
101. Remskar, M., Skraba, Z., Sanjines, R., Levy, F. MoS2 and WS2 nanotubes alloyed with gold and silver. *Surf. Rev. Lett.* **6**, 1283–1287 (1999).
102. Kralj-Iglič, V., Remskar, M., Iglič, A. Deviatoric elasticity as a mechanism describing stable shapes of nanotubes. Reimer, A. (Ed.). *Horizons in world physics.* **244**, Hauppauge (NY): Nova Science Publishers, 111–156 (2004).
103. Perrin, C., Cordier, S., Gulo, F., Perrin, A. The octahedral cluster compounds of early transition metals: an original class of dielectric materials. *Ferroelectrics* **254**, 83–90 (2001).
104. Wilson, J. A., Yoffe, A. D. The transition metal dichalcogenides discussion and interpretation of the observed optical, electrical and structural properties. *Adv. Phys.* **18**, 193–335 (1969).
105. Remskar, M. Inorganic nanotubes synthesized by chemical transport reactions. Schwarz, J. A. (Ed.). *Dekker encyclopedia of nanoscience and nanotechnology.* **Part 1**, New York, Basel: Publ. Marcel Dekker, 1457–1465 (2004).
106. Zhu, Y. Q., Hsu, W. K., Kroto, H. W., Walton, D. R. M. An Alternative route to NbS2 nanotubes. *J. Phys. Chem. B* **106**, 7623–7626 (2002).
107. Jaszczak, J. A. Graphite: Flat, Fibrous and Spherical. G. D.Mendenhall, J.Liebman, A. Greenberg (Eds.). *Mesomolecules: from molecules to materials.* **1**, New York: Chapman & Hall, 161–180 (1995)
108. Rosentsveig, R., Margolin, A., Feldman, Y., Popovitz-Biro, R., Tenne, R. Bundels and foils of WS2 nanotubes. *Appl. Phys. A* **74**, 367–369 (2002).
109. Kasuga, T., Hiramatsu, M., Hoson, A., Sekino, T., Niihara, K. Formation of titanium oxide nanotube. *Langmuir* **14**, 3160–3163 (1998).
110. Du, G. H., Chen, Q., Che, R. C., Yuan, Z. Y., Peng, L.-M. Preparation and structure analysis of titanium oxide nanotubes. *Appl. Phys. Lett.* **79**, 3702–3704 (2001).
111. Zhang, S., et al. Formation mechanism of H2Ti3O7 nanotubes. *Phys. Rev. Lett.* **91**, 256103 1–4 (2003).
112. Rafailov, P. M., Thomsen, C., Gartsman, K., Kaplan-Ashiri, I., Tenne, R. Orientation dependence of the polarizability of an individual WS2 nanotube by resonant Raman spectroscopy. *Phys. Rev. B* **72**, 205436 1–4 (2005).
113. Dobardzic, E., Daki, B., Damnjanovic, M., Milosevic, I. Zero m phonons in MoS2 nanotubes. *Phys. Rev. B* **71**, 121405 1–4 (2005).
114. Tenne, R. Inorganic nanotubes and fullerene-like nanoparticles. *J. Mater. Res.* **21**, 2726–2743 (2006).
115. Bar-Sadan, M., Kaplan-Ashiri, I., Tenne, R. Inorganic fullerenes and nanotubes: Wealth of materials and morphologies. *Eur. Phys. J. Special Topics* **149**, 71–101 (2007).
116. Chopra, N. G., Zettl, A. Measurement of the elastic modulus of a multi-wall boron nitride nanotube. *Solid State Commun.* **105**, 297–300 (1998).
117. Hung, S. C., Su, Y. K., Fang, T. H., Chang, S. J., Ji, L. W. Buckling instabilities in GaN nanotubes under uniaxial compression. *Nanotechnology* **16**, 2203–2208 (2005).
118. Kaplan-Ashiri, I., et al. Mechanical behavior of individual WS2 nanotubes. *J. Mater. Res.* **19**, 454–459 (2004).
119. Kaplan-Ashiri, I., et al. On the mechanical behavior of WS2 nanotubes under axial tension and compression. *Proc. Natl. Acad. Sci. USA.* **103**, 523–528 (2006).
120. Yu, M. F., Files, B. S., Arepalli, S., Ruoff, R. S. Tensile loading of ropes of single wall carbon nanotubes and their mechanical properties. *Phys. Rev. Lett.* **84**, 5552–5555 (2000).
121. Yu, M. F., et al. Strength and breaking mechanism of multiwalled carbon nanotubes under tensile load. *Science* **287**, 637–640 (2000).
122. Remskar, M., Mrzel, A. High-temperature fibres composed of transition metal inorganic nanotubes. *Curr. Opin. Solid State Mater. Sci.* **8**, 121–125 (2004).
123. Horváth, L., Gault, R. A. The mineralogy of Mont Saint-Hilaire, Quebec. *Mineralog. Record* **21**, 281–362 (1990).

Chapter 12
Synthesis Approaches of Inorganic Nanotubes

Mihaela Daub and Kornelius Nielsch

Abstract The fabrication and characterisation of nanotubes and nanotubes with multifunctional properties has attracted a lot of interest due to their potential applications ranging from diagnostics of cancer cells to nanoelectronics. In contrast to a conventional spherical nanoparticle, several physical or chemical properties can be implemented within a nanowire or nanotube, like optical information, magnetical properties, and chemical functionalisation of certain surface areas. This chapter reviews the development of internationally leading research groups of multifunctional nanowire and nanotubes based on inorganic materials for applications in the fields of biotechnology, magnetic media, sensors, thermoelectric materials and nanoelectronics.

12.1 Introduction

The first report on carbon nanotubes in 1991 by S. Iijima [1], followed in 1992 by the first study on inorganic (WS_2) nanotubes by R. Tenne [2], initiated intense experimental and theoretical research on this new type of nanostructures. In recent years, the number of reports on synthesis of different types of inorganic nanotubes has increased monthly, due to the importance of this new and fast developing research area of nanotechnology. The particular physical and chemical properties of inorganic nanotubes indicate promising applications in nanoscale devices, sensors, drug delivery or energy conversion. One-dimensional materials play an important role in understanding, for instance, the influence of dimensionality on optical, mechanical, electrical, and magnetic properties. The cylindrical geometry of the nanotubes leads to low mass density, high porosity and a large surface to weight ratio, properties which are especially relevant in applications such as gas sensors, catalysis, solid-state batteries and fuel cell electrodes.

Six categories of inorganic nanotubes have been synthesized until now [5]:

- transition metal chalcogenide nanotubes: MoS_2, $MoSe_2$, WS_2, WSe_2, NbS_2, $NbSe_2$, TaS_2, ZrS_2, HfS_2, TiS_2, ZnS, NiS, CdSe, CdS;

K. Nielsch (✉)
Max Planck Institute of Microstructure Physics, 06120 Halle, Germany; Institute of Applied Physics, University of Hamburg, 20355 Hamburg, Germany
e-mail: knielsch@physnet.uni-hamburg.de; knielsch@mpi-halle.de

O. Hayden, K. Nielsch (eds.), *Molecular- and Nano-Tubes*,
DOI 10.1007/978-1-4419-9443-1_12, © Springer Science+Business Media, LLC 2011

- oxide nanotubes: TiO_2, NiO, ZnO, GaO/ZnO, VO_x, $W_{18}O_{49}$, V_2O_5, Al_2O_3, Fe_3O_4, In_2O_3, Ga_2O_3, $BaTiO_3$, $PbTiO_3$, SiO_2, MoO_3, RuO_2, MgO, rare earth oxides: (Er, Tm, Yb, Lu) oxide;
- transition metal halogenide nanotubes: $NiCl_2$;
- mixed-phase and metal-doped nanotubes: $PbNb_nS_{2n+1}$, $Mo_{1-x}WS_2$, $W_xMo_yC_zS_z$, Nb-WS_2, WS_2-carbon nanotubes, NbS_2-carbon nanotubes, Au-WS_2, Ag-MoS_2, $Cu_{5.5}FeS_{6.5}$;
- boron- and silicon-based nanotubes: BN, BCN, Si;
- metal nanotubes: Au, Co, Fe, CoFe, FePt, Cu, Ni, Pb, Te, Bi, Se, Pt, Ru.

The latest review on inorganic nanotubes, which include an overview on their properties and theoretical calculations, was published in 2006 by R. Tenne [3, 4]. A review on the synthesis, morphology and applications of inorganic nanotubes was done by Remskar [5] in 2004. Other reviews focusing on the synthesis of inorganic nanotubes were published by Pokropivny [6], Ivanovskii [7], Rao and Nath [8] and Xia et al. [9]. This chapter will provide a brief overview over the main methods used for the synthesis of inorganic nanotubes, with a focus on the newest methods developed in the last few years, such as electrodeposition and atomic layer deposition in template systems.

12.2 Synthesis Methods

BN nanotubes were first synthesized in 1995 by *arc discharge* by Chopra et al. [10], followed by Loiseau et al. [11]. A high temperature plasma jet method was later proposed by Shimizu et al. [12] for the synthesis of BN nanotubes and coexisting amorphous phases. Cumings et al. [13] used a high-yield plasma-arc method to produce macroscopic amounts of pure BN nanotubes. Mixed BN-C nanotubes by Suenaga et al. [14] and Goldberg et al. [15] and $B_xC_yN_z$ nanotubes were also fabricated by using this method by Weng-Sieh et al. [16].

The method of *laser ablation* was used by Goldberg et al. [17, 18] to fabricate inclusion-free BN nanotubes, as well as by Yu et al. [19] for the synthesis of BN single-walled nanotubes. Coaxial multi-element nanocables of SiC-SiO_2-BN-C, consisting of SiC nanowires, an amorphous intermediate layer of SiO_2 and an outer layer of BN-C were produced by Zhang et al. [20] by the same method. *Chemical substitution reactions* were used by Han et al. [21] for the synthesis of multi-walled B-C-N nanotubes.

One of the earliest techniques for producing inorganic nanotubes of layered metal dichalcogenide (MX_2) compounds is the *vapour chemical transport method*: a powder of ME_2 is placed on the hot side of an evacuated quartz ampoule, in the presence of a transport agent (bromine or iodine). By maintaining a temperature gradient of 20–50°C, nanotubes can grow on the cold side of the ampoule. By this method, Remskar et al. fabricated nanotubes of WS_2 [22, 23], MoS_2 [24, 25] (Fig. 12.1) and $MoS_{2-x}I_y$ [26].

Fig. 12.1 Scanning electron and electron transmission images on different length scales of MoS_2 nanotubes. (**a**) Bundles appear to self-assemble into various different microscopic structures. (**b**) The bundles end in sharp points. (**c**) A split tip of a bundle terminating in strands 4 nm wide. (**d**) Expanded electron transmission view of a strand composed of only a few individual nanotubes. (Reproduced with permission from Ref. [25])

Another method for producing inorganic nanotubes is the *sol-gel technique*. By dissolving a metal organic compound together with a template-forming species in alcohol, a template nanotube structure is created. The addition of small amounts of water leads to a slow hydrolysis of the metal organic compound and to the formation of the metal-oxide sol, which can transform into a gel by tuning the pH. Thermal treatment and hydrogen reduction are further leading to the formation of nanotubes in the template. Cheng et al. [27] used the sol-gel chemistry and porous alumina templating to fabricate In_2O_3 and Ga_2O_3 nanotubes, which is shown in Fig. 12.2. Fe, Ni, Pb and CoFe nanotubes were also prepared by Hua et al. [28] by the sol-gel method, followed by hydrogen reduction of the nanotubes in the anodic alumina template. Other examples are vanadium oxide nanotubes by Krumeich et al. [29], $BaTiO_3$ and $PbTiO_3$ nanotubes by Hernandez et al. [30], silica nanotubes by Nakamura et al. [31], Mitchell et al. [32] and Gasparac et al. [33], TiO_2 nanotubes by Zhang et al. [34], ZrO_2 nanotubes by Bao et al. [35] and iron oxide nanotubes by Suber et al. [36].

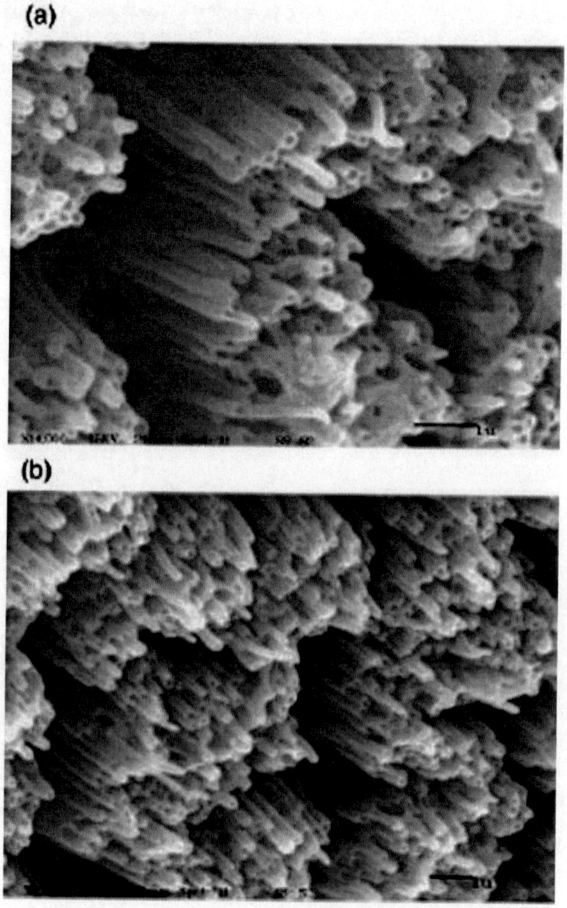

Fig. 12.2 SEM images of In_2O_3 and Ga_2O_3 nanotubes prepared by the sol-gel method; a 1 μm bar is indicated. (Reproduced with permission from Ref. [27])

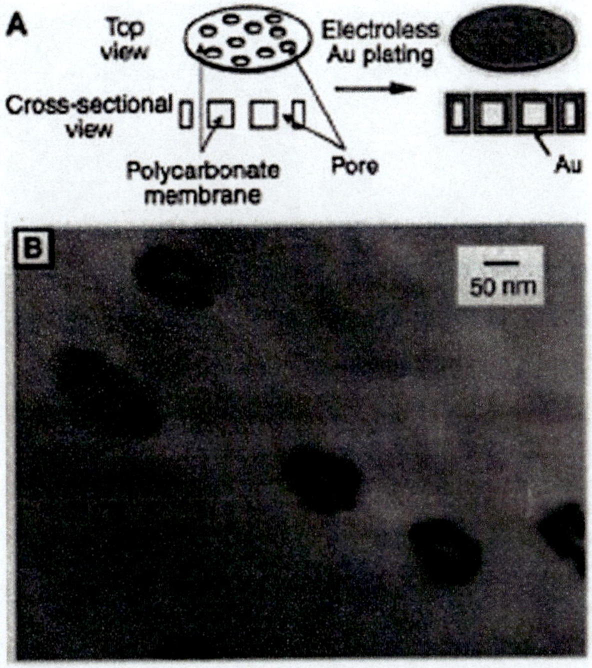

Fig. 12.3 (**a**) Schematic diagram of the electroless plating procedure; (**b**) Transmission electron micrograph showing a cross section of a Au-nanotubule membrane. (Reproduced with permission from Ref. [37])

In 1995, Nishizawa et al. [37] presented a general method for preparing nanotubes, which they named template synthesis (Fig. 12.3), consisting in the deposition of the desired material within the cylindrical pores of a nanoporous membrane. They used commercially available polycarbonate filtration membranes with pores of 30 nm in diameter as templates for the deposition of Au nanotubes. Gold was *electroless* deposited onto the pore walls, by applying first a catalyst to the surface of the membrane and then immersing the membrane into an electroless plating bath which containes a Au^{I} species and a chemical reducing agent. The thickness of the deposited Au layer was controlled by varying the deposition time. At longer plating times, membranes containing nanowires were obtained [37, 38]. By using the same method, Demoustier-Champagne et al. [39] obtained Au and Ag nanotubes inside the pores of polycarbonate membranes with mean pore size in the range of 100–1,000 nm and low pore size distribution [4].

Mertig et al. [40] fabricated nanotubes by electroless metal film deposition onto tubular biomolecular templates. They used microtubules, in vitro assembled protein filaments with a 25 nm outer diameter and a length of several micrometers, for the electroless deposition of magnetic nanotubes (Ni, Co), shown in Fig. 12.4. The metal deposition was initiated by direct absorption of noble metal catalysts (Pd, Pt) on the protein surface, under neutral pH and by using dimethylamine-borane as reducing agent of nickel and cobalt salts. Under physiological conditions, the morphology of the template was completely preserved.

Co, Ni and Cu nanotubes were also prepared by Wang et al. [41] by electroless deposition, using porous anodic aluminium oxide (AAO) as templates (Fig. 12.5).

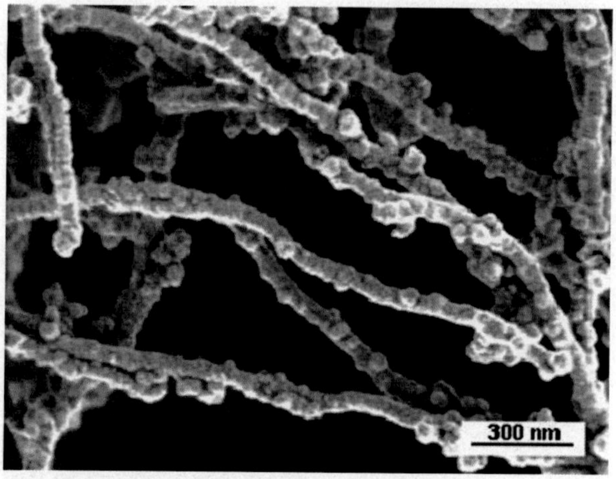

Fig. 12.4 SEM micrograph of nickel electroless plated microtubules. (Reproduced with permission from Ref. [40])

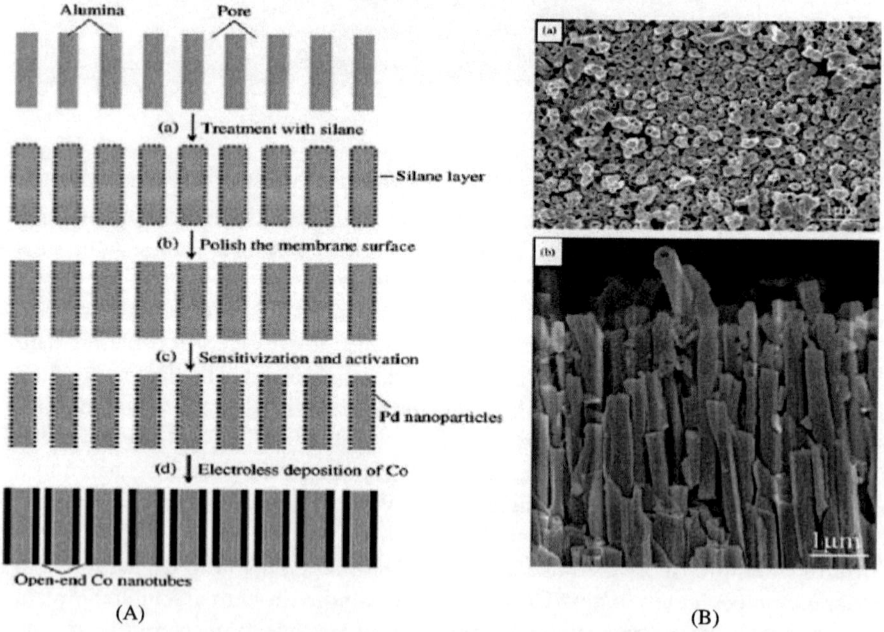

Fig. 12.5 (**A**) Scheme for the synthesis of metal nanotube arrays within the Al_2O_3 porous template. The membrane is (**a**) modified with silane, (**b**) polished to remove the silane layer on the surface, (**c**) sensitized with Sn^{2+} and then activated with Pd^{2+}, resulting in the deposition of Pd nanoparticles on the pore walls and (**d**) immersed in the metal plating solution to form metal nanotube arrays. (**B**) (**a**) *Top* view and (**b**) *side* view SEM images of Co nanotube arrays. (Reproduced with permission from Ref. [41])

Prior to the deposition, the alumina membrane was modified with silane. The electroless deposition was carried out in an aqueous solution, via the reaction of an oxidizer and a reductant in an electrolyte solution. The obtained nanotubes had a well-defined outer diameters determined by the size of the alumina pores, while the inner diameter can be tailored by adjusting the deposition time. The length of the nanotubes was determined by the thickness of the template.

Electrochemical deposition is a versatile method for obtaining metallic or semiconducting nanotubes, which combines low processing costs with ambient conditions. The structural properties of electrodeposited nanotubes are influenced by the electrodeposition parameters, which makes possible their fabrication with a high reproducibility and in a controlled manner in terms of their crystallographic properties.

Au nanotubes with a length of 2 μm were first obtained by Brumlik et al. [42, 43] using an alumina membrane coated with an organocyanide as molecular anchor to bind the electrochemically deposited Au to the inner walls of the template. Bao et al. [44] fabricated 35 μm long Ni nanotubes by electrodeposition in the pores of alumina membranes that were pretreated with an organoamine as a pore-wall modifying agent. The Ni nanotubes had an average outer diameter of 160 nm ± 20 nm and polycrystalline structure. In a further report, Bao et al. reported the fabrication of Co nanotubes in alumina membranes using the same organoamine as pore-wall modifying agent [45].

TiO_2 nanotubes were obtained by P. Hoyer [46] using electrochemical deposition, anodic alumina membranes as templates and a polymer (PMMA) negatype structure as mold. Corresponding micrographs are shown in Fig. 12.6. The titanium

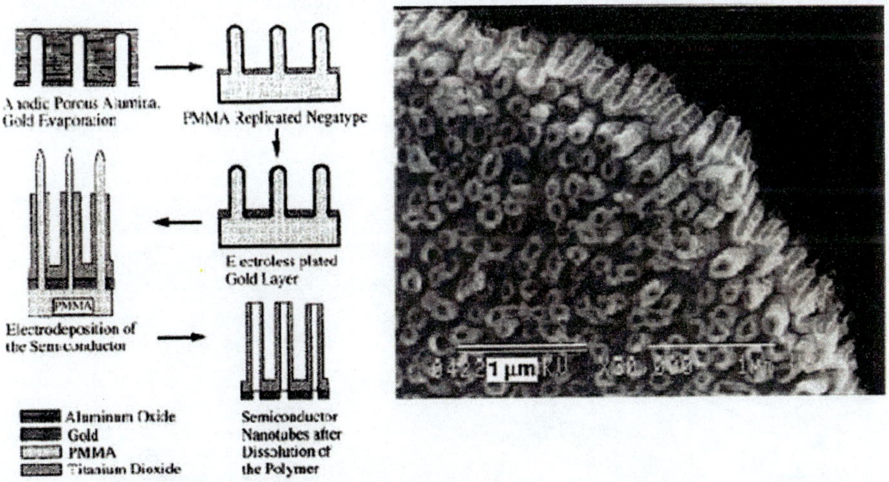

(A) (B)

Fig. 12.6 (**a**) Schematic view of a replication process using electrochemical deposition. (**b**) SEM image of the cross-section of the as-prepared film of titania with the *upper part* removed. (Reproduced with permission from Ref. [46])

dioxide was electrochemically deposited onto the polymer mold and, after the dissolution of the polymer matrix, nanotubes with both ends opened were obtained. The TiO_2 nanotubes had an outer diameter between 140 and 180 nm and an inner diameter of 70–100 nm. The as-deposited nanotubes were amorphous, but polycrystalline anatase samples of the same structure were obtained after heat treatment.

Tourillon et al. [47] obtained Co and Fe nanotubes by pulsed electrochemical deposition in commercially available track-etched polycarbonate membranes 6 μm thick and with a nominal pore diameter of 30 nm (Fig. 12.7). A thin Au layer was first evaporated on the bottom side of the membranes and was used as the working electrode in a standard three-electrode electrochemical cell together with a saturated calomel reference electrode and a Pt grid. The pulsed electrodeposition was performed at room temperature with an aqueous solution of 0.1 M $CoSO_4$ and 0.1 M H_3BO_3. The Co and Fe nanotubes had an outer diameter of 30 nm $\pm$ 20%, length of 6 μm and a wall thickness of 1–2 nm. In a further report by the same group [48], the Fe oxide nanotubes obtained by pulsed electrodeposition were proven to be built up of Fe_3O_4 nanocrystals with a common crystallographic axis, creating a pseudomonocrystalline wall in the nanotubes.

Other examples of inorganic nanotubes obtained by using anodic aluminium oxide or polycarbonate membranes as templates and electrochemical deposition are: $Ni(OH)_2$ and Fe-doped $Ni(OH)_2$ nanotubes, by Chou et al. [49], Ni nanotubes by Pi et al. [50], CoNiCu and Cu nanotubes by Davis et al. [51, 52], Ni nanotubes by Wu et al. [53] and Xue et al. [54], Pt nanotubes by Mu et al. [55] and Yoo et al. [56].

Multisegmented nanotubes were fabricated by Lee et al. [57] using a multi-step approach: metallic nanoparticles (Ag) were first immobilized on the pore walls of an anodic alumina membrane, then a thin Au layer was sputtered on one side of the membrane to serve as working electrode. Finally, the electrodeposition of Au was performed, resulting in the formation of Au nanotubes with surfaces decorated with

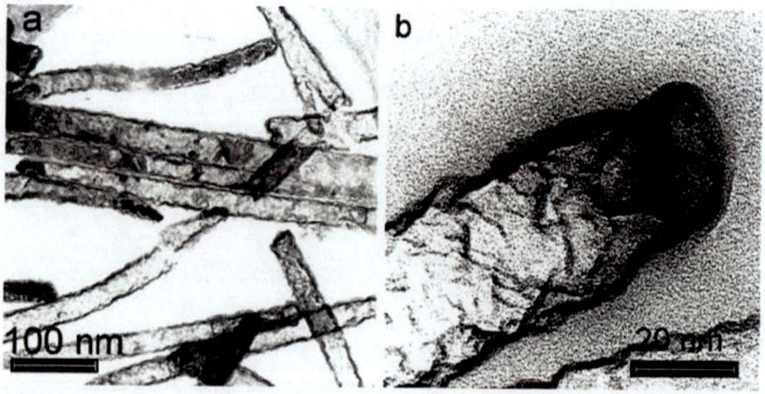

Fig. 12.7 TEM image of iron nanotubes obtained by pulsed electrochemical deposition. (Reproduced with permission from Ref. [47])

Ag nanoparticles. The preferential deposition of the metal along the wall surfaces of the nanopores allowed the preparation of multisegmented metallic nanotubes with a bimetallic stacking configuration: Au/Ni/Au/

Multilayered CoNiFe/Cu nanotubes were obtained by Davis et al. [52] under pulsed potential conditions from a single bath electrolyte, using commercially available nanoporous membranes as templates.

Chemical vapour deposition (CVD) is another method for obtaining inorganic nanotubes, which was developed mainly in the last years. Single-crystalline gallium nitride nanotubes with diameters of 30–200 nm and wall thicknesses of 5–50 nm were obtained in ZnO nanowires templates by Goldberger et al. [58]. Micrographs of the nanostructures are shown in Fig. 12.8. Two micrographs of these structures are displayed in Fig. 12.9. The ZnO nanowire arrays were placed inside a reaction tube for GaN chemical vapour deposition. Trimethylgallium and ammonia were used as precursors and argon or nitrogen as carrier gas. The ZnO nanowires were removed after the GaN deposition, by heating the samples at 600°C in 10% H_2. Si nanotubes were also fabricated by Sha et al. [59] with a similar method. Metal-organic chemical vapour deposition (MOCVD) was used for obtaining nickel oxide nanotubes by Malandrino et al. [60] and GaN nanotubes by Jung et al. [61].

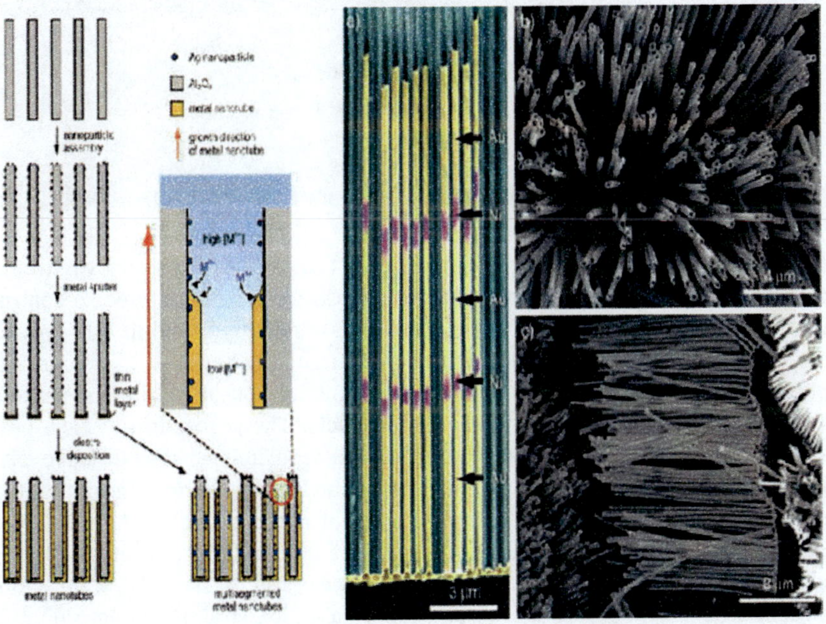

(A) (B)

Fig. 12.8 (**a**) Procedure for the preparation of multisegmented metal nanotubes and the proposed mechanism for metal nanotube growth; (**b**) SEM images of multisegmented metal nanotubes with a stacking configuration of Au-Ni-Au-Ni-Au along the tube axis. (Reproduced with permission from Ref. [57])

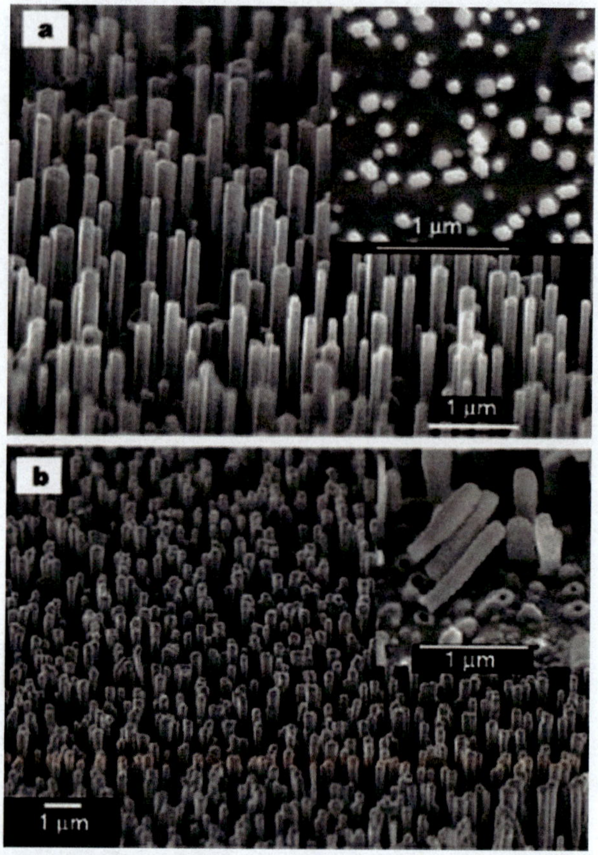

Fig. 12.9 Arrays of ZnO nanowires and GaN nanotubes. Shown are SEM images of the ZnO nanowire template arrays (**a**) and the resulting GaN nanotube array (**b**). *Inset* in (**a**) shows cross-sections of the ZnO nanowires. *Inset* in (**b**) shows the fractured interface between the GaN nanotubes and the substrate. (Reproduced with permission from Ref. [58])

Atomic layer deposition or atomic layer epitaxy (ALD or ALE) is a very new and promising method for growing very uniform nanostructures with a precise control of deposition thickness and characteristics. ALD is a self-limiting thin film growth process during which the substrate surface is alternatively exposed to precursors. The deposition process consists in alternating chemisorption of precursors, surface reaction and desorption of the gaseous side-products. The adsorption of each precursor is self-limiting, up to a full monolayer can be deposited per operational cycle. The important advantage of ALD is that the film thickness is dependent not on the kinetics of the reactions but only on the number of reaction cycles. Atomic layer deposition offers other several practical advantages in comparison to other deposition methods, including accurate and simple thickness control, large area and large batch capability, good conformability and reproducibility, straight-forward doping and scale-up, as well as the possibility to produce sharp and precisely defined interfaces. By atomic layer deposition on porous anodic alumina or silicon as templates, inorganic nanotubes can be obtained. One of the first reports explaining the process of ALD on porous substrates (alumina or silica) was published in 1994 by E.-L. Lakomaa [62].

Uniform arrays of TiO_2 nanotubes were fabricated by Sander et al. [63] using $TiCl_4$ and H_2O as ALD precursors and nanoporous alumina templates on a silicon substrate. After the deposition of TiO_2, the layer deposited on the top of the template was removed by gentle mechanical polish and the alumina was etched away to reveal arrays of titania nanotubes on the substrate. The process allowed the fabrication of nanotubes with well-defined and easy-to-tailor diameter and wall thickness.

Shin et al. [64] used commercially available polycarbonate membranes as templates for the deposition of TiO_2 and ZrO_2 nanotubes. Kemell et al. [65] fabricated TiO_2 nanotubes by depositing directly into commercially available alumina membranes or on electrodeposited Ni nanowires.

Ni and Co nanotubes were obtained by Daub et al. [66] in alumina membranes, using nickelocene, cobaltocene and water or ozone as precursors (Fig. 12.10). The fabrication process consisted in depositing first the nickel oxide and then further exposing the substrate during the same ALD cycle to hydrogen, or performing a heat treatment under Ar/H_2 atmosphere at the end of the deposition process. The tubes obtained by the second method, using ozone as a second ALD precursor,

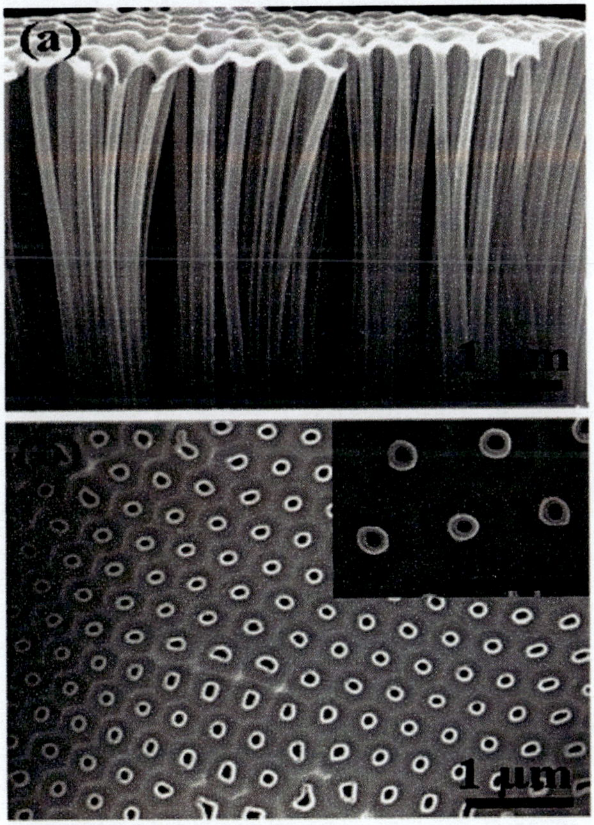

Fig. 12.10 TiO_2/Ni/TiO_2 nanotubes obtained by atomic layer deposition inside the pores of alumina templates. (**a**) The tubes were removed from the template; (**b**) The multilayer nanotubes are embedded in the alumina membrane, the ferromagnetic layer on top being removed by ion milling. (Reproduced with permission from Ref. [66])

presented a fine granular crystal structure, low surface roughness and significantly improved magnetic properties. Furthermore Bachmanne et al. [67] has recently synthesized Fe_3O_4 nanotubes based on low-temperature atomic layer deposition and have studied in detail the magnetic properties as function of the wall thickness.

Another possibility to obtain inorganic nanotubes by ALD is the use of carbon nanotubes as templates. Herrmann et al. [68] functionalized multi-walled carbon nanotubes by covering them with Al_2O_3. They also obtained multilayered coatings consisting of alternating layers of Al_2O_3 and W deposited on multi-walled carbon nanotubes. The functionalization of single-walled carbon nanotubes was reported by Farmer et al. [69].

Ruthenium oxide nanotubes were successfully fabricated by Min et al. [70] using carbon nanotube arrays as removable templates. Carbon nanotubes were first grown inside porous anodic alumina membranes by catalytically pyrolyzing acetylene in nitrogen carrier gas. Atomic layer deposition with $Ru(Et)_3$ and oxygen resulted in a metallic Ru thin film on the carbon nanotubes. During the process of removing of

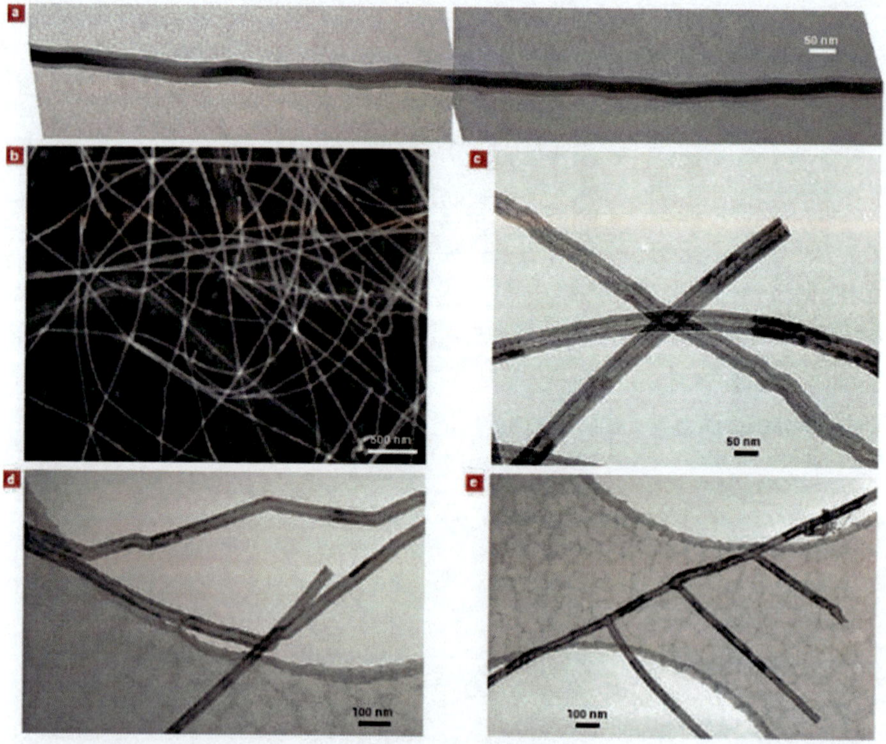

Fig. 12.11 Transformation of core-shell nanowires to nanotubes by means of the Kirkendall effect. (**a**) TEM image of an example ZnO-Al_2O_3 core-shell nanowire. (**b**) SEM image and (**c**)–(**e**) TEM images of $ZnAl_2O_4$ spinel nanotubes. Owing to the hollow interiors, the central part of the nanotubes becomes more transparent to the electron beam than the outer part, in contrast to the structure before annealing. (Reproduced with permission from Ref. [71])

the carbon nanotube template by ashing, the Ru tubes were oxidized and ruthenium oxide nanotube arrays were subsequently obtained.

Fan et al. [71] reported a new method for obtaining monocrystalline spinel nanotubes by using the Kirkendall effect (Fig. 12.11). Here, ZnO nanowires were conformally coated by ALD with Al_2O_3, using trimethylaluminium and water at 200°C. An interfacial solid-state reaction at an elevated annealing temperature of 700°C consumed the ZnO core, producing hollow spinel $ZnAlO_4$ nanotubes with a diameter of 40 nm and wall thickness of 10 nm.

Tobacco mosaic viruses were used by Knez et al. [72] as templates for the atomic layer deposition of TiO_2 and Al_2O_3. The smallest nanotubes realized so far are based on metal oxides and templated by the inner nanochannel of the Tobacco mosaic viruses (TMV) virus, which are presented in Fig. 12.12. They obtained by this method TiO_2 and Al_2O_3 nanotubes with an outer diameter of 4 nm and an inner diameter of 1.5 nm or less.

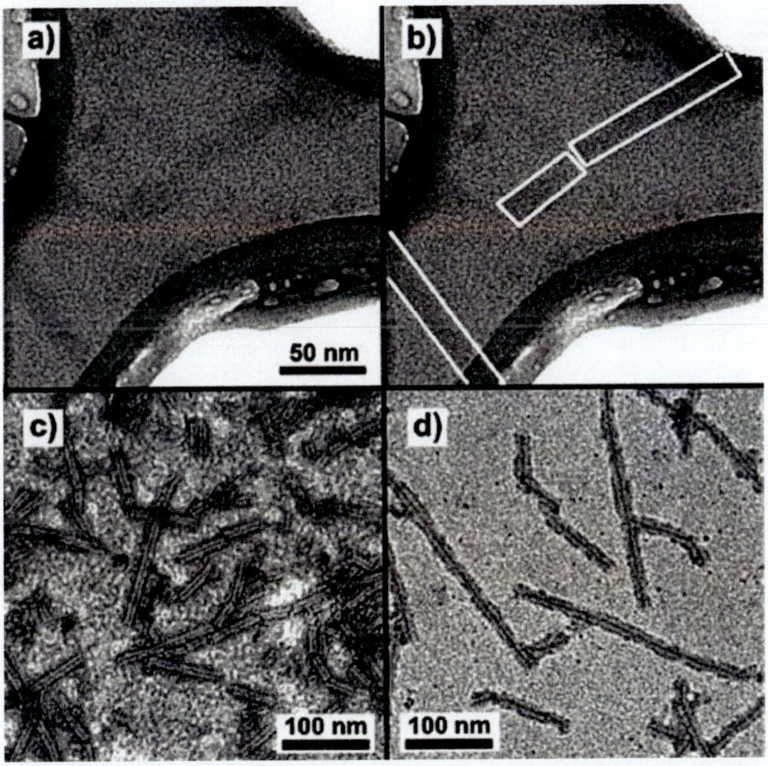

Fig. 12.12 *Upper images*: TEM image of untreated TMV. **(c)** TEM image of TMV treated with TiO_2 by ALD. The coverage of viruses with TiO_2 reaches 100%. **(d)** After ultrasonication the TiO_2 is partially removed from the outer surface and mainly the inner channel of the TMV remains covered with TiO_2. (Reproduced with permission from Ref. [72])

12.3 Concluding Remarks

This chapter presented an overview on the main methods for synthesizing inorganic nanotubes, with a focus on the most recent and promising methods developed in the recent years, such as electrodeposition and atomic layer deposition, which allows a precise tailoring of the layer thickness and a very conform and uniform deposition characteristics. Control over the dimensions and properties of nanotubes, is an important feature of the fabricating process and an essential step for the potential technological applications.

References

1. Iijima, S. Helical microtubules of graphitic carbon. *Nature* **354**, 54–56 (1991).
2. Tenne, R., Margulis, L., Genut, M. & Hodes, G. Polyhedral and cylindrical structures of tungsten disulfide. *Nature* **360**, 444–445 (1992).
3. Tenne, R. Inorganic nanotubes and fullerene-like nanoparticles. *Nat. Nanotechnol.* **1**, 103–111 (2006).
4. Tenne, R. Inorganic nanotubes and fullerene-like nanoparticles. *J. Mater. Res.* **21**, No. 11, 2726–2743 (2006).
5. Remskar, M. Inorganic nanotubes. *Adv. Mater.* **16**, No. 17, 1497–1504 (2004).
6. Pokropivny, V. V. Non-carbon nanotubes (Review). I. Synthesis methods. *Powder Metallurgy and Metal Ceramics* **40**, No. 9–10, 485–496 (2001).
7. Ivanovskii, A. L. Non-carbon nanotubes: synthesis and simulation. *Russian Chemical Reviews* **71**, No. 3, 175–194 (2002).
8. Rao, C. N. R. & Nath, M. Inorganic nanotubes. *Dalton Trans.* 1–24 (2003).
9. Xia, Y., Yang, P., Sun, Y., Wu, Y., Mayers, B., Gates, B., Yin, Y., Kim, F. & Yan, H. One-dimensional nanostructures: synthesis, characterization and applications. *Adv. Mater.* **15**, No. 5, 353–389 (2003).
10. Chopra, N. G., Luyken, R.J., Cherry, K., et al. Boron nitride nanotubes. *Science* **269**, 966–967 (1995).
11. Loiseau, A., Willaime, F., Demoncey, N., et al. Boron nitride nanotubes with reduced number of layers synthesised by arc discharge. *Phys. Rev. Lett.* **76**, No. 25, 4737–4740 (1996).
12. Shimizu, Y., Moriyoshi, Y., Tanaka, H. & Komatsu, S. BN nanotubes, webs, and coexisting amorphous phase formed by the plasma jet method. *Appl. Phys. Lett.* **75**, No. 7, 929–932 (1999).
13. Cumings, J. & Zettl, A. Mass-production of boron nitride double-wall nanotubes and nanococoons. *Chem. Phys. Lett.* **316**, 211–216 (2000).
14. Suenaga, K., Colliex, C., Demoncey, N., Loiseau, A., et al. Synthesis of nanoparticles and nanotubes with well separated layers of BN and carbon. *Science* **278**, 653–655 (1997).
15. Goldberg, D., Bando, Y., Eremetz, M., et al. BN nanotube growth defects and their annealing-out under electron irradiation. *Chem. Phys. Lett.* **279**, 191–196 (1997).
16. Weng-Sieh, Z., Cherrey, K., Chopra, N.G., et al. Synthesis of $B_xC_yN_z$ nanotubes. *Phys. Rev. B* **51**, No. 16, 11229–11231 (1995).
17. Goldberg, D., Bando, Y., Eremets, M., et al. Nanotubes in boron nitride laser heated at high pressure. *Appl. Phys. Lett.* **69**, No. 14, 2045–2047 (1996).
18. Goldberg, D., Bando, Y., Han, W., Kurashima, K. & Sato, N. Single-walled B-doped carbon, B/N-doped carbon, and BN nanotubes synthesized from SWCNTs through a substitution method. *Chem. Phys. Lett.* **308**, 337–342 (1999).

19. Yu, D. P., Sun, X.S., Lee, C.S., et al. Synthesis of boron nitride nanotubes by means of excimer laser ablation at high temperature. *Appl. Phys. Lett.* **16**, 1996–1968.
20. Zhang, Y., Suenaga, K., Colliex, C. & Iijima, S. Coaxial nanocable: silicon carbide and silicon oxide sheathed with boron nitride and carbon. *Science* **281**, 973–975 (1998).
21. Han, W., Banbo, Y., Kurashima, K. & Sato, T. Synthesis of boron nitride nanotubes from carbon nanotubes by a substitution reaction. *J. Appl. Phys. Lett.* **73**, No. 21, 3085–3087 (1998).
22. Remskar, M., Skraba, Z., Regula, M., Ballif, C., Sanjines, R. & Levy, F. New crystal structures of WS_2: microtubes, ribbons and ropes. *Adv. Mater.* **10**, No 3, 246–249 (1998).
23. Remskar, M., Skraba, Z., Sanjines, R. & Levy, F. Syntactic coalescence of WS_2 nanotubes. *Appl. Phys. Lett.* **74**, No. 24, 3633–3635 (1999).
24. Remskar, M., Skraba, Z., Stadelmann, P. & Levy, F. Structural stabilization of new compounds: MoS_2 and WS_2 micro- and nanotubes alloyed with gold and silver. *Adv. Mater.* **12**, No. 11, 814–818 (2000).
25. Remskar, M., Mrzel, A., Skraba, Z., Jesih, A., Ceh, M., Demsar, J., Stadelmann, P., Levy, F. & Mihailovic, D. Self-assembly of sub-nanometer-diameter single-wall MoS_2 nanotubes. *Science* **292**, 479–481 (2001).
26. Remskar, M., Mrzel, A., Sanjines, R. & Levy, F. Metallic Sub-nanometer $MoS_{2-x}I_y$ nanotubes. *Adv. Mater.* **15**, No. 3, 237–240 (2003).
27. Cheng, B. & Samulski, E. Fabrication and characterization of nanotubular semiconductor oxides In_2O_3 and Ga_2O_3. *J. Mater. Chem.* **11**, 2901–2902 (2001).
28. Hua, Z., Yang, S., Huang, H., Lv, L., Lu, M., Gu, B. & Du, Y. Metal nanotubes prepared by a sol-gel method followed by a hydrogen reduction procedure. *Nanotechnology* **17**, 5106–5110 (2006).
29. Krumeich, F., Muhr, H.-J., Niederberger, M., Bieri, F., Schnyder, B. & Nesper, R. Morphology and topochemical reactions of novel vanadium oxide nanotubes. *J. Am. Chem. Soc.* **121**, 8324–8331 (1999).
30. Hernandez, B. A., Chang, K.-S., Fisher, E.R. & Dorhout, P.K. Sol.gel template synthesis and characterization of $BaTiO_3$ and $PbTiO_3$ nanotubes. *Chem. Mater.* **14**, 480–482 (2002).
31. Nakamura, H. & Matsui, Y. Silica gel nanotubes obtained by the sol-gel method. *J. Am. Chem. Soc.* **117**, 2651–2652 (1995).
32. Mitchell, D. T., Lee, S.B., Trofin, l, Li, N., Nevanen, T.K., Soderlund, H. & Martin, C.R. Smart nanotubes for bioseparations and biocatalysis. *J. Am. Chem. Soc.* **124**, 11864–11865 (2002).
33. Gasparac, R., Kohli, P., Mota, M.O., Trofin, L. & Martin, C.R. Template synthesis of nanotest tubes. *Nano Lett.* **4**, No. 3, 513–516 (2004).
34. Zhang, M. & Wada, B.K. Sol-gel template preparation of TiO_2 nanotubes and nanorods. *J. Mater. Sci. Lett.* **20**, 167–170 (2001).
35. Bao, J., Tie, C., Xu, Z., Ma, Q., Hong, J., Sang, H. & Sheng, D. An array of concentric composite nanostructures of zirconia nanotubules/cobalt nanowires: preparation and magnetic properties. *Adv. Mater.* **14**, No. 1, 44–47 (2002).
36. Suber, L., Imperatori, P., Ausanio, G., Fabbri, F. & Hofmeister, H. Synthesis, morphology and magnetic characterization of iron oxide nanowires and nanotubes. *J. Phys. Chem. B* **109**, 7103–7109 (2005).
37. Nishizawa, M., Menon, V.P. & Martin, C.R. Metal nanotubule membranes with electrochemically switchable ion-transport selectivity. *Science* **268**, 700–702 (1995).
38. Wirtz, M. & Martin, C.R. Template-fabricated gold nanowires and nanotubes. *Adv. Mat.* **15**, No. 5, 455–458 (2003).
39. Demoustier-Champagne, S. & Delvaux, M. Preparation of polymeric and metallic nanostructures using a template-based deposition method. *Mat. Sc. Eng. C* **15**, 269–271 (2001).
40. Mertig, M. & Kirsch, R. Pompe & W. Biomolecular approach to nanotube fabrication. *Appl. Phys. A* **66**, S723–S727 (1998).

41. Wang, W., Li, N., Li, X., Geng, W. & Qiu, Z. Synthesis of metallic nanotube arrays in porous anodic aluminum oxide template through electroless deposition. *Mat. Res. Bull.* **41**, 1417–1423 (2006).
42. Brumlik, C. J. & Martin, C.R.J. *Am. Chem. Soc.* **113**, 3174 (1991).
43. Brumlik, C. J., Menon, V.P. & Martin, C.R.J. *Mater. Res.* **5**, 1174 (1994).
44. Bao, J., Tie, C., Xu, Z., Zhou, Q., Shen, D. & Ma, Q. Template synthesis of an array of nickel nanotubules and its magnetic behavior. *Adv. Mater.* **13**, No. 21, 1631–1633 (2001).
45. Bao, J., Xu, Z., Hong, J., Ma, X. & Lu, Z. Fabrication of cobalt nanostructures with different shapes in alumina template. *Scripta Mater.* **50**, 19–23 (2004).
46. Hoyer, P. Formation of a titanium dioxide nanotube array. *Langmuir* **12**, 1411–1413 (1996).
47. Tourillon, G., Pontonnier, L., Levy, J.P. & Langlais, V. Electrochemically synthesized Co and Fe nanowires and nanotubes. *Electrochem. Solid-State Lett.* **3**, 20–23 (2000).
48. Verbeeck, J., Lebedev, O.I., Van Tendeloo, G., Cagnon, L., Bougerol, C. & Tourillon, G. Fe and Co nanowires and nanotubes synthesized by template electrodeposition. *J. Electrochem. Soc.* **150**, No. 10, E468–E471 (2003).
49. Chou, S., Cheng, F. & Chen, J. Electrochemical deposition of $Ni(OH)_2$ and Fe-doped $Ni(OH)_2$ tubes. *Eur. J. Inorg. Chem.* 4035–4039 (2005).
50. Pi, Z., Tian, T., Tian, X., Yang, C., Zhang, S. & Zheng, J. A template-based electrochemical method for the synthesis of high dense nickel nanotube arrays. *J. Nanosci. Nanotechnol.* **7**, 673–676 (2007).
51. Davis, D. M. & Podlaha, E.J. CoNiCu and Cu nanotube electrodeposition. *Electrochem. Solid-State Lett.* **8**, No. 2, D1–D4 (2005).
52. Davis, D. M., Moldovan, M., Young, D.P., Henk, M., Xie, X. & Podlaha, E.J. Magnetoresistance in electrodeposited CoNiFe/Cu multilayered nanotubes. *Electrochem. Solid-State Lett.* **9**, No. 9, C153–C155 (2006).
53. Wu, M. T., Leu, I.C., Yen, J.H. & Hon, M.H. Novel electrodeposition behavior of Ni on porous anodic alumina templates without a conductive interlayer. *J. Phys. Chem. B* **109**, 9575–9580 (2005).
54. Xue, S., Cao, C. & Zhu, H. Electrochemically and template-synthesized nickel nanorod arrays and nanotubes. *J. Mater. Sci.* **41**, 5598–5601 (2006).
55. Mu, C., Yu, Y., Wang, R., Wu, K., Xu, D. & Guo, G. *Adv. Mater.* **16**, 1550 (2004).
56. Yoo, W. C. & Lee, J.K. *Adv. Mater.* **16**, 1097 (2004).
57. Lee, W., Scholz, R., Nielsch, K. & Goesele, U. A template-based electrochemical method for the synthesis of multisegmented metallic nanotubes. *Angew. Chem. Int. Ed.* **44**, 6050–6054 (2005).
58. Goldberger, J., He, R., Zhang, Y., Lee, S., Yan, H., Choi, H.-J. & Yang, P. Single-crystal gallium nitride nanotubes. *Nature* **422**, 599–602 (2003).
59. Sha, J., Niu, J., Ma, X., Xu, J., Zhang, X., Yang, Q. & Yang, D. Silicon nanotubes. *Adv. Mater.* **14**, No. 17, 1219–1221 (2002).
60. Malandrino, G., Perdicaro, L.M.S., Fragala, I.L., Nigro, R.L., Losurdo, M. & Bruno, G. MOCVD template approach to the fabrication of free-standing nickel(II) oxide nanotube arrays: structural, morphological, and optical properties characterization. *J. Phys. Chem. Lett. C* **111**, 3211–3215 (2007).
61. Jung, W. -G., Jung, S.-H., Kung, P. & Razeghi, M. Fabrication of GaN nanotubular material using MoCVD with an aluminium oxide membrane. *Nanotechnology* **17**, 54–59 (2006).
62. Lakomaa, E. -L. Atomic layer epitaxy (ALE) on porous substrates. *Appl. Surf. Sci.* **75**, 185–196 (1994).
63. Sander, M. S., Cote, M.J., Gu, W., Kile, B.M. & Tripp, C.P. Template-assisted fabrication of dense, aligned arrays of titania nanotubes with well-controlled dimensions on substrates. *Adv. Mater.* **16**, 2052–2057 (2004).
64. Shin, H., Jeong, J.K., Lee, J., Sung, M.M. & Kim, J. Formation of TiO_2 and ZrO_2 nanotubes using atomic layer deposition with ultraprecise control of the wall thickness. *Adv. Mater.* **16**, 1197–1200 (2004).

65. Kemell, M., Pore, V., Tupala, J., Ritala, M. & Leskela, M. Atomic layer deposition of nanostructured TiO_2 photocatalysts via template approach. *Chem. Mater.* **19**, 1816–1820 (2007).
66. Daub, M., Knez, M., Goesele, U. & Nielsch, K. Ferromagnetic nanotubes by atomic layer deposition in anodic alumina membranes. *J. Appl. Phys.* **101**, 09J111-1-3 (2007).
67. Bachmann, J., et al. *J. Am. Chem. Soc.* **129** (2007).
68. Herrmann, C. F., Fabreguette, F.H., Finch, D.S., Geiss, R. & George, S.M. *Appl. Phys. Lett.* **87**, 123110-1-3 (2005).
69. Farmer, D. B. & Gordon, R.G. Atomic layer deposition on suspended single-walled carbon nanotubes via gas-phase noncovalent functionalization. *Nanolett* **6**, No. 4, 699–703 (2006).
70. Min, Y. -S., Bae, E.J., Jeong, K.S., Cho, Y.J., Lee, J.-H., Choi, W.B. & Park, G.-S. Ruthenium oxide nanotube arrays fabricated by atomic layer deposition using a carbon nanotube template. *Adv. Mater.* **15**, No. 12, 1019–1022 (2003).
71. Fan, H. J., Knez, M., Scholz, R., Nielsch, K., Pippel, E., Hesse, D., Zacharias, M. & Goesele, U. Monocrystalline spinel nanotube fabrication based on the Kirkendall effect. *Nat. Mater.* **5**, 627–631 (2006).
72. Knez, M., Kadri, A., Wege, C., Goesele, U., Jeske, H. & Nielsch, K. Atomic layer deposition on biological macromolecules: metal oxide coating of tobacco mosaic virus and ferritin. *Nano Lett.* **6**, 1172–1177 (2006).

Chapter 13
Macroporous Silicon

Andreas Langner, Frank Müller, and Ulrich Gösele

Abstract Macroporous silicon is a material system ideally suited for the fabrication of tubular structures on the micrometer size scale. With nanometer precision on an arbitrarily sized area, millions to billions of identically shaped objects can be produced. In a modified way it is even possible to alter the shape of the structures in situ, i.e. to grow three-dimensionally modulated structures. The fabrication of macroporous silicon is based on a classical top down approach. Starting from bulk material, regular pore structures are formed in an electrochemical etching process. The underlying fabrication method is a parallel process which makes the production of macroporous silicon scalable to large areas and therefore economically attractive. In this review an introduction is given to the method of etching macroporous silicon while the growth of straight and modulated pores is explained in more detail. In conjunction with suitable post treatment steps a variety of applications for macroporous silicon has been realized from which a selection will be presented here.

The condemned live longer. And so does silicon. It is yet the material par excellence for making our modern live a complex one. It was often heard that the time has come for a new age beyond silicon. But meanwhile, silicon has learned numerous tricks, e.g. superconductivity [1, 2], spin transport [3], and even active emission of photons [4–7] and their modulation [8] is feasible. In the course of time and with all the efforts put in the field of silicon also a porous form of silicon was fabricated and investigated thoroughly. In the scope of this book we are interested in a material system that can provide us with regularly shaped and well controlled arrays of pores. The advantage of having a porous material based on silicon is obvious: All the knowledge and techniques established in silicon industry can be applied to this material system as well. Especially the ease of growing a native oxide on silicon was one key to make silicon the most popular material nowadays. In addition, silicon has the advantage to be nontoxic and biocompatible.

Macroporous silicon is a material system ideally suited for tubular structures on the micrometer size scale. With nanometer precision on an arbitrary area, millions to billions of identically shaped objects can be produced. In a modified way it is even

Ulrich Gösele (Deceased)

A. Langner (✉)
Max Planck Institute of Microstructure Physics, D-06120 Halle, Germany
e-mail: alangner@mpi-halle.mpg.de

O. Hayden, K. Nielsch (eds.), *Molecular- and Nano-Tubes*,
DOI 10.1007/978-1-4419-9443-1_13,

possible to alter the shape of the structures in situ, i.e. to grow three-dimensionally modulated structures. The fabrication of macroporous silicon is based on a classical top down approach. Starting with a bulk material, regular pore structures will be produced just by removing part of the material in an electrochemical etching process. Important to mention at this point is the fact that the underlying fabrication method is a parallel process. Thus macroporous silicon is scalable to large areas and therefore economically attractive.

The chapter is divided into two parts. At first, we give an introduction to the method of etching macroporous silicon. The growth of straight and modulated pores will be considered in detail, achievable material parameters are discussed, and different methods are presented suitable for modifying the obtained structures for certain applications. Secondly, we will concentrate on a selection of applications realized with macroporous silicon.

13.1 Introduction

The formation of porous silicon in an anodization setup is known since the 1950s from the pioneering work by Uhlir and Turner at Bell Labs [9, 10]. But it took more than 30 years when Lehmann and Föll invented an electrochemical etching process to produce arrays of ordered macropores in silicon [11]. Other groups followed, and this new topic was established as an important field in material science.

The etching of silicon and therefore the "growth" of pores into silicon covers several orders of magnitude. According to the IUPAC nomenclature for porous materials [12], structures with a pore width below two nanometers are called microporous. Their formation needs to be described quantum mechanically [13]. The mesoporous structures range from 2 to 50 nm. Bigger pores are referred as macropores.

The obtained pore morphology is dependent on the properties of the silicon (doping, resistivity, and orientation), the composition of the electrolyte (organic, anorganic, and concentration), and external parameters such as temperature, voltage and illumination. Beside numerous review articles that can be found in literature (e.g. [14, 15]), a book written by V. Lehmann is highly recommended for a more detailed study of the electrochemistry of silicon [16].

Among the different ways of fabricating porous silicon, macroporous structures etched in an aqueous solution of hydrofluoric acid (HF) allow to design the sample parameters specifically. This method has developed to a well understood and viable tool for device and template fabrication. One significant advantage is the possibility to produce ordered pore arrays with defined diameters and lengths. The pore diameters can be adjusted between a few hundred nanometers and several micrometers. Remarkably is the potential to etch pores of a few hundred micrometers in depth. Thus aspect ratios of 500:1 and even more can be obtained without loosing the ordering and the defined shape of the pores. The underlying mechanisms of the macropore formation in silicon will be discussed in more detail in the following.

13.2 Etching of Macroporous Silicon

13.2.1 Fundamentals

The basic ingredients for the fabrication of macroporous silicon can be defined as follows: Most obviously silicon is needed. For reasons that will be discussed later on, n-type doped silicon with an adapted resistivity is preferred. The wafers must be grown in a float-zone process. This ensures a charge carrier life time sufficiently long to realize diffusion lengths in the order of the wafer thickness. In addition, this raw material should receive a lithographically defined and etched prepatterning. Moreover, a highly doped backside contact is needed. Low concentrated HF with some surfactant added is used in the chemical process. Furthermore, an anodization setup is needed that provides control over voltage and backside illumination of the substrate.

13.2.1.1 Space Charge Region

Starting with the bulk material, structures are designed by dissolving part of the silicon – a typical top-down approach. In Fig. 13.1a a sketch of an anodization setup is shown. The silicon wafer is clamped between two parts of a PVC cell (inert to low concentrated HF). One side (hereinafter called front side) of the wafer is exposed to the HF. Such an intersection between a moderately doped semiconductor and an ionic liquid resembles a behavior similar to a Schottky contact. The electrochemical potential between both materials will adjust by forming a Helmholtz double layer. In the electrolyte, the charge carriers are mobile and can therefore screen the electric field within a few nanometers. In the semiconductor, free charge carriers can only be generated from the fixed donor-atoms. The density of these dopants is much lower

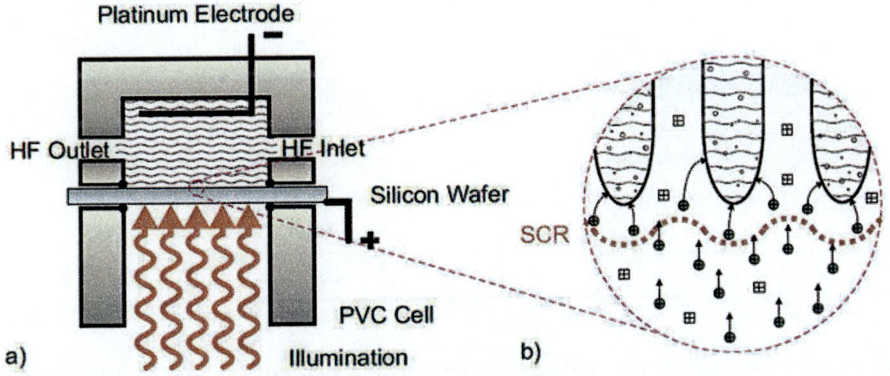

Fig. 13.1 (**a**) Sketch of an electrochemical etching setup for the growth of macroporous silicon. (**b**) An enlarged view of the situation at the bottom of the pores: The SCR extends to the *dotted line* in the silicon, so this region is depleted and the minority charge carriers are focused to the pore tips

than the charge carrier density in the electrolyte. Consequently, the electric field evolves inside the silicon and forms a space charge region (SCR) (Fig. 13.1b). For the case of a planar interface, the width of this region W_{SCR} can be estimated to

$$W_{SCR} = \sqrt{\frac{2\varepsilon\varepsilon_0 U}{eN_D}} \tag{13.1}$$

where ε is the dielectric constant of silicon, ε_0 the permittivity of the free space, e the elementary charge and N_D the density of dopants. The voltage U is the difference between the built-in potential of the silicon-HF contact and the external applied voltage. There is no reference electrode in this setup because the voltage drop over the SCR is significantly larger than the electrochemical potential. Furthermore, the chemical reaction is limited by the charge carriers generated due to the backside illumination.

Although Eq. (13.1) is only valid for a one-dimensional geometry, it is a good approximation for the width of the SCR for pores within the micrometer range. For example, a moderate doping density of 10^{15} cm^{-3} and a voltage of 2 V would result in a width of the SCR of 1.6 μm.

13.2.1.2 Electrochemistry

In Fig. 13.2 the current-voltage profile of an n-type silicon-electrolyte junction is depicted. The forward-biased regime (negative potential) is characterized by a charge carrier flow of majority charge carriers (electrons) from the silicon to the electrolyte. This case is of minor interest because it only leads to the reduction of H^+-ions and hence hydrogen is produced in the electrolyte.

In the anodic regime where the voltage is reversed, two different processes can be distinguished. The border between these two regions is defined by the critical current

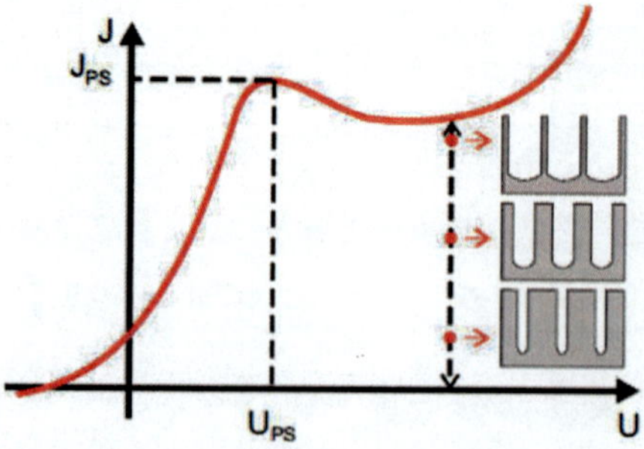

Fig. 13.2 Schematics of the I–V curve of an illuminated reverse-biased Si-HF contact. The critical current density J_{PS} marks the transition from divalent to tetravalent dissolution. The limitation of the overall current density to values below J_{PS} by illumination from the backside adjusts the diameter of the pores

density J_{PS}. It was found experimentally that J_{PS} is dependent on HF concentration and temperature [17]. For current densities below J_{PS} divalent dissolution of silicon takes place:

$$Si + 4HF_2^- + h^+ \rightarrow SiF_6^{2-} + 2HF + H_2 + e^- \quad (13.2)$$

Thereby, two charges per dissolved silicon atom can be measured in the electric circuit: One defect electron (h^+), moving from the silicon to the electrolyte and one electron (e^-), moving in the opposite direction through the interface. Current densities exceeding the critical value J_{PS} will involve four minority charge carriers. A tetravalent dissolution occurs:

$$Si + 2H_2O + 4h^+ \rightarrow SiO_2 + 4H^+ \quad (13.3)$$

The formed silicon dioxide is instantaneously removed by the HF from the electrolyte. Both, the divalent and the tetravalent reaction lead to a dissolution of silicon. But only the first reaction (Eq. 13.2) etches pores into silicon because the divalent dissolution shows a strong dependence on the crystal orientation.

The model presented here is sufficient to describe the etching process and the obtained results. Nevertheless, this continuum-like theory is based on macroscopical findings. A detailed description on an intrinsic time and length scale is given by the current-burst model. It can be found for instance in the review by Föll et al. [14].

13.2.1.3 Etching Speed

HF concentration and temperature are not only important for the magnitude of J_{PS}. They also determine the etching speed of the whole process. Increasing both quantities enhances the etching speed. However, it has turned out that moderate values will give the best results. For example, an etching speed of nearly 1 μm/min is obtained for an acid concentration of 5% and an electrolyte temperature of 10°C.

13.2.1.4 Porosity

As mentioned before, n-type doped silicon is used for this process. Therefore, the defect electrons which are needed in the dissolution process (Eqs. 2 and 3) are the minority charge carriers. Generated by, for example, backside illumination with an LED-array, the minority charge carriers are obtained from electron-hole pairs. Due to the applied voltage, the holes will diffuse towards the front side and take part in the dissolution of silicon.

For sufficiently high current densities the dissolution reaction is limited by the chemistry. This situation is represented by the curve in Fig. 13.2. Lower currents can be adjusted by a decreased backside illumination intensity. Experiments show that the etching speed is unaffected by an altered current density. With less charge carriers available, however, the etched area decreases and therefore also the porosity p of the sample. This correlation is summarized in the following equation:

$$p = \frac{J}{J_{PS}} \tag{13.4}$$

This means that the porosity p of a sample can be adjusted by the current density J. It is controlled by the backside illumination intensity independently from the voltage.

13.2.1.5 Pore Wall Passivation

The width of the SCR is a critical factor in the etching process. It prevents the pore walls from being post-etched. In the SCR there are no free charge carriers available. The silicon is depleted and therefore no charge transfer through the silicon-electrolyte interface is possible. As can be seen in Eq. (13.1) the width W_{SCR} can be adjusted by the voltage U and the doping density N_D. Controlling W_{SCR} by the voltage is limited to small changes only: A decreased voltage will diminish the focusing effect of the charge carriers towards the pore tips. In contrast, an increased voltage leads to a higher amount of dark currents that cannot be controlled. Consequently, the preferred way is to tune the W_{SCR} by the doping density of the material.

13.2.1.6 Lithography

In contrast to the anodization of metals (e.g. aluminum, titanium) the growth of macropores in silicon is not a self-ordering process. The pores would not form a periodic arrangement naturally. The pore position, however, can be predefined by lithography and subsequently etching in potassium hydroxide (KOH). The obtained etch pits then work as nucleation sites for the pore growth (Fig. 13.3a). The advantage of this process is that one is not only restricted to a hexagonal pore arrangement. Rather, the prepatterning can be varied in the pore arrangement and pore size, having a specific impact on several applications as shown below.

Although the etching process is not a self-ordering process, it is a self-organizing one. For a given doping density, applied voltage and backside illumination an average porosity and pore diameter will arise. Therefore, the lithography has to fit with the intrinsic material parameters. The passivation of the pore walls and thus the prevention of the pore walls from being post-etched is a consequence of the SCR. For a stabilized pore growth the remaining silicon between neighboring pores should be completely depleted from charge carriers. From this requirement a rule of thumb can be derived for the lattice constant of the lithography: The interpore distance a has to be chosen twice as large as the width W_{SCR}. It implies, that higher doped material is preferentially used for smaller interpore distances and vice versa.

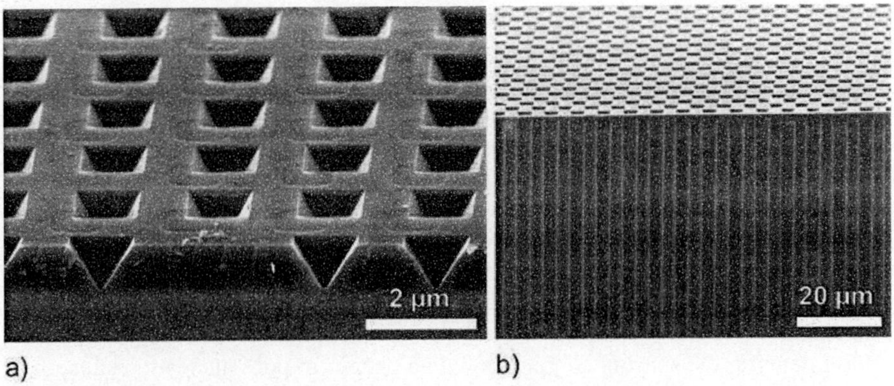

Fig. 13.3 SEM images of macroporous silicon: (**a**) Bird's eye view of KOH etch pits formed at the silicon surface prior to the etching process. (**b**) Straight pores with 50 μm in depth etched into silicon with a predefined hexagonal lattice of 6 μm pore-to-pore distance

13.2.2 Etching of Straight Tubular Structures

In the previous section the basic model of macroporous pore etching in n-type silicon was presented. The most important points should be summarized once again:

- for ordered pore composition, a prepatterning with a certain mask design is necessary
- the SCR is responsible for pore wall passivation
- the etching speed is dependent only on acid concentration and temperature
- the porosity and therefore the pore diameter is controlled by the backside illumination intensity

With these four points at hand and in mind one can understand the formation of macropores: A given sample with a defined mask is etched at a speed determined by temperature and acid concentration with an interpore distance fixed by the mask and a pore diameter given by the intensity of the backside illumination.

13.2.2.1 Stability

Concerning the stability of the etching process several points are relevant. Firstly, the etching speed is affected by the electrolyte temperature. To avoid inhomogeneities during the etching the temperature has to be kept constant within 0.1 K. Another important point is the fact that during the pore etching HF-molecules are consumed and hydrogen is produced. To avoid hydrogen bubbles to stick at the surface the electrolyte can be stirred or pumped. In addition, some surfactant should be added.

The consumption of HF-molecules can be neglected as long as the volume of the electrolyte is large compared to the dissolved volume of silicon. However, things are different at the micrometer scale inside the pores. While a pore is growing, the exchange of reactants with the HF-basin at the front side of the sample becomes more and more affected by diffusion processes. A concentration gradient establishes between the pore bottom and the opening of the pore at the sample surface. Consequently, with a lower concentration the etching speed slows down and also the critical current density J_{PS} diminishes. According to Eq. (13.4), a decreasing J_{PS} results in an increasing porosity p if the illumination intensity and thus the current density J is kept constant. The pores grow larger in diameter and finally – for $J_{PS} < J$ – electropolishing sets in. To avoid this situation one has to establish a correction parameter for the illumination intensity in dependence on the pore depth. In that way, a uniform pore pattern with aspect ratios of several hundreds to one can be obtained.

After defining the starting conditions with a pattern of pore nucleation sites, the etching process itself is self-stabilizing. The porosity is controlled by the current density J over the entire etched surface. Thus there is no direct control mechanism for each individual pore. The self-stabilizing effect is based on the fact, that the HF concentration determines the growth velocity. Let us assume a situation in which one pore is somewhat ahead of the surrounding ones. This pore will then collect more charge carriers because of its enlarged SCR. However, the excess of consumed charge carriers will also decrease the concentration of the HF in this pore. As a consequence, the growth velocity is reduced and the offset to neighboring pores is compensated. Analogous argument holds for pores staying behind the surrounding ones. Because of this feedback interaction the pore depth is the same for all pores (Fig. 13.3b).

13.2.2.2 Surface Roughness

The adjustment of the effective current density controlled by the backside illumination is done over the whole etched area. However, the local current density at the tip of the pore is always J_{PS}. Consequently, the bottom of the pore is electropolished and a very smooth surface is obtained (Fig. 13.4). In contrast, the pore walls exhibit a layer of microporous silicon with a certain roughness. The reason for this layer can be found in a kind of post etching process. While at the outermost pore tip electropolishing occurs the current drops gradually along the pore wall and therefore the already etched pore walls are being post-etched. Both, the divalent and the tetravalent dissolution of silicon as shown in Eqs. (13.2) and (13.3) take place simultaneously. This is experimentally confirmed by the fact that the number of charge carriers required for the dissolution of one silicon atom is around 2.7. Additionally, there is always a certain amount of dark currents, e.g. charge carriers generated by tunneling processes. Therefore, the thickness as well as the roughness of this microporous layer depends on the applied voltage as well as on the mixture of the electrolyte.

13.2.2.3 Limits

Naturally, the question arises what are the limits of this process? It appears in Eq. (13.1) that any width W_{SCR} can be obtained just by the proper doping of the

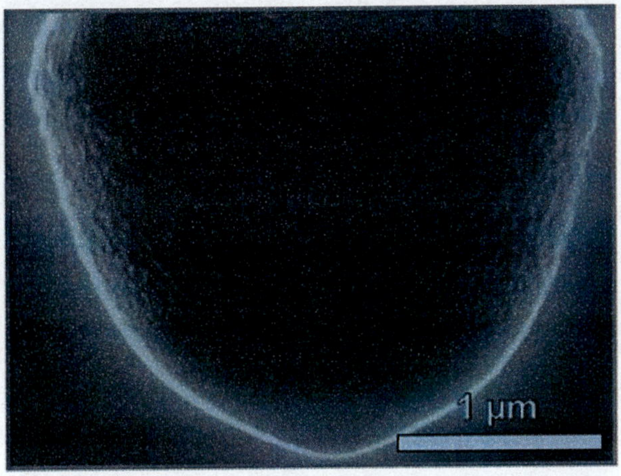

Fig. 13.4 SEM picture of the bottom of a pore. While the pore tip has a smooth polished surface the walls show a certain roughness [18]

material. In lowly doped material, large interpore distances and pore diameters can be realized, e.g. etching of pore diameters up to 100 µm were reported [19]. For small interpore distances, highly doped substrates are required. The pore formation process then is limited by electrical breakdown, because the critical voltage at which charge carriers are generated by electrical breakdown approaches only a few volt. It has to be considered that the local electric field strength at the pore bottom is increased due to the curvature of the pore tip. In this case the field strength is significantly higher than that for a planar interface. This effect is most pronounced in the small interpore distance region and limits the lowest lattice constant which can be realized to a value of about 500 nm.

The same argumentation is true for the pore diameters. With a fixed doping density and a given lattice constant pores with diameters close to the interpore distance are possible (Fig. 13.5a). The limit for diameters that are small compared to the interpore distance is determined by the critical local breakdown voltage. If the pores are too small in diameter the resulting small radius of curvature would locally boost the electrical field strength and unintentional charge carrier generation can no longer be prevented. Furthermore, charge carriers can bypass the region of the pore tip and post-etch the pore walls. Therefore, small pores with diameters of only a few hundred nanometers or even less can be grown, but the pore walls show a ragged and non-uniform surface (Fig. 13.5b). Mesoporous pore growth resulting in sponge-like structures occurs and the continuum model discussed so far is not valid any longer.

13.2.3 Pores with Modulated Diameter

So far, the formation of straight pores with a constant diameter was demonstrated. With regard to Eq. (13.4) we learned that the porosity of a sample can be tuned

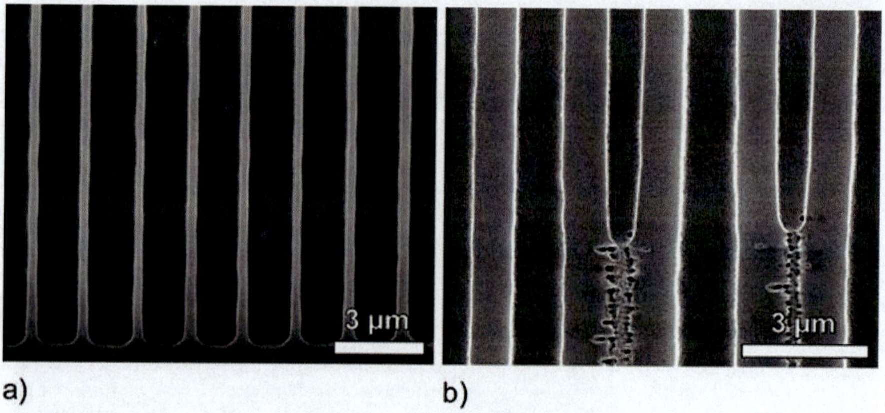

Fig. 13.5 Cross section SEM of (**a**) pores with a very large diameter close to the interpore distance and (**b**) dying pores when the diameter is not big enough. The amount of holes generated by electrical breakdown is increased leading to eroded pores

by the current density J. Moreover, we used this dependence to compensate for the concentration loss of the acid due to diffusion effects in the pores.

With rethinking this dependence, we can find a way to alter the pore shape while manufacturing. In contrast to the slow and steady adjustment necessary for diffusion compensation, the current density is now strongly modulated on a timescale of seconds to minutes. In combination with a periodic variation of the voltage, a pore shape with sharp kinks can be obtained. With this rather fast adjustment of the current as well as the voltage, the formation of diameter-modulated pores does no longer occur in the steady-state region of stable pore growth. Nevertheless, highly ordered three-dimensionally modulated structures can be obtained with this method. The mechanisms responsible for the pore growth will be presented in a detailed form in the following sections.

13.2.3.1 Current-Only Modulation

In a first approach a periodically modulated current profile is applied to the backside illumination LED array. In proportion to that, the minority charge carrier density is also modulated and therefore the amount of defect electrons at the pore tip changes. In Fig. 13.6a an applied sawtooth-like current profile is shown while the voltage is kept constant at 1.8 V. The obtained pore shape belonging to this profile is shown in Fig. 13.6b. The diameter of the pore changes in a sinusoidal form between 1 and 1.4 μm. The shape of the pores did not follow directly the applied current profile and the shape of the pores is very smooth without sharp spikes. The reason is the altered condition in the SCR. If the current is increased, charge carriers are accumulated in the region of the pore tip. However, since the pore cannot adjust its diameter to a certain value instantaneously, it has to increase. With these additional charges the

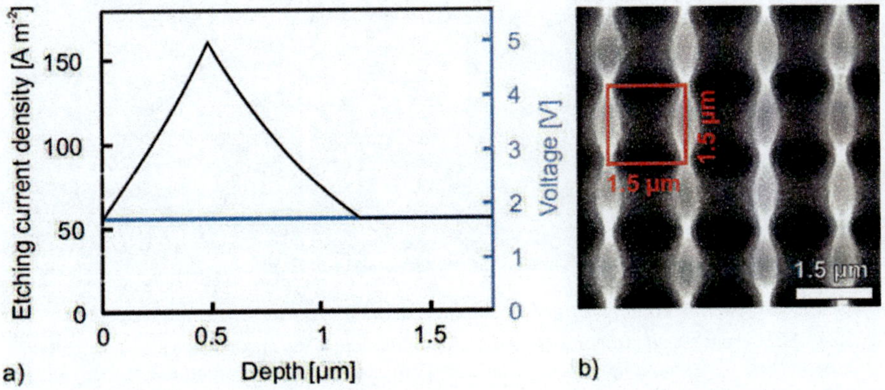

Fig. 13.6 (**a**) Sawtooth-like current profile (*black line*) for one modulation period with a constant voltage of 1.8 V (*blue line*). (**b**) SEM picture of the resulting structure with a unique lattice constant in all spatial dimensions [18]

focusing effect of the SCR is lowered and therefore the passivation of the pore walls is diminished [20]. The defect electrons do not only react at the tip of the pores but also in areas that were already etched before and the pore shape becomes smeared.

13.2.3.2 Voltage-Assisted Current Modulation

To overcome the impact of the modified charge carrier density, the SCR itself has to be adjusted as well. According to Eq. (13.1) the width of the SCR W_{SCR} – and therefore its capability of focusing incoming charges – is dependent on the voltage. A higher voltage will enlarge the width of the SCR and vice versa. However, an increased voltage brings the electrical condition at the pore tip closer to the limit of breakdown. This effect – dependent on the doping level of the silicon and the pore geometry – destabilizes the pore growth and is therefore not a general solution to this problem. Only a combination of both, stable and unstable etching steps is capable of producing uniform pores with a sharp modulation. A detailed description of strongly modulated pore etching is given in the following.

In Fig. 13.7 a three-dimensionally modulated structure with sharp kinks and a symmetric shape is shown. The square lattice has a lattice constant of a = 1.5 μm and the same modulation length. As an example we will observe the growth of one single modulation of this structure. The current-voltage profile that was applied while etching the structure is shown in Fig. 13.8b: The black curve is the etching current density, the blue one the applied voltage for one period. To visualize the progress of the pore formation the pore growth was stopped at seven different points of the modulation. For every point the sample cross section was looked at in an SEM (Fig. 13.8a). The obtained pore shape was redrawn and is schematically reproduced in Fig. 13.8c.

In the beginning (point 1) a new pore with a small diameter at the bottom should be grown. The mechanism used for this purpose is quite different from

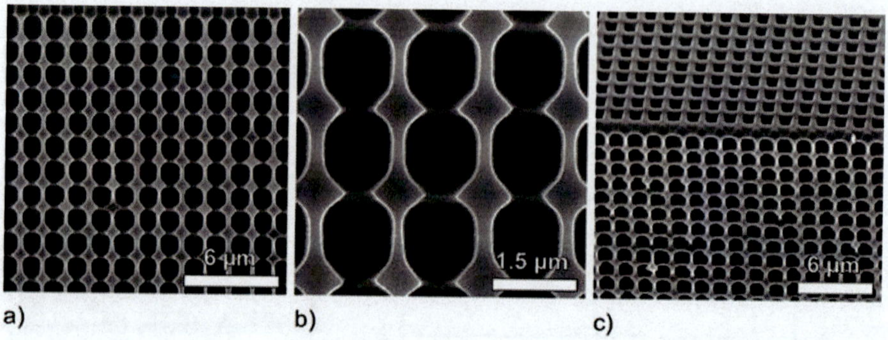

Fig. 13.7 SEM cross section of pores etched with modulation of the current as well as the voltage. (**a**) Cross section of 20 modulations. (**b**) Magnified area with 3 – by – 3 modulations. (**c**) Bird's eye view of the structure [21]

the macropore etching described above. As can be seen in Fig. 13.8b the voltage is increased by a large amount that locally exceeds the critical breakdown value. Additional charge carriers in the vicinity of the pore bottom are generated by tunneling. Combining both, the charge carriers generated by illumination and the charge carriers generated by tunneling due to the locally increased electric field strength, a nanometer scaled pore in the [100] – direction (point 2) is obtained. This process of pore generation by breakdown voltage is stable only for some 10 nm. After that, the voltage has to be reduced again. Otherwise the pores would start to branch or die and the ordering would be lost. Simultaneously, the current density is increased (points 3 up to 7). As expected the pore grows larger in diameter.

In Fig. 13.8c the contour of the pore is redrawn for every point. It can be seen that the etching takes place not only at the outer region of the pore tip. Also regions already etched get widened in their diameter. Furthermore, the sharp kink that marks the beginning of a new pore is propagating slightly with increasing pore depth. This can be seen best comparing the shapes of point 2 and point 7 in Fig. 13.8c. While for the last steps (points 5–7) the position of this kink remains stable, it has changed significantly during the formation and etching of the small pore (points 2–4). During the formation of the small pore more charge carriers can propagate between neighboring pores. Despite the increased width of the SCR not all of them are directed to the tip of the evolving pore and therefore they hit the surface at regions beyond the etch front. As a result, the already etched parts are widened and the position of the kink moves.

The current density has reached its maximum value at point 7. The voltage was raised again while approaching the maximum current density. This can help to protect the side walls from being etched too much because the majority of the charge carriers is focused to the pore tip. After this final point the current density is switched to zero for a short time. During this stage remaining charge carriers can be consumed. In addition, time is given for the diffusion of reactants and the relaxation of the electrolyte to normal concentration.

In contrast to the sinusoidally shaped pores (Fig. 13.6), strongly modulated pores with sharp kinks require a modulation of the current density and an adaptation of the

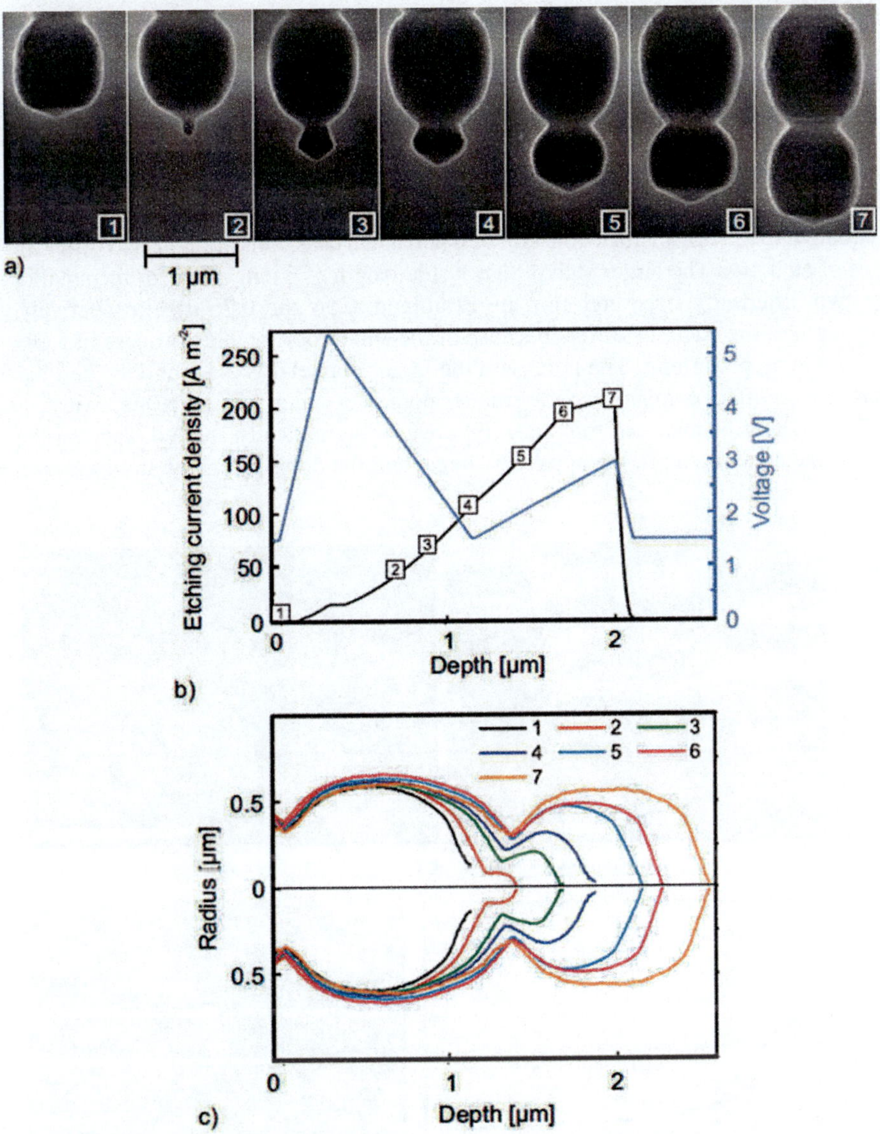

Fig. 13.8 (**a**) SEM pictures of the evolution of one etched pore at seven different points of the modulation profile. It starts with a tiny pore that is widened afterwards. (**b**) The current-voltage profile for one modulation, the *black curve* denotes the current density, the *blue curve* the voltage. (**c**) Visualized pore shape acquired from the pictures in (**a**) [20]

SCR via the applied voltage. Thereby, the steady-state regime is left and even unstable conditions are used for the pore formation. A detailed understanding of these processes nevertheless allows the fabrication of highly ordered diameter-modulated pore structures [21].

13.2.4 Nonuniform Pore Growth

The structures presented so far were equally sized in interpore distance and pore diameter. In the following, the controlled introduction of a nonuniformity should emphasize the complexity but also the versatility of the macroporous silicon etching process.

The prestructured mask of the sample under consideration is shown in Fig. 13.9a: A square lattice with a lattice constant of 6 μm and a base consisting of two different sizes of etch pits. The larger etch pit has a side length of 3 μm while for the smaller one two differently sized etch pits are examined: 0.56 and 0.76 μm, respectively. The depth of the etch pits differs because of the anisotropic etching process in KOH during the prepatterning. The pore with the larger diameter is 2.1 μm deep and has therefore an offset compared to the smaller pore of 1.7 and 1.6 μm, respectively.

The etching behavior of this mask type was systematically studied, varying the size of the small etch pit, the applied voltage, and the doping density. It was shown

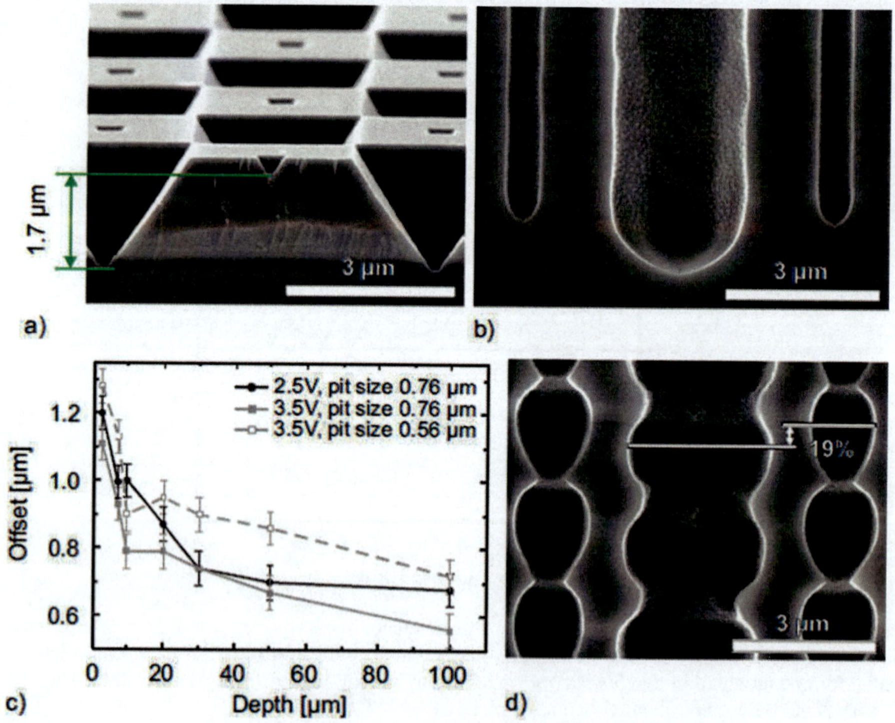

Fig. 13.9 (**a**) Bird's eye view of a *square* lattice with an alternating sequence of two differently sized pores (3 and 0.56 μm, respectively) and an offset between larger and smaller etch pits. (**b**) Stable pore growth could be observed for at least 100 μm resulting in pores with different diameters and lengths. (**c**) Measured offset depending on the pore depth, the applied voltage, and the initial pit size of the small pore. (**d**) SEM image of a modulated macroporous silicon sample with two-pore lattice. The maxima in diameter for the thick pore occur at a different depth than that for the thin pore [22]

that a stable pore growth of at least 100 μm in depth is possible [22]. In Fig. 13.9b a more detailed view is given for the bottom part of a sample with a small etch pit of 0.76 μm side length: There are thicker as well as thinner pores and there is an offset in the length between them. Interestingly, the ratio of the pore diameters equals the side length ratio of the etch pits. Roughly speaking, the large etch pits start with a larger diameter and thus collect more charge carriers than the small ones. Therefore, the diameter ratio of the small and large pores can be adjusted by the side length of the etch pits.

In Fig. 13.9c the evolution of the offset between the two types of pores is shown for different parameters. For an applied voltage of 3.5 V the offset strongly diminishes during the first ten micrometers and then steadily decreases at a rate of 0.23 μm per 100 μm. The difference between the two curves for 3.5 V and different depths of the etch pits (0.56 and 0.76 μm) is equal to the difference in the depth of the etch pits. For a voltage of 2.5 V and a size of 0.76 μm for the small pits it is worth noting that finally a region is reached in which the offset remains nearly constant at 0.7 μm (black curve in Fig. 13.9c).

The material used had a resistivity of 1 Ωcm. Comparable experiments were performed with lower doped material of 5 Ωcm resistivity. For the latter one no growth of the small pores could be observed. Only the pores with the large diameter grew. From this result it can be concluded that in the case of the lower doped material no charge carriers were transported to the small pores. Because the SCR extends broader into the lowly doped material, the small pores are shielded from incoming charge carriers, i.e. they cannot grow. In the higher doped material the form of the SCR is more closely adapted to the pore shape. Charge carriers are focused to the large pores as well as to the small ones.

It was also possible to modulate the diameter during the etching process. Applying a strongly modulated current-voltage profile similar to the one discussed in detail in the preceding section leads to a diameter modulation as shown in Fig. 13.9d. The offset is visible in the different positions of the maxima in diameter between small and large pores.

To summarize this section it was shown that not only straight pores with a constant diameter can be grown over large depths. Also pores modulated in diameter can be obtained when the porosity is altered in situ. Generally, smoothly modulated pores are obtained. In extension of this process it was shown that an additional adjustment of the SCR via the applied voltage can lead to pores with a strong modulation and sharp kinks. In the application section, further pore shapes will be presented and their applications will be discussed.

13.3 Post Treatment

The range of applications can be significantly expanded by proper post treatment of the etched samples. On one hand the pore surface can be modified to obtain specific surface properties. On the other hand post etching steps can be applied in order to alter the structure itself, i.e. to modify the volume or shape of the macropores.

The treatment of the surface is a very complex topic that is strongly dependent on the desired application. After taking the samples out of the electrolyte and rinsing them in clean water their surface is hydrogen-terminated. In ambient atmosphere this surface coverage is not stable and after some hours a silicon dioxide layer forms. The thickness of this layer is about 3 nm and remains stable. Due to the water in the air the dangling bonds at the surface are saturated with hydroxyl molecules.

When macroporous silicon is used as a template material (see Chapter 5 by Steinhart), in catalysis, chemisorption or in microfluidics, its surface has to be modified with precursors to remove the hydroxyl groups and alter the surface functionality. To give an example, the wettability of the pore walls can be tuned: The hydrogen-terminated silicon surface that is present directly after the etching process is hydrophobic. In contrast, a pure SiO_2 surface and also a surface terminated with Si–OH (silanol) groups is hydrophilic.

13.3.1 Isotropic Form Treatment

Beyond the modification of the surface conditions the ease of forming an oxide layer is one of the most valuable properties of silicon. Silicon dioxide, also known as a principal component of glass, is nontoxic and biocompatible. Therefore, it is well suited for interactions with organic molecules. Beside the native oxide on the surface, an oxide layer can also be grown thermally, either in a wet or dry atmosphere. Thermal oxidation of silicon is typically performed at temperatures between 800 and 1,200°C. In a dry oxidation process the environment contains only oxygen. The grown films are very uniform and denser than films grown under wet conditions (i.e. in the presence of hydrogen). The growth rate of a dry oxide, however, is much slower than that for wet oxidation.

About silicon and its oxide numerous publications can be found. In here, only a special feature related to the pore shape will be discussed. The growth rates given in literature are mostly obtained for planar surfaces. The oxidation of pores on the other hand is – dependent on the diameter of the pores – retarded compared to flat interface conditions. The reason is the strain induced during the growth of the oxide layer. The volume of silicon dioxide is 2.25 times larger than that of the original silicon. Inside a pore the grown silicon dioxide layer cannot relax in the same way as on a planar surface. This additional induced strain significantly lowers the diffusion of oxygen through the silicon dioxide layer towards the silicon interface where the oxide growth takes place. Hence, also the growth rate is reduced and the oxidation process – especially for thick silicon dioxide layers of several tens to hundred nanometers – is retarded.

The counterpart of the oxidation – the removal of the oxide layer – is performed in HF-containing acids such as aqueous HF or NH_4F. As shown in Fig. 13.4 the pore walls appear to be rough after the etching. This surface roughness can be decreased by at least one order of magnitude when the macroporous silicon is shortly oxidized followed by a dip in an HF-containing acid to remove the oxide layer.

During the oxidation of silicon not only the surface is affected. With longer oxidation times the amount of silicon that is oxidized increases and the oxide layer grows deeper into the entire structure. Despite the retarded oxidation of macroporous silicon due to the curved surface of the pores it is possible to fully oxidize the silicon [23]. This can be interesting for applications, e.g. in life sciences: While silicon is only transparent for wavelengths larger than 1.1 μm, silicon dioxide is also transparent in the visible spectrum of light.

Beside these two extrema, the surface modification and the complete oxidation of the structure, a partial oxidation can be used to tune the pore diameter after the etching or even to fit a special design as shown in the application section. The dry oxidation in a controlled environment, e.g. in a tube furnace with stabilized heating zones and mass flow controller, gives a very reproducible and homogeneous oxide thickness. In combination with a subsequent etching step in HF or NH_4F, the structure is widened isotropically. The advantage of this process compared to an isotropic widening in, e.g., a mixture of HF/HNO_3 is the precise thickness control: The amount of silicon removed by this procedure is determined by the thickness of the grown oxide. This allows for tuning of the pore diameter after the etching.

Though this process seems to be straightforward, some restrictions have to be kept in mind. For thin oxide layers (short oxidation times) the oxide growth speed is very high. Hence, statistical fluctuations during the growth can have an undesired effect and the uniform shape of the structure could suffer during this procedure. In the case of thick oxide layers the dominating factor is the induced stress. Therefore, the possibility of damaging the structure has to be taken into account. An oxide layer thickness of 50–100 nm is an optimum value avoiding both, inhomogeneity and cracking. If thicker layers have to be removed a splitting of this procedure into several sequential oxidation and oxide removal steps is recommended.

13.3.2 Anisotropic Form Treatment

Another possibility to alter the pore shape after the etching is the treatment in an alkaline solution such as potassium hydroxide (KOH). The {100} – and the {110} – planes are preferentially etched and the etch rate is one to two orders of magnitude larger than that for the {111} – planes [24]. Treating a three-dimensional macroporous silicon structure with this process can result in a variety of different shapes (Fig. 13.10).

Furthermore, it can be useful to have a porous structure with both ends of the pores opened. This means that the remaining silicon from the backside up to the pore bottoms has to be removed. Therefore, a further property of KOH is considered: The etch ratio between silicon and silicon oxide can reach a factor of several thousands. Based on this property the porous structure is first oxidized (a few 10 nm) and afterwards the backside of the silicon is etched in KOH. When the pore bottoms are reached, the silicon oxide protects the porous structure from the KOH for a couple of minutes, giving enough time to stop the process. Subsequently, the oxide is etched away in an HF-containing solution and a porous membrane is obtained (Fig. 13.11).

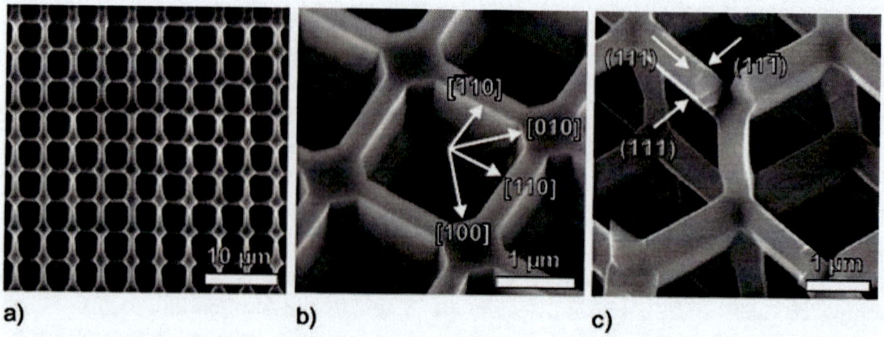

Fig. 13.10 (**a**) The original three-dimensional macroporous silicon structure shown in cross section. (**b**) Top view of the sample after etching in lowly concentrated (2 wt%) KOH: The structure possesses crystallographically defined side walls. (**c**) For longer etching times a highly porous scaffold structure with interconnected microbars is obtained [25]

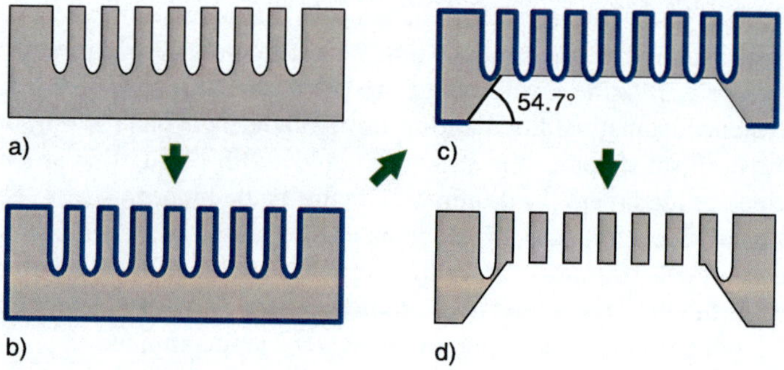

Fig. 13.11 Schematic representation of membrane fabrication: The etched macroporous silicon (**a**) is oxidized (**b**) and subsequently etched in KOH (**c**). The angle between the (100) – oriented surface and the remaining (111) pore wall is 54.7°. The silicon dioxide can be removed with an HF-dip and a porous membrane is obtained (**d**)

13.4 Applications

The versatile and complex etching process of macroporous silicon leads to a variety of different pore shapes and surface conditions. This fact is also reflected in numerous applications. In the Chapter 5 by Steinhart several methods are presented in which macroporous silicon is used as a template material. For that purpose the etched samples get a surface treatment well suited for the subsequent filling of the pores. After removal of the silicon an inverted copy of the macroporous silicon sample is obtained. A specific application of this technique is proposed in [26], where diameter-modulated gold microwires are fabricated and their use as microbarcodes is suggested. The utilization as template material is by far not the only application of macroporous silicon. A selection of applications based on macroporous silicon will be presented in the following sections.

13.4.1 Photonic Crystals

A photonic crystal (PC) is a periodic structure with an alternating sequence of materials with different permittivities and a periodicity in the order of the wavelength of the electromagnetic wave. Similar to Bragg-like x-ray scattering at atomic crystal lattices, waves of larger wavelength are scattered at PCs [27, 28]. Macroporous silicon with feature sizes on a micrometer scale and its highly ordered arrangement of pores is an ideal candidate for such applications. A precisely designed PC is capable of guiding light in a specified way or to dramatically reduce the speed of light in the medium. Beyond that, a PC can suppress electromagnetic waves from propagation in a certain energy range. This effect is associated with a photonic band gap and was one of the driving forces in the design of PCs.

Typically, two-dimensional (2D) structures are used as waveguides, or frequency-, direction-, and polarization-filters. More effort is required when PCs are used to control the emission of light. Since a classic dipole emits in all three spatial dimensions also three-dimensionally (3D) modulated structures are required to control and manipulate such a behavior. Furthermore, only 3D PCs can confine light in all spatial dimensions using a complete photonic band gap. Thus the propagation of light is prohibited for certain frequencies in any direction of the PC.

13.4.1.1 Two-Dimensional Photonic Crystals

Although they cannot confine light in all three dimensions, 2D PCs can exhibit large band gaps in two dimensions. The most common types of 2D PCs have a quadratic or hexagonal arrangement of pores. It can be distinguished between two polarization directions of the electromagnetic wave with respect to the pore geometry: While in the transversal electric (TE) mode the electric field oscillates

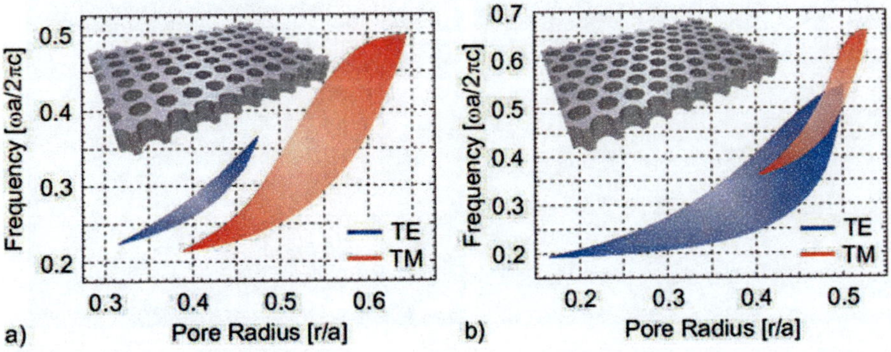

Fig. 13.12 Size of the lowest photonic band gap in dependence on the pore diameter for (**a**) square and (**b**) triangular lattices with a dielectric constant of $\varepsilon = 12$. The pore radius and the frequency are given in units of the lattice constant. The values were calculated with the MIT photonic bands package [29]

perpendicular to the pores, in the transversal-magnetic (TM) mode the electric field is parallel to the pore axis.

In Fig. 13.12 for both of these lattice types the frequency range of the lowest photonic band gap in dependence on the pore radius is given. For the case of the square lattice (Fig. 13.12a) the photonic band gaps for the TE (blue area) and TM (red area) polarization occur in separated frequency and pore radii ranges. In a hexagonal lattice there is an overlap of both ranges resulting in a 2D complete photonic band gap.

For the experimental characterization of the photonic band gap a 2D PC was grown with a hexagonal lattice and a pore-to-pore distance of a = 1.5 μm. Thin bars of only a few crystal layers were prepared (Fig. 13.13) using a microstructuring technique [32].

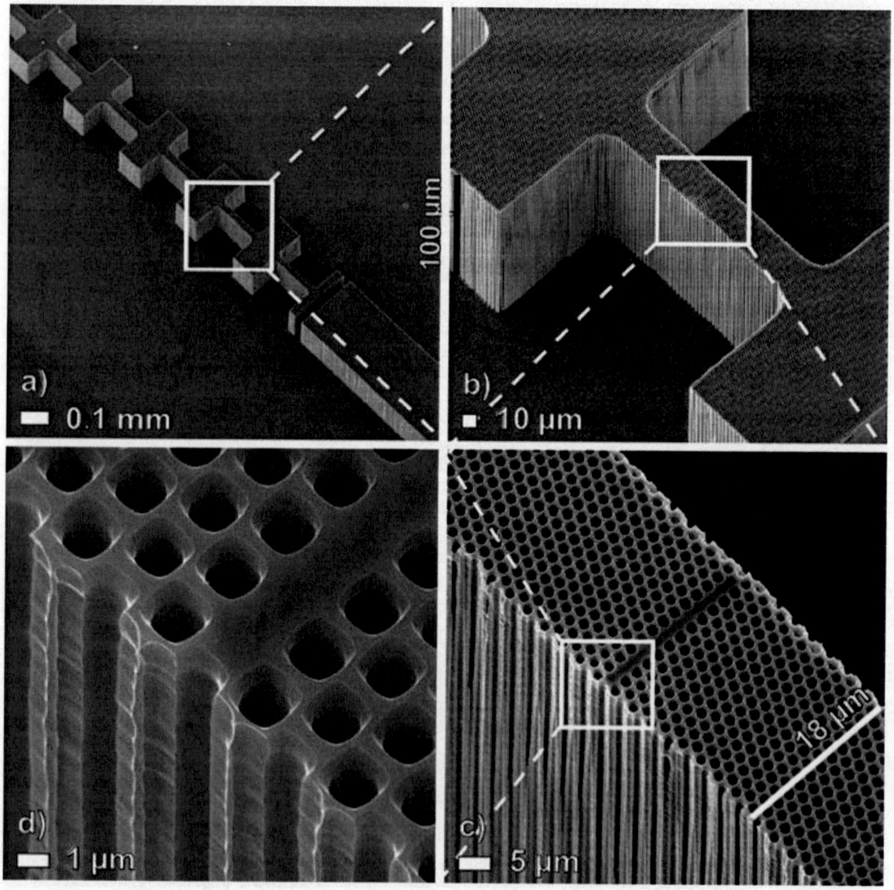

Fig. 13.13 SEM pictures of 2D macroporous silicon with a hexagonal lattice and a lattice constant of a = 1.5 μm. The thinner part with its surface along the Γ – M direction consists of 13 pore rows. The thicker part is made for reasons of stability. In addition, this structure also features a line defect as can be seen in (**c**) and (**d**) [30, 31]

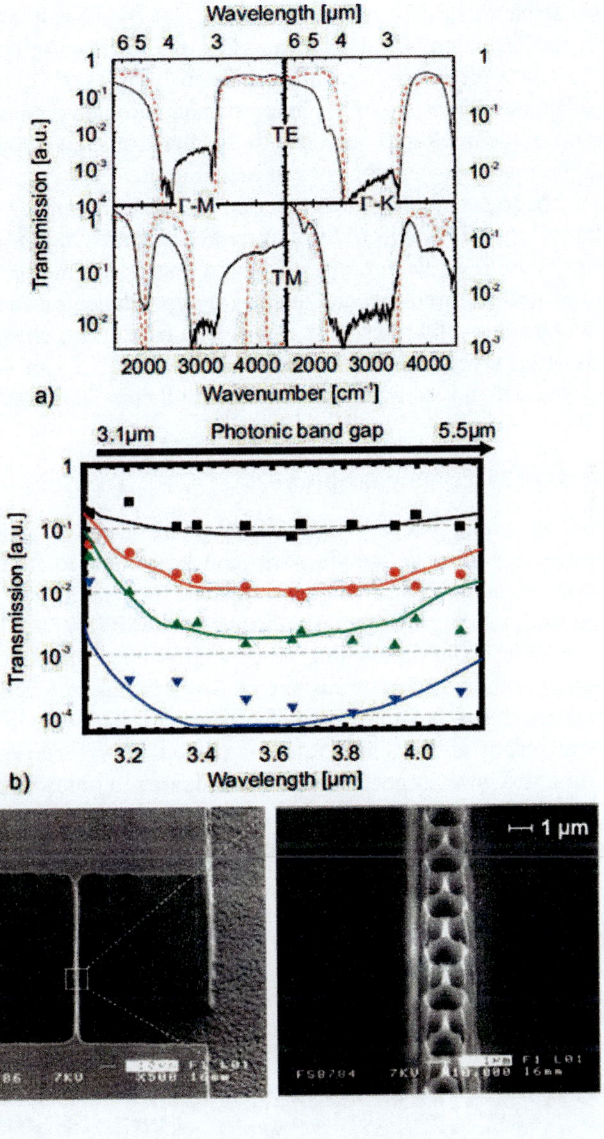

Fig. 13.14 (**a**) Measured (*solid line*) and calculated (*dashed line*) transmission spectra of a 2D hexagonal PC for TE (H-polarized) and TM (E-polarized) polarization and the interface cut along the $\Gamma - \text{M}$ and $\Gamma - \text{K}$ direction. The sample consists of 13 pore rows with a lattice constant of a $= 1.5\ \mu\text{m}$ and a pore radius of $r = 0.46\text{a}$. The photonic band gap is centered around $\lambda = 3.2\ \mu\text{m}$ [30]. (**b**) Measured (*points*) and calculated (*solid lines*) transmissivity for samples with a thickness of 1 (■), 2 (•), 3 (▲), and 4 (▼) crystal layers [33]. (**c**) SEM picture of a single pore row as used for the measurement shown in (**b**)

The transmission was measured in the vicinity of the photonic band gap for a crystal with its surface along the $\Gamma - M$ and $\Gamma - K$ direction for both polarization modes (Fig. 13.14a). The loss in transmission intensity becomes evident for at least three orders of magnitude around the photonic band gap (centered at $\lambda = 3.2\ \mu m$). The comparison of the measured curves with the theoretical calculated ones reveals a very good agreement between theory and experiment.

The photonic band gap is a volume property of a PC. This can be verified by varying the number of crystal rows [33]. Similar to the structures in Fig. 13.13 PCs with one up to four crystal rows were prepared. The transmission measurement of these structures in the mid-infrared revealed clearly a dependence on the thickness and thus on the number of crystal rows (Fig. 13.14b). In fact, an exponential attenuation of 10 dB per crystal layer within the photonic band gap ($3.1\ \mu m < \lambda < 5.5\ \mu m$) could be observed which is in good agreement with theoretical calculations.

13.4.1.2 Defects in Macroporous Silicon

In extension of the etching process of 2D pore structures the periodicity of the lattice can be perturbed. For instance, single pores can be prevented from being etched or their position or diameter is changed compared to the surrounding ones. This can be realized with a lithographic mask designed accordingly. A missing pore can be thought of as a defect in the etching process. No charge carriers are consumed during the etching at the position of this defect. Since in this self-stabilizing etching process the carrier density is homogeneously dispersed, the pores in the vicinity of this defect have to consume this additional number of charge carriers. As a result, the surrounding pores grow larger in diameter compared to pores further away from this defect. This effect is less developed for small pore diameters than for larger ones since the charge carrier density rises with the square of the pore diameter. Therefore, the pores are preferentially etched with a smaller diameter and afterwards widened isotropically as described in the post treatment section.

As mentioned in the previous section, in a 2D PC a photonic band gap can exist. In the picture of the photonic band structure a defect (for instance a point defect as shown in Fig. 13.15a) breaks the symmetry of the lattice and therefore changes the

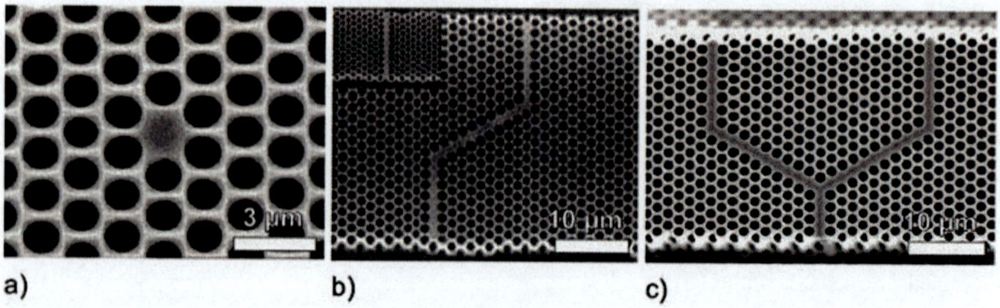

Fig. 13.15 SEM pictures of defect structures realized in a 2D PC with an $a = 1.5\ \mu m$ hexagonal lattice. (**a**) A single missing pore. (**b**) A waveguide that can bend the light sharply around corners and even split one beam in two parts (**c**) [34, 35]

properties of the PC. With a properly designed defect in a PC it is even possible to generate states inside the photonic band gap. An electromagnetic wave package is localized at this location if it is in resonance with the eigenfrequency of such a defect state [36]. Because of the photonic band gap the wave package cannot escape through the surrounding pores since it is backscattered. Consequently, the light is stored at the location of the defect within the photonic band gap.

Combining several point defects to a line defect, such resonant modes can be guided through a PC. These so called waveguides are as narrow as the wavelength of the transmitted light or even less. Therefore, PC-based waveguides are much narrower than conventional structures used for this purpose based on internal reflections.

The defect structures shown in Fig. 13.15b, c consist of a combination of several missing pores giving a line defect. For example, light can be sharply bent around corners (Fig. 13.15b) or split into two separate beams (Fig. 13.15c). There are no radiation losses in an ideal 2D geometry because of the surrounding material and the band gap. Even in the presence of scattering the radiation losses into the third dimension are low and a high output quality can be reached.

13.4.1.3 Three-Dimensional Photonic Crystals

A 3D structure in simple cubic geometry with a complete photonic band gap was proposed by Leonard [37]. In the band structure (Fig. 13.16a) a region can be seen where no propagation of light is possible for any direction in the PC. The model

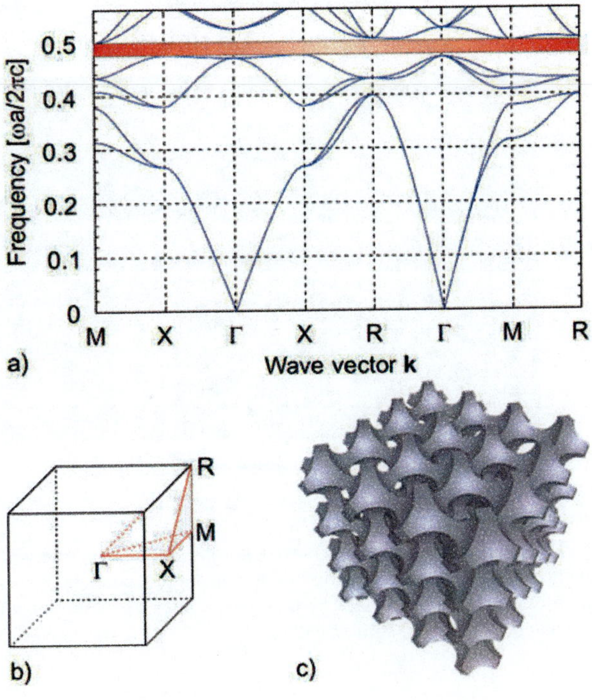

Fig. 13.16 A 3D simple cubic PC in silicon ($\varepsilon = 11.7$) with a complete photonic band gap as suggested by Leonard [37]. (**a**) The photonic band structure reveals that there is no propagation possible in any direction for a range of frequencies, i.e. a photonic band gap (*grey bar*) exists. (**b**) The reciprocal lattice cell with the points of high symmetry. (**c**) A model structure of the simple cubic crystal with overlapping air spheres of radius $r = 0.605\,a$

structure for this theoretical prediction is shown in Fig. 13.16c. It is a simple cubic lattice of silicon with inserted overlapping empty spheres. For the highest possible width of the complete photonic band gap the radius of the spheres is $r = 0.605\,a$ (a is the lattice constant). Consequently, the spheres are overlapping each other and thus all pores are connected to each other.

This structure was realized experimentally within the framework of the macroporous silicon material system. Starting with a square lattice of $a = 1.5\ \mu m$ lattice constant a strongly modulated structure was grown (Fig. 13.17a). Thereby, the modulation period equals the lattice constant. The pores are connected in the vertical direction due to the etching process – but not yet in the lateral direction. In order

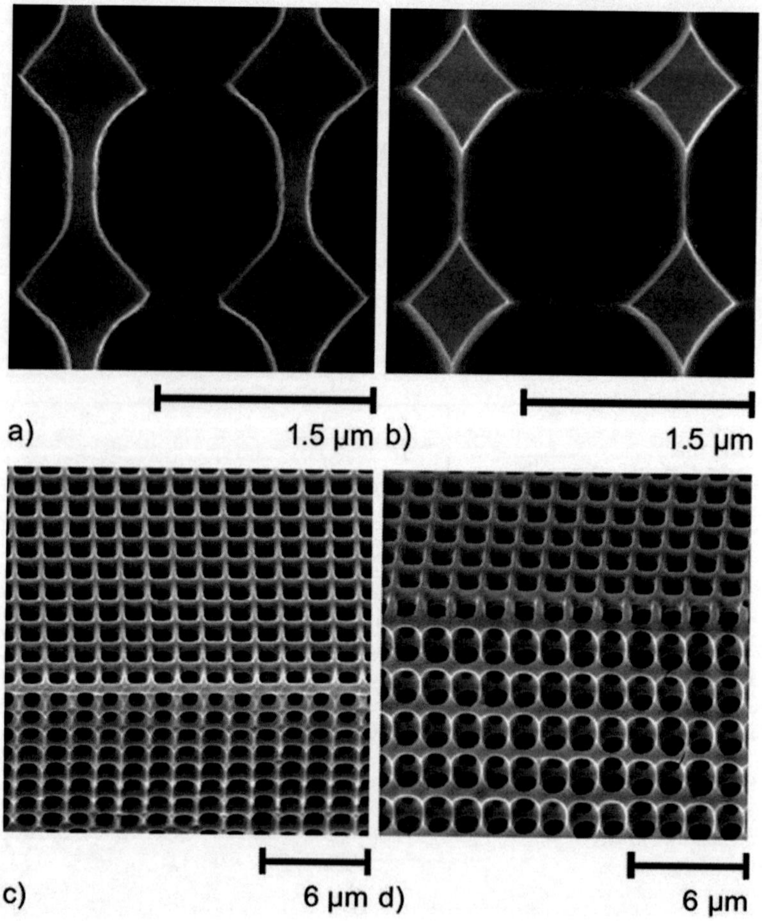

Fig. 13.17 SEM images of a 3D macroporous silicon structure with a square lattice of $a = 1.5\ \mu m$ lattice constant. (**a**) Cross section of the strongly modulated grown structure with a modulation length of $1.5\ \mu m$. (**b**) Symmetric pore shape after the homogeneous pore widening. (**c, d**) Bird's eye view of this 3D PC structure [21]

to realize the overlapping spheres, the procedure of homogeneous pore widening was applied as described in the post treatment section. All in all four iterations of growing an oxide layer (900°C, 2 h) and removing it in NH_4F (12.5%, 1 h) were carried out. A film of about 70 nm thickness was removed in every step (Fig. 13.17b). In the area of the large pore diameter the walls were opened and neighboring pores became connected to each other. Thus an ordered porous structure was fabricated capable of acting as a PC with a complete photonic band gap in 3D (Fig. 13.17c, d) [21].

The material requirements for applications as PCs are high: The periodicity of the structure has to be on the order of the desired wavelength with a precision in the range of 1/20 of the wavelength. While a couple of preparation techniques are already known for the fabrication of 2D structures, manufacturing 3D structures is still a challenging task. PCs based on macroporous silicon offer this precision. Moreover, they can be fabricated in a large scale process.

13.4.2 Microfluidics

Macroporous silicon cannot only be applied to alter the propagation of light. It can also be used to control the flow of liquids. The material parameters of interest are different from that required for photonic applications. The formation of channels and their shape are the important points to consider. In microfluidics, materials are required with a well defined surface and shape. Macroporous silicon is an excellent candidate for this aim as demonstrated in the following example of a ratchet device.

It was proposed based on theoretical work [38] that a ratchet-shaped pore profile can separate particles of different size – dispensed in a liquid – without a net transport of the liquid itself. An upper and a lower basin are separated by a membrane with asymmetrically shaped pores (Fig. 13.18a). The liquid with the dispersed microparticles is periodically pumped between upper and lower basin with an average net liquid flow of zero.

For the experimental verification of this experiment a macroporous silicon sample with an asymmetric pore profile was grown and afterwards the remaining silicon was removed from the backside with KOH (cf. Fig. 13.11). The pore geometry can be seen in Fig. 13.18c, d: It is a triangular pattern with 6 μm pore-to-pore distance, 150 μm in depth, and the maximum (minimum) diameter of the modulation is 4.8 μm (2.5 μm). Particles of 0.32 μm diameter could be transported to the upper basin with such a membrane when a periodic pressure oscillation was applied.

The density of the spheres in the upper basin was measured via photoluminescence (PL). In Fig. 13.18b some experimental data are shown. The pressure oscillations were toggled on and off every 60 s to exclude long-term drifts. During this period the intensity in the upper basin – and therefore the concentration of particles – starts from a homogeneous particle distribution and increases during the pump-on times (curve U). The effect of the asymmetry of the pore shape can be seen when turning the membrane upside down. Now the intensity decreases ($U_{reversed}$). It can be

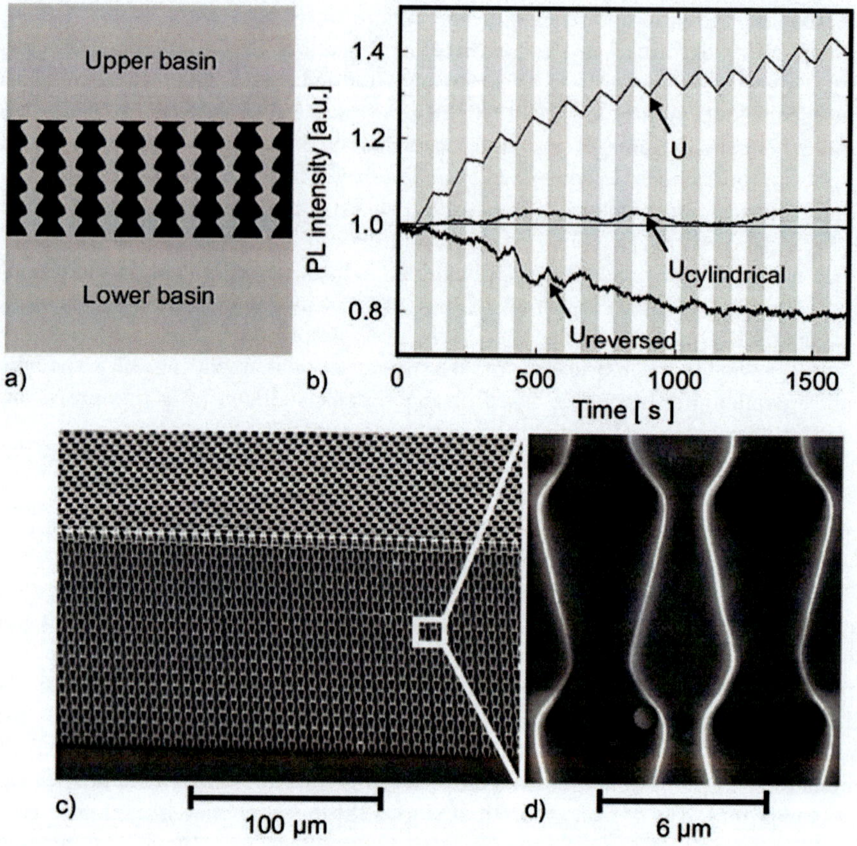

Fig. 13.18 (**a**) Schematics of the horizontally mounted and asymmetric diameter-modulated membrane separating two basins. (**b**) Measured photoluminescence (PL) intensity in the upper basin (U), for the reverse mounted membrane ($U_{reversed}$), and a cylindrical pore shape for comparison ($U_{cylindrical}$). (**c**) SEM picture of a cleaved macroporous silicon membrane used for the experiment. (**d**) Magnified version with a colloidal sphere of 1 μm in diameter inside the pore sticking to the pore wall [39]

seen in the curve $U_{cylindrical}$ that this effect is clearly related to the asymmetric shape of the membrane: The measurement was performed with a sample of cylindrical pore shape (diameter 2.4 μm) but no significant particle transport could be observed.

13.4.3 Materials Science

This last section will briefly highlight a few additional ideas realized with macroporous silicon. In Fig. 13.19a an area of silicon spikes can be seen. Since silicon is a very hard material it can be used as a micrometer sized structuring tool, e.g. as a stamp for imprint lithography. This structure was prepared using a 2D

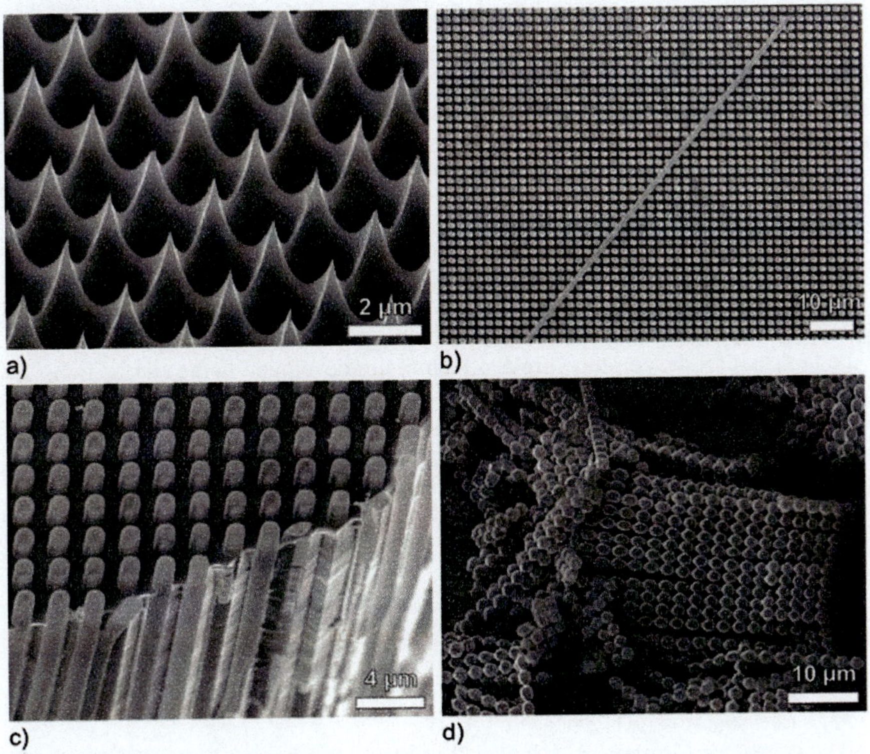

Fig. 13.19 Certain forms of appearance of structures based on macroporous silicon. (**a**) Pattern of sharp silicon spikes in a square lattice of 2 µm. (**b**) One silicon dioxide tube on top of an array of oxidized macroporous silicon pores after partly removing the silicon from the backside. (**c**) Bird's eye view of the sample in (**b**) which shows an edge and uncovered silicon dioxide tubes. (**d**) Diameter-modulated titanium dioxide tubes, fabricated by replicating a macroporous silicon sample with atomic layer deposition

macroporous silicon sample in which the pore diameter was strongly increased at the end. Thus neighboring pores got connected to each other and the porous silicon layer can be removed easily. The remaining pore bottom with the spikes has the same lattice geometry as the lithographically defined etch pits.

In Fig. 13.19b a bottom view of a macroporous silicon membrane is shown. The pores were grown in a normal way and afterwards oxidized. With the KOH etching used for the membrane preparation the bulk silicon from the backside was removed. On top of the sample a completely released silicon dioxide microtube can be seen. The bright shining domes are SiO_2 tubes with a closed bottom, partly released from the silicon (Fig. 13.19c). A suggested application as a window for electron beams in high vacuum chambers is given in [40].

Also the modulated pore structures can be used as a template. In Fig. 13.19d modulated titanium dioxide tubes are shown: The minimum diameter is 0.4 and

1.6 μm the maximum one. They were fabricated by atomic layer deposition of titanium dioxide in the pores of a modulated macroporous silicon sample. Afterwards, the silicon was selectively removed in KOH. Applications of these biocompatible titanium dioxide spheres may be found in life science, e.g. for drug delivery purposes.

13.5 Summary

A basic introduction to the fabrication as well as applied aspects of macroporous silicon were presented. The enormous adaptability of this material system is based on the fact, that all the common process techniques developed in the silicon industry can be applied. The basic features were presented in this chapter and should give the necessary informations whether a closer look to this material system is worthwhile for the reader. Nevertheless, the most difficult part is still open to the reader: No realization without imagination.

Acknowledgment We thank J. V. Wittemann for carefully reading and correcting the manuscript.

References

1. E. Bourgeois and X. Blase. Superconductivity in doped cubic silicon: An ab initio study. *Appl. Phys. Lett.*, 90(14):142511, 2007.
2. E. Bustarret, C. Marcenat, P. Achatz, J. Kačmarčik, F. Lévy, A. Huxley, L. Ortéga, E. Bourgeois, X. Blase, D. Débarre, and J. Boulmer Superconductivity in doped cubic silicon. *Nature*, 444(7118):465–468, 2006.
3. I. Appelbaum, B. Huang and D. J. Monsma Electronic measurement and control of spin transport in silicon. *Nature*, 447(7142):295–298, 2007.
4. S. Prucnal, J. M. Sun, W. Skorupa and M. Helm. Switchable two-color electroluminescence based on a Si metal-oxide-semiconductor structure doped with Eu. *Appl. Phys. Lett.*, 90(18):181121, 2007.
5. H. Rong, R. Jones, A. Liu, O. Cohen, D. Hak, A. Fang and M. Paniccia A continuous-wave Raman silicon laser. *Nature*, 433(7027):725–728, 2005.
6. H. Rong, A. Liu, R. Jones, O. Cohen, D. Hak, R. Nicolaescu, A. Fang and M. Paniccia An all-silicon Raman laser. *Nature*, 433(7023):292–294, 2005.
7. H. Rong, S. Xu, Y. -H. Kuo, V. Sih, O. Cohen, O. Raday and M. Paniccia Low-threshold continuous-wave Raman silicon laser. *Nat. Photon*, 1(4):232–237, 2007.
8. A. Liu, R. Jones, L. Liao, D. Samara-Rubio, D. Rubin, O. Cohen, R. Nicolaescu and M. Paniccia A high-speed silicon optical modulator based on a metal-oxide-semiconductor capacitor. *Nature*, 427(6975):615–618, 2004.
9. D. R. Turner. Electropolishing Silicon in Hydrofluoric Acid Solutions. *J. Electrochem. Soc*, 105(7):402–408, 1958.
10. A. Uhlir. Electrolytic Shaping of Germanium and Silicon. *Bell System Tech. J.*, 35:333–347, 1956.
11. V. Lehmann and H. Föll. Formation Mechanism and Properties of Electrochemically Etched Trenches in n-Type Silicon. *J. Electrochem. Soc*, 137(2):653–659, 1990.
12. D. H. Everett. IUPAC, Manual of Symbol and Terminology for Physicochemical Quantities and Units, Appendix, Definitions, Terminology and Symbols in Colloid and Surface Chemistry, Part I. *Pure Appl. Chem.*, 31(4):577, 1972.

13. V. Lehmann and U. Gösele. Porous silicon formation: A quantum wire effect. *Appl. Phys. Lett.*, 58(8):856–858, 1991.
14. H. Föll, M. Christophersen, J. Carstensen and G. Hasse. Formation and application of porous silicon. *Mater. Sci. Eng. R*, 39(4):93–141, 2002.
15. X. G. Zhang. Morphology and Formation Mechanisms of Porous Silicon. *J. Electrochem. Soc*, 151(1):C69–C80, 2004.
16. V. Lehmann. *Electrochemistry of Silicon: Instrumentation, Science, Materials and Applications*. Wiley-VCH Verlag GmbH, 2002.
17. V. Lehmann. The Physics of Macropore Formation in Low Doped n-Type Silicon. *J. Electrochem. Soc.*, 140(10):2836–2843, 1993.
18. S. Matthias. *Herstellung und Charakterisierung von 3D-photonischen Kristallen aus makroporösem Silizium*. PhD thesis Martin-Luther-Universität Halle-Wittenberg, 2005.
19. P. Kleimann, J. Linnros and S. Petersson Formation of wide and deep pores in silicon by electrochemical etching. *Mater. Sci. Eng. B*, 69–70:29–33, 2000.
20. S. Matthias, F. Müller, J. Schilling and U. Gösele. Pushing the limits of macroporous silicon etching. *Appl. Phys. A: Mater. Sci. Process.*, 80(7):1391–1396, 2005.
21. S. Matthias, F. Müller, C. Jamois, R. B. Wehrspohn and U. Gösele. Large-Area Three-Dimensional Structuring by Electrochemical Etching and Lithography. *Adv. Mater.*, 16(23–24):2166–2170, 2004.
22. S. Matthias, F. Müller and U. Gösele Controlled nonuniformity in macroporous silicon pore growth. *Appl. Phys. Lett.*, 87(22):224106, 2005.
23. M. H. Klühr, A. Sauermann, C. A. Elsner, K. H. Thein and S. K. Dertinger Partially Oxidized Macroporous Silicon: A Three-Dimensional Photonic Matrix for Microarray Applications. *Adv. Mater.*, 18(23):3135–3139, 2006.
24. H. Seidel, L. Csepregi, A. Heuberger and H. Baumgärtel. Anisotropic etching of crystalline silicon in alkaline solutions. *J. Electrochem. Soc.*, 137(11):3612–3632, 1990.
25. S. Matthias, F. Müller and U. Gösele Simple cubic three-dimensional photonic crystals based on macroporous silicon and anisotropic posttreatment. *J. Appl. Phys.*, 98(2):023524, 2005.
26. S. Matthias, J. Schilling, K. Nielsch, F. Müller, R. B. Wehrspohn and U. Gösele Monodisperse diameter-modulated gold microwires. *Adv. Mater.*, 14(22):1618–1621, 2002.
27. S. John. Strong localization of photons in certain disordered dielectric superlattices. *Phys. Rev. Lett.*, 58(23):2486–2489, 1987.
28. E. Yablonovitch. Inhibited Spontaneous Emission in Solid-State Physics and Electronics. *Phys. Rev. Lett.*, 58(20):2059–2062, 1987.
29. S. G. Johnson and J. D. Joannopoulos. Block-iterative frequency-domain methods for Maxwell's equations in a planewave basis. *Opt. Express*, 8(3):173–190, 2001.
30. A. Birner, A. P. Li, F. Müller, U. Gösele, P. Kramper, V. Sandoghdar, J. Mlynek, K. Busch and V. Lehmann Transmission of a microcavity structure in a two-dimensional photonic crystal based on macroporous silicon. *Mater. Sci. Semicond. Process*, 3(5–6):487–491, 2000.
31. S. W. Leonard, H. M. van Driel, A. Birner, U. Gösele and P. R. Villeneuve. Single-mode transmission in two-dimensional macroporous silicon photonic crystal waveguides. *Opt. Lett.*, 25(20):1550–1552, 2000.
32. S. Ottow, V. Lehmann and H. Föll. Processing of Three-Dimensional Microstructures Using Macroporous n-Type Silicon. *J. Electrochem. Soc*, 143(1):385–390, 1996.
33. S. W. Leonard, H. M. van Driel, K. Busch, S. John, A. Birner, A. -P. Li, F. Müller, U. Gösele and V. Lehmann. Attenuation of optical transmission within the band gap of thin two-dimensional macroporous silicon photonic crystals. *Appl. Phys. Lett.*, 75(20):3063–3065, 1999.
34. A. Birner, R. B. Wehrspohn, U. M. Gösele and K. Busch. Silicon-Based Photonic Crystals. *Adv Mater*, 13(6):377–388, 2001.
35. A. Birner, K. Busch and F. Müller Photonische Kristalle. *Phys. Bl*, 55(4):27–33, 1999.

36. P. Kramper, M. Agio, C. M. Soukoulis, A. Birner, F. Müller, R. B. Wehrspohn, U. Gösele and V. Sandoghdar. Highly Directional Emission from Photonic Crystal Waveguides of Subwavelength Width. *Phys. Rev. Lett.*, 92(11):113903, 2004.
37. S. W. Leonard. Complete three-dimensional band gap in macroporous silicon photonic crystals. *Appl. Phys. Lett.*, 81(16):2917–2919, 2002.
38. C. Kettner, P. Reimann, P. Hänggi and F. Müller Drift ratchet. *Phys. Rev. E*, 61(1):312–323, 2000.
39. S. Matthias and F. Müller Asymmetric pores in a silicon membrane acting as massively parallel brownian ratchets. *Nature*, 424(6944):53–57, 2003.
40. J. Schilling, A. Scherer, U. Gösele and M. Kolbe. Macroporous silicon membranes as electron and x-ray transmissive windows. *Appl. Phys. Lett.*, 85(7):1152–1154, 2004.

Index

O. Hayden, K. Nielsch (eds.), *Molecular- and Nano-Tubes*,
DOI 10.1007/978-1-4419-9443-1, © Springer Science+Business Media, LLC 2011

M